Arithmetic Operations:

$$ab + ac = a(b+c)$$

$$\frac{a}{b} + \frac{c}{d} = \frac{ad+bc}{bd}$$

$$\frac{a+b}{c} = \frac{a}{c} + \frac{b}{c}$$

$$\frac{\left(\frac{a}{b}\right)}{\left(\frac{c}{d}\right)} = \frac{ad}{bc}$$

$$a\left(\frac{b}{c}\right) = \frac{ab}{c}$$

$$\frac{a-b}{c-d} = \frac{b-a}{d-c}$$

$$\frac{ab+ac}{a} = b+c,\ a \neq 0$$

$$\frac{\left(\frac{a}{b}\right)}{c} = \frac{a}{bc}$$

$$\frac{a}{\left(\frac{b}{c}\right)} = \frac{ac}{b}$$

Exponents and Radicals:

$$a^0 = 1,\ a \neq 0$$

$$\frac{a^x}{a^y} = a^{x-y}$$

$$\left(\frac{a}{b}\right)^x = \frac{a^x}{b^x}$$

$$\sqrt[n]{a^m} = a^{m/n} = (\sqrt[n]{a})^m$$

$$a^{-x} = \frac{1}{a^x}$$

$$(a^x)^y = a^{xy}$$

$$\sqrt{a} = a^{1/2}$$

$$\sqrt[n]{ab} = \sqrt[n]{a}\,\sqrt[n]{b}$$

$$a^x a^y = a^{x+y}$$

$$(ab)^x = a^x b^x$$

$$\sqrt[n]{a} = a^{1/n}$$

$$\sqrt[n]{\left(\frac{a}{b}\right)} = \frac{\sqrt[n]{a}}{\sqrt[n]{b}}$$

Algebraic Errors to Avoid:

$$\frac{a}{x+b} \neq \frac{a}{x} + \frac{a}{b}$$ (To see this error, let $a = b = x = 1$.)

$$\sqrt{x^2 + a^2} \neq x + a$$ (To see this error, let $x = 3$ and $a = 4$.)

$$a - b(x-1) \neq a - bx - b$$ (Remember to distribute negative signs. The equation should be $a - b(x-1) = a - bx + b$.)

$$\frac{\left(\frac{x}{a}\right)}{b} \neq \frac{bx}{a}$$ (To divide fractions, invert and multiply. The equation should be

$$\frac{\frac{x}{a}}{b} = \frac{\frac{x}{a}}{\frac{b}{1}} = \left(\frac{x}{a}\right)\left(\frac{1}{b}\right) = \frac{x}{ab}.)$$

$$\sqrt{-x^2 + a^2} \neq -\sqrt{x^2 - a^2}$$ (We can't factor a negative sign outside of the square root.)

$$\frac{\cancel{a} + bx}{\cancel{a}} \neq 1 + bx$$ (This is one of many examples of incorrect cancellation. The equation should be $\frac{a+bx}{a} = \frac{a}{a} + \frac{bx}{a} = 1 + \frac{bx}{a}.)$

$$\frac{1}{x^{1/2} - x^{1/3}} \neq x^{-1/2} - x^{-1/3}$$ (This error is a sophisticated version of the first error.)

$$(x^2)^3 \neq x^5$$ (The equation should be $(x^2)^3 = x^2 x^2 x^2 = x^6.)$

Conversion Table:

1 centimeter =	0.394 inches	1 joule =	0.738 foot-pounds	1 mile = 1.609 kilometers
1 meter =	39.370 inches	1 gram =	0.035 ounces	1 gallon = 3.785 liters
=	3.281 feet	1 kilogram =	2.205 pounds	1 pound = 4.448 newtons
1 kilometer =	0.621 miles	1 inch =	2.540 centimeters	1 foot-lb = 1.356 joules
1 liter =	0.264 gallons	1 foot =	30.480 centimeters	1 ounce = 28.350 grams
1 newton =	0.225 pounds	=	0.305 meters	1 pound = 0.454 kilograms

Daniel D. Bonar, Denison University
Richard Cutts, University of Wisconsin—Stout
John E. Bruha, University of Northern Iowa
H. Eugene Hall, DeKalb Community College
Randal Hoppens, Blinn College
E. John Hornsby, Jr., University of New Orleans
William B. Jones, University of Colorado
Jimmie D. Lawson, Louisiana State University
Peter J. Livorsi, Oakton Community College
Wade T. Macey, Appalachian State University
Jerome L. Paul, University of Cincinnati
Marilyn Schiermeier, North Carolina State University
George W. Schultz, St. Petersburg Junior College
Edith Silver, Mercer County Community College
Shirley C. Sorensen, University of Maryland
Charles Stone, DeKalb Community College
Bruce Williamson, University of Wisconsin—River Falls

The mathematicians listed below responded to a survey conducted by D. C. Heath and Company that helped us outline our revision.

Holli Adams, Portland Community College
Marion Baumler, Niagara County Community College
Diane Blansett, Delta State University
Derek I. Bloomfield, Orange County Community College
Daniel D. Bonar, Denison University
John E. Bruha, University of Northern Iowa
William L. Campbell, University of Wisconsin—Platteville
John Caraluzzo, Orange County Community College
William E. Chatfield, University of Wisconsin—Platteville
Robert P. Finley, Mississippi State University
August J. Garver, University of Missouri—Rolla
Sue Goodman, University of North Carolina
Louis Hoelzle, Bucks County Community College
Randal Hoppens, Blinn College
Moana Karsteter, Florida State University
Robert C. Limburg, St. Louis Community College at Florissant Valley
Peter J. Livorsi, Oakton Community College
John Locker, University of North Alabama
Wade T. Macey, Appalachian State University
J. Kent Minichiello, Howard University
Terry Mullen, Carroll College
Richard Nation, Palomar College

Supplements

• For students, the *Study and Solutions Guide* by Dianna L. Zook contains detailed solutions of many of the odd-numbered exercises in the text. Each of these is indicated in the text by a box surrounding the exercise number. (Approximately 40% of the odd-numbered exercises in the text are solved in the *Study and Solutions Guide*.) This guide also contains summaries of important concepts for each section and self-tests for each chapter.

• For students, *Computer Activities for Precalculus* is an IBM-PC®* package that offers activities based on programs that enhance the learning of such topics as linear functions, quadratic functions, exponential and logarithmic functions, and trigonometric functions. Exploratory practice and directed tutorials are included as well as a function grapher.

• For instructors, the *Complete Solutions Guide* contains solutions for all of the exercises in the text.

• For instructors, the *Instructor's Guide* by Meredythe M. Burrows contains sample tests for each chapter in the text and also contains suggestions for class room instruction.

• For instructors, we have prepared test-generating software called *HeathTest* to accompany the text. This software will run on an IBM-PC® that has at least 256K of memory. (It will also run on many IBM compatibles.) To print the tests, the software requires an IBM graphics-compatible dot matrix printer. An Apple II®† version is also available.

• For instructors, we have prepared a package containing 25 two-color transparencies.

Acknowledgements

We would like to thank the many people who have helped us at various stages of preparing the First and Second Editions of this text. Their encouragement, criticisms, and suggestions have been invaluable to us.

Special thanks goes to the reviewers of the First and Second Editions.

Hollie Baker, Norfolk State University
Derek I. Bloomfield, Orange County Community College
Ben P. Bockstage, Broward Community College

---

*IBM is a registered trademark of International Business Machines Corp.

†Apple is a registered trademark of Apple Computer, Inc.

Graphics   The ability to visualize a problem is a critical part of a student's ability to solve the problem. This text contains over 1075 figures. Of these, approximately 300 are in the exercise sets and approximately 340 are in the odd-numbered answers in the back of the text. The art package for the Second Edition is completely new. Every graph in the text was computer generated for the greatest possible accuracy.

Applications   Throughout the Second Edition we have included many applied problems that give students insight about the usefulness of trigonometry in a wide variety of fields including business, economics, biology, engineering, chemistry, and physics.

Examples   The Second Edition contains 450 examples, each carefully chosen to illustrate a particular concept or problem-solving technique. Each example is titled for quick reference, and many examples include color side comments to justify or explain the steps in the solution.

Exercises   Over 3900 exercises are included in the Second Edition. These are designed to build competence, skill, and understanding. Each exercise set is graded in difficulty to allow students to gain confidence as they progress. To help students develop skills in analytic geometry, we stress a graphical approach in many sections and have included numerous graphs in the exercises.

Warm up Exercises   [New in the Second Edition] We have found that students in trigonometry can benefit greatly from reinforcement of previously learned concepts. Most sections in the text contain a set of ten Warm up exercises that give students practice in the "old skills" that are necessary to master the "new skills" presented in the section. *All* of the Warm up exercises are answered in the back of the text.

Calculators   Hints and instructions for working with calculators occur in many places in the Second Edition. Because calculators have become commonplace, we no longer identify exercises that require decimal approximations.

Algebra of Calculus   Special emphasis has been given to algebraic skills that are needed in calculus. Many examples in the Second Edition discuss trigonometric techniques that arise in calculus. These examples are clearly identified.

Remarks   In the Second Edition we include special instructional notes to students called *Remarks*. These appear after definitions, theorems, or examples and are designed to give additional insight, help avoid common errors, or describe generalizations.

Preface

Success in college level mathematics courses is enhanced by a good understanding of trigonometry. The goal of this text is to help students develop this understanding. Although we review some of the basic concepts in algebra, we assume that most students in this course will have completed two years of high school algebra.

What's New in the Second Edition

Many users of the first edition of the text have given us suggestions for improving the text. We appreciate this type of input very much and have incorporated most of the suggestions into the Second Edition. *Every* section in the text has been revised and many sections were completely rewritten. Most sections in the Second Edition have more exercises than were in the First Edition. The major changes are as follows.

In Chapter 1 we added a new section on solving equations (Section 1.2). The introduction of functions in Section 1.5 was completely rewritten. The first three sections of Chapter 2 were expanded to four sections to improve the flow from trigonometric functions of real numbers to right-triangle trigonometry. In Chapter 3, we now have an earlier coverage of solving trigonometric equations (Section 3.3), and the coverage of multiple-angle formulas and product-sum formulas has been condensed to one section (Section 3.5). Chapter 6 was extensively reorganized and rewritten. More emphasis is now given to solving exponential and logarithmic equations (Section 6.4), and we included a new section on applications of exponential and logarithmic equations (Section 6.5).

Features of the Second Edition

The Second Edition contains many features that we have found help students improve their skills and acquire an understanding of the material.

Senior Acquisitions Editor: Mary Lu Walsh
Developmental and Senior Production Editor: Cathy Cantin
Senior Designer: Sally Steele
Editorial Assistant: Carolyn Johnson
Production Manager: Mike O'Dea
Composition: Jonathan Peck Typographers
Technical Art: Folium
Cover: Martucci Studio

Published simultaneously in Canada.

Printed in the United States of America.

International Standard Book Number: 0-669-16266-3

Library of Congress Catalog Card Number: 88-81730

5   6   7   8   9   10

Trigonometry

SECOND EDITION

Roland E. Larson
Robert P. Hostetler
**The Pennsylvania State University
The Behrend College**

With the assistance of
**David E. Heyd
The Pennsylvania State University
The Behrend College**

D.C. Heath and Company
Lexington, Massachusetts   **Toronto**

Trigonometry

William Paul, Appalachian State University
Richard A. Quint, Ventura College
Charles T. Scarborough, Mississippi State University
Shannon Schumann, University of Wyoming
Arthur E. Schwartz, Mercer County Community College
Joseph Sharp, West Georgia College
Burla J. Sims, University of Arkansas at Little Rock
James R. Smith, Appalachian State University
B. Louise Whisler, San Bernardino Valley College
Bruce Williamson, University of Wisconsin—River Falls

We would also like to thank all of the people at D. C. Heath and Company who worked with us in the development of the Second Edition, especially Mary Lu Walsh, Senior Mathematics Acquisitions Editor; Cathy Cantin, Developmental Editor and Senior Production Editor; Sally Steele, Designer; Carolyn Johnson, Editorial Assistant; and Mike O'Dea, Production Manager.

Several other people also worked on this project. David E. Heyd assisted us with the text, Dianna Zook wrote the *Study and Solutions Guide*, Helen Medley proofread the manuscript and worked the exercises, and Meredythe Burrows wrote the *Instructor's Guide*. Timothy R. Larson prepared the art and worked the exercises. Linda L. Matta proofread the galleys and typed the *Instructor's Guide*. Linda M. Bollinger proofread the galleys and typed the text manuscript, the *Study and Solutions Guide*, and the *Complete Solutions Guide*. Randall Hammond and Lisa Bickel worked the exercises.

We are grateful to our wives, Deanna Gilbert Larson and Eloise Hostetler, for their love, patience, and understanding.

If you have suggestions for improving the text, please feel free to write to us. Over the past several years we have received many useful comments from both instructors and students and we value this very much.

Roland E. Larson
Robert P. Hostetler

The Larson and Hostetler Precalculus Series

To accommodate the different methods of teaching college algebra, trigonometry, and analytic geometry, we have prepared four volumes. Each has its own supplement package. These four titles are described below.

College Algebra, Second Edition

This text is designed for a one-term course covering standard topics such as algebraic functions and their graphs, exponential and logarithmic functions, systems of equations, matrices, determinants, sequences, series, and probability.

Trigonometry, Second Edition

This text is used in a one-term course covering the trigonometric functions and their graphs, exponential and logarithmic functions, and analytic geometry (including polar coordinates and parametric equations).

Algebra and Trigonometry, Second Edition

This book combines the content of the two texts mentioned above (with the exception of polar coordinates and parametric equations). It is comprehensive enough for a two-term course, or, with careful selection, may be used in a one-term course.

Precalculus, Second Edition

With this book, students cover the algebraic, exponential and logarithmic, and trigonometric functions and their graphs, as well as analytic geometry in preparation for a course in calculus. This may be used in a one- or two-term course.

Contents

Introduction to Calculators

This text includes several examples and exercises that use a scientific calculator. As we encounter each new calculator application, we will give instructions for using a calculator efficiently. These instructions are somewhat general and may not agree precisely with the steps required by your calculator.

For use with this text, we recommend a calculator with the following features.

1. At least 8-digit display
2. Four arithmetic operations: $\boxed{+}$, $\boxed{-}$, $\boxed{\times}$, $\boxed{\div}$
3. Change sign key: $\boxed{+/-}$
4. Memory key and Recall key: $\boxed{\textbf{STO}}$, $\boxed{\textbf{RCL}}$
5. Parentheses: $\boxed{(}$, $\boxed{)}$
6. Exponential key: $\boxed{y^x}$
7. Natural logarithmic key: $\boxed{\textbf{ln } x}$
8. Pi and Degree-Radian conversion: $\boxed{\pi}$, $\boxed{\textbf{DRG}}$
9. Inverse, reciprocal, square root: $\boxed{\textbf{INV}}$, $\boxed{1/x}$, $\boxed{\sqrt{x}}$
10. Trigonometric functions: $\boxed{\textbf{sin}}$, $\boxed{\textbf{cos}}$, $\boxed{\textbf{tan}}$

One of the basic differences in calculators is their order of operations. Some calculators use an order of operations called RPN (for Reverse Polish Notation). In this text, however, all calculator steps will be given using *algebraic logic*. For example, the calculation

$$4.69[5 + 2(6.87 - 3.042)]$$

can be performed with the following steps.

$$4.69 \boxed{\times} \boxed{(} 5 \boxed{+} 2 \boxed{\times} \boxed{(} 6.87 \boxed{-} 3.042 \boxed{)} \boxed{)} \boxed{=}$$

This yields the value 59.35664. Without parentheses, we would work from the inside out with the sequence

6.87 $\boxed{-}$ 3.042 $\boxed{=}$ $\boxed{\times}$ 2 $\boxed{+}$ 5 $\boxed{=}$ $\boxed{\times}$ 4.69 $\boxed{=}$

to obtain the same result.

Rounding Numbers

For all their usefulness, calculators do have a problem representing numbers because they are limited to a finite number of digits. For instance, what does your calculator display when you compute 2 ÷ 3? Some calculators simply truncate (drop) the digits that exceed their display range and display .66666666. Others will round the number and display .66666667. Although the second display is more accurate, *both* of these decimal representations of 2/3 contain a rounding error.

When rounding decimals, we use the following rules.

1. Determine the number of digits of accuracy you wish to keep. The digit in the last position you keep is called the **rounding digit,** and the digit in the first position you discard is called the **decision digit.**
2. If the decision digit is 5 or greater, round up by adding 1 to the rounding digit.
3. If the decision digit is 4 or less, round down by leaving the rounding digit unchanged.

Here are some examples. Note that we round down in the first example because the decision digit is 4 or less, and we round up in the other two examples because the decision digit is 5 or greater.

Number	*Rounded to three decimal places*	
(a) $\sqrt{2} = 1.4142136 \ldots$	1.414	*Round down*
(b) $\pi = 3.1415927 \ldots$	3.142	*Round up*
(c) $\dfrac{7}{9} = 0.77777777 \ldots$	0.778	*Round up*

One of the best ways to minimize error due to rounding is to leave numbers in your calculator until your calculations are complete. If you want to save a number for future use, store it in your calculator's memory.

CHAPTER 1

Prerequisites for Trigonometry

1.1 The Real Number System

We begin our review of prerequisites for trigonometry with a look at the **real number system.** Real numbers are used in everyday life to describe quantities like age, miles per gallon, container size, population, and so on. To represent real numbers we use symbols such as

$$9, \quad -5, \quad 0, \quad \frac{4}{3}, \quad 0.6666\ldots, \quad 28.21, \quad \sqrt{2}, \quad \pi, \quad \text{and} \quad \sqrt[3]{-32}.$$

The set of real numbers contains some important subsets with which you need to be familiar (see Figure 1.1). For instance, the numbers

$$\ldots, \quad -3, \quad -2, \quad -1, \quad 0, \quad 1, \quad 2, \quad 3, \quad \ldots$$

are called **integers.** A real number is called **rational** if it can be written as the ratio p/q of two integers, where $q \neq 0$. For instance, the numbers

$$\frac{1}{3} = 0.3333\ldots, \quad \frac{1}{8} = 0.125, \quad \text{and} \quad \frac{125}{111} = 1.126126\ldots$$

are rational. The decimal representation of a rational number either repeats (as in $3.1454545\ldots$) or terminates (as in $1/2 = 0.5$). Real numbers that

1

Prerequisites for Trigonometry

cannot be written as the ratio of two integers are called **irrational.** For instance, the numbers

$$\sqrt{2} \approx 1.4142136 \quad \text{and} \quad \pi \approx 3.1415927$$

are irrational. (The symbol $\approx$ means "approximately equal to.")

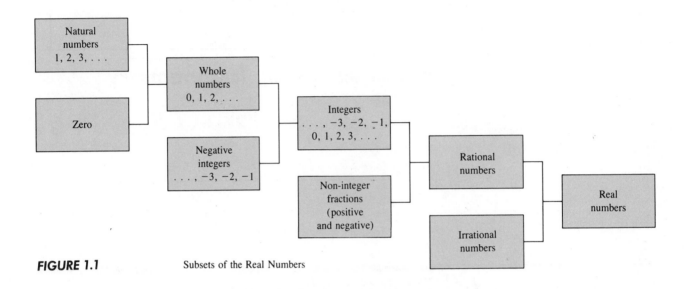

FIGURE 1.1   Subsets of the Real Numbers

The model we use to represent the real number system is called the **real number line.** It consists of a horizontal line with an arbitrary point (the **origin**) labeled 0. Numbers to the right of the origin are positive and numbers to the left of the origin are negative, as shown in Figure 1.2. We use the term **nonnegative** to describe a number that is either positive or zero.

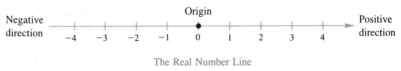

The Real Number Line

FIGURE 1.2

Each point on the real number line corresponds to one and only one real number and *each real number corresponds to one and only one point on the real number line*. This type of relationship is called a **one-to-one correspondence,** as shown in Figure 1.3.

The Real Number System

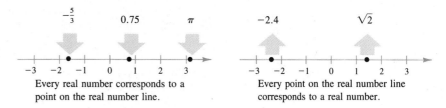

| Every real number corresponds to a | Every point on the real number line |
| point on the real number line. | corresponds to a real number. |

One-to-One Correspondence

FIGURE 1.3

The number associated with a point on the real number line is called the **coordinate** of the point. For example, in Figure 1.3, $-\frac{5}{3}$ is the coordinate of the left-most point and $\sqrt{2}$ is the coordinate of the right-most point.

Ordering the Real Numbers

One important property of real numbers is that they are **ordered.**

Definition of Order on the Real Number Line

If a and b are real numbers, then a is **less than** b if $b - a$ is positive. We denote this order by the **inequality**

$a < b.$

The symbol $a \leq b$ means that a is **less than or equal to** b.

Geometrically, this definition implies that $a < b$ if and only if a lies to the *left* of b on the real number line, as shown in Figure 1.4. For example, $1 < 2$ because 1 lies to the left of 2 on the real number line.

$a < b$ if and only if a lies to the left of b.

FIGURE 1.4

Inequalities are useful in denoting subsets of the real numbers, as shown in Examples 1 and 2.

EXAMPLE 1 *Interpreting Inequalities*

(a) The inequality $x \leq 2$ denotes all real numbers less than or equal to 2, as shown in Figure 1.5(a).

(b) The inequality $-2 \leq x < 3$ means that $x \geq -2$ *and* $x < 3$. This "double" inequality denotes all real numbers between -2 and 3, including -2 but *not* including 3, as shown in Figure 1.5(b).

(c) The inequality $x > -5$ denotes all real numbers greater than -5, as shown in Figure 1.5(c).

(a)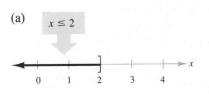

(b)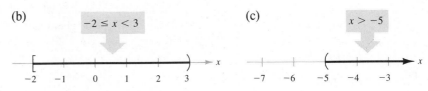

(c)

FIGURE 1.5

EXAMPLE 2 *Using Inequalities to Represent Sets of Real Numbers*

Use inequality notation to describe each of the following.

(a) c is nonnegative.
(b) b is at most 5.
(c) d is negative and greater than -3.
(d) x is positive but not more than 6.

SOLUTION

(a) "c is nonnegative" means that c is greater than or equal to zero, and we write $c \geq 0$.

(b) "b is at most 5" can be written as $b \leq 5$.

(c) "d is negative" can be written as $d < 0$, and "d is greater than -3" can be written as $-3 < d$. Combining these two inequalities produces $-3 < d < 0$.

(d) "x is positive" can be written as $0 < x$, and "x is not more than 6" can be written as $x \leq 6$. Combining these two inequalities produces $0 < x \leq 6$.

The following property of real numbers is called the **Law of Trichotomy.** It tells us that for any two real numbers a and b, *precisely* one of three orders is possible.

$$a = b, \quad a < b, \quad \text{or} \quad a > b \qquad \qquad \textit{Law of Trichotomy}$$

The Real Number System

Absolute Value and Distance

By the **absolute value** of a real number, we mean its *magnitude* (its value disregarding its sign).

Definition of Absolute Value

If a is a real number, then the **absolute value** of a is given by

$$|a| = \begin{cases} a, & \text{if } a \geq 0 \\ -a, & \text{if } a < 0. \end{cases}$$

Be sure you see that the absolute value of a number can never be negative. For instance, if $a = -5$, then

$$|a| = |-5| = -a = -(-5) = 5$$

because $-5 < 0$. Similarly, $|0| = 0$ because $0 \geq 0$, and

$$|2 - \pi| = -(2 - \pi) = \pi - 2$$

because $(2 - \pi) < 0$.

EXAMPLE 3   *Evaluating the Absolute Value of a Number*

Evaluate the fraction

$$\frac{|x|}{x}$$

for (a) $x > 0$ and (b) $x < 0$.

SOLUTION

(a) If $x > 0$, then $|x| = x$ and we have

$$\frac{|x|}{x} = \frac{x}{x} = 1.$$

For instance, $|4|/4 = 1$.

(b) If $x < 0$, then $|x| = -x$ and we have

$$\frac{|x|}{x} = \frac{-x}{x} = -1.$$

For instance, $|-4|/(-4) = 4/(-4) = -1$.

Note that when $x = 0$, the expression $|x|/x$ is undefined.

The following list gives four useful properties of absolute value. When you see a list like this, try to formulate verbal descriptions of the properties. For instance, the third property tells us that the absolute value of a product of two numbers is equal to the product of the absolute values of the two numbers.

Properties of Absolute Value

Let a and b be real numbers. Then the following properties are true.

1. $|a| \geq 0$ 2. $|-a| = |a|$

3. $|ab| = |a||b|$ 4. $\left|\dfrac{a}{b}\right| = \dfrac{|a|}{|b|}, \quad b \neq 0$

Absolute value can be used to define the distance between two numbers on the real number line. To see how this is done, consider the numbers -3 and 4, shown in Figure 1.6. To find the distance between these two points, we subtract *either* number from the other and then take the absolute value of the difference. For instance,

$$\text{distance} = |-3 - 4| = |-7| = 7$$

or equivalently,

$$\text{distance} = |4 - (-3)| = |7| = 7.$$

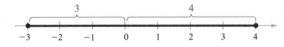

FIGURE 1.6   The distance between -3 and 4 is 7.

Distance Between Two Points on the Real Line

Let a and b be real numbers. The **distance between a and b** is

$$d(a, b) = |b - a| = |a - b|.$$

EXAMPLE 4 *Distance and Absolute Value*

(a) The distance between $\sqrt{7}$ and 4 is given by
$$d(\sqrt{7}, 4) = |4 - \sqrt{7}| = 4 - \sqrt{7}.$$

(b) The statement "the distance between c and -2 is at least 7" can be written
$$d(c, -2) = |c + 2| \geq 7.$$

(c) The statement "the distance between x and 2.3 is less than 1" can be written

$$d(x, 2.3) = |x - 2.3| < 1.$$

(d) The distance between -4 and the origin is given by

$$d(-4, 0) = |-4 - 0| = |-4| = 4.$$

Intervals on the Real Number Line

Subsets of real numbers are sometimes expressed in the **interval** forms shown in the following table.

Interval Notation for Subsets of Real Numbers

The following intervals on the real number line are called **bounded intervals.**

Notation	Interval Type	Inequality	Graph
$[a, b]$	closed	$a \leq x \leq b$	
(a, b)	open	$a < x < b$	
$[a, b)$	half-open	$a \leq x < b$	
$(a, b]$	half-open	$a < x \leq b$	

The following intervals on the real number line are called **unbounded intervals.**

Notation	Interval Type	Inequality	Graph
$[a, \infty)$	half-open	$x \geq a$	
(a, ∞)	open	$x > a$	
$(-\infty, b]$	half-open	$x \leq b$	
$(-\infty, b)$	open	$x < b$	
$(-\infty, \infty)$	entire real line		

Remark: The symbols ∞ (**positive infinity**) and −∞ (**negative infinity**) are not real numbers. They are simply convenient symbols used to denote the *unboundedness* of an interval such as $(1, \infty)$. Note also that an **open interval** (a, b) excludes its endpoints a and b, whereas a **closed interval** includes them. **Half-open intervals** include just one of the endpoints.

EXAMPLE 5   *Intervals and Inequalities*

Write an inequality to represent each of the following intervals and state whether the interval is bounded or unbounded.

(a) $(-3, 5]$ (b) $(-3, \infty)$ (c) $[0, 2]$

SOLUTION

(a) $(-3, 5]$ corresponds to $-3 < x \le 5$.   *Bounded*
(b) $(-3, \infty)$ corresponds to $-3 < x$.   *Unbounded*
(c) $[0, 2]$ corresponds to $0 \le x \le 2$.   *Bounded*

The following list gives two basic types of inequalities involving absolute values.

Two Basic Types of Inequalities Involving Absolute Value

Let a be a positive real number.

$|x| < a$ if and only if $-a < x < a$.
$|x| > a$ if and only if $x < -a$ or $x > a$.

Inequality	Interpretation	Graph
$|x| < a$	All numbers x whose distance from 0 is *less* than a	$-a < x < a$
$|x| > a$	All numbers x whose distance from 0 is *greater* than a	$x < -a$ $x > a$

Remark: Note that $-a < x < a$ means that $-a < x$ *and* $x < a$.

EXAMPLE 6 An Inequality Involving Absolute Value

Sketch the graph of the interval on the real line represented by the following inequalities.

(a) $|x - 5| < 2$ (b) $|x + 3| \geq 7$

SOLUTION

(a) This inequality represents the set of all numbers x whose distance from 5 is less than 2 units, as shown in Figure 1.7(a).

$$|x - 5| < 2 \qquad \textit{Given}$$
$$-2 < x - 5 < 2 \qquad \textit{Interpret absolute value}$$
$$3 < x < 7 \qquad \textit{Add 5}$$

Thus, the interval is (3, 7).

(b) This inequality represents the set of all numbers x whose distance from -3 is at least 7 units, as shown in Figure 1.7(b).

$$|x + 3| \geq 7 \qquad \textit{Given}$$
$$x + 3 \leq -7 \quad \text{or} \quad x + 3 \geq 7 \qquad \textit{Interpret absolute value}$$
$$x \leq -10 \qquad\qquad x \geq 4 \qquad \textit{Subtract 3}$$

Thus, the intervals are $(-\infty, -10]$ and $[4, \infty)$.

(a)

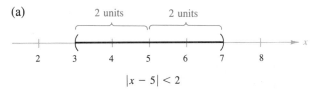

$$|x - 5| < 2$$

(b)

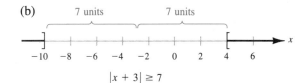

$$|x + 3| \geq 7$$

FIGURE 1.7

EXERCISES 1.1

In Exercises 1–6, plot the two real numbers on the real number line and place the appropriate inequality sign (< or >) between them.

1. $\frac{3}{2}$, 7

2. -3.5, 1

3. -4, -8

4. 1, $-\frac{16}{3}$

*__5.__ $\frac{5}{6}$, $\frac{2}{3}$

6. $-\frac{8}{7}$, $-\frac{3}{7}$

In Exercises 7–12, use inequality notation to denote the given expression.

7. x is negative.

8. y is greater than 5 and less than or equal to 12.

9. Burt's age, A, is at least 30.

10. The yield, Y, is no more than 45 bushels per acre.

*A boxed number indicates that a detailed solution can be found in the *Student Solution Guide*.

11. The annual rate of inflation, R, is expected to be at least 3.5%, but no more than 6%.

12. The price, P, of unleaded gasoline is not expected to go above $1.25 per gallon during the coming year.

In Exercises 13–20, write the given number without absolute value signs.

13. $|-10|$

14. $|0|$

15. $|3 - \pi|$

16. $|4 - \pi|$

17. $\dfrac{-5}{|-5|}$

18. $-3|-3|$

19. $-3 - |-3|$

20. $|-1| - |-2|$

In Exercises 21–28, find the distance between a and b.

21. $a = -1$ $b = 3$

22. $a = -4$ $b = -\dfrac{3}{2}$

23. $a = -\dfrac{5}{2}$ $b = 0$

24. $a = \dfrac{1}{4}$ $b = \dfrac{11}{4}$

25. $a = 126,\ b = 75$

26. $a = -126,\ b = -75$

27. $a = 9.34,\ b = -5.65$

28. $a = \dfrac{16}{5},\ b = \dfrac{112}{75}$

In Exercises 29–34, use absolute value notation to describe the given expression.

29. The distance between x and 5 is no more than 3.

30. The distance between x and -10 is at least 6.

31. The distance between z and $\dfrac{3}{2}$ is greater than 1.

32. The distance between z and 0 is less than 8.

33. y is at least 6 units from 0.

34. y is at most 2 units from a.

35. A rational number expressed in decimal form is either a terminating decimal or a nonterminating decimal with a repeating pattern. Use your calculator to find the decimal form of each of the following rational numbers. If it is a nonterminating decimal, give the repeating pattern.

(a) $\dfrac{5}{8}$ (b) $\dfrac{1}{3}$ (c) $\dfrac{41}{333}$

[*Note:* For some rational numbers you need to have more decimal places than shown on a typical calculator before you can observe the repeating pattern. For example, try to find the decimal representation of $\dfrac{4}{23}$.]

36. Repeat Exercise 35 for the following rational numbers.

(a) $\dfrac{6}{11}$ (b) $\dfrac{85}{750}$ (c) $\dfrac{1}{7}$

37. (a) Use a calculator to order the following numbers from smallest to largest.

$$\dfrac{7071}{5000}, \quad \dfrac{584}{413}, \quad \sqrt{2}, \quad \dfrac{47}{33}, \quad \dfrac{127}{90}$$

(b) Which of the rational numbers in part (a) is closest to $\sqrt{2}$?

38. Use a calculator to order the following numbers from smallest to largest.

$$\dfrac{26}{15}, \quad \sqrt{3}, \quad 1.73\overline{20}, \quad \dfrac{381}{220}, \quad \sqrt{10} - \sqrt{2}$$

In Exercises 39–46, match the given inequality with its graph. [The graphs are labeled (a)–(h).]

39. $x < 4$ **40.** $x \geq 6$

41. $-2 < x \leq 5$ **42.** $0 \leq x \leq \dfrac{7}{2}$

43. $|x| < 4$ **44.** $|x| > 3$

45. $|x - 5| > 2$ **46.** $|x + 6| < 3$

(a)

(b)

(c)

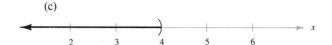

(d)

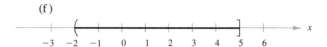

(e)

(f)

(g)

(h)

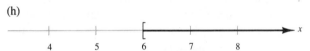

In Exercises 47–54, use absolute value notation to define each interval (or pair of intervals) on the real line.

47.

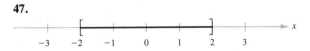

48.

49.

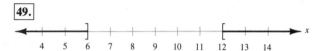

50.

51. All real numbers within 10 units of 12.

52. All real numbers at least 5 units from 8.

53. All real numbers whose distances from −3 are more than 5.

54. All real numbers whose distances from −6 are more than 7.

1.2 Solving Equations

An **equation** is a statement that two expressions are equal. Some examples of equations in x are

$$3x - 5 = 7, \qquad x^2 - x - 6 = 0, \qquad \text{and} \qquad \sqrt{2x} = 4.$$

To **solve** an equation in x means that we find all values of x for which the equation is true. Such values are called **solutions.** For instance, $x = 4$ is a solution of the equation $3x - 5 = 7$, because $3(4) - 5 = 7$ is a true statement.

The solutions of an equation depend upon the kinds of numbers being considered. For instance, in the set of rational numbers the equation $x^2 = 10$ has no solution because there is no rational number whose square is 10. However, in the set of real numbers this equation has the two solutions $\sqrt{10}$ and $-\sqrt{10}$ because $(\sqrt{10})^2 = 10$ and $(-\sqrt{10})^2 = 10$.

A **polynomial equation** in x is of the form

$$a_n x^n + a_{n-1} x^{n-1} + \cdots + a_2 x^2 + a_1 x + a_0 = 0, \qquad a_n \neq 0$$

where n is the **degree** of the equation. For instance, the degree of the polynomial equation

$$-x^3 + 2x - 1 = 0$$

is 3.

An equation that is true for *every* real number in the domain of the variable is called an **identity.** Two examples of identities are

$$x^2 - 9 = (x + 3)(x - 3) \qquad \text{and} \qquad \frac{x}{3x^2} = \frac{1}{3x}, \qquad x \neq 0.$$

The first equation is an identity because it is a true statement for any real value of x. The second equation is an identity because it is true for any nonzero real value of x.

An equation that is true for just *some* (or even none) of the real numbers in the domain of the variable is called a **conditional equation.** For example, the equation $x^2 - 9 = 0$ is conditional because $x = 3$ and $x = -3$ are the only values in the domain that satisfy the equation.

To solve a conditional equation we generate a sequence of **equivalent** (and usually simpler) equations, each having the same solution(s) as the original equation. The operations that yield equivalent equations are given in the following statement.

Generating Equivalent Equations

A given equation is transformed into an *equivalent equation* by one or more of the following steps.

	Given Equation	*Equivalent Equation*
1. Remove symbols of grouping, combine like terms, or reduce fractions on both sides of the equation.	$2x - x = 4$	$x = 4$
2. Add or subtract the same quantity to both sides of the equation.	$x + 1 = 6$	$x = 5$
3. Multiply or divide both sides of the equation by the same nonzero quantity.	$2x = 6$	$x = 3$
4. Interchange the two sides of the equation.	$2 = x$	$x = 2$

Solving Linear Equations

A **linear equation** in variable x is an equation that can be written in the form

$$ax + b = 0$$

where a and b are real numbers with $a \neq 0$. (Note that a linear equation in x is simply a first-degree polynomial equation in x.) To solve a linear equation in x, we *isolate* the variable x by using operations that generate equivalent equations.

EXAMPLE 1 *Solving a Linear Equation*

Solve $3x - 6 = 0$ for x.

SOLUTION

$$3x - 6 = 0 \qquad\qquad \textit{Given}$$
$$3x = 6 \qquad\qquad \textit{Add 6 to both sides}$$
$$x = 2 \qquad\qquad \textit{Divide both sides by 3}$$

When solving an equation, it is a good idea to **check each solution** in the *original* equation. For instance, in Example 1, we can check that $x = 2$ is a solution by substituting in the original equation $3x - 6 = 0$ to obtain

$$3(2) - 6 = 6 - 6 = 0. \qquad\qquad \textit{Check}$$

To solve an equation involving *fractional expressions*, we multiply every term of the equation by the lowest common denominator, LCD, of its terms. This procedure clears the equation of fractions, as shown in the following example.

EXAMPLE 2 *Solving an Equation Involving Fractional Expressions*

Solve the following equation for x.

$$\frac{1}{x - 2} = \frac{3}{x + 2} - \frac{4x}{x^2 - 4}$$

SOLUTION

In this case, the LCD is $x^2 - 4$ or $(x + 2)(x - 2)$. Multiplying every term by this LCD and reducing produces

$$\frac{1}{x - 2}(x + 2)(x - 2) = \frac{3}{x + 2}(x + 2)(x - 2) - \frac{4x}{x^2 - 4}(x + 2)(x - 2)$$
$$x + 2 = 3(x - 2) - 4x$$
$$x + 2 = 3x - 6 - 4x$$
$$2x = -8$$
$$x = -4. \qquad\qquad \textit{Solution}$$

When multiplying or dividing by a *variable* quantity as we did in Example 2, it is possible to introduce an **extraneous** solution—one that does not satisfy the original equation. For instance, to solve the equation

$$\frac{1}{x - 2} = \frac{3}{x + 2} - \frac{6x}{x^2 - 4}$$

we would proceed as in Example 2 and obtain

$$\frac{1}{x - 2}(x + 2)(x - 2) = \frac{3}{x + 2}(x + 2)(x - 2) - \frac{6x}{x^2 - 4}(x + 2)(x - 2)$$

$$x + 2 = 3(x - 2) - 6x$$
$$x + 2 = 3x - 6 - 6x$$
$$4x = -8$$
$$x = -2. \qquad \textit{Extraneous solution}$$

This value of *x* is *not* a solution of the original equation because it produces a denominator of zero in one of the terms in the original equation. Hence, the original equation has *no solution*. This example points out the fact that you should get in the habit of checking all solutions by substituting in the original equation.

Factoring Polynomials

To solve a polynomial equation in *x* of degree 2 or more, we write the equation in standard form (with the polynomial on the left side of the equation and 0 on the right) and factor the polynomial. Then we use the Factorization Principle, which states that:

If $ab = 0$, then $a = 0$ or $b = 0$.   *Factorization Principle*

For instance, $25x^2 - 9$ factors as $(5x + 3)(5x - 3)$; hence, the solution to $25x^2 - 9 = 0$ can be obtained as follows.

$$25x^2 - 9 = 0 \qquad \textit{Standard form}$$
$$(5x + 3)(5x - 3) = 0 \qquad \textit{Factored form}$$
$$5x + 3 = 0, \qquad 5x - 3 = 0 \qquad \textit{Set factors equal to zero}$$
$$5x = -3 \qquad 5x = 3 \qquad \textit{Isolate x-terms}$$
$$x = -\frac{3}{5} \qquad x = \frac{3}{5} \qquad \textit{Solutions}$$

To assist you in the factorization process, we list the following special product formulas from algebra.

Factoring Special Polynomial Forms

Factored Form	*Example*

$u^2 - v^2 = (u + v)(u - v)$

$\begin{aligned} 9x^4 - 4 &= (3x^2)^2 - 2^2 \\ &= (3x^2 + 2)(3x^2 - 2) \end{aligned}$

$u^2 + 2uv + v^2 = (u + v)^2$

$\begin{aligned} x^2 + 6x + 9 &= x^2 + 2(x)(3) + 3^2 \\ &= (x + 3)^2 \end{aligned}$

$u^2 - 2uv + v^2 = (u - v)^2$

$\begin{aligned} x^2 - 6x + 9 &= x^2 - 2(x)(3) + 3^2 \\ &= (x - 3)^2 \end{aligned}$

$u^3 + v^3 = (u + v)(u^2 - uv + v^2)$

$\begin{aligned} x^3 + 8 &= x^3 + 2^3 \\ &= (x + 2)(x^2 - 2x + 4) \end{aligned}$

$u^3 - v^3 = (u - v)(u^2 + uv + v^2)$

$\begin{aligned} 27x^3 - 1 &= (3x)^3 - 1^3 \\ &= (3x - 1)(9x^2 + 3x + 1) \end{aligned}$

The first step in factoring a polynomial is to check for a common factor. Once the common factor is removed, it is often possible to recognize patterns that were not obvious at first glance. Watch for this in some of the examples that follow.

Solving Quadratic Equations

A **quadratic equation** in x is a second-degree polynomial equation that can be written in the standard form

$$ax^2 + bx + c = 0$$

where a, b, and c are real numbers and $a \neq 0$. We can solve a quadratic equation either by factoring or by the quadratic formula. First, we demonstrate solution by factoring.

EXAMPLE 3 **A Quadratic Equation Involving the Difference of Squares**

Solve the equation $3 - 12x^2 = 0$.

SOLUTION

First, we factor out the common factor 3, and then use the difference of two squares formula $u^2 - v^2 = (u + v)(u - v)$ with $u = 1$ and $v = 2x$.

$$3 - 12x^2 = 0 \qquad \textit{Standard form}$$
$$3(1 - 4x^2) = 0 \qquad \textit{Common factor}$$
$$3(1 + 2x)(1 - 2x) = 0 \qquad \textit{Factored form}$$
$$1 + 2x = 0, \qquad 1 - 2x = 0 \qquad \textit{Set factors to zero}$$
$$2x = -1 \qquad 2x = 1$$
$$x = -\frac{1}{2} \qquad x = \frac{1}{2} \qquad \textit{Solutions}$$

Remark: In Example 3, note that the common constant factor, 3, does not affect the solution.

If the two factors of a quadratic expression are identical, then the corresponding solution is called a **double** or **repeated** solution. This occurs in the next example.

EXAMPLE 4 A Quadratic Equation Involving a Perfect Square Trinomial

Solve the equation $16x^2 + 8x + 1 = 0$.

SOLUTION

We recognize this equation as a perfect square trinomial that fits the formula $u^2 + 2uv + v^2 = (u + v)^2$, where $u = 4x$ and $v = 1$.

$$16x^2 + 8x + 1 = 0 \qquad \textit{Standard form}$$
$$(4x + 1)^2 = 0 \qquad \textit{Factored form}$$
$$4x + 1 = 0, \qquad 4x + 1 = 0 \qquad \textit{Set factors to zero}$$
$$x = -\frac{1}{4} \qquad x = -\frac{1}{4} \qquad \textit{Repeated solution}$$

Remark: Remember that the Factorization Principle works *only* for equations written in standard form. Therefore, all terms must be collected to one side *before* factoring.

EXAMPLE 5 A Quadratic Equation Involving a General Trinomial

Solve the equation $2x^2 + x = 15$.

SOLUTION

We first write the equation in the standard form

$$2x^2 + x - 15 = 0.$$

Next, we attempt to factor the trinomial $2x^2 + x - 15$ as a product $(ax + b)(cx + d)$ of two binomials where a and c are factors of 2, and b and d are factors of -15. The eight possible factorizations are

$$(2x - 1)(x + 15) \qquad (2x + 1)(x - 15)$$
$$(2x - 3)(x + 5) \qquad (2x + 3)(x - 5)$$
$$(2x - 5)(x + 3) \qquad (2x + 5)(x - 3)$$
$$(2x - 15)(x + 1) \qquad (2x + 15)(x - 1).$$

By testing each sum of the *inner* and *outer* products, we find the correct factorization to be

Outer

$$2x^2 + x - 15 = (2x - 5)(x + 3).$$

Inner

Therefore, the solutions are obtained as follows.

$$(2x - 5)(x + 3) = 0 \qquad\qquad \textit{Factored form}$$
$$2x - 5 = 0, \qquad x + 3 = 0 \qquad \textit{Set factors to zero}$$
$$x = \frac{5}{2} \qquad\qquad x = -3 \qquad\qquad \textit{Solutions}$$

Quadratic equations can also be solved by the **quadratic formula.**

The Quadratic Formula

The solutions of the quadratic equation $ax^2 + bx + c = 0$, $a \neq 0$, are given by the **quadratic formula:**

$$x = \frac{-b \pm \sqrt{b^2 - 4ac}}{2a}.$$

The quantity under the radical sign, $b^2 - 4ac$, is called the **discriminant** of the quadratic expression $ax^2 + bx + c$. It is used to determine the nature of the solutions of a quadratic equation, as listed next.

1. If $b^2 - 4ac > 0$, then there are *two distinct real solutions.*
2. If $b^2 - 4ac = 0$, then there is *one repeated solution.*
3. If $b^2 - 4ac < 0$, then there are *no real solutions.*

The third case (no real solutions) will be studied in Section 5.2.

EXAMPLE 6 Using the Quadratic Formula: Two Distinct Solutions

Use the quadratic formula to solve $x^2 + 3x = 9$.

SOLUTION

In standard form, the equation is

$$x^2 + 3x - 9 = 0$$

where $a = 1$, $b = 3$, and $c = -9$. By the quadratic formula, we have

$$x = \frac{-b \pm \sqrt{b^2 - 4ac}}{2a} = \frac{-3 \pm \sqrt{(3)^2 - 4(1)(-9)}}{2(1)}$$

$$= \frac{-3 \pm \sqrt{45}}{2} = \frac{-3 \pm 3\sqrt{5}}{2}.$$

Thus, the solutions are

$$x = \frac{-3 + 3\sqrt{5}}{2} \qquad \text{and} \qquad x = \frac{-3 - 3\sqrt{5}}{2}.$$

EXAMPLE 7 Using the Quadratic Formula: One Repeated Solution

Use the quadratic formula to solve $8x^2 - 24x + 18 = 0$.

SOLUTION

To simplify the calculations, we begin by dividing both sides of the equation by 2.

$$8x^2 - 24x + 18 = 0 \qquad \qquad \textit{Given}$$
$$4x^2 - 12x + 9 = 0 \qquad \qquad \textit{Divide both sides by 2}$$

Thus, $a = 4$, $b = -12$, and $c = 9$, and by the quadratic formula we obtain

$$x = \frac{-b \pm \sqrt{b^2 - 4ac}}{2a} = \frac{-(-12) \pm \sqrt{(-12)^2 - 4(4)(9)}}{2(4)}$$

$$= \frac{12 \pm \sqrt{0}}{8} = \frac{3}{2}.$$

Note that there is only one (repeated) solution. This occurs because the discriminant is zero.

The discriminant in Example 7 is a perfect square (zero in this case), and we could have *factored* the quadratic as

$$4x^2 - 12x + 9 = (2x - 3)^2 = 0$$

to conclude that the solution is $x = \frac{3}{2}$. Since factoring is easier than applying the quadratic formula, try factoring first. If, however, factors cannot be readily found, then use the quadratic formula. For instance, try solving the quadratic equation $x^2 - x - 12 = 0$ in two ways—by factoring and by the quadratic formula—to see that you get the same solutions either way.

Solving Polynomial Equations of Higher Degree

A polynomial equation of degree greater than 2 can often be solved by factoring the polynomial into products of linear and quadratic factors. For instance, the third-degree equation $x^3 - 7x^2 + 12x = 0$ can be factored as

$$x(x^2 - 7x + 12) = x(x - 3)(x - 4) = 0.$$

By the Factorization Principle, the resulting solutions are $x = 0$, $x = 3$, and $x = 4$. In the next three examples, we show techniques commonly used to solve polynomial equations.

EXAMPLE 8 A Fourth-Degree Difference of Squares Equation

Solve the equation $x^4 - 16 = 0$.

SOLUTION

We can write this equation as the difference of two squares and obtain the solutions as follows.

$$
\begin{aligned}
x^4 - 16 &= 0 &&\text{\textit{Standard form}}\\
(x^2 + 4)(x^2 - 4) &= 0 &&\text{\textit{Factor}}\\
(x^2 + 4)(x + 2)(x - 2) &= 0 &&\text{\textit{Factor again}}
\end{aligned}
$$

The equation $x^2 + 4 = 0$ yields no real solutions because its discriminant, $b^2 - 4ac = 0^2 - 4(1)(4) = -16$, is less than zero. Therefore, the only real solutions are $x = -2$ and $x = 2$, which come from the factors $(x + 2)$ and $(x - 2)$.

EXAMPLE 9 *A Fourth-Degree Trinomial Equation*

Solve the equation $3x^4 - 6x^3 - 12x^2 = 0$.

SOLUTION

In this case, we first factor out the common monomial factor, $3x^2$, and obtain $3x^2(x^2 - 2x - 4) = 0$. Setting both factors to zero and using the quadratic formula for the second equation, we obtain the following.

$$3x^2 = 0, \qquad x^2 - 2x - 4 = 0$$

$$x = 0 \qquad\qquad x = \frac{-(-2) \pm \sqrt{4 - 4(1)(-4)}}{2(1)}$$

$$= \frac{2 \pm \sqrt{20}}{2} = 1 \pm \sqrt{5}$$

Therefore, the solutions are $x = 0$ (repeated), $x = 1 + \sqrt{5}$, and $x = 1 - \sqrt{5}$.

Sometimes polynomials with more than three terms can be factored by a method called **factoring by grouping.**

EXAMPLE 10 *Solving an Equation Using Factoring by Grouping*

Solve the equation $x^3 - 2x^2 - 3x + 6 = 0$.

SOLUTION

By grouping the polynomial as $(x^3 - 2x^2) - (3x - 6)$, we obtain the following results.

$$
\begin{array}{ll}
x^3 - 2x^2 - 3x + 6 = 0 & \textit{Standard form} \\
(x^3 - 2x^2) - (3x - 6) = 0 & \textit{Group terms} \\
x^2(x - 2) - 3(x - 2) = 0 & \textit{Factor groups} \\
(x - 2)(x^2 - 3) = 0 & \textit{Common factor} \\
x - 2 = 0, \qquad x^2 - 3 = 0 & \textit{Set factors to zero} \\
x = 2 \qquad\qquad x^2 = 3 & \\
\qquad\qquad\quad x = \pm\sqrt{3} &
\end{array}
$$

Remark: In Example 10, we solved the equation $x^2 - 3 = 0$ by writing it in the form $x^2 = 3$ and taking the square root of both sides.

EXERCISES 1.2

In Exercises 1–4, determine whether the given value of x is a solution of the equation.

1. $5x - 3 = 3x + 5$
   (a) $x = 0$ (b) $x = -5$
   (c) $x = 4$ (d) $x = 10$

2. $7 - 3x = 5x - 17$
   (a) $x = -3$ (b) $x = 0$
   (c) $x = 8$ (d) $x = 3$

3. $3x^2 + 2x - 5 = 2x^2 - 2$
   (a) $x = -3$ (b) $x = 1$
   (c) $x = 4$ (d) $x = -5$

4. $5x^3 + 2x - 3 = 4x^3 + 2x - 11$
   (a) $x = 2$ (b) $x = -2$
   (c) $x = 0$ (d) $x = 10$

In Exercises 5–26, solve the given equation (if possible) and check your answer.

5. $2(x + 5) - 7 = 3(x - 2)$

6. $2(13t - 15) + 3(t - 19) = 0$

7. $\dfrac{5x}{4} + \dfrac{1}{2} = x - \dfrac{1}{2}$

8. $\dfrac{x}{5} - \dfrac{x}{2} = 3$

9. $0.25x + 0.75(10 - x) = 3$

10. $0.60x + 0.40(100 - x) = 50$

11. $x + 8 = 2(x - 2) - x$

12. $-3(x + 3) = 5(1 - x) - 1$

13. $\dfrac{100 - 4u}{3} = \dfrac{5u + 6}{4} + 6$

14. $\dfrac{17 + y}{y} + \dfrac{32 + y}{y} = 100$

15. $\dfrac{5x - 4}{5x + 4} = \dfrac{2}{3}$ **16.** $\dfrac{10x + 3}{5x + 6} = \dfrac{1}{2}$

17. $10 - \dfrac{13}{x} = 4 + \dfrac{5}{x}$ **18.** $\dfrac{15}{x} - 4 = \dfrac{6}{x} + 3$

19. $\dfrac{1}{x - 3} + \dfrac{1}{x + 3} = \dfrac{10}{x^2 - 9}$

20. $\dfrac{1}{x - 2} + \dfrac{3}{x + 3} = \dfrac{4}{x^2 + x - 6}$

21. $\dfrac{7}{2x + 1} - \dfrac{8x}{2x - 1} = -4$

22. $\dfrac{4}{u - 1} + \dfrac{6}{3u + 1} = \dfrac{15}{3u + 1}$

23. $(x + 2)^2 + 5 = (x + 3)^2$

24. $(x + 1)^2 + 2(x - 2) = (x + 1)(x - 2)$

25. $4 - 2(x - 2b) = ax + 3$

26. $5 + ax = 12 - bx$

In Exercises 27–34, solve the quadratic equation by factoring.

27. $6x^2 + 3x = 0$ **28.** $9x^2 - 1 = 0$

29. $x^2 - 2x - 8 = 0$ **30.** $x^2 + 10x + 25 = 0$

31. $16x^2 + 56x + 49 = 0$ **32.** $3 + 5x - 2x^2 = 0$

33. $2x^2 = 19x + 33$ **34.** $(x + a)^2 - b^2 = 0$

In Exercises 35–38, solve the equation by taking the square root of both sides.

35. $3x^2 = 36$ **36.** $9x^2 = 25$

37. $(x - 12)^2 = 18$ **38.** $(x + 13)^2 = 21$

In Exercises 39–52, use the quadratic formula to solve the equation.

39. $2x^2 + x - 1 = 0$ **40.** $2x^2 - x - 1 = 0$

41. $16x^2 + 8x - 3 = 0$ **42.** $25x^2 - 20x + 3 = 0$

43. $2 + 2x - x^2 = 0$ **44.** $x^2 - 10x + 22 = 0$

45. $12x - 9x^2 = -3$ **46.** $16x^2 + 22 = 40x$

47. $4x^2 + 4x = 7$ **48.** $16x^2 - 40x + 5 = 0$

49. $(y - 5)^2 = 2y$ **50.** $(z + 6)^2 = -2z$

51. $\dfrac{1}{x} - \dfrac{1}{x + 1} = 3$ **52.** $\dfrac{x}{x^2 - 4} + \dfrac{1}{x + 2} = 3$

In Exercises 53–56, use a calculator to solve the equation. Round your answers to three decimal places.

53. $5.1x^2 - 1.7x - 3.2 = 0$

54. $10.4x^2 + 8.6x + 1.2 = 0$

55. $7.06x^2 - 4.85x + 0.50 = 0$

56. $-0.005x^2 + 0.101x - 0.193 = 0$

In Exercises 57–76, find all real solutions of the given equation. Check your answer in the original equation.

57. $4x^4 - 18x^2 = 0$

58. $20x^3 - 125x = 0$

59. $x^3 - 2x^2 - 3x = 0$

60. $2x^4 - 15x^3 + 18x^2 = 0$

61. $x^4 - 81 = 0$

62. $x^6 - 64 = 0$

63. $5x^3 + 30x^2 + 45x = 0$

64. $9x^4 - 24x^3 + 16x^2 = 0$

65. $x^3 - 3x^2 - x + 3 = 0$

66. $x^3 + 2x^2 + 3x + 6 = 0$

67. $x^4 - x^3 + x - 1 = 0$

68. $x^4 + 2x^3 - 8x - 16 = 0$

69. $x^4 - 10x^2 + 9 = 0$

70. $x^4 - 29x^2 + 100 = 0$

71. $x^4 + 5x^2 - 36 = 0$

72. $x^4 - 4x^2 + 3 = 0$

73. $4x^4 - 65x^2 + 16 = 0$

74. $36t^4 + 29t^2 - 7 = 0$

75. $x^6 + 7x^3 - 8 = 0$

76. $x^6 + 3x^3 + 2 = 0$

1.3 The Cartesian Plane

Just as we can represent real numbers by points on the real line, we can represent ordered pairs of real numbers by points in a plane. This plane is called the **rectangular coordinate system,** or the **Cartesian plane,** after the French mathematician René Descartes (1596–1650).

The Cartesian plane is formed by two real lines intersecting at right angles, as shown in Figure 1.8. The horizontal real line is usually called the **x-axis,** and the vertical real line is usually called the **y-axis.** The point of intersection of these two axes is called the **origin,** and the axes divide the plane into four parts called **quadrants.**

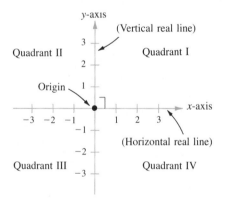

FIGURE 1.8   The Cartesian Plane

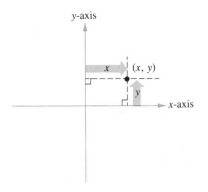

y-axis

x-axis

FIGURE 1.9

Each point in the plane corresponds to an **ordered pair** (x, y) of real numbers x and y, called **coordinates** of the point. The **x-coordinate** (or **abscissa**) represents the directed distance from the *y*-axis to the point, and the **y-coordinate** (or **ordinate**) represents the directed distance from the *x*-axis to the point, as shown in Figure 1.9.

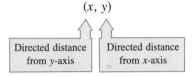

(x, y)

| Directed distance from *y*-axis | Directed distance from *x*-axis |

Remark: It is customary to use the notation (x, y) to denote both a point in the plane and an open interval on the real line. The nature of a specific problem will show which of the two we are talking about.

EXAMPLE 1 *Plotting Points in the Cartesian Plane*

Locate the points $(-1, 2)$, $(3, 4)$, $(0, 0)$, $(3, 0)$, and $(-2, -3)$ in the Cartesian plane.

SOLUTION

To plot the point $(-1, 2)$ we envision a vertical line through -1 on the *x*-axis and a horizontal line through 2 on the *y*-axis. The intersection of these two lines is the point $(-1, 2)$, as shown in Figure 1.10. The other four points can be plotted in a similar way.

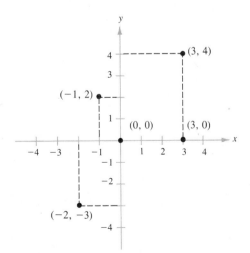

FIGURE 1.10

The value of the rectangular coordinate system is that it allows us to visualize relationships between the variables x and y. Today, Descartes's ideas are in common use in virtually every scientific and business-related field.

EXAMPLE 2 *An Application of the Rectangular Coordinate System*

The prime interest rate in the United States from 1976 to 1986 is given in Table 1.1. Plot these points on a rectangular coordinate system.

TABLE 1.1

Year	1976	1977	1978	1979	1980	1981
Rate	6.84	6.83	9.06	12.67	15.26	18.87

Year	1982	1983	1984	1985	1986
Rate	14.86	10.79	11.51	9.93	8.33

SOLUTION

The points are shown in Figure 1.11. The break in the x-axis indicates that we have omitted the numbers between 0 and 1976.

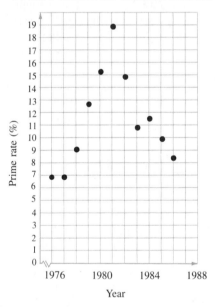

FIGURE 1.11   Prime Interest Rate

The Distance Between Two Points in the Plane

We know from Section 1.1, that the distance d between two points a and b on a number line is simply

$$d = |a - b|.$$

This same rule is used to find the distance between two points that lie on the same *vertical* or *horizontal* line in the plane.

EXAMPLE 3   *Finding Horizontal and Vertical Distances*

(a) Find the distance between the points $(1, -1)$ and $(1, 4)$.
(b) Find the distance between the points $(-3, -1)$ and $(1, -1)$.

SOLUTION

(a) Because the x-coordinates are equal, we visualize a vertical line through the points $(1, -1)$ and $(1, 4)$, as shown in Figure 1.12. The distance between these two points is given by the absolute value of the difference of their y-coordinates. That is,

$$\text{vertical distance} = |4 - (-1)| = 5. \qquad \textit{Subtract y-coordinates}$$

(b) Because the y-coordinates are equal, we visualize a horizontal line through the points $(-3, -1)$ and $(1, -1)$, as shown in Figure 1.12. The distance between these two points is given by the absolute value of the difference of their x-coordinates. That is,

$$\text{horizontal distance} = |1 - (-3)| = 4 \qquad \textit{Subtract x-coordinates}$$

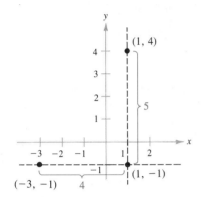

FIGURE 1.12

The technique used in Example 3 can be used to develop a general formula for finding the distance between two points in the plane. This general formula will work for any two points, even if they do not lie on the same vertical or horizontal line. To develop the formula, we use the Pythagorean Theorem, which says that for a right triangle with hypotenuse c and sides a and b, we have the relationship $a^2 + b^2 = c^2$, as shown in Figure 1.13. (The converse is also true. That is, if $a^2 + b^2 = c^2$, then the triangle is a right triangle.)

Now, suppose we want to determine the distance d between the two points (x_1, y_1) and (x_2, y_2) in the plane. With these two points, a right triangle can be formed, as shown in Figure 1.14. Note that the third vertex of the triangle is (x_1, y_2). Since (x_1, y_1) and (x_1, y_2) lie on the same vertical line, the length of the vertical side of the triangle is $|y_2 - y_1|$. Similarly, the length of the horizontal side is $|x_2 - x_1|$. Thus, by the Pythagorean Theorem, we have

$$d^2 = |x_2 - x_1|^2 + |y_2 - y_1|^2.$$

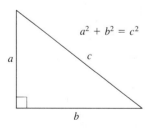

Pythagorean Theorem

FIGURE 1.13

Prerequisites for Trigonometry

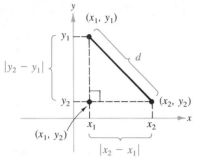

Distance Between Two Points

FIGURE 1.14

Since the distance d must be positive, we choose the positive square root and write

$$d = \sqrt{|x_2 - x_1|^2 + |y_2 - y_1|^2}.$$

Finally, replacing $|x_2 - x_1|^2$ and $|y_2 - y_1|^2$ by the equivalent expressions $(x_2 - x_1)^2$ and $(y_2 - y_1)^2$ gives us the following formula for the distance between any two points in a coordinate plane.

The Distance Formula

The distance d between the points (x_1, y_1) and (x_2, y_2) in the coordinate plane is

$$d = \sqrt{(x_2 - x_1)^2 + (y_2 - y_1)^2}.$$

Remark: Note that for the special case in which the two points lie on the same vertical or horizontal line, the Distance Formula still works. For instance, applying the Distance Formula to the points $(1, -1)$ and $(1, 4)$, we obtain

$$d = \sqrt{(1 - 1)^2 + [4 - (-1)]^2} = \sqrt{5^2} = 5$$

which is the same result we obtained in Example 3.

EXAMPLE 4 Finding the Distance Between Two Points

Find the distance between the points $(-2, 1)$ and $(3, 4)$.

SOLUTION

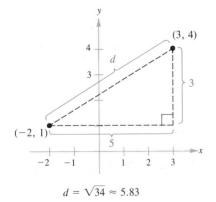

$d = \sqrt{34} \approx 5.83$

FIGURE 1.15

Letting $(x_1, y_1) = (-2, 1)$ and $(x_2, y_2) = (3, 4)$, we apply the Distance Formula to obtain

$$\begin{aligned}
d &= \sqrt{[3 - (-2)]^2 + (4 - 1)^2} \\
&= \sqrt{5^2 + 3^2} \\
&= \sqrt{25 + 9} \\
&= \sqrt{34} \approx 5.83.
\end{aligned}$$

See Figure 1.15.

In Example 4, the figure provided was not essential to the solution of the problem. *Nevertheless*, we recommend that you include graphs with your problem solutions.

The Cartesian Plane

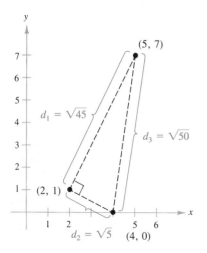

FIGURE 1.16

EXAMPLE 5 An Application of the Distance Formula

Show that the points $(2, 1)$, $(4, 0)$, and $(5, 7)$ are the vertices of a right triangle.

SOLUTION

The three points are plotted in Figure 1.16. Using the Distance Formula, we find the lengths of the three sides of the triangle.

$$d_1 = \sqrt{(5 - 2)^2 + (7 - 1)^2} = \sqrt{9 + 36} = \sqrt{45}$$
$$d_2 = \sqrt{(4 - 2)^2 + (0 - 1)^2} = \sqrt{4 + 1} = \sqrt{5}$$
$$d_3 = \sqrt{(5 - 4)^2 + (7 - 0)^2} = \sqrt{1 + 49} = \sqrt{50}$$

Since $d_1^2 + d_2^2 = 45 + 5 = 50 = d_3^2$, we can conclude from the Pythagorean Theorem that the triangle is a right triangle.

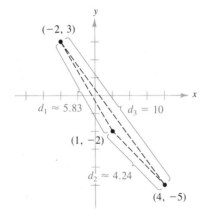

FIGURE 1.17

EXAMPLE 6 Testing for Collinearity

Determine whether the points $(-2, 3)$, $(1, -2)$, and $(4, -5)$ are collinear (lie on the same line) or form the vertices of a triangle.

SOLUTION

By the Distance Formula, we have

$$d_1 = \sqrt{(1 + 2)^2 + (-2 - 3)^2} = \sqrt{34} \approx 5.83$$
$$d_2 = \sqrt{(4 - 1)^2 + (-5 + 2)^2} = \sqrt{18} \approx 4.24$$
$$d_3 = \sqrt{(4 + 2)^2 + (-5 - 3)^2} = \sqrt{100} = 10.$$

Since no two distances add up to the third, the points are not collinear. Thus, they form a triangle, as shown in Figure 1.17.

EXAMPLE 7 Finding Points at a Specified Distance from a Given Point

Find x so that the distance between $(x, 3)$ and $(2, -1)$ is 5.

SOLUTION

As we begin this problem we do not know how many values of x satisfy the given requirements. Even so, we can use the Distance Formula to find the distance to be

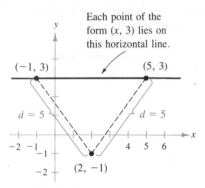

FIGURE 1.18

$$\sqrt{(x-2)^2 + (3+1)^2} = 5.$$

Squaring both sides of this equation, we obtain

$$(x^2 - 4x + 4) + 16 = 25$$
$$x^2 - 4x - 5 = 0$$
$$(x - 5)(x + 1) = 0$$
$$x = 5 \text{ or } -1.$$

We see that there are two solutions, and we conclude that both of the points $(5, 3)$ and $(-1, 3)$ lie five units from the point $(2, -1)$, as shown in Figure 1.18.

The Midpoint Formula

Next we introduce a formula for finding the midpoint of a line segment joining two points. The coordinates of the midpoint are simply the average values of the corresponding coordinates of the two endpoints.

The Midpoint Formula

The **midpoint** of the line segment joining the points (x_1, y_1) and (x_2, y_2) is

$$\left(\frac{x_1 + x_2}{2}, \frac{y_1 + y_2}{2} \right).$$

Midpoint Formula

FIGURE 1.19

PROOF

Using Figure 1.19, we need to show that

$$d_1 = d_2 \quad \text{and} \quad d_1 + d_2 = d_3.$$

By the Distance Formula, we obtain

$$d_1 = \sqrt{\left(\frac{x_1 + x_2}{2} - x_1\right)^2 + \left(\frac{y_1 + y_2}{2} - y_1\right)^2} = \frac{1}{2}\sqrt{(x_2 - x_1)^2 + (y_2 - y_1)^2}$$

$$d_2 = \sqrt{\left(x_2 - \frac{x_1 + x_2}{2}\right)^2 + \left(y_2 - \frac{y_1 + y_2}{2}\right)^2} = \frac{1}{2}\sqrt{(x_2 - x_1)^2 + (y_2 - y_1)^2}$$

$$d_3 = \sqrt{(x_2 - x_1)^2 + (y_2 - y_1)^2}.$$

Thus, it follows that $d_1 = d_2$ and $d_1 + d_2 = d_3$.

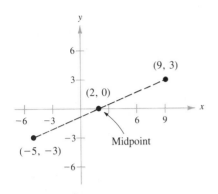

FIGURE 1.20

EXAMPLE 8 *Finding the Midpoint of a Line Segment*

Find the midpoint of the line segment joining the points $(-5, -3)$ and $(9, 3)$.

SOLUTION

Figure 1.20 shows the two given points and their midpoint. By the Midpoint Formula, we have

$$\text{midpoint} = \left(\frac{-5 + 9}{2}, \frac{-3 + 3}{2} \right)$$

$$= (2, 0).$$

WARM UP

Simplify the given expressions.

1. $\sqrt{(2 - 6)^2 + [1 - (-2)]^2}$

2. $\sqrt{(1 - 4)^2 + (-2 - 1)^2}$

3. $\dfrac{4 + (-2)}{2}$

4. $\dfrac{-1 + (-3)}{2}$

5. $\sqrt{18} + \sqrt{45}$

6. $\sqrt{12} + \sqrt{44}$

Solve for x or y.

7. $\sqrt{(4 - x)^2 + (5 - 2)^2} = \sqrt{58}$

8. $\sqrt{(8 - 6)^2 + (y - 5)^2} = 2\sqrt{5}$

9. $\dfrac{x + 3}{2} = 7$

10. $\dfrac{-2 + y}{2} = 1$

EXERCISES 1.3

In Exercises 1–4, sketch the polygon with the indicated vertices.

1. Triangle: $(-1, 1)$, $(2, -1)$, $(3, 4)$

2. Triangle: $(0, 3)$, $(-1, -2)$, $(4, 8)$

3. Square: $(2, 4)$, $(5, 1)$, $(2, -2)$, $(-1, 1)$

4. Parallelogram: $(5, 2)$, $(7, 0)$, $(1, -2)$, $(-1, 0)$

In Exercises 5–8, find the distance between the given points. [*Note:* In each case the two points lie on the same horizontal or vertical line.]

5. $(6, -3)$, $(6, 5)$

6. $(1, 4)$, $(8, 4)$

7. $(-3, -1)$, $(2, -1)$

8. $(-3, -4)$, $(-3, 6)$

In Exercises 9–12, (a) find the length of the two sides of the right triangle and use the Pythagorean Theorem to find the length of the hypotenuse, and (b) use the Distance Formula to find the length of the hypotenuse of the triangle.

9.

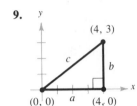

10.

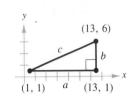

11.

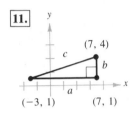

12.

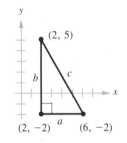

In Exercises 13–24, (a) plot the points, (b) find the distance between the points, and (c) find the midpoint of the line segment joining the points.

13. $(1, 1)$, $(9, 7)$

14. $(1, 12)$, $(6, 0)$

15. $(-4, 10)$, $(4, -5)$

16. $(-7, -4)$, $(2, 8)$

17. $(-1, 2)$, $(5, 4)$

18. $(2, 10)$, $(10, 2)$

19. $\left(\frac{1}{2}, 1\right)$, $\left(-\frac{5}{2}, \frac{4}{3}\right)$

20. $\left(-\frac{1}{3}, -\frac{1}{3}\right)$, $\left(-\frac{1}{6}, -\frac{1}{2}\right)$

21. $(6.2, 5.4)$, $(-3.7, 1.8)$

22. $(-16.8, 12.3)$, $(5.6, 4.9)$

23. $(-36, -18)$, $(48, -72)$

24. $(1.451, 3.051)$, $(5.906, 11.360)$

In Exercises 25–28, show that the given points form the vertices of the indicated polygon. (A rhombus is a parallelogram whose sides are all of the same length.)

25. Right triangle: $(4, 0)$, $(2, 1)$, $(-1, -5)$

26. Isosceles triangle: $(1, -3)$, $(3, 2)$, $(-2, 4)$

27. Rhombus: $(0, 0)$, $(1, 2)$, $(2, 1)$, $(3, 3)$

28. Parallelogram: $(0, 1)$, $(3, 7)$, $(4, 4)$, $(1, -2)$

In Exercises 29 and 30, find x so that the distance between the points is 13.

29. $(1, 2)$, $(x, -10)$

30. $(-8, 0)$, $(x, 5)$

In Exercises 31 and 32, find y so that the distance between the points is 17.

31. $(0, 0)$, $(8, y)$

32. $(-8, 4)$, $(7, y)$

In Exercises 33 and 34, find a relationship between x and y so that (x, y) is equidistant from the two given points.

33. $(4, -1)$, $(-2, 3)$

34. $\left(3, \frac{5}{2}\right)$, $(-7, -1)$

In Exercises 35–42, determine the quadrant(s) in which (x, y) is located so that the given conditions are satisfied.

35. $x > 0$ and $y < 0$

36. $x < 0$ and $y < 0$

37. $x > 0$ and $y > 0$

38. $x < 0$ and $y > 0$

39. $x = -4$ and $y > 0$

40. $x > 2$ and $y = 3$

41. $y < -5$

42. $x > 4$

43. A line segment has (x_1, y_1) as one endpoint and (x_m, y_m) as its midpoint. Find the other endpoint (x_2, y_2) of the line segment in terms of x_1, y_1, x_m, and y_m.

44. Use the result of Exercise 43 to find the endpoint of a line segment if the other endpoint and midpoint are, respectively,
(a) $(-3, 2)$ and $(1, 7)$ (b) $(-5, 11)$ and $(2, 4)$.

45. Use the Midpoint Formula twice to find the three points that divide the line segment joining (x_1, y_1) and (x_2, y_2) into four parts.

46. Use the result of Exercise 45 to find the points that divide the line segment joining the given points into four equal parts.
(a) $(1, -2)$, $(4, -1)$ (b) $(-2, -3)$, $(0, 0)$

In Exercises 47 and 48, use the Midpoint Formula to estimate the sales of a company for 1983, given the sales in 1980 and 1986. Assume the sales followed a linear pattern.

47.

Year	1980	1986
Sales	$520,000	$740,000

48.

Year	1980	1986
Sales	$4,200,000	$5,650,000

49. Find the price of corn for the following dates (see figure).
    (a) May 19   (b) July 7
    (c) August 11   (d) September 15

50. Find the *decrease* in corn prices from May 19 to September 15.

FIGURE FOR 49 and 50

51. Find the trade deficit for the following months (see figure).
    (a) February   (b) December

52. Find the percentage *increase* in the trade deficit from October to November.

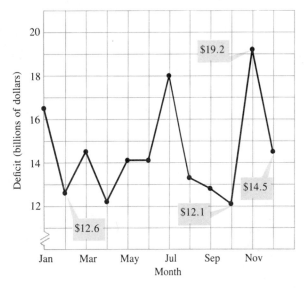

U.S. Trade Deficit

FIGURE FOR 51 and 52

1.4 Graphs of Equations

News magazines frequently show graphs comparing the rate of inflation, the gross national product, wholesale prices, or the unemployment rate to the time of year. Industrial firms and businesses use graphs to report their monthly production and sales statistics. Such graphs provide a simple geometric picture of the way one quantity changes with respect to another.

Frequently, the relationship between two quantities is expressed in the form of an equation. In this section, we introduce the basic procedure for determining the geometric picture associated with an algebraic equation.

For an equation in variables x and y, a point (a, b) is a **solution point** if the substitution of $x = a$ and $y = b$ satisfies the equation. Of course, most equations have many solution points. For example, the equation

$$3x + y = 5$$

has solution points $(0, 5)$, $(1, 2)$, $(2, -1)$, $(3, -4)$, and so on. The set of all solution points for a given equation is called the **graph** of the equation.

Prerequisites for Trigonometry

The Point-Plotting Method of Graphing

To sketch a graph of an equation by point plotting, use the following steps.

1. If possible, rewrite the equation by isolating one of the variables.
2. Make up a table of several solution points.
3. Plot these points in the coordinate plane.
4. Connect the points with a smooth curve.

EXAMPLE 1   *Sketching the Graph of an Equation*

Sketch a graph of the equation $3x + y = 5$.

SOLUTION

In this case we isolate variable y, to get

$$y = 5 - 3x.$$

Using negative, zero, and positive values for x, we obtain the following table of values (solution points).

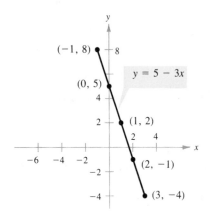

x		−1	0	1	2	3
$y = 5 - 3x$		8	5	2	−1	−4

Next, we plot these points and connect them as shown in Figure 1.21. It appears that the graph is a straight line. (We will study lines extensively in Sections 7.1 and 7.2.)

FIGURE 1.21

Step 4 of the point-plotting method can be difficult. For instance, how would you connect the four points in Figure 1.22? Without further information about the equation, any one of the three graphs in Figure 1.23 would be reasonable.

Graphs of Equations

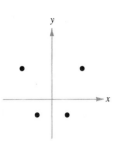

FIGURE 1.22

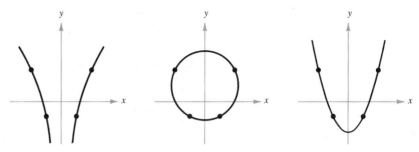

FIGURE 1.23

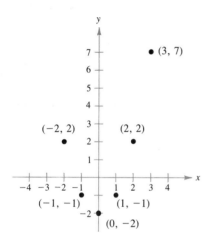

Plot several points.

FIGURE 1.24

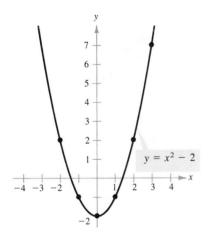

Connect points with a smooth curve.

FIGURE 1.25

With too few solution points, we could grossly misrepresent the graph of a given equation. Just how many points should be plotted? For a straight-line graph, two points are sufficient. For more complicated graphs, we need many more points. More sophisticated techniques will be discussed in later sections, but for now plot enough points so as to reveal the essential behavior of the graph. A programmable calculator is useful for determining the many solution points needed for an accurate graph.

EXAMPLE 2   *Sketching the Graph of an Equation*

Sketch the graph of the equation $y = x^2 - 2$.

SOLUTION

First, we make a table of values by choosing several convenient values of x and calculating the corresponding values of y.

x	-2	-1	0	1	2	3
$y = x^2 - 2$	2	-1	-2	-1	2	7

Next, we plot these points, as shown in Figure 1.24. Finally, we connect the points by a smooth curve, as shown in Figure 1.25.

Intercepts of a Graph

When choosing points to plot, we suggest you start with those that are easiest to calculate. Two such points have zero as either their x- or y-coordinate.

These points are called **intercepts,** because they are points at which the graph intersects the *x*- or *y*-axis.

Definition of Intercepts

The point $(a, 0)$ is called an **x-intercept** of the graph of an equation if it is a solution point of the equation. To find the *x*-intercepts, let *y* be zero and solve the equation for *x*.

The point $(0, b)$ is called a **y-intercept** of the graph of an equation if it is a solution point of the equation. To find the *y*-intercepts, let *x* be zero and solve the equation for *y*.

Remark: Some texts denote the *x*-intercept as the *x*-coordinate of the point $(a, 0)$ rather than the point itself. Unless it is necessary to make a distinction, we will use "intercept" to mean either the point or the coordinate.

Of course, it is possible that a particular graph will have no intercepts or several intercepts. For instance, consider the three graphs in Figure 1.26.

FIGURE 1.26

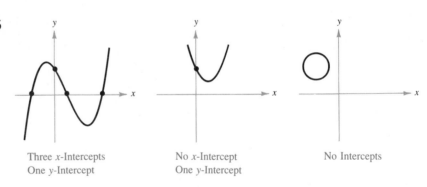

Three *x*-Intercepts
One *y*-Intercept

No *x*-Intercept
One *y*-Intercept

No Intercepts

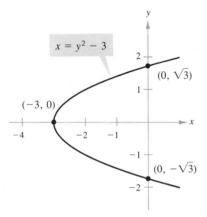

FIGURE 1.27

EXAMPLE 3 *Finding x- and y-Intercepts*

Find the *x*- and *y*-intercepts for the graph of $y^2 - 3 = x$.

SOLUTION

Let $y = 0$. Then $-3 = x$.

    x-intercept: $(-3, 0)$

Let $x = 0$. Then $y^2 - 3 = 0$ has solutions $y = \pm\sqrt{3}$.

    y-intercepts: $(0, \sqrt{3}), (0, -\sqrt{3})$

See Figure 1.27.

Graphs of Equations

Symmetry

The graph shown in Figure 1.27 is said to be "symmetric" with respect to the *x*-axis. This means that if the Cartesian plane were folded along the *x*-axis, the portion of the graph above the *x*-axis would coincide with the portion below the *x*-axis. Symmetry with respect to the *y*-axis can be described in a similar manner.

Knowing the symmetry of a graph *before* attempting to sketch it is beneficial because we then need only half as many solution points as we would otherwise. We define the three basic types of symmetry as follows. (See Figure 1.28.)

Definition of Symmetry with Respect to the Coordinate Axes and the Origin

A graph is said to be **symmetric with respect to the y-axis** if, whenever (x, y) is on the graph, $(-x, y)$ is also on the graph.

A graph is said to be **symmetric with respect to the x-axis** if, whenever (x, y) is on the graph, $(x, -y)$ is also on the graph.

A graph is said to be **symmetric with respect to the origin** if, whenever (x, y) is on the graph, $(-x, -y)$ is also on the graph.

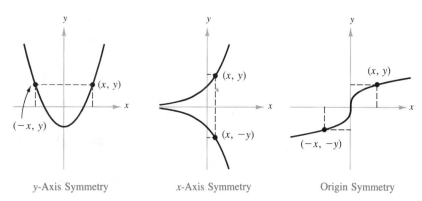

| *y*-Axis Symmetry | *x*-Axis Symmetry | Origin Symmetry |

FIGURE 1.28

Suppose we apply this definition of symmetry to the graph of the equation $y = x^2 - 2$. Replacing x with $-x$ produces

$$y = (-x)^2 - 2$$
$$y = x^2 - 2.$$

Prerequisites for Trigonometry

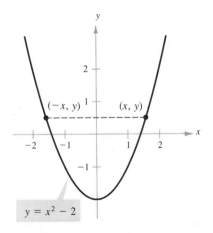

$y = x^2 - 2$

y-Axis Symmetry

FIGURE 1.29

Since this substitution does not change the equation, it follows that if (x, y) is a solution point of the equation, then $(-x, y)$ must also be a solution point. Thus, the graph of $y = x^2 - 2$ is symmetric with respect to the y-axis. (See Figure 1.29.)

A similar test can be made for symmetry with respect to the x-axis or to the origin. These three tests are summarized as follows.

Tests for Symmetry

1. The graph of an equation is symmetric with respect to the *y-axis* if replacing x with $-x$ yields an equivalent equation.
2. The graph of an equation is symmetric with respect to the *x-axis* if replacing y with $-y$ yields an equivalent equation.
3. The graph of an equation is symmetric with respect to the *origin* if replacing x with $-x$ *and* y with $-y$ yields an equivalent equation.

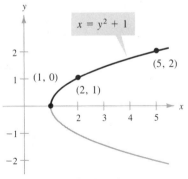

$x = y^2 + 1$

First plot the points above the x-axis, then use symmetry to complete the graph.

FIGURE 1.30

EXAMPLE 4 *Using Intercepts and Symmetry as Sketching Aids*

Use intercepts and symmetry to sketch the graph of $x - y^2 = 1$.

SOLUTION

Intercepts: Letting $x = 0$, we see that $-y^2 = 1$ or $y^2 = -1$ has no real solutions. Hence, there are no y-intercepts. Letting $y = 0$, we obtain $x = 1$. Thus, the x-intercept is $(1, 0)$.

Symmetry: Replacing y with $-y$ yields

$$x - (-y)^2 = 1$$
$$x - y^2 = 1$$

which means that the graph is symmetric with respect to the x-axis.

Using symmetry, we need only find solution points above the x-axis and then reflect them to obtain the desired graph (see Figure 1.30).

y	0	1	2
$x = y^2 + 1$	1	2	5

Graphs of Equations

EXAMPLE 5 *Using Intercepts and Symmetry as Sketching Aids*

Use intercepts and symmetry to sketch the graph of $y = x^3 - 4x$.

SOLUTION

Intercepts: Letting $y = 0$, we have

$$0 = x^3 - 4x = x(x^2 - 4).$$

Setting both factors equal to zero and solving for x, we obtain $x = 0$ and $x = \pm 2$. Thus, the x-intercepts are $(0, 0)$, $(2, 0)$, and $(-2, 0)$. Letting $x = 0$, we obtain $y = 0$, which tells us that the y-intercept is $(0, 0)$.

Symmetry: Replacing x with $-x$ and y with $-y$ yields

$$-y = (-x)^3 - 4(-x)$$
$$-y = -x^3 + 4x$$
$$y = x^3 - 4x \qquad\qquad \textit{Multiply by } -1$$

which is the original equation. Thus, the graph of $y = x^3 - 4x$ is symmetric with respect to the origin.

Using the intercepts, symmetry, and the following table of values, we obtain the graph shown in Figure 1.31.

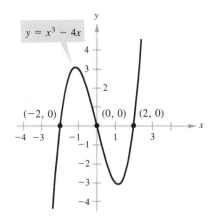

FIGURE 1.31

x	0	1	2	3
$y = x^3 - 4x$	0	-3	0	15

Not all equations have graphs that are symmetric with respect to one of the axes or the origin. For instance, the graph of the equation shown in Example 6 has none of these three types of symmetry.

EXAMPLE 6 *Sketching the Graph of an Equation*

Sketch the graph of $y = |4x - x^2|$.

SOLUTION

Intercepts: Letting $x = 0$ yields $y = 0$, which means that $(0, 0)$ is a y-intercept. Letting $y = 0$ yields $x = 0$ and $x = 4$, which means that $(0, 0)$ and $(4, 0)$ are x-intercepts.

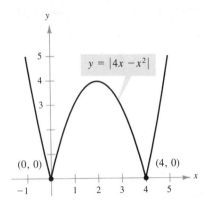

$y = |4x - x^2|$

(0, 0)

(4, 0)

FIGURE 1.32

Symmetry: This equation fails all three tests for symmetry and consequently its graph is not symmetric with respect to either axis or to the origin.

The absolute value sign indicates that y is always nonnegative, and we obtain the following table of values.

x	-1	0	1	2	3	4	5
$y = |4x - x^2|$	5	0	3	4	3	0	5

The graph is shown in Figure 1.32.

Circles

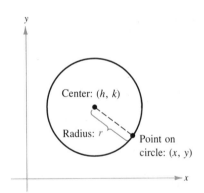

Center: (h, k)

Radius: r

Point on circle: (x, y)

FIGURE 1.33

So far in this section we have studied the point-plotting method and two additional concepts (intercepts and symmetry) that can be used to streamline the graphing procedure.

Another graphing aid is *equation recognition*, the ability to recognize the general shape of a graph simply by looking at its equation. A circle is one type of graph whose equation is easily recognized.

Figure 1.33 shows a circle of radius r with center at (h, k). The point (x, y) is on this circle if and only if its distance from the center (h, k) is r. This means that a **circle** consists of the set of all points (x, y) that are at a given positive distance r from a fixed point (h, k). Expressing this relationship by means of the Distance Formula, we have

$$\sqrt{(x - h)^2 + (y - k)^2} = r.$$

By squaring both sides of this equation, we obtain the following definition.

Standard Form of the Equation of a Circle

The **standard form of the equation of a circle** with radius r and center at (h, k) is

$$(x - h)^2 + (y - k)^2 = r^2.$$

Remark: The standard form of the equation of a circle whose *center is the origin* is simply

$$x^2 + y^2 = r^2.$$

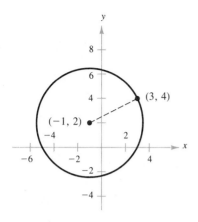

FIGURE 1.34

EXAMPLE 7 *Finding an Equation of a Circle*

The point $(3, 4)$ lies on a circle whose center is at $(-1, 2)$, as shown in Figure 1.34. Find an equation for the circle.

SOLUTION

The radius r of the circle is the distance between $(-1, 2)$ and $(3, 4)$. Thus, we have

$$\begin{aligned} r &= \sqrt{[3 - (-1)]^2 + (4 - 2)^2} \\ &= \sqrt{16 + 4} \\ &= \sqrt{20}. \end{aligned}$$

We conclude that the standard equation for this circle is

$$[x - (-1)]^2 + (y - 2)^2 = (\sqrt{20})^2$$
$$(x + 1)^2 + (y - 2)^2 = 20.$$

If we remove parentheses in the standard equation in Example 7, we obtain

$$(x + 1)^2 + (y - 2)^2 = 20 \qquad \textit{Standard form}$$
$$x^2 + 2x + 1 + y^2 - 4y + 4 = 20$$
$$x^2 + y^2 + 2x - 4y - 15 = 0 \qquad \textit{General form}$$

where the last equation is in the **general form of the equation of a circle:**

$$Ax^2 + Ay^2 + Dx + Ey + F = 0, \qquad A \neq 0.$$

The general form of the equation of a circle is less useful than the standard form. For instance, it is not immediately apparent from the general equation of the circle in Example 7 that the center is at $(-1, 2)$ and the radius is $\sqrt{20}$. To graph the equation of a circle, it is best to write the equation in standard form. We do this by **completing the square,** as demonstrated in the following example.

EXAMPLE 8 *Sketching a Circle*

Sketch the circle whose general equation is

$$4x^2 + 4y^2 + 20x - 16y + 37 = 0.$$

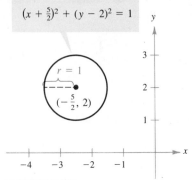

$(x + \frac{5}{2})^2 + (y - 2)^2 = 1$

$r = 1$

$(-\frac{5}{2}, 2)$

FIGURE 1.35

SOLUTION

To complete the square we first divide by 4 so that the coefficients of x^2 and y^2 are both 1.

$$4x^2 + 4y^2 + 20x - 16y + 37 = 0 \qquad \text{\textit{General form}}$$

$$x^2 + y^2 + 5x - 4y + \frac{37}{4} = 0 \qquad \text{\textit{Divide by 4}}$$

$$(x^2 + 5x + \quad) + (y^2 - 4y + \quad) = -\frac{37}{4} \qquad \text{\textit{Group terms}}$$

$$\left(x^2 + 5x + \frac{25}{4}\right) + (y^2 - 4y + 4) = -\frac{37}{4} + \frac{25}{4} + 4 \qquad \begin{array}{l}\text{\textit{Add 25/4 and 4}}\\\text{\textit{to both sides}}\end{array}$$

$$\underbrace{\qquad}_{(\text{half})^2} \qquad \underbrace{\qquad}_{(\text{half})^2}$$

$$\left(x + \frac{5}{2}\right)^2 + (y - 2)^2 = 1 \qquad \text{\textit{Standard form}}$$

Therefore, the circle is centered at $(-5/2, 2)$, and its radius is 1, as shown in Figure 1.35.

Remark: The general equation $Ax^2 + Ay^2 + Dx + Ey + F = 0$ may not always represent a circle. Such an equation will have no solution points if the procedure of completing the square yields the *impossible* result

$$(x - h)^2 + (y - k)^2 = \text{negative number.}$$

Moreover, the general equation $Ax^2 + Ay^2 + Dx + Ey + F = 0$ will have exactly one solution point if the procedure of completing the square yields the result $(x - h)^2 + (y - k)^2 = 0$. The point (h, k) is the only solution point for this equation.

The Unit Circle

In Chapter 2 we will use the circle given by

$$x^2 + y^2 = 1$$

to introduce the trigonometric functions. We call the graph of this equation the **unit circle.** It has a radius of 1, and its center is at the origin. There are several points on the unit circle that play a special role in trigonometry. We discuss some of the properties of these special points in the next two examples. Study both examples carefully.

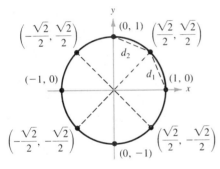

FIGURE 1.36

EXAMPLE 9 Points on the Unit Circle

Use the Distance Formula to show that the points shown in Figure 1.36 divide the unit circle into eight equal parts.

SOLUTION

Referring to Figure 1.36, we use the the Distance Formula as follows.

$$d_1 = \sqrt{\left(\frac{\sqrt{2}}{2} - 1\right)^2 + \left(\frac{\sqrt{2}}{2} - 0\right)^2}$$

$$= \sqrt{\frac{2}{4} - \sqrt{2} + 1 + \frac{2}{4}} = \sqrt{2 - \sqrt{2}}$$

$$d_2 = \sqrt{\left(\frac{\sqrt{2}}{2} - 0\right)^2 + \left(\frac{\sqrt{2}}{2} - 1\right)^2}$$

$$= \sqrt{\frac{2}{4} + \frac{2}{4} - \sqrt{2} + 1} = \sqrt{2 - \sqrt{2}}$$

In a similar way, we could show that the length of the chord connecting any two adjacent points is $\sqrt{2 - \sqrt{2}}$. From this we can conclude that these eight points divide the unit circle into eight equal parts.

EXAMPLE 10 Points on the Unit Circle

Use the Distance Formula to show that the points shown in Figure 1.37 divide the unit circle into 12 equal parts.

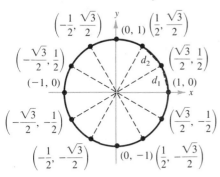

FIGURE 1.37

SOLUTION

Referring to Figure 1.37, we use the Distance Formula as follows.

$$d_1 = \sqrt{\left(\frac{\sqrt{3}}{2} - 1\right)^2 + \left(\frac{1}{2} - 0\right)^2}$$

$$= \sqrt{\frac{3}{4} - \sqrt{3} + 1 + \frac{1}{4}} = \sqrt{2 - \sqrt{3}}$$

$$d_2 = \sqrt{\left(\frac{1}{2} - \frac{\sqrt{3}}{2}\right)^2 + \left(\frac{\sqrt{3}}{2} - \frac{1}{2}\right)^2}$$

$$= \sqrt{\frac{1}{4} - \frac{\sqrt{3}}{2} + \frac{3}{4} + \frac{3}{4} - \frac{\sqrt{3}}{2} + \frac{1}{4}} = \sqrt{2 - \sqrt{3}}$$

In a similar way, we could show that the length of the chord connecting any two adjacent points is $\sqrt{2 - \sqrt{3}}$. From this we can conclude that these 12 points divide the unit circle into 12 equal parts.

WARM UP

Solve for y in terms of x.

1. $3x - 5y = 2$ 　　　　　　　　　　　　　**2.** $x^2 - 4x + 2y - 5 = 0$

Solve for x.

3. $x^2 - 4x + 4 = 0$ 　　　　　　　　　　　**4.** $(x - 1)(x + 5) = 0$
5. $x^3 - 9x = 0$ 　　　　　　　　　　　　　**6.** $x^4 - 8x^2 + 16 = 0$

Simplify the equations.

7. $-y = (-x)^3 + 4(-x)$ 　　　　　　　　　**8.** $(-x)^2 + (-y)^2 = 4$
9. $y = 4(-x)^2 + 8$ 　　　　　　　　　　　**10.** $(-y)^2 = 3(-x)^2 + 4$

EXERCISES 1.4

In Exercises 1–6, determine whether the indicated points lie on the graph of the given equation.

1. $y = \sqrt{x + 4}$
　(a) $(0, 2)$ 　　　　　　　(b) $(5, 3)$

2. $y = x^2 - 3x + 2$
　(a) $(2, 0)$ 　　　　　　　(b) $(-2, 8)$

3. $2x - y - 3 = 0$
　(a) $(1, 2)$ 　　　　　　　(b) $(1, -1)$

4. $x^2 + y^2 = 20$
　(a) $(3, -2)$ 　　　　　　(b) $(-4, 2)$

5. $x^2 y - x^2 + 4y = 0$
　(a) $\left(1, \frac{1}{5}\right)$ 　　　　　　(b) $\left(2, \frac{1}{2}\right)$

6. $y = \dfrac{1}{x^2 + 1}$
　(a) $(0, 0)$ 　　　　　　　(b) $(3, 0.1)$

In Exercises 7–10, find the constant C such that the given ordered pair is a solution point of the equation.

7. $y = x^2 + C$, 　$(2, 6)$
8. $y = Cx^3$, 　$(-4, 8)$

9. $y = C\sqrt{x + 1}$, 　$(3, 8)$
10. $x + C(y + 2) = 0$, 　$(4, 3)$

In Exercises 11–18, find the x- and y-intercepts of the graph of the given equation.

11. $y = x - 5$ 　　　　　　**12.** $y = (x - 1)(x - 3)$
13. $y = x^2 + x - 2$ 　　　**14.** $y = 4 - x^2$
15. $y = x\sqrt{x + 2}$ 　　　　**16.** $xy = 4$
17. $xy - 2y - x + 1 = 0$ 　**18.** $x^2 y - x^2 + 4y = 0$

In Exercises 19–26, check for symmetry with respect to both axes and the origin.

19. $x^2 - y = 0$ 　　　　　　**20.** $xy^2 + 10 = 0$
21. $x - y^2 = 0$ 　　　　　　**22.** $y = \sqrt{9 - x^2}$
23. $y = x^3$ 　　　　　　　　**24.** $xy = 4$
25. $y = \dfrac{x}{x^2 + 1}$ 　　　　　**26.** $y = x^4 - x^2 + 3$

In Exercises 27–32, match the given equation with its graph. [The graphs are labeled (a)–(f).]

27. $y = 4 - x$

28. $y = x^2 + 2x$

29. $y = \sqrt{4 - x^2}$

30. $y = \sqrt{x}$

31. $y = x^3 - x$

32. $y = |x| - 2$

(a)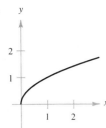

(b)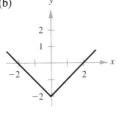

(c)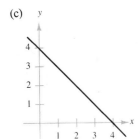

(d)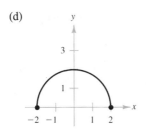

(e)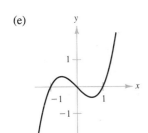

(f)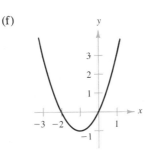

In Exercises 33–52, sketch the graph of the given equation. Identify any intercepts and test for symmetry.

33. $y = -3x + 2$

34. $y = 2x - 3$

35. $y = 1 - x^2$

36. $y = x^2 - 1$

37. $y = x^2 - 4x + 3$

38. $y = -x^2 - 4x$

39. $y = x^3 + 2$

40. $y = x^3 - 1$

41. $y = x(x - 2)^2$

42. $y = \dfrac{4}{x^2 + 1}$

43. $y = \sqrt{x - 3}$

44. $y = \sqrt{1 - x}$

45. $y = \sqrt[3]{x}$

46. $y = \sqrt[3]{x + 1}$

47. $y = |x - 2|$

48. $y = 4 - |x|$

49. $x = y^2 - 1$

50. $x = y^2 - 4$

51. $x^2 + y^2 = 4$

52. $x^2 + y^2 = 16$

In Exercises 53–60, find the standard form of the equation of the specified circle.

53. Center: $(0, 0)$, radius: 3

54. Center: $(0, 0)$, radius: 5

55. Center: $(2, -1)$, radius: 4

56. Center: $(0, \frac{1}{3})$, radius: $\frac{1}{3}$

57. Center: $(-1, 2)$, solution point: $(0, 0)$

58. Center: $(3, -2)$, solution point: $(-1, 1)$

59. Endpoints of a diameter: $(0, 0)$, $(6, 8)$

60. Endpoints of a diameter: $(-4, -1)$, $(4, 1)$

In Exercises 61–68, write the given equation of the circle in standard form and sketch its graph.

61. $x^2 + y^2 - 2x + 6y + 6 = 0$

62. $x^2 + y^2 - 2x + 6y - 15 = 0$

63. $x^2 + y^2 - 2x + 6y + 10 = 0$

64. $3x^2 + 3y^2 - 6y - 1 = 0$

65. $2x^2 + 2y^2 - 2x - 2y - 3 = 0$

66. $4x^2 + 4y^2 - 4x + 2y - 1 = 0$

67. $16x^2 + 16y^2 + 16x + 40y - 7 = 0$

68. $x^2 + y^2 - 4x + 2y + 3 = 0$

In Exercises 69–71, (a) sketch a graph to compare the given data and the model for that data, and (b) use the model to predict y for the year 1990.

69. The following table gives the consumer price index (CPI) for selected years from 1970 to 1985. In the base year of 1967, CPI = 100.

Year	1970	1972	1974	1976	1978	1980
CPI	116.3	125.3	147.7	170.5	195.3	247.0

Year	1981	1982	1983	1984	1985	1990
CPI	272.3	288.6	297.4	311.1	322.2	?

A mathematical model for the CPI during this period is

$$y = 15.21t + 247.45$$

where y represents the CPI and t represents time in years with $t = 0$ corresponding to 1980.

70. The following table gives the life expectancy of a child (at birth) for selected years from 1920 to 1980.

Year	1920	1930	1940	1950
Life Expectancy	54.1	59.7	62.9	68.2

Year	1960	1970	1980	1990
Life Expectancy	69.7	70.8	73.7	?

A mathematical model for the life expectancy during this period is

$$y = 0.31t + 65.59$$

where y represents the life expectancy and t represents time in years with $t = 0$ corresponding to 1950.

71. The following table gives the per capita public debt for the United States for selected years from 1950 to 1985.

Year	1950	1960	1970
Per Capita Debt	$1688.30	$1572.31	$1807.09

Year	1980	1985	1990
Per Capita Debt	$3969.55	$7614.15	?

A mathematical model for the per capita debt during this period is

$$y = 9.84t^2 - 200.2t + 1970.4$$

where y represents the per capita debt and t is time in years with $t = 0$ corresponding to 1950.

72. Use the Distance Formula to show that the points shown in the accompanying figure divide the unit circle into three equal parts.

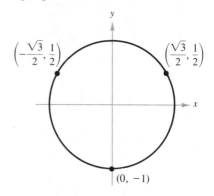

FIGURE FOR 72

73. Use the Distance Formula to show that the points shown in the accompanying figure divide the unit circle into four equal parts.

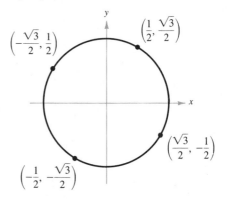

FIGURE FOR 73

1.5 Functions

Many everyday phenomena involve two quantities that are related to each other by some rule of correspondence. Here are some examples.

1. The simple interest I earned on $1000 for one year is related to the annual percentage rate r by the formula $I = 1000r$.
2. The distance d traveled on a bicycle in two hours is related to the speed s of the bicycle by the formula $d = 2s$.
3. The area A of a circle is related to its radius r by the formula $A = \pi r^2$.

Not all correspondences between two quantities have simple mathematical formulas. For instance, we commonly match up quantities such as NFL starting quarterbacks with touchdown passes, days of the year with the Dow-Jones Industrial Average, and hours of the day with temperature. In each of these cases, however, there is some rule of correspondence that matches each item from one set with exactly one item from a different set. Such a rule of correspondence is called a **function.**

Definition of a Function

A **function** f from a set A to a set B is a rule of correspondence that assigns to each element x in the set A exactly one element y in the set B.

Remark: The set A is called the **domain** (or set of inputs) of the function f, and the set B contains the **range** (or set of outputs).

To get a better idea of this definition, look at the function illustrated in Figure 1.38. This function can be represented more efficiently by the following set of ordered pairs:

$$\{(1, 9°),\quad (2, 13°),\quad (3, 15°),\quad (4, 15°),\quad (5, 12°),\quad (6, 4°)\}$$

where the first coordinate is the input and the second is the output. From this list and Figure 1.38, we note the following characteristics of a function.

1. Each element in A must be matched with an element in B.
2. Some elements in B may not be matched with any element in A.
3. Two or more elements of A may be matched with the same element of B.

The converse of the third statement is not true. That is, an element of A (the domain) cannot be matched with two different elements of B. This is illustrated in Example 1.

Prerequisites for Trigonometry

Hours of the day   Celsius temperature

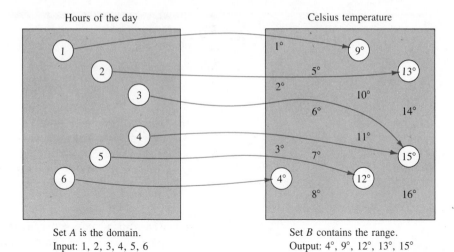

Set *A* is the domain.   Set *B* contains the range.
Input: 1, 2, 3, 4, 5, 6   Output: 4°, 9°, 12°, 13°, 15°

FIGURE 1.38   Function from Set *A* to Set *B*

EXAMPLE 1 *Testing for Functions*

Let $A = \{a, b, c\}$ and $B = \{1, 2, 3, 4, 5\}$. Determine which of the following sets of ordered pairs or figures represents a function from set *A* to set *B*.

(a) $\{(a, 2), (b, 3), (c, 4)\}$ (b) $\{(a, 4), (b, 5)\}$

(c)   (d)

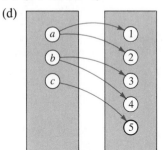

SOLUTION

(a) This collection of ordered pairs *does* represent a function from *A* to *B*. Each element of *A* is matched with exactly one element of *B*.

(b) This collection of ordered pairs *does not* represent a function from *A* to *B*. Not all elements of *A* are matched with an element of *B*.

(c) This figure *does* represent a function from *A* to *B*. It does not matter that each element of *A* is matched with the same element in *B*.

(d) This figure *does not* represent a function from *A* to *B*. The element *a* in *A* is matched with *two* elements, 1 and 2, of *B*. This is also true of the element *b*.

Representing functions by sets of ordered pairs is a common practice in the study of *discrete mathematics*. In algebra or in calculus, however, it is more common to represent functions by equations or formulas involving two variables. For instance, the equation

$$y = x^2$$

represents the variable y as a function of the variable x. We call x the **independent variable** and y the **dependent variable.** In this context, the domain of the function is the set of all values taken on by the independent variable x, and the range of the function is the set of all values taken on by the dependent variable y.

EXAMPLE 2 *Testing for Functions Represented by Equations*

Determine which of the following equations represent y as a function of x.

(a) $x^2 + y = 1$

(b) $-x + y^2 = 1$

SOLUTION

In each case, to determine whether y is a function of x, it is helpful to solve for y in terms of x.

(a) Solving for y, we obtain

$$
\begin{aligned}
x^2 + y &= 1 \qquad\qquad & \textit{Given equation}\\
y &= 1 - x^2. \qquad\qquad & \textit{Solve for } y
\end{aligned}
$$

To each value of x there corresponds just one value for y; hence y *is* a function of x.

(b) In this case, solving for y yields

$$
\begin{aligned}
-x + y^2 &= 1 \qquad\qquad & \textit{Given equation}\\
y^2 &= 1 + x\\
y &= \pm\sqrt{1 + x}. \qquad\qquad & \textit{Solve for } y
\end{aligned}
$$

The $\pm$ indicates that to a given value of x there correspond *two* values for y. Hence y *is not* a function of x.

Function Notation

When using an equation to represent a function it is convenient to name the function so that it can be easily referenced. For example, we know that the equation $y = 1 - x^2$, Example 2(a), describes y as a function of x. Suppose we give this function the name "f." Then we can use the following **function notation.**

Input	*Output*	*Equation*
x	$f(x)$	$f(x) = 1 - x^2$

The symbol $f(x)$ is read as the **value of f at x** or simply "f of x." This corresponds to the y-value for a given x. Thus, we can (and often do) write $y = f(x)$.

Keep in mind that f is the *name* of the function, while $f(x)$ is the *value* of the function at x. For instance, the function given by

$$f(x) = 3 - 2x$$

has *function values* denoted by $f(-1)$, $f(0)$, $f(2)$, and so on. To find these values, we substitute the specified input values into the given equation, as follows.

For $x = -1$, $f(-1) = 3 - 2(-1) = 3 + 2 = 5.$
For $x = 0$, $f(0) = 3 - 2(0) = 3 + 0 = 3.$
For $x = 2$, $f(2) = 3 - 2(2) = 3 - 4 = -1.$

Although we generally use f as a convenient function name and x as the independent variable, we can use other letters. For instance, $f(x) = x^2 - 4x + 7$, $f(t) = t^2 - 4t + 7$, and $g(s) = s^2 - 4s + 7$ all define the same function. In fact, the role of the independent variable in a function is simply that of a "placeholder." Consequently, the above function can be properly described by the form

$$f(\ ) = (\ )^2 - 4(\ ) + 7$$

where parentheses are used in place of a letter. Therefore, to evaluate $f(-2)$, we simply place -2 in each set of parentheses:

$$f(-2) = (-2)^2 - 4(-2) + 7$$
$$= 4 + 8 + 7$$
$$= 19.$$

Similarly, the value of $f(3x)$ is

$$f(3x) = (3x)^2 - 4(3x) + 7$$
$$= 9x^2 - 12x + 7.$$

EXAMPLE 3 *Evaluating Functions*

Let $g(x) = 4x - x^2$ and find the following.

(a) $g(2)$ (b) $g(x + 2)$

SOLUTION

(a) Replacing x with 2, we obtain

$$g(2) = 4(2) - (2)^2 = 8 - 4 = 4.$$

(b) Replacing x with $x + 2$, we obtain

$$
\begin{aligned}
g(x + 2) &= 4(x + 2) - (x + 2)^2 \\
&= 4x + 8 - (x^2 + 4x + 4) \\
&= -x^2 + 4.
\end{aligned}
$$

Remark: Example 3 shows that $g(x + 2) \neq g(x) + g(2)$ because $-x^2 + 4 \neq 4x - x^2 + 4$. In general, $g(u + v)$ is not equal to $g(u) + g(v)$.

Sometimes a function is defined using more than one equation. An illustration is given in Example 4.

EXAMPLE 4 *A Function Defined by Two Equations*

Evaluate the function given by

$$
f(x) = \begin{cases} x^2 + 1, & x < 0 \\ x - 1, & x \geq 0 \end{cases}
$$

at $x = -1, 0$, and 1.

SOLUTION

Since $x = -1 < 0$, we use $f(x) = x^2 + 1$ to obtain

$$f(-1) = (-1)^2 + 1 = 2.$$

For $x = 0$, we use $f(x) = x - 1$ to obtain

$$f(0) = (0) - 1 = -1.$$

For $x = 1$, we use $f(x) = x - 1$ to obtain

$$f(1) = (1) - 1 = 0.$$

Finding the Domain of a Function

The domain of a function may be explicitly described along with the function, or it may be *implied* by the expression used to define the function. The **implied domain** is the set of all real numbers for which the expression is defined. For instance, the function given by

$$f(x) = \frac{1}{x^2 - 4}$$

has an implied domain which consists of all real x other than $x = \pm 2$. These two values are excluded from the domain because division by zero is undefined. Another common type of implied domain is that used to avoid even roots of negative numbers. For example, the function given by

$$f(x) = \sqrt{x}$$

is defined only for $x \geq 0$. Hence, its implied domain is the interval $[0, \infty)$.

The *ranges* of such functions are more difficult to find, and can best be obtained from their graphs, which we will study in Section 1.6.

EXAMPLE 5 Finding the Domain of a Function

Find the domain of each of the functions.

(a) f: $\{(-3, 0), (-1, 4), (0, 2), (2, 2), (4, -1)\}$

(b) Volume of a sphere: $V = \frac{4}{3}\pi r^3$

(c) $g(x) = \dfrac{1}{x + 5}$ (d) $h(x) = \sqrt{4 - x^2}$

SOLUTION

(a) The domain of f consists of all first coordinates in the set of ordered pairs, and is therefore the set

$$\text{domain} = \{-3, -1, 0, 2, 4\}.$$

(b) For the volume of a sphere we must choose nonnegative values for the radius (independent variable) r. Thus, the domain is the set of all real numbers r such that $r \geq 0$.

(c) Excluding x-values that yield zero in the denominator, the domain of g is the set of all real numbers $x \neq -5$.

(d) We choose x-values for which $4 - x^2 \geq 0$. This implies that $-2 \leq x \leq 2$. Thus, the domain is the interval $[-2, 2]$.

Remark: In Example 5(b), note that the domain of a function may be implied by the physical context. For instance, from the equation $V = \frac{4}{3}\pi r^3$, we would have no reason to restrict r to nonnegative values, but the physical context tells us that a sphere cannot have a negative radius.

Applications

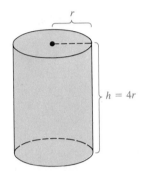

FIGURE 1.39

EXAMPLE 6   The Dimensions of a Container

The marketing manager of a soft drink company wants to design a can with a height that is four times the radius of the can.

(a) Express the volume of the can as a function of the radius r.
(b) Express the volume of the can as a function of the height h.

SOLUTION

The volume of a right circular cylinder is given by the formula

$$V = \pi(\text{radius})^2(\text{height}).$$

(a) If the radius is r, then the height is $4r$, as shown in Figure 1.39. The volume as a function of r is given by

$$V = \pi r^2(4r) = 4\pi r^3.$$

(b) If the height is h and $h = 4r$, then $r = h/4$. The volume as a function of h is

$$V = \pi\left(\frac{h}{4}\right)^2 h = \frac{\pi h^3}{16}.$$

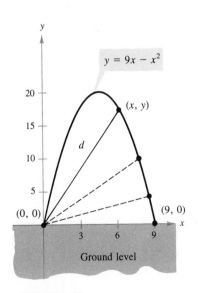

FIGURE 1.40

EXAMPLE 7   The Path of a Projectile

A projectile moves in a parabolic path given by $y = 9x - x^2$, as shown in Figure 1.40.

(a) For any point (x, y) on the path, express the distance d from (x, y) to the launch point $(0, 0)$ as a function of x.
(b) Find the domain of d.

SOLUTION

(a) By the Distance Formula, we know that

$$d = \sqrt{(x_2 - x_1)^2 + (y_2 - y_1)^2} = \sqrt{(x - 0)^2 + (y - 0)^2}.$$

Since $y = 9x - x^2$, it follows that as a function of x, d is given by

$$d = \sqrt{x^2 + (9x - x^2)^2} = \sqrt{x^4 - 18x^3 + 82x^2}.$$

(b) The domain of d is restricted to the x-values between the launch point $(0, 0)$ and the point of impact (the other x-intercept). Solving

$$0 = 9x - x^2 = x(9 - x)$$

we find the other x-intercept to be $(9, 0)$. Consequently, $0 \le x \le 9$, and the domain of d is the interval $[0, 9]$.

One of the basic definitions in calculus employs the ratio

$$\frac{f(x + h) - f(x)}{h}, \qquad h \ne 0$$

called a **difference quotient.** The next example shows how such an expression can be evaluated.

EXAMPLE 8 *Evaluating a Difference Quotient*

For the function given by $f(x) = x^2 - 4x + 7$, find

$$\frac{f(x + h) - f(x)}{h}.$$

SOLUTION

$$
\begin{aligned}
\frac{f(x + h) - f(x)}{h} &= \frac{[(x + h)^2 - 4(x + h) + 7] - [x^2 - 4x + 7]}{h} \\
&= \frac{x^2 + 2xh + h^2 - 4x - 4h + 7 - x^2 + 4x - 7}{h} \\
&= \frac{2xh + h^2 - 4h}{h} \\
&= \frac{h(2x + h - 4)}{h} \\
&= 2x + h - 4
\end{aligned}
$$

Summary of Function Terminology

Function

A relationship between two variables such that to each value of the independent variable there corresponds exactly one value of the dependent variable.

Function Notation $y = f(x)$

f is the **name** of the function.
y is the **dependent variable.**
x is the **independent variable.**
$f(x)$ is the **value of the function at x.**

Domain

The set of all values (inputs) of the independent variable for which the function is defined. If x is in the domain of f, we say that f is **defined** at x. If x is not in the domain of f, we say that f is **undefined** at x.

Range

The set of all values (outputs) assumed by the dependent variable (that is, the set of all function values).

Implied Domain

If f is defined by an algebraic expression and the domain is not specified, then the implied domain consists of all real numbers for which the expression is defined.

WARM UP

Simplify the expression.

1. $2(-3)^3 + 4(-3) - 7$

2. $4(-1)^2 - 5(-1) + 4$

3. $(x + 1)^2 + 3(x + 1) - 4 - (x^2 + 3x - 4)$

4. $(x - 2)^2 - 4(x - 2) - (x^2 - 4)$

Solve for y in terms of x.

5. $2x + 5y - 7 = 0$

6. $y^2 = x^2$

Solve the inequality.

7. $x^2 - 4 \geq 0$

8. $9 - x^2 \geq 0$

9. $x^2 + 2x + 1 \geq 0$

10. $x^2 - 3x + 2 \geq 0$

EXERCISES 1.5

In Exercises 1–4, fill in the blanks using the specified function and the given value of the independent variable. (The symbol Δx represents a single variable and is read "delta x." This symbol is commonly used in calculus to denote a small change in x.)

1. $f(x) = 6 - 4x$
  (a) $f(3) = 6 - 4(\quad)$
  (b) $f(-7) = 6 - 4(\quad)$
  (c) $f(t) = 6 - 4(\quad)$
  (d) $f(c + 1) = 6 - 4(\quad)$

2. $g(x) = x^2 - 2x$
  (a) $g(2) = (\quad)^2 - 2(\quad)$
  (b) $g(-3) = (\quad)^2 - 2(\quad)$
  (c) $g(t + 1) = (\quad)^2 - 2(\quad)$
  (d) $g(x + \Delta x) = (\quad)^2 - 2(\quad)$

3. $f(s) = \dfrac{1}{s + 1}$

  (a) $f(4) = \dfrac{1}{(\quad) + 1}$

  (b) $f(0) = \dfrac{1}{(\quad) + 1}$

  (c) $f(4x) = \dfrac{1}{(\quad) + 1}$

  (d) $f(x + h) = \dfrac{1}{(\quad) + 1}$

4. $f(t) = \sqrt{25 - t^2}$
  (a) $f(3) = \sqrt{25 - (\quad)^2}$
  (b) $f(5) = \sqrt{25 - (\quad)^2}$
  (c) $f(x + 5) = \sqrt{25 - (\quad)^2}$
  (d) $f(2 + h) = \sqrt{25 - (\quad)^2}$

In Exercises 5–16, evaluate the function at the specified value of the independent variable and simplify the results.

5. $f(x) = 2x - 3$
  (a) $f(1)$
  (b) $f(-3)$
  (c) $f(x - 1)$
  (d) $f\left(\frac{1}{4}\right)$

6. $g(y) = 7 - 3y$
  (a) $g(0)$
  (b) $g\left(\frac{7}{3}\right)$
  (c) $g(s)$
  (d) $g(s + 2)$

7. $h(t) = t^2 - 2t$
  (a) $h(2)$
  (b) $h(-1)$
  (c) $h(x + 2)$
  (d) $h(1.5)$

8. $V(r) = \frac{4}{3}\pi r^3$
  (a) $V(3)$
  (b) $V(0)$
  (c) $V\left(\frac{3}{2}\right)$
  (d) $V(2r)$

9. $f(y) = 3 - \sqrt{y}$
  (a) $f(4)$
  (b) $f(100)$
  (c) $f(4x^2)$
  (d) $f(0.25)$

10. $f(x) = \sqrt{x + 8} + 2$
  (a) $f(-8)$
  (b) $f(1)$
  (c) $f(x - 8)$
  (d) $f(h + 8)$

11. $q(x) = \dfrac{1}{x^2 - 9}$
  (a) $q(4)$
  (b) $q(0)$
  (c) $q(3)$
  (d) $q(y + 3)$

12. $q(t) = \dfrac{2t^2 + 3}{t^2}$
  (a) $q(2)$
  (b) $q(0)$
  (c) $q(x)$
  (d) $q(-x)$

13. $f(x) = \dfrac{|x|}{x}$
  (a) $f(2)$
  (b) $f(-2)$
  (c) $f(x^2)$
  (d) $f(x - 1)$

14. $f(x) = |x| + 4$
  (a) $f(2)$
  (b) $f(-2)$
  (c) $f(x^2)$
  (d) $f(x + \Delta x) - f(x)$

15. $f(x) = \begin{cases} 2x + 1, & x < 0 \\ 2x + 2, & x \geq 0 \end{cases}$
  (a) $f(-1)$
  (b) $f(0)$
  (c) $f(1)$
  (d) $f(2)$

16. $f(x) = \begin{cases} x^2 + 2, & x \leq 1 \\ 2x^2 + 2, & x > 1 \end{cases}$
  (a) $f(-2)$
  (b) $f(0)$
  (c) $f(1)$
  (d) $f(2)$

In Exercises 17–22, find the indicated difference quotient and simplify your answer.

17. $f(x) = x^2 - x + 1$

$$\dfrac{f(2 + h) - f(2)}{h}$$

18. $f(x) = 5x - x^2$

$$\dfrac{f(5 + h) - f(5)}{h}$$

19. $f(x) = x^3$

$$\dfrac{f(x + \Delta x) - f(x)}{\Delta x}$$

20. $f(x) = 2x$

$$\dfrac{f(x + \Delta x) - f(x)}{\Delta x}$$

21. $g(x) = 3x - 1$

$$\dfrac{g(x) - g(3)}{x - 3}$$

22. $f(t) = \dfrac{1}{t}$

$$\dfrac{f(t) - f(1)}{t - 1}$$

In Exercises 23–28, find all real values of x such that $f(x) = 0$.

23. $f(x) = 15 - 3x$

24. $f(x) = \dfrac{3x - 4}{5}$

25. $f(x) = x^2 - 9$

26. $f(x) = x^3 - x$

27. $f(x) = \dfrac{3}{x - 1} + \dfrac{4}{x - 2}$

28. $f(x) = a + \dfrac{b}{x}$

In Exercises 29–38, find the domain of the function.

29. $f(x) = 5x^2 + 2x - 1$

30. $g(x) = 1 - 2x^2$

31. $h(t) = \dfrac{4}{t}$

32. $s(y) = \dfrac{3y}{y + 5}$

33. $g(y) = \sqrt{y - 10}$

34. $f(t) = \sqrt[3]{t + 4}$

35. $f(x) = \sqrt[4]{1 - x^2}$

36. $h(x) = \dfrac{10}{x^2 - 2x}$

37. $g(x) = \dfrac{1}{x} - \dfrac{3}{x + 2}$

38. $f(s) = \dfrac{\sqrt{s - 1}}{s - 4}$

In Exercises 39–48, identify the equations that determine y as a function of x.

39. $x^2 + y^2 = 4$

40. $x = y^2$

41. $x^2 + y = 4$

42. $x + y^2 = 4$

43. $2x + 3y = 4$

44. $x^2 + y^2 - 2x - 4y + 1 = 0$

45. $y^2 = x^2 - 1$

46. $y = \sqrt{x + 5}$

47. $x^2y - x^2 + 4y = 0$

48. $xy - y - x - 2 = 0$

In Exercises 49 and 50, determine which of the sets of ordered pairs represents a function from A to B. Give reasons for your answers.

49. $A = \{0, 1, 2, 3\}$ and $B = \{-2, -1, 0, 1, 2\}$
  (a) $\{(0, 1), (1, -2), (2, 0), (3, 2)\}$
  (b) $\{(0, -1), (2, 2), (1, -2), (3, 0), (1, 1)\}$
  (c) $\{(0, 0), (1, 0), (2, 0), (3, 0)\}$
  (d) $\{(0, 2), (3, 0), (1, 1)\}$

50. $A = \{\alpha, \beta, \gamma\}$ and $B = \{w, x, y, z\}$
  (a) $\{(\alpha, x), (\gamma, y), (\gamma, z), (\beta, z)\}$
  (b) $\{(\alpha, x), (\beta, y), (\gamma, z)\}$
  (c) $\{(x, \alpha), (w, \alpha), (y, \gamma), (z, \beta)\}$
  (d) $\{(\gamma, w), (\beta, w), (\alpha, z)\}$

In Exercises 51–54, assume that the domain of f is the set $A = \{-2, -1, 0, 1, 2\}$. Determine the set of ordered pairs representing the function f.

51. $f(x) = x^2$

52. $f(x) = \dfrac{2x}{x^2 + 1}$

53. $f(x) = \sqrt{x + 2}$

54. $f(x) = |x + 1|$

In Exercises 55–58, find the value(s) of x for which $f(x) = g(x)$.

55. $f(x) = x^2$, $g(x) = x + 2$

56. $f(x) = x^2 + 2x + 1$, $g(x) = 3x + 3$

57. $f(x) = \sqrt{3x} + 1$, $g(x) = x + 1$

58. $f(x) = x^4 - 2x^2$, $g(x) = 2x^2$

59. Express the volume V of a cube as a function of the length e of one of its edges.

60. Express the circumference C of a circle as a function of its (a) radius r, and (b) diameter d.

61. Express the area A of a circle as a function of its circumference C.

62. Express the area A of an equilateral triangle as a function of the length s of one of its sides.

63. An open box is to be made from a square piece of material, 12 inches on a side, by cutting equal squares from each corner and turning up the sides (see figure). Write the volume V of the box as a function of x.

64. A rectangular package to be sent by a postal service can have a maximum combined length and girth (perimeter of a cross section) of 108 inches (see figure). Write the volume of the package as a function of x.

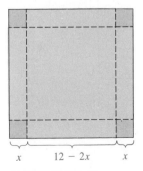

x $12 - 2x$ x

FIGURE FOR 63

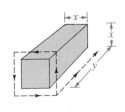

FIGURE FOR 64

65. A right triangle is formed in the first quadrant by the x- and y-axes and a line through the point $(1, 2)$ (see figure). Write the area of the triangle as a function of x, and determine the domain of the function.

66. A rectangle is bounded by the x-axis and the semicircle $y = \sqrt{25 - x^2}$ (see figure). Write the area of the rectangle as a function of x, and determine the domain of the function.

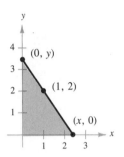

FIGURE FOR 65

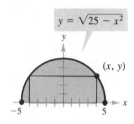

$y = \sqrt{25 - x^2}$

FIGURE FOR 66

67. A balloon carrying a transmitter ascends vertically from a point 2000 feet from the receiving station (see figure). Let d be the distance between the balloon and the receiving station. Express the height of the balloon as a function of d.

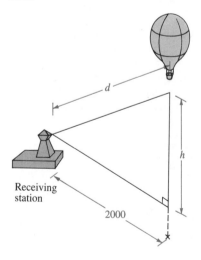

FIGURE FOR 67

68. A mechanical arm is used to fill containers as they move on a conveyor belt (see figure). Express the length L of the arm as a function of x, as labeled in the figure.

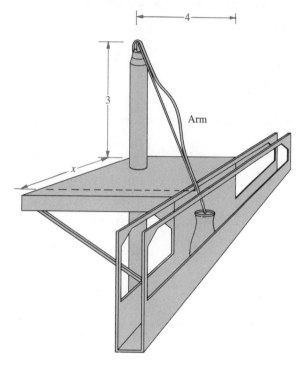

FIGURE FOR 68

69. A company produces a product for which the variable cost is $12.30 per unit and the fixed costs are $98,000. The product sells for $17.98. Let x be the number of units produced.
  (a) Write the total cost C as a function of the number of units produced.
  (b) Write the revenue R as a function of the number of units produced.
  (c) Write the profit P as a function of the number of units produced. (*Note: $P = R - C$.*)

70. The inventor of a new game believes that the variable cost for producing the game is $0.95 per unit and the fixed costs are $6000. The inventor sells each game for $1.69. Let x be the number of games sold.
  (a) Write the total cost C as a function of the number of games sold.
  (b) Write the average cost per unit $\bar{C} = C/x$ as a function of x.

71. For groups of 80 or more people, a charter bus company determines the rate per person according to the following formula:

Rate $= \$8.00 - \$0.05(n - 80),\quad n \geq 80$

where n is the number of persons.

(a) Express the total revenue R for the bus company as a function of n.

(b) Complete the following table by using the function from part (a).

n	90	100	110	120	130	140	150
$R(n)$							

72. Assume that the amount of money deposited in a bank is proportional to the square of the interest rate the bank pays on the money. That is, $d = kr^2$, where d is the total deposit, r is the interest rate, and k is the proportionality constant. Assuming the bank can reinvest the money for a return of 18%, write the bank's profit P as a function of the interest rate r.

73. The force F (in tons) of water pressure against the face of a dam is a function of the depth y of the water given by

$$F(y) = 149.76\sqrt{10}y^{5/2}.$$

Complete the following table by finding the force for various water levels.

y	5	10	20	30	40
$F(y)$					

74. The work W (in foot-pounds) required to fill a storage tank to a depth of y feet is given by

$$W(y) = 25\pi\left(4y^2 - \frac{y^3}{3}\right).$$

Complete the following table by finding the work required to fill the tank to various levels.

y	1	2	4	6	8
$W(y)$					

1.6 Graphs of Functions

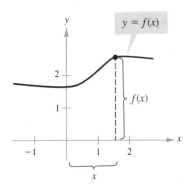

FIGURE 1.41

In Section 1.5 we discussed functions from an algebraic (or analytic) point of view. Here we look at functions from a geometric perspective. The **graph of a function** f is the collection of ordered pairs $(x, f(x))$ such that x is in the domain of f. As you study this section, remember that

$$x = \text{the directed distance from the } y\text{-axis}$$
$$f(x) = \text{the directed distance from the } x\text{-axis}$$

as shown in Figure 1.41.

We noted in Section 1.5 that the *range* (set of values assumed by the dependent variable) of a function is often more easily determined from its graph than from its equation. This technique is illustrated in Example 1.

EXAMPLE 1 *Finding Domain and Range from the Graph of a Function*

Use the graph of the function f, shown in Figure 1.42, to find (a) the domain of f, (b) $f(-1)$ and $f(2)$, and (c) the range of f.

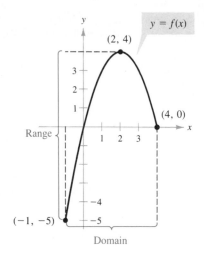

FIGURE 1.42

SOLUTION

(a) Because the graph does not extend beyond $x = -1$ (on the left) and $x = 4$ (on the right), the domain of f is all x in the interval $[-1, 4]$.

(b) Since $(-1, -5)$ is a point on the graph, it follows that

$$f(-1) = -5.$$

Similarly, since $(2, 4)$ is a point on the graph, it follows that

$$f(2) = 4.$$

(c) Because the graph does not extend below $f(-1) = -5$ nor above $f(2) = 4$, the range of f is the interval $[-5, 4]$.

Remark: Note in Figure 1.42 that the solid dots representing the points $(-1, -5)$ and $(4, 0)$ indicate that the graph terminates at these points.

By the definition of a function, at most one y-value corresponds to a given x-value. It follows, then, that a vertical line can intersect the graph of a function at most once. This observation provides us with a convenient visual test for functions.

Vertical Line Test for Functions

A set of points in a coordinate plane is the graph of y as a function of x if and only if no vertical line intersects the graph at more than one point.

(a)

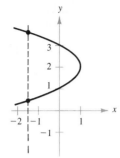

EXAMPLE 2 *Vertical Line Test for Functions*

Which of the graphs in Figure 1.43 represents y as a function of x?

SOLUTION

(a) This *is not* a graph of y as a function of x because we can find a vertical line that intersects the graph twice.

(b) This *is* a graph of y as a function of x because every vertical line intersects the graph at most once.

(b)

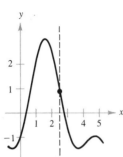

(c) This *is* a graph of y as a function of x. (Note that if a vertical line does not intersect the graph, it simply means that the function is undefined for this particular value of x.)

(c)

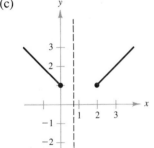

The more we know about the graph of a function, the more we know about the function itself. Consider the graph shown in Figure 1.44. As we move *from left to right*, this graph falls for values of x between -2 and 0, is constant from $x = 0$ to $x = 2$, and then rises for values of x between 2 and 4. From these observations, we say that the function is:

decreasing on the interval $(-2, 0)$,

constant on the interval $(0, 2)$, and

increasing on the interval $(2, 4)$.

FIGURE 1.43

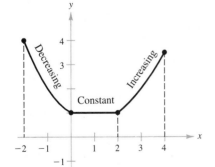

FIGURE 1.44

Definition of Increasing, Decreasing, and Constant Functions

A function f is **increasing** on an interval if, for any x_1 and x_2 in the interval,

$$x_1 < x_2 \quad \text{implies} \quad f(x_1) < f(x_2).$$

A function f is **decreasing** on an interval if, for any x_1 and x_2 in the interval,

$$x_1 < x_2 \quad \text{implies} \quad f(x_1) > f(x_2).$$

A function f is **constant** on an interval if, for every x_1 and x_2 in the interval,

$$f(x_1) = f(x_2).$$

Prerequisites for Trigonometry

EXAMPLE 3 Increasing and Decreasing Functions

In Figure 1.45 determine the open intervals on which each function is increasing, decreasing, or constant.

SOLUTION

(a) Although it might appear that there is an interval about zero over which this function is constant, we see that if $x_1 < x_2$, then $f(x_1) = x_1^3 < x_2^3 = f(x_2)$, and we conclude that the function is increasing for all x.

(b) This function is

  increasing on the interval $(-\infty, -1)$,
  decreasing on the interval $(-1, 1)$, and
  increasing on the interval $(1, \infty)$.

(c) This function is

  increasing on the interval $(-\infty, 0)$,
  constant on the interval $(0, 2)$, and
  decreasing on the interval $(2, \infty)$.

(a)

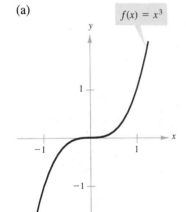

(b)

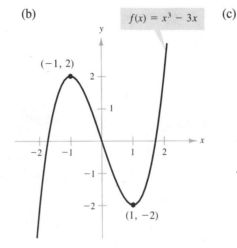

(c)

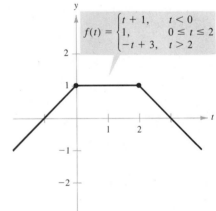

FIGURE 1.45

Graphs of Functions

Remark: The points at which a function changes its increasing, decreasing, or constant behavior are especially important in producing an accurate graph of the function. These points often identify *maximum* or *minimum* values of the function. Techniques for finding the exact location of these special points are developed in calculus.

EXAMPLE 4 *The Greatest Integer Function*

The **greatest integer function** is denoted by $[x]$ and is defined by

$$f(x) = [x] = \text{the greatest integer less than or equal to } x.$$

The graph of this function is shown in Figure 1.46. Note that the graph of f jumps vertically one unit at each integer and is constant (a horizontal line segment) between each pair of consecutive integers. Some values of this function are

$$[-1] = -1, \quad [-0.5] = -1, \quad [1] = 1, \quad \text{and} \quad [1.5] = 1.$$

The range of the greatest integer function is the set of all integers.

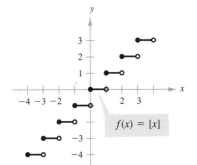

Greatest Integer Function

FIGURE 1.46

$f(x) = [x]$

Aids for Sketching Graphs of Functions

Intercepts, symmetry, and equation recognition, which we used for sketching the graph of an equation, can also be used to sketch the graph of a function.

For instance, recall that an *x*-intercept of a graph is a point at which the graph intersects the *x*-axis. When the graph of a function intersects the *x*-axis, we say that the function has a zero at that point. That is, the function f has a **zero** at $x = a$ if $f(a) = 0$. For instance, $x = 4$ is a zero of the function given by $f(x) = x - 4$ because $f(4) = 4 - 4 = 0$.

In Section 1.4, we also discussed different types of symmetry. In the terminology of functions, we say that a function is **even** if its graph is symmetric with respect to the *y*-axis, and a function is **odd** if its graph is symmetric with respect to the origin. Thus, the symmetry tests given in Section 1.4 yield the following tests for even and odd functions.

Test for Even and Odd Functions

A function given by $y = f(x)$ is **even** if, for each x in the domain of f,

$$f(-x) = f(x).$$

A function given by $y = f(x)$ is **odd** if, for each x in the domain of f,

$$f(-x) = -f(x).$$

EXAMPLE 5 *Even and Odd Functions*

Determine whether the following functions are even, odd, or neither.

(a) $g(x) = x^3 - x$ (b) $h(x) = x^2 + 1$ (c) $f(x) = x^3 - 1$

SOLUTION

(a) This function is odd because

$$g(-x) = (-x)^3 - (-x) \qquad \textit{Substitute } -x \textit{ for } x$$
$$= -x^3 + x$$
$$= -(x^3 - x)$$
$$= -g(x).$$

(b) This function is even because

$$h(-x) = (-x)^2 + 1 = x^2 + 1 = h(x).$$

(c) By substituting $-x$ for x , we have

$$f(-x) = (-x)^3 - 1 = -x^3 - 1.$$

Because $f(x) = x^3 - 1$ and $-f(x) = -x^3 + 1$, we conclude that $f(-x) \neq f(x)$ and $f(-x) \neq -f(x)$. Hence, the function is neither even nor odd.

The graphs of these three functions are shown in Figure 1.47.

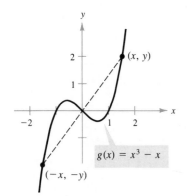

(a) Odd function
     (symmetric to origin)

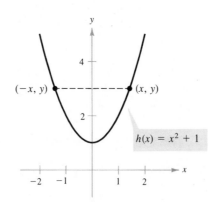

(b) Even function
     (symmetric to y-axis)

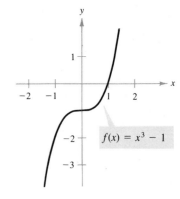

(c) Neither even nor odd

FIGURE 1.47

Graphs of Functions

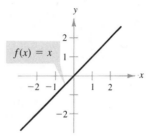

(a)   Constant function

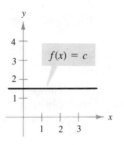

(b)   Identity function

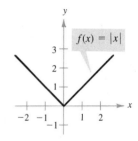

(c)   Absolute value function

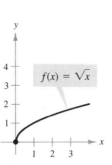

(d)   Square root function

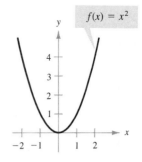

(e)   Squaring function

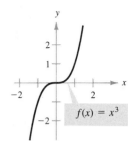

(f)   Cubing function

FIGURE 1.48

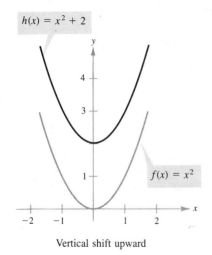

Vertical shift upward

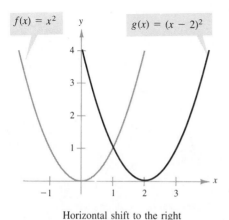

Horizontal shift to the right

FIGURE 1.49

Figure 1.48 shows the graphs of some basic functions that occur frequently in applications. You need to be familiar with these graphs.

Shifting, Reflecting, and Stretching Graphs

Many functions have graphs that are simple transformations of familiar graphs like those in Figure 1.48. For example, we can obtain the graph of $h(x) = x^2 + 2$ by shifting the graph of $f(x) = x^2$ upward two units, as shown in Figure 1.49. In function notation, h and f are related as follows.

$$h(x) = x^2 + 2 = f(x) + 2 \qquad \textit{Upward shift of 2}$$

Similarly, we can obtain the graph of $g(x) = (x - 2)^2$ by shifting the graph of $f(x) = x^2$ to the right two units. In this case, the functions g and f have the following relationship.

$$g(x) = (x - 2)^2 = f(x - 2) \qquad \textit{Right shift of 2}$$

We summarize these two types of **horizontal** and **vertical shifts** of the graphs of functions as follows.

Prerequisites for Trigonometry

Vertical and Horizontal Shifts

Let c be a positive real number. **Vertical** and **horizontal shifts** in the graph of $y = f(x)$ are represented as follows.

Vertical shift c units **upward:**	$h(x) = f(x) + c$
Vertical shift c units **downward:**	$h(x) = f(x) - c$
Horizontal shift c units to the **right:**	$h(x) = f(x - c)$
Horizontal shift c units to the **left:**	$h(x) = f(x + c)$

Some graphs can be obtained from a *combination* of vertical and horizontal shifts. This is demonstrated in part (c) of the next example.

EXAMPLE 6   Shifts in the Graph of a Function

Use the graph of $f(x) = x^3$ to sketch the graph of each of the following functions.

(a) $g(x) = x^3 + 1$ (b) $h(x) = (x - 1)^3$

(c) $k(x) = (x + 2)^3 + 1$

SOLUTION

Relative to the graph of $f(x) = x^3$, the graph of $g(x) = x^3 + 1$ is an upward shift of one unit, the graph of $h(x) = (x - 1)^3$ is a right shift of one unit, and the graph of $k(x) = (x + 2)^3 + 1$ involves a left shift of two units and an upward shift of one unit. The graphs are shown in Figure 1.50.

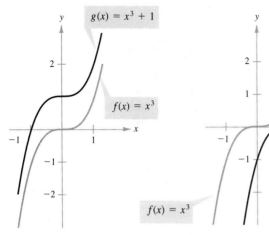

(a)   Vertical shift: 1 up

(b)   Horizontal shift: 1 right

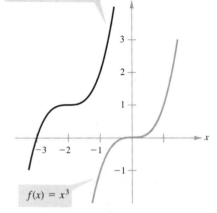

(c)   Horizontal shift: 2 left
      Vertical shift: 1 up

FIGURE 1.50

Graphs of Functions

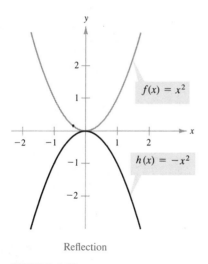

$f(x) = x^2$

$h(x) = -x^2$

Reflection

FIGURE 1.51

The second common type of transformation is called a **reflection.** For instance, if we assume that the x-axis represents a mirror, then the graph of $h(x) = -x^2$ is the mirror image (or reflection) of the graph of $f(x) = x^2$, as shown in Figure 1.51.

Reflections in the Coordinate Axes

Reflections, in the coordinate axes, of the graph of $y = f(x)$ are represented as follows.

Reflection in the x-axis: $h(x) = -f(x)$
Reflection in the y-axis: $h(x) = f(-x)$

EXAMPLE 7   Reflections and Shifts

Sketch a graph of each of the following functions.

(a) $g(x) = -\sqrt{x}$ (b) $h(x) = \sqrt{-x}$ (c) $k(x) = -\sqrt{x + 2}$

SOLUTION

(a) Relative to the graph of $f(x) = \sqrt{x}$, the graph of g is a reflection in the x-axis because

$$g(x) = -\sqrt{x} = -f(x).$$

(b) The graph of h is a reflection of the graph of $f(x) = \sqrt{x}$ in the y-axis because

$$h(x) = \sqrt{-x} = f(-x).$$

(c) From the equation

$$k(x) = -\sqrt{x + 2} = -f(x + 2)$$

we conclude that the graph of k is, first, a left shift of two units, followed by a reflection in the x-axis.

The graphs of all three functions are shown in Figure 1.52.

Prerequisites for Trigonometry

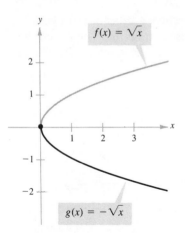

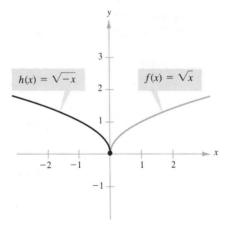

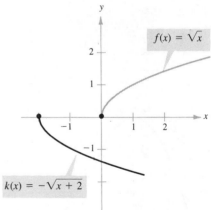

(a) Reflection in x-axis   (b) Reflection in y-axis   (c) Shift and reflection

FIGURE 1.52

(a)

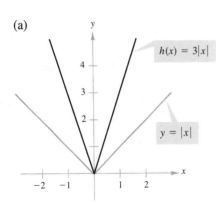

(b)

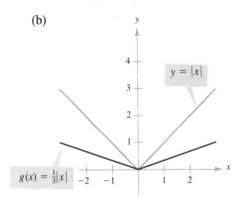

FIGURE 1.53

Horizontal shifts, vertical shifts, and reflections are called **rigid** transformations because the basic shape of the graph is unchanged. These transformations change only the *position* of the graph in the xy-plane. **Nonrigid** transformations are those which cause a *distortion*—a change in the shape of the original graph. For instance, a nonrigid transformation of the graph of $y = f(x)$ is represented by $y = cf(x)$, where the transformation is a **vertical stretch** if $c > 1$ and a **vertical shrink** if $0 < c < 1$.

EXAMPLE 8 *Nonrigid Transformations*

Sketch a graph of each of the following.

(a) $h(x) = 3|x|$ (b) $g(x) = \dfrac{1}{3}|x|$

SOLUTION

(a) Relative to the graph of $f(x) = |x|$, the graph of

$$h(x) = 3|x| = 3f(x)$$

is a vertical stretch (multiply each y-value by 3) of the graph of f.

(b) Similarly, the equation

$$g(x) = \frac{1}{3}|x| = \frac{1}{3}f(x)$$

indicates that the graph of h is a vertical shrink of the graph of f.

The graphs of both functions are shown in Figure 1.53.

1. Find $f(2)$ for $f(x) = -x^3 + 5x$.

3. Find $f(-x)$ for $f(x) = \dfrac{3}{x}$.

2. Find $f(6)$ for $f(x) = x^2 - 6x$.

4. Find $f(-x)$ for $f(x) = x^2 + 3$.

Solve for x.

5. $x^3 - 16x = 0$

6. $2x^2 - 3x + 1 = 0$

Find the domain of the function.

7. $g(x) = \dfrac{4}{x - 4}$

8. $f(x) = \dfrac{2x}{x^2 - 9x + 20}$

9. $h(t) = \sqrt[4]{5 - 3t}$

10. $f(t) = t^3 + 3t - 5$

EXERCISES 1.6

In Exercises 1–6, determine the domain and range of the given function.

1. $f(x) = \sqrt{x - 1}$

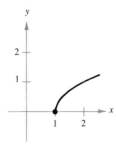

2. $f(x) = 4 - x^2$

5. $f(x) = \sqrt{25 - x^2}$

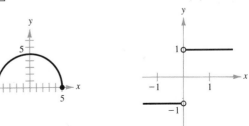

6. $f(x) = |x|/x$

In Exercises 7–12, use the vertical line test to determine if y is a function of x.

3. $f(x) = \sqrt{x^2 - 4}$

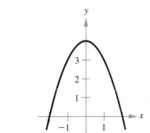
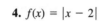

4. $f(x) = |x - 2|$

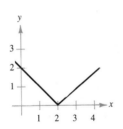

7. $y = x^2$

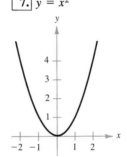

8. $y = x^3 - 1$

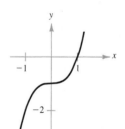

9. $x - y^2 = 0$

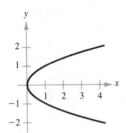

10. $x^2 + y^2 = 9$

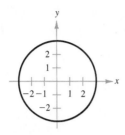

17. $f(x) = 3x^4 - 6x^2$

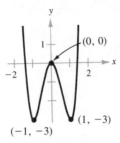

18. $f(x) = x^{2/3}$

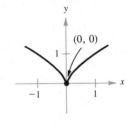

11. $x^2 = xy - 1$

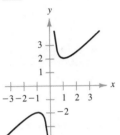

12. $x = |y|$

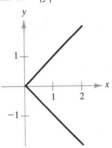

19. $f(x) = x\sqrt{x + 3}$

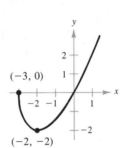

20. $f(x) = |x + 1| + |x - 1|$

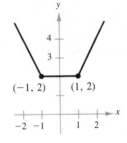

In Exercises 13–20, (a) determine the intervals over which the function is increasing, decreasing, or constant, and (b) determine if the function is even, odd, or neither.

13. $f(x) = 2x$

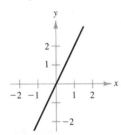

14. $f(x) = x^2 - 2x$

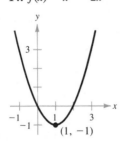

In Exercises 21–26, determine whether the given function is even, odd, or neither.

21. $f(x) = x^6 - 2x^2 + 3$

22. $h(x) = x^3 - 5$

23. $g(x) = x^3 - 5x$

24. $f(x) = x\sqrt{1 - x^2}$

25. $f(t) = t^2 + 2t - 3$

26. $g(s) = 4s^{2/3}$

15. $f(x) = x^3 - 3x^2$

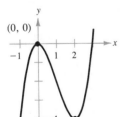

16. $f(x) = \sqrt{x^2 - 4}$

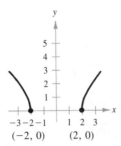

In Exercises 27–38, sketch the graph of the given function and determine whether the function is even, odd, or neither.

27. $f(x) = 3$

28. $g(x) = x$

29. $f(x) = 5 - 3x$

30. $h(x) = x^2 - 4$

31. $g(s) = \dfrac{s^3}{4}$

32. $f(t) = -t^4$

33. $f(x) = \sqrt{1 - x}$

34. $f(x) = x^{3/2}$

35. $g(t) = \sqrt[3]{t-1}$

36. $f(x) = |x+2|$

37. $f(x) = \begin{cases} x+3, & x \le 0 \\ 3, & 0 < x \le 2 \\ 2x-1, & x > 2 \end{cases}$

38. $f(x) = \begin{cases} 2x+1, & x \le -1 \\ x^2-2, & x > -1 \end{cases}$

In Exercises 39–48, sketch the graph of the function and determine the interval(s) (if any) on the real axis for which $f(x) \ge 0$.

39. $f(x) = 4 - x$

40. $f(x) = 4x + 2$

41. $f(x) = x^2 - 9$

42. $f(x) = x^2 - 4x$

43. $f(x) = 1 - x^4$

44. $f(x) = \sqrt{x+2}$

45. $f(x) = x^2 + 1$

46. $f(x) = -(1 + |x|)$

47. $f(x) = -5$

48. $f(x) = \frac{1}{2}(2 + |x|)$

49. Sketch (on the same set of coordinate axes) a graph of f for $c = -2, 0,$ and 2.
  (a) $f(x) = \frac{1}{2}x + c$
  (b) $f(x) = \frac{1}{2}(x - c)$
  (c) $f(x) = \frac{1}{2}(cx)$

50. Sketch (on the same set of coordinate axes) a graph of f for $c = -2, 0,$ and 2.
  (a) $f(x) = x^3 + c$
  (b) $f(x) = (x - c)^3$
  (c) $f(x) = (x - 2)^3 + c$

51. Use the graph of $f(x) = \sqrt{x}$ (see figure) to sketch the graph of each of the following.
  (a) $y = \sqrt{x} + 2$
  (b) $y = -\sqrt{x}$
  (c) $y = \sqrt{x} - 2$
  (d) $y = \sqrt{x} + 3$
  (e) $y = 2 - \sqrt{x} - 4$
  (f) $y = \sqrt{2x}$

52. Use the graph of $f(x) = \sqrt[3]{x}$ (see figure) to sketch the graph of each of the following.
  (a) $y = \sqrt[3]{x} - 1$ (b) $y = \sqrt[3]{x} + 1$
  (c) $y = \sqrt[3]{x-1}$ (d) $y = -\sqrt[3]{x} - 2$
  (e) $y = \sqrt[3]{x+1} - 1$ (f) $y = \frac{1}{2}\sqrt[3]{x}$

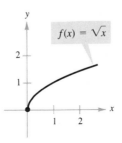
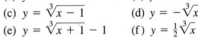

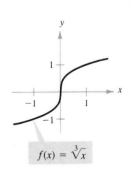

FIGURE FOR 51   **FIGURE FOR 52**

53. Use the graph of $f(x) = x^2$ [see Figure 1.48(e)] to write formulas for the functions whose graphs are shown in parts (a) and (b).
  (a)   (b)

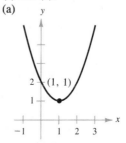
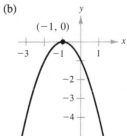

54. Use the graph of $f(x) = x^3$ [see Figure 1.48(f)] to write formulas for the functions whose graphs are shown in parts (a) and (b).
  (a)   (b)

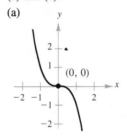
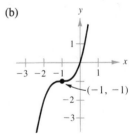

In Exercises 55–58, write the height *h* of the given rectangle as a function of *x*.

In Exercises 59 and 60, write the length *L* of the given rectangle as a function of *y*.

55.

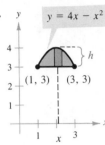

56.

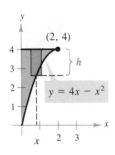

59.

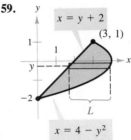

60.

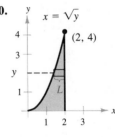

57.

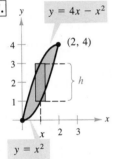

58.

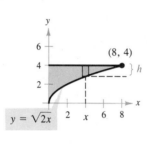

61. Prove that a function of the form

$$f(x) = a_{2n+1}x^{2n+1} + a_{2n-1}x^{2n-1} + \cdots + a_3x^3 + a_1x$$

is odd.

62. Prove that a function of the form

$$f(x) = a_{2n}x^{2n} + a_{2n-2}x^{2n-2} + \cdots + a_2x^2 + a_0$$

is even.

1.7 Combinations of Functions and Inverse Functions

Just as two real numbers can be combined by the operations of addition, subtraction, multiplication, and division to form other real numbers, two functions can be combined to create new functions. For example, if

$$f(x) = 2x - 3 \quad \text{and} \quad g(x) = x^2 - 1$$

we can form the sum, difference, product, and quotient of *f* and *g* as follows.

$$f(x) + g(x) = (2x - 3) + (x^2 - 1) = x^2 + 2x - 4 \qquad \textit{Sum}$$

$$f(x) - g(x) = (2x - 3) - (x^2 - 1) = -x^2 + 2x - 2 \qquad \textit{Difference}$$

$$f(x)g(x) = (2x - 3)(x^2 - 1) = 2x^3 - 3x^2 - 2x + 3 \qquad \textit{Product}$$

$$\frac{f(x)}{g(x)} = \frac{2x - 3}{x^2 - 1}, \qquad x \neq \pm 1 \qquad \textit{Quotient}$$

The domain of an arithmetic combination of functions f and g consists of all real numbers *common* to the domains of f and g. In the case of $f(x)/g(x)$, there is the further restriction that $g(x) \neq 0$.

Definition of Sum, Difference, Product, and Quotient of Functions

Let f and g be two functions with overlapping domains. Then, for all x common to both domains, the sum, difference, product, and quotient of f and g are defined as follows.

1. *Sum:* $(f + g)(x) = f(x) + g(x)$
2. *Difference:* $(f - g)(x) = f(x) - g(x)$
3. *Product:* $(fg)(x) = f(x) \cdot g(x)$
4. *Quotient:* $\left(\dfrac{f}{g}\right)(x) = \dfrac{f(x)}{g(x)}, \qquad g(x) \neq 0$

EXAMPLE 1   *Evaluating the Sum and Difference of Two Functions*

Given $f(x) = 2x + 1$ and $g(x) = x^2 + 2x - 1$, find the following.

(a) $(f + g)(2)$ (b) $(f - g)(2)$

SOLUTION

(a) Since

$$(f + g)(x) = f(x) + g(x)$$
$$= (2x + 1) + (x^2 + 2x - 1)$$
$$= x^2 + 4x$$

it follows that

$$(f + g)(2) = 2^2 + 4(2) = 12.$$

(b) Since

$$(f - g)(x) = f(x) - g(x)$$
$$= (2x + 1) - (x^2 + 2x - 1)$$
$$= -x^2 + 2$$

it follows that

$$(f - g)(2) = -(2)^2 + 2 = -2.$$

In Example 1, both f and g have domains that consist of all real numbers. Thus, the domain of their sum and difference is also the set of all real numbers. Remember that any restrictions on the domains of f or g must be taken into account when forming the sum, difference, product, or quotient of f and g. For instance, the domain of $f(x) = 1/x$ is all $x \neq 0$, and the domain of $g(x) = \sqrt{x}$ is $[0, \infty)$. This implies that the domain of $f + g$ is $(0, \infty)$.

EXAMPLE 2 The Quotient of Two Functions

Find $(f/g)(x)$ and $(g/f)(x)$ for the functions

$$f(x) = \sqrt{x} \quad \text{and} \quad g(x) = \sqrt{4 - x^2}.$$

Then find the domains of f/g and g/f.

SOLUTION

The quotient of f and g is given by

$$\left(\frac{f}{g}\right)(x) = \frac{f(x)}{g(x)} = \frac{\sqrt{x}}{\sqrt{4 - x^2}}$$

and the quotient of g and f is given by

$$\left(\frac{g}{f}\right)(x) = \frac{g(x)}{f(x)} = \frac{\sqrt{4 - x^2}}{\sqrt{x}}.$$

The domain of f is $[0, \infty)$ and the domain of g is $[-2, 2]$. The intersection of these two domains is $[0, 2]$. Thus, we have the following domains for f/g and g/f.

Domain of $\dfrac{f}{g}$: $[0, 2)$

Domain of $\dfrac{g}{f}$: $(0, 2]$

Can you see why these two domains differ slightly?

Composition of Functions

Another way of combining two functions is to form the **composition** of one with the other. For instance, if $f(x) = x^2$ and $g(x) = x + 1$, then the composition of f with g is given by

$$f(g(x)) = f(x + 1) = (x + 1)^2.$$

We denote this **composite function** as $f \circ g$.

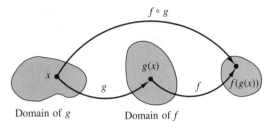

FIGURE 1.54

Definition of Composition of Two Functions

The **composition of the functions f and g** is given by

$(f \circ g)(x) = f(g(x))$.

The domain of $f \circ g$ is the set of all x in the domain of g such that $g(x)$ is in the domain of f. (See Figure 1.54.)

EXAMPLE 3 Forming the Composition of f with g

Find $(f \circ g)(x)$ for

$f(x) = \sqrt{x}, \qquad x \geq 0$

and

$g(x) = x - 1, \qquad x \geq 1.$

If possible, find $(f \circ g)(2)$ and $(f \circ g)(0)$.

SOLUTION

To find $(f \circ g)(x)$, we write

$$\begin{aligned}(f \circ g)(x) &= f(g(x)) && \text{\textit{Definition of} } f \circ g \\ &= f(x - 1) && \text{\textit{Definition of} } g(x) \\ &= \sqrt{x - 1}. && \text{\textit{Definition of} } f(x)\end{aligned}$$

The domain of $f \circ g$ is $[1, \infty)$. Thus,

$(f \circ g)(2) = \sqrt{2 - 1} = 1$

is defined, but $(f \circ g)(0)$ is not defined because 0 is not in the domain of $f \circ g$.

The composition of f with g is generally *not* the same as the composition of g with f. This is illustrated by the fact that for the functions in Example 3

$$(g \circ f)(x) = g(f(x)) = g(\sqrt{x}) = \sqrt{x} - 1$$

which is not equal to $(f \circ g)(x)$. (Try letting $x = 4$ to verify that $\sqrt{x} - 1 \neq \sqrt{x - 1}$.)

EXAMPLE 4 A Case in Which f ∘ g = g ∘ f

Given $f(x) = 2x + 3$ and $g(x) = \frac{1}{2}(x - 3)$, find the following.

(a) $(f \circ g)(x)$ (b) $(g \circ f)(x)$

SOLUTION

(a) $(f \circ g)(x) = f(g(x)) = f(\frac{1}{2}(x - 3))$
$$= 2[\tfrac{1}{2}(x - 3)] + 3 = x - 3 + 3 = x$$

(b) $(g \circ f)(x) = g(f(x)) = g(2x + 3)$
$$= \tfrac{1}{2}[(2x + 3) - 3] = \tfrac{1}{2}(2x) = x$$

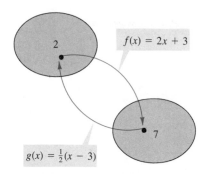

$f(x) = 2x + 3$

$g(x) = \frac{1}{2}(x - 3)$

FIGURE 1.55

Inverse Functions

In Example 4, note that both composite functions $f \circ g$ and $g \circ f$ equal the identity function. That is,

$$f(g(x)) = x = g(f(x)).$$

The functions f and g have the effect of "undoing" each other, as shown in Figure 1.55. We call such functions **inverses** of each other.

Definition of Inverse Functions

Two functions f and g are **inverses** of each other if

$$f(g(x)) = x \qquad \text{for every } x \text{ in the domain of } g$$

and

$$g(f(x)) = x \qquad \text{for every } x \text{ in the domain of } f.$$

We denote g by f^{-1} (read "f inverse"). Thus, $f(f^{-1}(x)) = x$ and $f^{-1}(f(x)) = x$.

Remark: For inverse functions f and g, the range of g must be *equal to* the domain of f, and vice versa.

Don't be confused by the use of -1 to denote the inverse function f^{-1}. Whenever we write $f^{-1}(x)$, we will *always* be referring to the inverse of the function f and *not* to the reciprocal of f.

EXAMPLE 5   Testing for Inverse Functions

Which of the functions

$$g(x) = \frac{x - 2}{5} \quad \text{and} \quad h(x) = \frac{5}{x} + 2$$

is the inverse of the function

$$f(x) = \frac{5}{x - 2}?$$

SOLUTION

The composition of f with g is

$$f(g(x)) = f\left(\frac{x - 2}{5}\right) = \frac{5}{[(x - 2)/5] - 2}$$

$$= \frac{25}{(x - 2) - 10} = \frac{25}{x - 12} \neq x.$$

The composition of f with h is

$$f(h(x)) = f\left(\frac{5}{x} + 2\right) = \frac{5}{(5/x) + 2 - 2}$$

$$= \frac{5}{5/x}$$

$$= x.$$

Thus, it appears that f and h are inverses of each other, which is confirmed by the fact that

$$h(f(x)) = h\left(\frac{5}{x - 2}\right) = \frac{5}{5/(x - 2)} + 2$$

$$= x - 2 + 2$$

$$= x.$$

Check to see that the domain of f is the same as the range of h and vice versa.

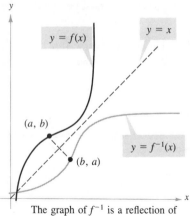

The graph of f^{-1} is a reflection of the graph of f in the line $y = x$.

FIGURE 1.56

The graphs of f and f^{-1} are related to each other in the following way. If the point (a, b) lies on the graph of f, then the point (b, a) lies on the graph of f^{-1} and vice versa. This means that the graph of f^{-1} is a reflection of the graph of f in the line $y = x$, as shown in Figure 1.56.

A function need not have an inverse. For instance, the function $f(x) = x^2$ has no inverse [assuming a domain of $(-\infty, \infty)$]. In order for a function to have an inverse, it is necessary that the function be **one-to-one,** which means that no two elements in the domain of f correspond to the same element in the range of f.

Definition of One-to-One Function

A function f is **one-to-one** if, for a and b in its domain,

$f(a) = f(b)$ implies that $a = b$.

The function $f(x) = x + 1$ *is* one-to-one because

$$a + 1 = b + 1$$

implies that a and b must be equal. However, the function $f(x) = x^2$ is *not* one-to-one because

$$a^2 = b^2$$

does not imply that $a = b$. For instance, $(-1)^2 = 1^2$ and yet $-1 \neq 1$.

The following theorem tells us that a function has an inverse if and only if the function is one-to-one.

Existence of an Inverse Function

A function f has an inverse function f^{-1} if and only if f is one-to-one.

From its graph, it is easy to tell whether a function of x is one-to-one. We simply check to see that every *horizontal* line intersects the graph of the function at most once. For instance, in Figure 1.57(b), note that for the graph of $g(x) = x^2 - x$ it is possible to find a horizontal line that intersects the graph twice. Hence, the function g is not one-to-one. The other two functions shown in Figure 1.57 are one-to-one.

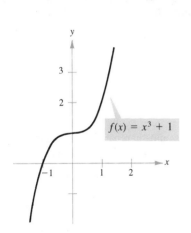

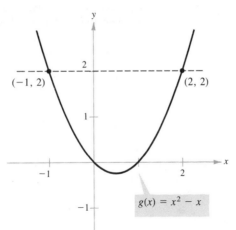

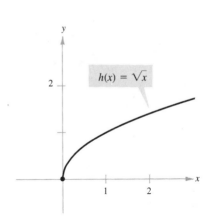

(a) f is one-to-one.   (b) g is not one-to-one.   (c) h is one-to-one.

FIGURE 1.57

Two special types of functions that do pass the **horizontal line test** are those that are increasing or decreasing on their entire domains.

1. If f is *increasing* on its entire domain, then f is one-to-one.
2. If f is *decreasing* on its entire domain, then f is one-to-one.

Finding the Inverse of a Function

Now that we have a test to determine whether a function has an inverse, we need a procedure for *finding* the inverse.

Finding the Inverse of a Function

To find the inverse of f, use the following steps.

1. Test to see that f is one-to-one.
2. Write the function in the form $y = f(x)$ and solve for x in terms of y to obtain $x = f^{-1}(y)$.
3. Check to see that the domain of f is the range of f^{-1}, and that the domain of f^{-1} is the range of f.

In the following example, remember that any letter can be used to represent the independent variable. For instance,

$$f^{-1}(x) = x^2 + 4 \qquad \text{and} \qquad f^{-1}(y) = y^2 + 4$$

represent the same function.

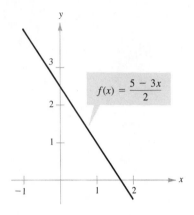

FIGURE 1.58

EXAMPLE 6 Finding the Inverse of a Function

Find the inverse (if it exists) of

$$f(x) = \frac{5 - 3x}{2}.$$

SOLUTION

From Figure 1.58 we see that f is one-to-one, and therefore has an inverse. To begin, we write the function in the form $y = f(x)$. Then we solve for x in terms of y, as follows.

$$y = \frac{5 - 3x}{2} \qquad \textit{Write in the form } y = f(x)$$

$$2y = 5 - 3x$$

$$3x = 5 - 2y$$

$$x = \frac{5 - 2y}{3} \qquad \textit{Solve for x}$$

Therefore, the inverse of f is

$$f^{-1}(y) = \frac{5 - 2y}{3}.$$

Or, using x as the independent variable, f^{-1} can be written as

$$f^{-1}(x) = \frac{5 - 2x}{3}.$$

The domain and range of both f and f^{-1} consist of all real numbers.

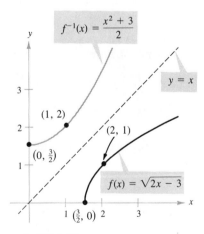

FIGURE 1.59

EXAMPLE 7 Finding the Inverse of a Function

Find the inverse of the function

$$f(x) = \sqrt{2x - 3}$$

and sketch the graphs of f and f^{-1}.

SOLUTION

The graph of f is shown in Figure 1.59. To find the inverse of f, we proceed as follows.

Combinations of Functions and Inverse Functions

$$y = \sqrt{2x - 3}$$
$$y^2 = 2x - 3$$
$$y^2 + 3 = 2x$$
$$x = \frac{y^2 + 3}{2} \qquad \text{\textit{Solve for } x}$$
$$f^{-1}(y) = \frac{y^2 + 3}{2} \qquad y \geq 0$$

Or, using x as the independent variable, f^{-1} can be written as

$$f^{-1}(x) = \frac{x^2 + 3}{2}. \qquad x \geq 0$$

The graph of f^{-1} is the reflection of the graph of f in the line $y = x$, as shown in Figure 1.59. Note that the domain of f is the interval $[3/2, \infty)$ and the range of f is the interval $[0, \infty)$. Moreover, the domain of f^{-1} is the interval $[0, \infty)$, and the range of f^{-1} is the interval $[3/2, \infty)$. ▬

The problem of finding the inverse of a function can be difficult (or even impossible) for two reasons. First, given $y = f(x)$, it may be algebraically difficult to solve for x in terms of y. Second, if f is not one-to-one, then f^{-1} does not exist.

In later chapters we will study two important classes of inverse functions: inverse trigonometric functions and logarithmic functions.

WARM UP

Find the domain of each function.

1. $f(x) = \sqrt[3]{x + 1}$

2. $f(x) = \sqrt{x + 1}$

3. $g(x) = \dfrac{2}{x^2 - 2x}$

4. $h(x) = \dfrac{x}{3x + 5}$

Simplify the expressions.

5. $2\left(\dfrac{x + 5}{2}\right) - 5$

6. $7 - 10\left(\dfrac{7 - x}{10}\right)$

7. $\sqrt[3]{2\left(\dfrac{x^3}{2} - 2\right) + 4}$

8. $(\sqrt[5]{x + 2})^5 - 2$

Solve for x in terms of y.

9. $y = \dfrac{2x - 6}{3}$

10. $y = \sqrt[3]{2x - 4}$

EXERCISES 1.7

In Exercises 1–6, find (a) $(f + g)(x)$, (b) $(f - g)(x)$, (c) $(fg)(x)$, and (d) $(f/g)(x)$. What is the domain of f/g?

1. $f(x) = x + 1,$ $g(x) = x - 1$
2. $f(x) = 2x - 5,$ $g(x) = 1 - x$
3. $f(x) = x^2,$ $g(x) = 1 - x$
4. $f(x) = 2x - 5,$ $g(x) = 5$
5. $f(x) = \dfrac{1}{x},$ $g(x) = \dfrac{1}{x^2}$
6. $f(x) = \dfrac{x}{x + 1},$ $g(x) = x^3$

In Exercises 7–16, evaluate the indicated function for $f(x) = x^2 + 1$ and $g(x) = x - 4$.

7. $(f + g)(3)$ 8. $(f - g)(-2)$
9. $(f - g)(2t)$ 10. $(f + g)(t - 1)$
11. $(fg)(4)$ 12. $(fg)(-6)$
13. $\left(\dfrac{f}{g}\right)(5)$ 14. $\left(\dfrac{f}{g}\right)(0)$
15. $(f - g)(0)$ 16. $(f + g)(1)$

In Exercises 17–20, find (a) $f \circ g$, (b) $g \circ f$, and (c) $f \circ f$.

17. $f(x) = x^2,$ $g(x) = x - 1$
18. $f(x) = 3x,$ $g(x) = 2x + 1$
19. $f(x) = 3x + 5,$ $g(x) = 5 - x$
20. $f(x) = x^3,$ $g(x) = \dfrac{1}{x}$

In Exercises 21–24, use the graphs of f and g (see figure) to evaluate the indicated functions.

21. (a) $(f + g)(3)$ (b) $\left(\dfrac{f}{g}\right)(2)$
22. (a) $(f - g)(1)$ (b) $(fg)(4)$
23. (a) $(f \circ g)(2)$ (b) $(g \circ f)(2)$
24. (a) $(f \circ g)(1)$ (b) $(g \circ f)(3)$

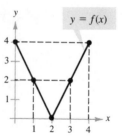
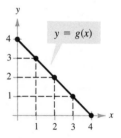

FIGURE FOR 21–24

In Exercises 25–34, (a) show that f and g are inverse functions by showing that $f(g(x)) = x$ and $g(f(x)) = x$, and (b) graph f and g on the same set of coordinate axes.

25. $f(x) = 2x,$ $g(x) = \dfrac{x}{2}$
26. $f(x) = x - 5,$ $g(x) = x + 5$
27. $f(x) = 5x + 1,$ $g(x) = \dfrac{x - 1}{5}$
28. $f(x) = 3 - 4x,$ $g(x) = \dfrac{3 - x}{4}$
29. $f(x) = x^3,$ $g(x) = \sqrt[3]{x}$
30. $f(x) = \dfrac{1}{x},$ $g(x) = \dfrac{1}{x}$
31. $f(x) = \sqrt{x - 4},$ $g(x) = x^2 + 4, x \geq 0$
32. $f(x) = 9 - x^2, x \geq 0,$ $g(x) = \sqrt{9 - x}, x \leq 9$
33. $f(x) = 1 - x^3,$ $g(x) = \sqrt[3]{1 - x}$
34. $f(x) = \dfrac{1}{1 + x}, x \geq 0,$ $g(x) = \dfrac{1 - x}{x}, 0 < x \leq 1$

In Exercises 35–44, determine whether the function is one-to-one.

35.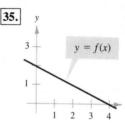

36.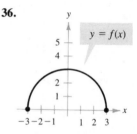

Exercises

37.

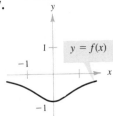

38.

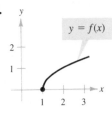

39. $g(x) = \dfrac{4 - x}{6}$

40. $f(x) = 10$

41. $h(x) = |x + 4|$

42. $g(x) = (x + 5)^3$

43. $f(x) = -\sqrt{16 - x^2}$

44. $f(x) = (x + 2)^2$

In Exercises 45–54, find the inverse of the one-to-one function f. Then graph both f and f^{-1} on the same coordinate plane.

45. $f(x) = 2x - 3$

46. $f(x) = 3x$

47. $f(x) = x^5$

48. $f(x) = x^3 + 1$

49. $f(x) = \sqrt{x}$

50. $f(x) = x^2, \quad x \geq 0$

51. $f(x) = \sqrt{4 - x^2}, \quad 0 \leq x \leq 2$

52. $f(x) = \dfrac{4}{x}$

53. $f(x) = \sqrt[3]{x - 1}$

54. $f(x) = x^{3/5}$

In Exercises 55–70, determine whether the given function is one-to-one. If it is, find its inverse.

55. $f(x) = x^4$

56. $f(x) = \dfrac{1}{x^2}$

57. $g(x) = \dfrac{x}{8}$

58. $f(x) = 3x + 5$

59. $p(x) = -4$

60. $f(x) = \dfrac{3x + 4}{5}$

61. $f(x) = (x + 3)^2, \quad x \geq -3$

62. $q(x) = (x - 5)^2$

63. $h(x) = \dfrac{1}{x}$

64. $f(x) = |x - 2|, \quad x \leq 2$

65. $f(x) = \sqrt{2x + 3}$

66. $f(x) = \sqrt{x - 2}$

67. $g(x) = x^2 - x^4$

68. $f(x) = \dfrac{x^2}{x^2 + 1}$

69. $f(x) = 25 - x^2, \quad -\infty < x \leq 0$

70. $f(x) = ax + b, \quad a \neq 0$

In Exercises 71 and 72, use the graph of the function f to complete the table and to sketch the graph of f^{-1}.

71.

x	0	1	2	3	4
$f^{-1}(x)$					

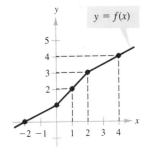

72.

x	0	2	4	6
$f^{-1}(x)$				

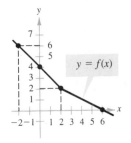

In Exercises 73–76, use the functions

$$f(x) = x + 4 \quad \text{and} \quad g(x) = 2x - 5$$

to find the indicated functions.

73. $g^{-1} \circ f^{-1}$

74. $f^{-1} \circ g^{-1}$

75. $(f \circ g)^{-1}$

76. $(g \circ f)^{-1}$

77. Prove that if f and g are one-to-one functions, then $(f \circ g)^{-1}(x) = (g^{-1} \circ f^{-1})(x)$.

CHAPTER 1 REVIEW EXERCISES

In Exercises 1–4, plot the two real numbers on the real number line and place the appropriate inequality sign (< or >) between them.

1. $-4, \frac{4}{3}$
2. $\frac{9}{4}, \frac{5}{3}$
3. $-4, -6$
4. $-\frac{5}{2}, \pi$

In Exercises 5–8, write the given number without absolute value signs.

5. $|5| - |-3|$
6. $\dfrac{6}{|6|}$
7. $-2|-5|$
8. $|\sqrt[3]{-27}|$

In Exercises 9–12, graph the inequality on the real number line.

9. $x \geq 2$
10. $-3 < x < \frac{3}{2}$
11. $|x| > 2$
12. $|x - 3| \leq 2$

In Exercises 13–16, use absolute value notation to describe the given statement.

13. The distance between x and 7 is at least 4.
14. The distance between x and 25 is no more than 10.
15. All the real numbers x within 2 units of 5.
16. All the real numbers y more than 2 units from $\frac{1}{2}$.

In Exercises 17–30, solve the given equation.

17. $3x - 2(x + 5) = 10$
18. $4x + 2(7 - x) = 5$
19. $5x^4 - 12x^3 = 0$
20. $4x^3 - 6x = 0$
21. $6x = 3x^2$
22. $2x = \dfrac{2}{x^3}$
23. $3\left(1 - \dfrac{1}{5t}\right) = 0$
24. $3x^2 + 1 = 0$
25. $2 - x^{-2} = 0$
26. $-t^2 + 6t = 4$
27. $4t^3 - 12t^2 + 8t = 0$
28. $12t^3 - 84t^2 + 120t = 0$
29. $(x - 1)(2x - 3) + (x^2 - 3x + 2) = 0$
30. $\dfrac{1}{x - 2} = 3$

In Exercises 31–34, find (a) the distance between the two points, (b) the coordinates of the midpoint of the line segment between the two points, and (c) an equation of the circle whose diameter is the line segment between the two points.

31. $(0, 0), (0, 10)$
32. $(-1, 4), (2, 0)$
33. $(2, 1), (14, 6)$
34. $(-2, 2), (3, -10)$

In Exercises 35–40, find the intercepts of the given graph and check for symmetry with respect to each of the coordinate axes and the origin.

35. $2y^2 = x^3$

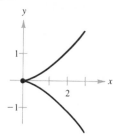

36. $x^2 + (y + 2)^2 = 4$

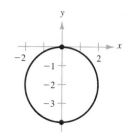

37. $y = \dfrac{x^4}{4} - 2x^2$

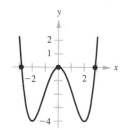

38. $y = \dfrac{x^3}{4} - 3x$

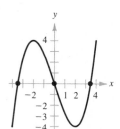

39. $y = x\sqrt{4 - x^2}$

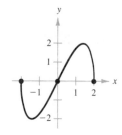

40. $y = x\sqrt{x + 3}$

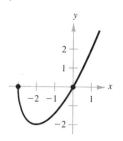

In Exercises 41–44, determine the center and radius of the circle. Then sketch its graph.

41. $x^2 + y^2 - 12x - 8y + 43 = 0$

42. $x^2 + y^2 - 20x - 10y + 100 = 0$

43. $4x^2 + 4y^2 - 4x - 40y + 92 = 0$

44. $5x^2 + 5y^2 - 14y = 0$

In Exercises 45–54, sketch the graph of the equation.

45. $y - 2x - 3 = 0$ **46.** $3x + 2y + 6 = 0$

47. $x - 5 = 0$ **48.** $y = \sqrt{x + 2}$

49. $y + 2x^2 = 0$ **50.** $y = x^2 - 4x$

51. $y = \sqrt{25 - x^2}$ **52.** $y = -(x - 4)^2$

53. $y = \frac{1}{4}(x + 1)^3$ **54.** $y = 4 - (x - 4)^2$

In Exercises 55 and 56, evaluate the function at the specified values of the independent variable. Simplify your answers.

55. $f(x) = x^2 + 1$
 (a) $f(2)$ (b) $f(-4)$
 (c) $f(t^2)$ (d) $f(x + 2)$

56. $h(x) = 6 - 5x^2$
 (a) $h(2)$ (b) $h(x + 3)$
 (c) $\dfrac{h(4) - h(2)}{4 - 2}$ (d) $h(x + 2) - h(x)$

In Exercises 57–60, determine the domain of the function.

57. $f(x) = \sqrt{25 - x^2}$ **58.** $f(x) = 3x + 4$

59. $g(s) = \dfrac{5}{3s - 9}$ **60.** $h(t) = \sqrt{4 - t}$

In Exercises 61–66, (a) find f^{-1}, (b) sketch the graphs of f and f^{-1} on the same coordinate axes, and (c) verify that $f^{-1}(f(x)) = x = f(f^{-1}(x))$.

61. $f(x) = \frac{1}{2}x - 3$ **62.** $f(x) = 5x - 7$

63. $f(x) = \sqrt{x + 1}$ **64.** $f(x) = x^3 + 2$

65. $f(x) = x^2 - 5, \quad x \geq 0$ **66.** $f(x) = \sqrt[3]{x + 1}$

In Exercises 67–70, restrict the domain of the function f to an interval where the function is increasing and determine f^{-1} over that interval.

67. $f(x) = 2(x - 4)^2$ **68.** $f(x) = \dfrac{x^2}{2}$

69. $f(x) = \sqrt{x^2 - 4}$ **70.** $f(x) = x^2 - 4$

In Exercises 71–76, let

$$f(x) = 3 - 2x, \quad g(x) = \sqrt{x}, \quad \text{and} \quad h(x) = 3x^2 + 2$$

and find the indicated value.

71. $(f - g)(4)$ **72.** $(fh)(1)$

73. $(h \circ g)(7)$ **74.** $\left(\dfrac{g}{h}\right)(1)$

75. $g^{-1}(3)$ **76.** $(h \circ f^{-1})(1)$

77. A wire 24 inches long is to be cut into four pieces to form a rectangle whose shortest side has a length of x. Express the area A of the rectangle as a function of x. Determine the domain of the function and sketch its graph over that domain.

78. A company produces a product for which the variable cost is $5.35 per unit and the fixed costs are $16,000. The company sells the product for $8.20, and can sell all it produces.
 (a) Find the total cost as a function of x, the number of units produced.
 (b) Find the profit as a function of x.

79. A group of farmers agree to share equally in the cost of a $48,000 piece of machinery. If they could find two more farmers to join the group, each person's share of the cost would decrease by $4000. How many farmers are presently in the group?

80. Each week a salesperson must make a 180-mile trip to pick up supplies. If she were to increase her average speed by 5 miles per hour, the trip would take 24 minutes less than usual. Find her usual average speed.

C H A P T E R 2

Trigonometry

2.1 Radian and Degree Measure

As derived from the Greek language, the word **trigonometry** means "measurement of triangles." Initially, trigonometry dealt with relationships among the sides and angles of triangles. As such, it was used in the development of astronomy, navigation, and surveying.

With the advent of calculus in the seventeenth century, and a resulting expansion of knowledge in the physical sciences, a different perspective arose—one that viewed the classic trigonometric relationships as *functions* with the set of real numbers as their domains. Consequently, the applications of trigonometry expanded to include a vast number of physical phenomena involving rotations, or vibrations. These include sound waves, light rays, planetary orbits, vibrating strings, pendulums, and orbits of atomic particles.

Our approach to trigonometry incorporates *both* perspectives, with the goal that you can become proficient at using this very practical branch of mathematics.

Angles

An **angle** is determined by rotating a ray (half-line) about its endpoint. The starting position of the ray is called the **initial side** of the angle, and the position after rotation is called the **terminal side,** as shown in Figure 2.1. The endpoint of the ray is called the **vertex** of the angle. This perception of an angle fits nicely into a coordinate system in which the origin is the vertex

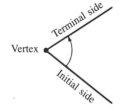

FIGURE 2.1

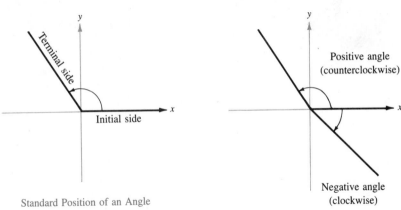

Standard Position of an Angle

FIGURE 2.2

FIGURE 2.3

and the initial side coincides with the positive *x*-axis. Such an angle is said to be in **standard position,** as shown in Figure 2.2. **Positive angles** are generated by counterclockwise rotation and **negative angles** by clockwise rotation, as shown in Figure 2.3.

To label angles in trigonometry, we use the Greek letters α (alpha), β (beta), and θ (theta), as well as upper-case letters A, B, and C. In Figure 2.4, note that angles α and β have the same initial and terminal sides. Such angles are called **coterminal.**

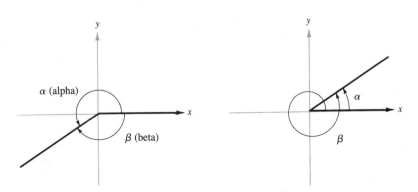

FIGURE 2.4   Coterminal Angles

Radian Measure

The **measure of an angle** is determined by the amount of rotation from the initial to the terminal side. One way to measure angles is in radians. This type of measure is needed in calculus. To define a radian we use a **central angle** of a circle, one whose vertex is the center of the circle, as shown in Figure 2.5.

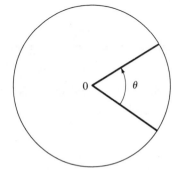

Central Angle θ

FIGURE 2.5

Trigonometry

Definition of a Radian

One **radian** is the measure of a central angle θ that subtends (intercepts) an arc s equal in length to the radius r of the circle. [See Figure 2.6(a).]

(a)   (b)

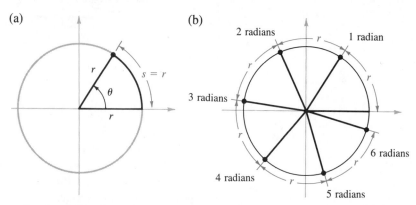

arc length = radius when θ = 1 radian

FIGURE 2.6

Since the circumference of a circle is $2\pi r$, it follows that a central angle of one full revolution (counterclockwise) corresponds to an arc length of $s = 2\pi r$. Moreover, because each radian intercepts an arc of length r, we conclude that one full revolution corresponds to an angle of $(2\pi r)/r = 2\pi$ radians. Note that since $2\pi \approx 6.28$, there are a little more than six radii lengths in a full circle, as shown in Figure 2.6(b).

In general, the radian measure of a central angle θ is obtained by dividing the arc length s by r. That is,

$$\frac{s}{r} = \theta$$

where θ *is measured in radians.* Because the units of measure for s and r are the same, this ratio is unitless—it is simply a real number.

EXAMPLE 1 Radian Measure of an Angle

(a) Because the arc length of a half circle of radius r is $s = \pi r$, the radian measure of one-half revolution is

$$\frac{\pi r}{r} = \pi. \qquad\qquad \textit{One-half revolution}$$

(b) The radian measure of one-quarter revolution is

$$\frac{\pi r/2}{r} = \frac{\pi}{2}. \qquad\qquad \textit{One-quarter revolution}$$

FIGURE 2.7

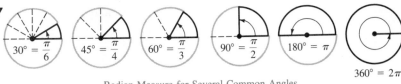

Radian Measure for Several Common Angles

Figure 2.7 shows these two angles together with the radian measure of some other common angles.

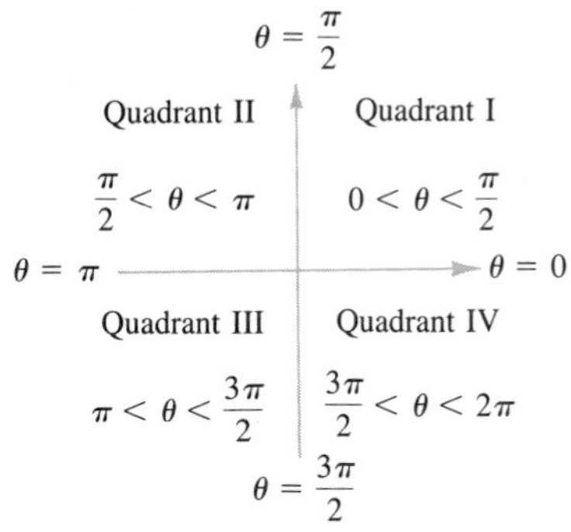

$\theta = \dfrac{\pi}{2}$

Quadrant II	Quadrant I

$\dfrac{\pi}{2} < \theta < \pi \qquad 0 < \theta < \dfrac{\pi}{2}$

$\theta = \pi \longrightarrow \theta = 0$

Quadrant III	Quadrant IV

$\pi < \theta < \dfrac{3\pi}{2} \qquad \dfrac{3\pi}{2} < \theta < 2\pi$

$\theta = \dfrac{3\pi}{2}$

Quadrant Location of Angles

FIGURE 2.8

Recall that the four quadrants in a coordinate system are numbered counterclockwise as I, II, III, and IV. Figure 2.8 shows which angles between 0 and 2π lie in each of the four quadrants.

We can find an angle that is coterminal to a given angle θ by adding or subtracting 2π (one revolution), as demonstrated in Example 2. (Note that a given angle has many coterminal angles. For instance, $\theta = \pi/6$ is coterminal with both $13\pi/6$ and $-11\pi/6$.)

EXAMPLE 2 *Sketching and Finding Coterminal Angles*

Sketch each of the following angles in standard position and find a coterminal angle for each.

(a) $\theta = \dfrac{13\pi}{6}$ 　　　　(b) $\theta = \dfrac{3\pi}{4}$ 　　　　(c) $\theta = -\dfrac{2\pi}{3}$

SOLUTION

(a) For the positive angle $\theta = 13\pi/6$, we subtract 2π and obtain the coterminal angle

$$\frac{13\pi}{6} - 2\pi = \frac{\pi}{6}.$$

Thus, the terminal side of θ lies in Quadrant I. Its sketch is shown in Figure 2.9(a).

(b) Again subtracting 2π, an angle coterminal with $3\pi/4$ is

$$\frac{3\pi}{4} - 2\pi = -\frac{5\pi}{4}.$$

Since $\pi/2 < \theta < \pi$ the terminal side of θ lies in Quadrant II. Moreover, since $\theta = 3\pi/4$ is $\pi/4$ less than π, it follows that θ lies in Quadrant II, $\pi/4$ radians up from the horizontal axis. A sketch is shown in Figure 2.9(b).

Trigonometry

(c) For the negative angle $\theta = -2\pi/3$, we add 2π to obtain the coterminal angle

$$\theta = -\frac{2\pi}{3} + 2\pi = \frac{4\pi}{3}$$

as shown in Figure 2.9(c).

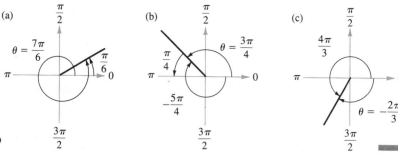

FIGURE 2.9

Figure 2.10 shows several common angles with their radian measures. Note that we classify angles between 0 and $\pi/2$ radians as **acute** and angles between $\pi/2$ and π as **obtuse**.

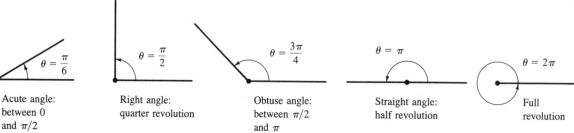

Acute angle: between 0 and $\pi/2$

Right angle: quarter revolution

Obtuse angle: between $\pi/2$ and π

Straight angle: half revolution

Full revolution

FIGURE 2.10

Two *positive* angles α and β are said to be **complementary** (or complements of each other) if their sum is $\pi/2$. For example, $\pi/6$ and $\pi/3$ are complementary angles because $(\pi/6) + (\pi/3) = \pi/2$. Two positive angles are **supplementary** (or supplements of each other) if their sum is π. For example, $2\pi/3$ and $\pi/3$ are supplementary angles because $(2\pi/3) + (\pi/3) = \pi$.

EXAMPLE 3 *Complementary and Supplementary Angles*

If possible, find the complementary and the supplementary angle for each of the following angles.

(a) $\frac{2\pi}{5}$

(b) $\frac{4\pi}{5}$

(a)

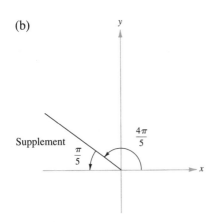

(b)

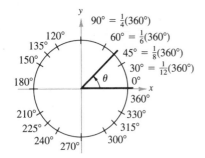

FIGURE 2.11

SOLUTION

(a) The complement of $\theta = 2\pi/5$ is

$$\frac{\pi}{2} - \frac{2\pi}{5} = \frac{5\pi}{10} - \frac{4\pi}{10} = \frac{\pi}{10}.$$

The supplement of $\theta = 2\pi/5$ is

$$\pi - \frac{2\pi}{5} = \frac{3\pi}{5}.$$

See Figure 2.11(a).

(b) Because $\theta = 4\pi/5$ is greater than $\pi/2$, it has no complement. (Remember, we use only *positive* angles for complements.) The supplement of $\theta = 4\pi/5$ is

$$\pi - \frac{4\pi}{5} = \frac{\pi}{5}.$$

See Figure 2.11(b).

Degree Measure

A second way to measure angles is in terms of degrees. A measure of **one degree (1°)** is equivalent to $1/360$ of a complete revolution about the vertex. To measure angles in degrees, it is convenient to mark degrees on the circumference of a circle as shown in Figure 2.12. Thus, a full revolution (counterclockwise) corresponds to 360°, a half revolution to 180°, and a quarter revolution to 90°.

Since 2π radians is the measure of an angle of one complete revolution, degrees and radians are related by the equations

$$360° = 2\pi \text{ rad}$$

and

$$180° = \pi \text{ rad}.$$

From the latter equation, we obtain

$$1° = \frac{\pi}{180} \text{ rad}$$

and

$$1 \text{ rad} = \left(\frac{180}{\pi}\right)°$$

which lead to the following conversion rules.

y

90° = $\frac{1}{4}$(360°)

120°
135°
150°

60° = $\frac{1}{6}$(360°)
45° = $\frac{1}{8}$(360°)
30° = $\frac{1}{12}$(360°)

180°
0°
x
360°

210°
225°
240°
270°

330°
315°
300°

θ

Degree Measure of an Angle

FIGURE 2.12

Trigonometry

Conversions: Degrees ⟺ Radians

1. To convert degrees to radians, multiply degrees by $\dfrac{\pi \text{ rad}}{180°}$.

2. To convert radians to degrees, multiply radians by $\dfrac{180°}{\pi \text{ rad}}$.

To apply these two conversion rules, you simply need to remember the basic relationship π rad $= 180°$, as demonstrated in Examples 4 and 5.

EXAMPLE 4 Converting from Degrees to Radians

Convert the following angles from degree to radian measure.

(a) 135° (b) 540° (c) −270°

SOLUTION

(a) $135° = (135 \text{ deg})\left(\dfrac{\pi \text{ rad}}{180 \text{ deg}}\right) = \dfrac{3\pi}{4}$ rad

(b) $540° = (540 \text{ deg})\left(\dfrac{\pi \text{ rad}}{180 \text{ deg}}\right) = 3\pi$ rad

(c) $-270° = (-270 \text{ deg})\left(\dfrac{\pi \text{ rad}}{180 \text{ deg}}\right) = -\dfrac{3\pi}{2}$ rad

EXAMPLE 5 Converting from Radians to Degrees

Convert the following radian measures to degree measures.

(a) $-\dfrac{\pi}{2}$ (b) $\dfrac{9\pi}{2}$ (c) 2

SOLUTION

(a) $-\dfrac{\pi}{2}$ rad $= \left(-\dfrac{\pi}{2} \text{ rad}\right)\left(\dfrac{180 \text{ deg}}{\pi \text{ rad}}\right) = -90°$

(b) $\dfrac{9\pi}{2}$ rad $= \left(\dfrac{9\pi}{2} \text{ rad}\right)\left(\dfrac{180 \text{ deg}}{\pi \text{ rad}}\right) = 810°$

(c) 2 rad $= (2 \text{ rad})\left(\dfrac{180 \text{ deg}}{\pi \text{ rad}}\right) = \dfrac{360}{\pi}$ deg $\approx 114.59°$

If you have a calculator with a radian-to-degree conversion key, try using it to verify the results.

Remark: Note that when no units of angle measure are specified, *radian measure is implied.* For instance, if we write $\theta = \pi$ or $\theta = 2$, we mean $\theta = \pi$ radians or $\theta = 2$ radians.

Degrees, Minutes, and Seconds

With calculators it is convenient to use *decimal* degrees to denote fractional parts of degrees. Historically, however, fractional parts of degrees were expressed in *minutes* and *seconds*, using the prime (') and double prime (") notations, respectively. That is,

$$1' = \text{one minute} = \frac{1}{60}(1°)$$

$$1'' = \text{one second} = \frac{1}{60}(1')$$

$$= \frac{1}{3600}(1°).$$

Consequently, an angle of 64 degrees, 32 minutes, and 47 seconds is represented by $\theta = 64°\ 32'\ 47''$.

Many calculators have special keys for converting an angle in degrees, minutes, and seconds ($D°\ M'\ S''$) into decimal degree form, and conversely. If your calculator does not have these special keys, you can use the techniques demonstrated in the next two examples to make the conversions.

EXAMPLE 6 *Converting an Angle from D° M′ S″ to Decimal Form*

Convert $152°\ 15'\ 29''$ to decimal degree form.

SOLUTION

Since

$$1' = \left(\frac{1}{60}\right)° \qquad \text{and} \qquad 1'' = \left(\frac{1}{60}\right)\left(\frac{1}{60}\right)° = \left(\frac{1}{3600}\right)°$$

we have

$$152°\ 15'\ 29'' = 152° + \left(\frac{15}{60}\right)° + \left(\frac{29}{3600}\right)°$$

$$\approx 152° + 0.25° + 0.00806°$$

$$= 152.25806°$$

EXAMPLE 7 *Converting an Angle to D° M' S" Form*

Convert 0.86492 radians to $D° M' S"$ form.

SOLUTION

First we convert to decimal degrees.

$$0.86492 = 0.86492 \left(\frac{180°}{\pi} \right) \approx 49.55627°$$

Then we have

$$
\begin{aligned}
49.55627° &= 49° + 0.55627° \\
&= 49° + (0.55627)(60') && \text{\textit{Because } } 1° = 60' \\
&= 49° + 33.3762' \\
&= 49° + 33' + 0.3762' \\
&= 49° + 33' + (0.3762)(60") && \text{\textit{Because } } 1' = 60" \\
&= 49° + 33' + 22.572".
\end{aligned}
$$

Consequently,

$$0.86492 \text{ radian} \approx 49° 33' 23".$$

Applications

The *radian measure* formula, $\theta = s/r$, can be used to measure arc length along a circle. Specifically, for a circle of radius r, a central angle θ subtends an arc of length s given by

$$s = r\theta \qquad\qquad \textit{Length of circular arc}$$

where θ is measured in radians.

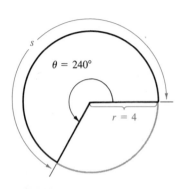

FIGURE 2.13

EXAMPLE 8 *Finding Arc Length*

A circle has a radius of 4 inches. Find the length of the arc cut off (subtended) by a central angle of 240°, as shown in Figure 2.13.

SOLUTION

To use the formula $s = r\theta$, we must first convert 240° to radian measure.

$$240° = (240 \text{ deg}) \left(\frac{\pi \text{ rad}}{180 \text{ deg}} \right) = \frac{4\pi}{3} \text{ rad}$$

Radian and Degree Measure

Then, using a radius of $r = 4$ inches, we find the arc length to be

$$s = r\theta = 4\left(\frac{4\pi}{3}\right) = \frac{16\pi}{3} \approx 16.76 \text{ inches.}$$

Note that the units for $r\theta$ are determined by the units for r because θ has no units.

The formula for the length of a circular arc can be used to analyze the motion of a particle moving at a *constant speed* along a circular path. Assume that the particle is moving at a constant speed along a circular path (of radius r). If s is the length of the arc traveled in time t, then we say that the **speed** of the particle is

$$\text{speed} = \frac{\text{distance}}{\text{time}} = \frac{s}{t}.$$

Moreover, if θ is the angle (in radian measure) corresponding to the arc length s, then the **angular speed** of the particle is

$$\text{angular speed} = \frac{\theta}{t}.$$

EXAMPLE 9 *Finding the Speed of an Object*

The second hand on a clock is 4 inches long, as shown in Figure 2.14. Find the speed of the tip of this second hand.

SOLUTION

The time required for the second hand to make one full revolution is

$$t = 60 \text{ seconds} = 1 \text{ minute.}$$

Moreover, the distance traveled by the tip of the second hand in one revolution is

$$s = 2\pi(\text{radius}) = 2\pi(4) = 8\pi \text{ inches.}$$

Therefore, the speed of the tip of the second hand is

$$\text{speed} = \frac{s}{t} = \frac{8\pi \text{ inches}}{60 \text{ seconds}} \approx 0.419 \text{ in/sec.}$$

FIGURE 2.14

Trigonometry

EXAMPLE 10 *Finding Angular Speed and Speed*

A lawn roller that is 30 inches in diameter makes 1.2 revolutions per second, as shown in Figure 2.15.

(a) Find the angular speed of the roller in radians per second.
(b) How fast is the roller moving across the lawn?

SOLUTION

(a) Since each revolution generates 2π radians, it follows that the roller turns $(1.2)(2\pi) = 2.4\pi$ radians per second. Thus, the angular speed is

$$\text{angular speed} = \frac{\theta}{t} = \frac{2.4\pi \text{ radians}}{1 \text{ second}} = 2.4\pi \text{ rad/sec.}$$

(b) To find the speed of the roller, we use the fact that its diameter is 30 inches. Thus, its radius is 15 inches and we have $s = 2\pi r = 2\pi(15) = 30\pi$. Since the roller makes 1.2 revolutions per second, its speed is

$$\text{speed} = \left(\frac{1.2 \text{ rev}}{1 \text{ sec}}\right)\left(\frac{30\pi \text{ in}}{1 \text{ rev}}\right) = 36\pi \text{ in/sec} \approx 113.1 \text{ in/sec.}$$

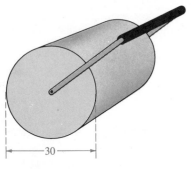

|← 30 →|

FIGURE 2.15

WARM UP

Solve for x.

1. $x + 135 = 180$

2. $790 = 720 + x$

3. $\pi = \frac{5\pi}{6} + x$

4. $2\pi - x = \frac{5\pi}{3}$

5. $\frac{45}{180} = \frac{x}{\pi}$

6. $\frac{240}{180} = \frac{x}{\pi}$

7. $\frac{\pi}{180} = \frac{x}{20}$

8. $\frac{180}{\pi} = \frac{330}{x}$

9. $\frac{x}{60} = \frac{3}{4}$

10. $\frac{x}{3600} = 0.0125$

EXERCISES 2.1

In Exercises 1–6, determine the quadrant in which the given angle lies. (The angle measure is given in radians.)

1. (a) $\dfrac{\pi}{5}$ (b) $\dfrac{7\pi}{5}$

2. (a) $\dfrac{5\pi}{4}$ (b) $\dfrac{7\pi}{4}$

3. (a) $-\dfrac{\pi}{12}$ (b) $-\dfrac{11\pi}{9}$

4. (a) -1 (b) -2

5. (a) 3.5 (b) 2.25

6. (a) 5.63 (b) -2.25

In Exercises 7–10, determine the quadrant in which the given angle lies.

7. (a) $130°$ (b) $285°$

8. (a) $8.3°$ (b) $257° \ 30'$

9. (a) $-132° \ 50'$ (b) $-336°$

10. (a) $-260°$ (b) $-3.4°$

In Exercises 11–16, sketch the given angle in standard position.

11. (a) $\dfrac{5\pi}{4}$ (b) $\dfrac{2\pi}{3}$

12. (a) $405°$ (b) $-480°$

13. (a) $-\dfrac{7\pi}{4}$ (b) $-\dfrac{5\pi}{2}$

14. (a) $\dfrac{11\pi}{6}$ (b) 7π

15. (a) $30°$ (b) $150°$

16. (a) $-270°$ (b) $-120°$

In Exercises 17–20, determine two coterminal angles (one positive and one negative) for the given angle. Give the answers in radians.

17. (a) $\theta = \dfrac{\pi}{9}$

(b) $\theta = \dfrac{4\pi}{3}$

18. (a) $\theta = \dfrac{11\pi}{6}$

(b) $\theta = -\dfrac{7\pi}{6}$
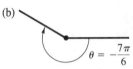

19. (a) $\theta = -\dfrac{9\pi}{4}$

(b) $\theta = -\dfrac{2\pi}{15}$

20. (a) $\theta = \dfrac{8\pi}{9}$

(b) $\theta = \dfrac{8\pi}{45}$
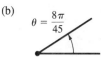

In Exercises 21–24, determine two coterminal angles (one positive and one negative) for the given angle. Give the answers in degrees.

21. (a) $\theta = 36°$ (b) $\theta = -45°$

22. (a) $\theta = -120°$ (b) $\theta = 390°$

23. (a) $\theta = 300°$ (b) $\theta = 740°$

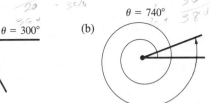

24. (a) $\theta = -420°$ (b) $\theta = 230°$

In Exercises 25–28, express the given angle in radian measure as a multiple of π. (Do not use a calculator.)

25. (a) 30°   (b) 150°

26. (a) 315°   (b) 120°

27. (a) −20°   (b) −240°

28. (a) −270°   (b) 144°

In Exercises 29–32, express the given angle in degree measure. (Do not use a calculator.)

29. (a) $\dfrac{3\pi}{2}$ (b) $\dfrac{7\pi}{6}$

30. (a) $-\dfrac{7\pi}{12}$ (b) $\dfrac{\pi}{9}$

31. (a) $\dfrac{7\pi}{3}$ (b) $-\dfrac{11\pi}{30}$

32. (a) $\dfrac{11\pi}{6}$ (b) $\dfrac{34\pi}{15}$

In Exercises 33–40, convert the angle from degrees to radian measure. List your answers to three decimal places.

33. 115°   **34.** 87.4°

35. −216.35°   **36.** −48.27°

37. 532°   **38.** 0.54°

39. −0.83°   **40.** 345°

In Exercises 41–48, convert the angle from radian to degree measure. List your answers to three decimal places.

41. $\dfrac{\pi}{7}$ **42.** $\dfrac{5\pi}{11}$

43. $\dfrac{15\pi}{8}$ **44.** 6.5π

45. -4.2π **46.** 4.8

47. −2   **48.** −0.57

In Exercises 49–52, convert the angle measurement to decimal form.

49. (a) 54° 45′   (b) −128° 30′

50. (a) 245° 10′   (b) 2° 12′

51. (a) 85° 18′ 30″   (b) 330° 25″

52. (a) −135° 36″   (b) −408° 16′ 25″

In Exercises 53–56, convert the angle measurement to D° M′ S″ form.

53. (a) 240.6°   (b) −145.8°

54. (a) −345.12°   (b) 0.45

55. (a) 2.5   (b) −3.58

56. (a) −0.355   (b) 0.7865

In Exercises 57–60, find the radian measure of the central angle using the given radius and arc length.

57. $r = 10$ inches, $s = 4$ inches

58. $r = 16$ feet, $s = 10$ feet

59. $r = 14.5$ cm, $s = 25$ cm

60. $r = 80$ km, $s = 160$ km

In Exercises 61–64, find the length of the arc on the circle of radius r subtended by the central angle θ.

61. $r = 15$ inches, $\theta = \pi$ radians

62. $r = 9$ feet, $\theta = \dfrac{\pi}{3}$ radians

63. $r = 6$ m, $\theta = 2$ radians

64. $r = 40$ cm, $\theta = \dfrac{3\pi}{4}$ radians

In Exercises 65–68, find the distance between the two cities of given latitudes. Assume that the earth is a sphere of radius 4000 miles and that the cities are on the same meridian (one city is due north of the other).

65. Dallas   32° 47′ 9″ N
Omaha   41° 15′ 42″ N

66. San Francisco   37° 46′ 39″ N
Seattle   47° 36′ 32″ N

67. Miami   25° 46′ 37″ N
Erie   42° 7′ 15″ N

68. Johannesburg, South Africa   26° 10′ S
Jerusalem, Israel   31° 47′ N

69. Assuming that the earth is a sphere of radius 4000 miles, what is the difference in latitude of two cities, one of which is 325 miles due north of the other?

70. The pointer on a voltmeter is 2 inches in length (see figure). Find the angle through which the pointer rotates when it moves $\frac{1}{2}$ inch on the scale.

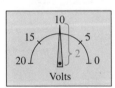

FIGURE FOR 70

71. An electric hoist is being used to lift a piece of equipment (see figure). The diameter of the drum on the hoist is 8 inches and the equipment must be raised one foot. Find the number of degrees through which the drum must rotate.

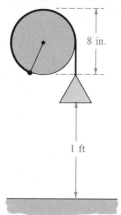

8 in.

1 ft

FIGURE FOR 71

72. Find the number of degrees through which a wheel will turn in 25 milliseconds ($\frac{25}{1000}$ seconds) for an angular speed of 3 radians per second.

73. A car is moving at the rate of 50 miles per hour, and the diameter of each of its wheels is 2.5 feet.
(a) Find the number of revolutions per minute that the wheels are rotating.
(b) Find the *angular speed* of the wheels.

74. Repeat Exercise 73 for the wheels on a truck that have a diameter of 3 feet.

75. A 2-inch-diameter pulley on an electric motor that runs at 1700 revolutions per minute is connected by a belt to a 4-inch-diameter pulley on a saw arbor.
(a) Find the angular speed (radians per minute) of each pulley.
(b) Find the revolutions per minute of the saw.

76. How long will it take a pulley rotating at 12 radians per second to make 100 complete revolutions?

2.2 *The Trigonometric Functions and the Unit Circle*

The two historical perspectives of trigonometry incorporate different methods for introducing the trigonometric functions. Our first introduction to these functions is based on the unit circle.

The Unit Circle

Consider the **unit circle** given by $x^2 + y^2 = 1$, as shown in Figure 2.16. Imagine that the real number line is wrapped around this circle, with positive numbers corresponding to a counterclockwise wrapping and negative numbers corresponding to a clockwise wrapping, as shown in Figure 2.17.

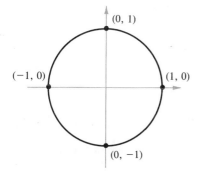

Unit Circle: $x^2 + y^2 = 1$

FIGURE 2.16

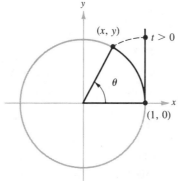

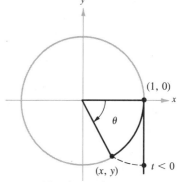

FIGURE 2.17   Positive numbers   Negative numbers

Trigonometry

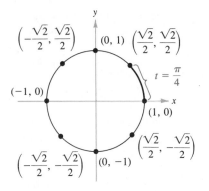

Unit Circle Divided into 8 Equal Arcs

FIGURE 2.18

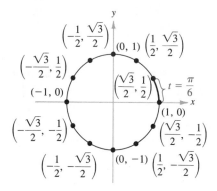

Unit Circle Divided into 12 Equal Arcs

FIGURE 2.19

As the real number line is wrapped around the unit circle, each real number t will correspond with a point (x, y) on the circle. For example, the real number 0 corresponds to the point $(1, 0)$. Moreover, because the unit circle has a circumference of 2π, the real number 2π will also correspond to the point $(1, 0)$.

In general, each real number t also corresponds to a central angle θ (in standard position) whose radian measure is t. With this interpretation of t, the arc length formula $s = r\theta$ (with $r = 1$) indicates that the real number t is the length of the arc subtended by the angle θ.

In Figure 2.18, the unit circle has been divided into eight equal arcs, corresponding to t-values of

$$0, \frac{\pi}{4}, \frac{\pi}{2}, \frac{3\pi}{4}, \pi, \frac{5\pi}{4}, \frac{3\pi}{2}, \frac{7\pi}{4}, \text{ and } 2\pi.$$

Similarly, in Figure 2.19, the unit circle has been divided into 12 equal arcs, corresponding to t-values of

$$0, \frac{\pi}{6}, \frac{\pi}{3}, \frac{\pi}{2}, \frac{2\pi}{3}, \frac{5\pi}{6}, \pi, \frac{7\pi}{6}, \frac{4\pi}{3}, \frac{3\pi}{2}, \frac{5\pi}{3}, \frac{11\pi}{6}, \text{ and } 2\pi.$$

The Trigonometric Functions

From the preceding discussion, it follows that the coordinates x and y are two functions of the real variable t. We use these coordinates to define the six trigonometric functions of t.

sine	cosecant
cosine	secant
tangent	cotangent

These six functions are normally abbreviated as sin, csc, cos, sec, tan, and cot, respectively.

Definition of Trigonometric Functions

Let t be a real number and (x, y) the point on the unit circle corresponding to t.

$$\sin t = y \qquad\qquad \csc t = \frac{1}{y}, \quad y \neq 0$$

$$\cos t = x \qquad\qquad \sec t = \frac{1}{x}, \quad x \neq 0$$

$$\tan t = \frac{y}{x}, \quad x \neq 0 \qquad \cot t = \frac{x}{y}, \quad y \neq 0$$

Remark: As an aid to memorizing these definitions, note that the functions in the second column are the *reciprocals* of the corresponding functions in the first column.

EXAMPLE 1 Matching Points with Real Numbers

Find the points on the unit circle that match the following real numbers.

(a) $t = \dfrac{3\pi}{4}$

(b) $t = \dfrac{7\pi}{6}$

(c) $t = \dfrac{5\pi}{3}$

SOLUTION

(a) Using Figure 2.18, we move counterclockwise around the unit circle, marking off an arc length of $t = 3\pi/4$ to obtain the second quadrant point

$$(x, y) = \left(-\frac{\sqrt{2}}{2}, \frac{\sqrt{2}}{2}\right).$$

(b) Using Figure 2.19, we move counterclockwise around the unit circle, marking off an arc length of $t = 7\pi/6$ to obtain the third quadrant point

$$(x, y) = \left(-\frac{\sqrt{3}}{2}, -\frac{1}{2}\right).$$

(c) Again using Figure 2.19, we move counterclockwise around the unit circle, marking off an arc length of $t = 5\pi/3$ to obtain the fourth quadrant point

$$(x, y) = \left(\frac{1}{2}, -\frac{\sqrt{3}}{2}\right).$$

Using the (x, y) coordinates found in Example 1, we can easily evaluate the trigonometric functions for the t-values given in the example. For instance, since $t = 3\pi/4$ corresponds to $(x, y) = (-\sqrt{2}/2, \sqrt{2}/2)$, we can write the following.

$$\sin \frac{3\pi}{4} = y = \frac{\sqrt{2}}{2} \qquad\qquad \csc \frac{3\pi}{4} = \frac{1}{y} = \frac{2}{\sqrt{2}} = \sqrt{2}$$

$$\cos \frac{3\pi}{4} = x = -\frac{\sqrt{2}}{2} \qquad\qquad \sec \frac{3\pi}{4} = \frac{1}{x} = -\frac{2}{\sqrt{2}} = -\sqrt{2}$$

$$\tan \frac{3\pi}{4} = \frac{y}{x} = \frac{\sqrt{2}/2}{-\sqrt{2}/2} = -1 \qquad\qquad \cot \frac{3\pi}{4} = \frac{x}{y} = \frac{-\sqrt{2}/2}{\sqrt{2}/2} = -1$$

Trigonometry

EXAMPLE 2 *Evaluating Trigonometric Functions of Real Numbers*

Evaluate sin t, cos t, and tan t at the following numbers.

(a) $t = 0$ (b) $t = \dfrac{\pi}{6}$ (c) $t = \dfrac{5\pi}{4}$

SOLUTION

(a) Since $t = 0$ corresponds to the point $(x, y) = (1, 0)$, it follows that

$$\sin 0 = y = 0$$
$$\cos 0 = x = 1$$
$$\tan 0 = \frac{y}{x} = \frac{0}{1} = 0.$$

(b) Since $t = \pi/6$ corresponds to the first quadrant point $(x, y) = (\sqrt{3}/2, 1/2)$, it follows that

$$\sin \frac{\pi}{6} = y = \frac{1}{2}$$

$$\cos \frac{\pi}{6} = x = \frac{\sqrt{3}}{2}$$

$$\tan \frac{\pi}{6} = \frac{y}{x} = \frac{1/2}{\sqrt{3}/2} = \frac{1}{\sqrt{3}}.$$

(c) Since $t = 5\pi/4$ corresponds to the third quadrant point $(x, y) = (-\sqrt{2}/2, -\sqrt{2}/2)$, it follows that

$$\sin \frac{5\pi}{4} = y = -\frac{\sqrt{2}}{2}$$

$$\cos \frac{5\pi}{4} = x = -\frac{\sqrt{2}}{2}$$

$$\tan \frac{5\pi}{4} = \frac{y}{x} = \frac{-\sqrt{2}/2}{-\sqrt{2}/2} = 1.$$

    In the definition of the trigonometric functions note that we do not define the tangent or secant if $x = 0$. For instance, because $t = \pi/2$ corresponds to $(x, y) = (0, 1)$, it follows that

$$\tan \frac{\pi}{2} \quad \text{and} \quad \sec \frac{\pi}{2} \qquad\qquad \textit{Undefined}$$

are *undefined*. Similarly, we do not define the cotangent or cosecant if $y = 0$. For instance, because $t = 0$ corresponds to $(x, y) = (1, 0)$, cot 0 and csc 0 are *undefined*.

The Trigonometric Functions and the Unit Circle

EXAMPLE 3 *Evaluating Trigonometric Functions of Real Numbers*

Evaluate the six trigonometric functions at the following real numbers.

(a) $t = -\dfrac{\pi}{3}$

(b) $t = \pi$

SOLUTION

(a) Moving *clockwise* around the unit circle we find that $t = -\pi/3$ corresponds to the point $(x, y) = (1/2, -\sqrt{3}/2)$. Hence, we have the following.

$$\sin\left(-\frac{\pi}{3}\right) = -\frac{\sqrt{3}}{2} \qquad \csc\left(-\frac{\pi}{3}\right) = -\frac{2}{\sqrt{3}}$$

$$\cos\left(-\frac{\pi}{3}\right) = \frac{1}{2} \qquad \sec\left(-\frac{\pi}{3}\right) = 2$$

$$\tan\left(-\frac{\pi}{3}\right) = -\sqrt{3} \qquad \cot\left(-\frac{\pi}{3}\right) = -\frac{1}{\sqrt{3}}$$

(b) Since $t = \pi$ corresponds to the point $(x, y) = (-1, 0)$ on the unit circle, we have the following.

$$\sin \pi = 0$$
$$\cos \pi = -1$$
$$\tan \pi = 0$$
$$\csc \pi \text{ is undefined}$$
$$\sec \pi = -1$$
$$\cot \pi \text{ is undefined}$$

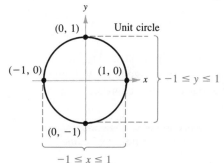

FIGURE 2.20

The *domain* of the sine and cosine functions is the set of all real numbers. To determine the *range* of these two functions, consider the unit circle shown in Figure 2.20. Since $r = 1$, it follows that $\sin t = y$ and $\cos t = x$. Moreover, because (x, y) is on the unit circle we know that $-1 \le y \le 1$ and $-1 \le x \le 1$, and it follows that the values of the sine and cosine also range between -1 and 1. That is,

$$-1 \le y \le 1 \qquad\qquad -1 \le x \le 1$$
$$-1 \le \sin t \le 1 \qquad \text{and} \qquad -1 \le \cos t \le 1.$$

Trigonometry

Suppose we add 2π to each value of t in the interval $[0, 2\pi]$, thus completing a second revolution around the unit circle, as shown in Figure 2.21. The values of $\sin(t + 2\pi)$ and $\cos(t + 2\pi)$ correspond to those of $\sin t$ and $\cos t$. Similar results can be obtained for repeated revolutions (positive or negative) on the unit circle. This leads to the general result

$$\sin(t + 2\pi n) = \sin t \qquad \text{and} \qquad \cos(t + 2\pi n) = \cos t$$

for any integer n and real number t. Functions that behave in such a repetitive (or cyclic) manner are called **periodic.**

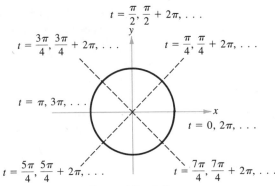

FIGURE 2.21

Repeated Revolutions
on the Unit Circle

Definition of a Periodic Function

A function f is **periodic** if there exists a positive real number c such that

$$f(t + c) = f(t)$$

for all t in the domain of f. The least number c for which f is periodic is called the **period** of f.

From this definition it follows that the sine and cosine functions are periodic and have a period of 2π. The other four trigonometric functions are also periodic, and we will say more about that in Section 2.6.

EXAMPLE 4 *Using the Period to Evaluate the Sine and Cosine*

Evaluate $\sin t$ and $\cos t$ at the following numbers.

(a) $t = \dfrac{13\pi}{6}$

(b) $t = -\dfrac{7\pi}{2}$

SOLUTION

(a) Using the fact that the sine and cosine functions have a period of 2π, together with the fact that

$$\frac{13\pi}{6} = 2\pi + \frac{\pi}{6},$$

we can write

$$\sin\frac{13\pi}{6} = \sin\left(2\pi + \frac{\pi}{6}\right)$$

$$= \sin\frac{\pi}{6} = \frac{1}{2}$$

$$\cos\frac{13\pi}{6} = \cos\left(2\pi + \frac{\pi}{6}\right)$$

$$= \cos\frac{\pi}{6} = \frac{\sqrt{3}}{2}.$$

(b) For $t = -7\pi/2$, we use the fact that

$$-\frac{7\pi}{2} = -4\pi + \frac{\pi}{2}$$

to write

$$\sin\left(-\frac{7\pi}{2}\right) = \sin\left(-4\pi + \frac{\pi}{2}\right)$$

$$= \sin\frac{\pi}{2} = 1$$

$$\cos\left(-\frac{7\pi}{2}\right) = \cos\left(-4\pi + \frac{\pi}{2}\right)$$

$$= \cos\frac{\pi}{2} = 0.$$

Recall from Section 1.6 that a function f is called *even* if $f(-t) = f(t)$ and is called *odd* if $f(-t) = -f(t)$. Of the six trigonometric functions, two are even and four are odd, as stated in the following theorem.

Even and Odd Trigonometric Functions

The cosine and secant functions are *even*.

$$\cos(-t) = \cos t \qquad \sec(-t) = \sec t$$

The sine, cosecant, tangent, and cotangent functions are *odd*.

$$\sin(-t) = -\sin t \qquad \csc(-t) = -\csc t$$
$$\tan(-t) = -\tan t \qquad \cot(-t) = -\cot t$$

Evaluating Trigonometric Functions with a Calculator

At the beginning of this section, we mentioned that each real number t can be viewed as a central angle θ given in radian measure. Thus, when we are evaluating trigonometric functions, *it doesn't make any difference whether we consider t to be a real number or an angle given in radians.*

A scientific calculator can be used to obtain decimal approximations of the values of the trigonometric functions of any real number. Before entering the number and function, we need to set a switch to the desired *mode* of measurement (degrees or radians). For instance, to find the value of $\tan(\pi/12)$, we can use the following keystroke sequence.

Radian Mode: π ÷ 12 = tan   *Display* 0.2679492

Most calculators do not have keys for cosecant, secant, or cotangent. To evaluate these functions, we use the $\boxed{1/x}$ key with their respective reciprocal functions sine, cosine, and tangent. For example, to evaluate $\csc(\pi/8)$, we use the fact that

$$\csc \frac{\pi}{8} = \frac{1}{\sin(\pi/8)}$$

and enter the following keystroke sequence.

Radian Mode: π ÷ 8 = sin 1/x   *Display* 2.6131259

EXAMPLE 5 *Using a Calculator to Evaluate Trigonometric Functions*

Use a calculator to evaluate each of the following.

(a) $\sin \dfrac{\pi}{6}$

(b) $\cot 1.5$

(c) $\sec 0.7$

SOLUTION

Function	Mode	Keystrokes	Display
(a) $\sin \dfrac{\pi}{6}$	Radian	π ÷ 6 = sin	0.5
(b) $\cot 1.5$	Radian	1.5 tan 1/x	0.0709148
(c) $\sec 0.7$	Radian	0.7 cos 1/x	1.3074593

Simplify the expressions.

1. $\dfrac{\sqrt{3}/2}{1/2}$

2. $\dfrac{\sqrt{2}/2}{-\sqrt{2}/2}$

3. $\dfrac{1/2}{-\sqrt{3}/2}$

4. $\dfrac{-\sqrt{2}/2}{-\sqrt{2}/2}$

In each case, find an angle $0 \le \theta < 2\pi$ that is coterminal with the given angle.

5. $\dfrac{8\pi}{3}$

6. $\dfrac{13\pi}{4}$

7. $-\dfrac{\pi}{4}$

8. $-\dfrac{2\pi}{3}$

9. What is the circumference of a circle whose radius is 1?

10. What is the arc length of a semicircle of radius 1?

EXERCISES 2.2

In Exercises 1–8, find the point (x, y) on the unit circle that corresponds to the given real number.

1. $t = \dfrac{\pi}{4}$

2. $t = \dfrac{\pi}{3}$

3. $t = \dfrac{5\pi}{6}$

4. $t = \dfrac{5\pi}{4}$

5. $t = \dfrac{4\pi}{3}$

6. $t = \dfrac{11\pi}{6}$

7. $t = \dfrac{3\pi}{2}$

8. $t = \pi$

In Exercises 9–20, evaluate the given trigonometric function.

9. (a) $\sin \dfrac{\pi}{4}$

   (b) $\cos \dfrac{\pi}{4}$

   (c) $\tan \dfrac{\pi}{4}$

10. (a) $\sin\left(-\dfrac{\pi}{4}\right)$

    (b) $\cos\left(-\dfrac{\pi}{4}\right)$

    (c) $\tan\left(-\dfrac{\pi}{4}\right)$

11. (a) $\sin\left(-\dfrac{\pi}{6}\right)$

    (b) $\cos\left(-\dfrac{\pi}{6}\right)$

    (c) $\tan\left(-\dfrac{\pi}{6}\right)$

12. (a) $\sin \dfrac{\pi}{3}$

    (b) $\cos \dfrac{\pi}{3}$

    (c) $\tan \dfrac{\pi}{3}$

13. (a) $\sin \pi$

    (b) $\cos \pi$

    (c) $\tan \pi$

14. (a) $\sin 2\pi$

    (b) $\cos 2\pi$

    (c) $\tan 2\pi$

15. (a) $\sin\left(-\dfrac{5\pi}{4}\right)$

    (b) $\cos\left(-\dfrac{5\pi}{4}\right)$

    (c) $\tan\left(-\dfrac{5\pi}{4}\right)$

16. (a) $\sin\left(-\dfrac{5\pi}{6}\right)$

    (b) $\cos\left(-\dfrac{5\pi}{6}\right)$

    (c) $\tan\left(-\dfrac{5\pi}{6}\right)$

17. (a) $\sin \dfrac{11\pi}{6}$

    (b) $\cos \dfrac{11\pi}{6}$

    (c) $\tan \dfrac{11\pi}{6}$

18. (a) $\sin \dfrac{2\pi}{3}$

    (b) $\cos \dfrac{2\pi}{3}$

    (c) $\tan \dfrac{2\pi}{3}$

19. (a) $\sin \dfrac{4\pi}{3}$

(b) $\cos \dfrac{4\pi}{3}$

(c) $\tan \dfrac{4\pi}{3}$

20. (a) $\sin \dfrac{7\pi}{4}$

(b) $\cos \dfrac{7\pi}{4}$

(c) $\tan \dfrac{7\pi}{4}$

In Exercises 21–26, evaluate (if possible) the six trigonometric functions for the given value of *t*.

21. $t = \dfrac{\pi}{4}$

22. $t = -\dfrac{2\pi}{3}$

23. $t = \dfrac{\pi}{2}$

24. $t = \dfrac{3\pi}{2}$

25. $t = -\dfrac{4\pi}{3}$

26. $t = -\dfrac{11\pi}{6}$

In Exercises 27–34, use the periodic nature of the sine and cosine to evaluate the given trigonometric function.

27. $\sin 3\pi$

28. $\cos 3\pi$

29. $\cos \dfrac{8\pi}{3}$

30. $\sin \dfrac{9\pi}{4}$

31. $\cos \dfrac{19\pi}{6}$

32. $\sin\left(-\dfrac{13\pi}{6}\right)$

33. $\sin\left(-\dfrac{9\pi}{4}\right)$

34. $\cos\left(-\dfrac{8\pi}{3}\right)$

In Exercises 35–38, use the value of the given trigonometric function to evaluate the indicated functions.

35. $\sin t = \tfrac{1}{3}$
(a) $\sin(-t)$

(b) $\csc(-t)$

36. $\cos t = -\tfrac{3}{4}$
(a) $\cos(-t)$

(b) $\sec(-t)$

37. $\cos(-t) = -\tfrac{7}{8}$
(a) $\cos t$

(b) $\sec(-t)$

38. $\sin(-t) = \tfrac{2}{5}$
(a) $\sin t$

(b) $\csc t$

In Exercises 39–50, use a calculator to evaluate the given trigonometric function. (Set your calculator in radian mode and round your answer to four decimal places.)

39. $\sin \dfrac{\pi}{4}$

40. $\tan \pi$

41. $\cos(-3)$

42. $\cot 1$

43. $\tan \dfrac{\pi}{10}$

44. $\cos\left(-\dfrac{\pi}{5}\right)$

45. $\cos(-1.7)$

46. $\csc 2.3$

47. $\csc 0.8$

48. $\sec 1.8$

49. $\sec 22.8$

50. $\sin(-0.9)$

51. The displacement from equilibrium of an oscillating weight suspended by a spring is given by

$$y(t) = \tfrac{1}{4} \cos 6t$$

where *y* is the displacement in feet and *t* is time in seconds. Find the displacement for (a) $t = 0$, (b) $t = \tfrac{1}{4}$, and (c) $t = \tfrac{1}{2}$.

52. Repeat Exercise 51 for the model given by

$$y(t) = \tfrac{1}{4}e^{-t} \cos 6t.$$

This model includes the damping effect of friction.

2.3 Trigonometric Functions of an Acute Angle

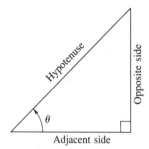

FIGURE 2.22

Our second look at the trigonometric functions is from a *right triangle* perspective. Note that the three sides of the right triangle shown in Figure 2.22 are labeled the **hypotenuse,** the **opposite side** (the side opposite the angle θ), and the **adjacent side** (the side adjacent to the angle θ). Using the lengths of these three sides we can form six ratios that define the six trigonometric functions of the acute angle θ.

In the following definition it is important to see that $0° < \theta < 90°$, and for such angles the value of each of the six trigonometric functions is *positive*.

Right Triangle Definition of Trigonometric Functions

Let θ be an *acute* angle of a right triangle. Then the six trigonometric functions of the angle θ are defined as follows.

$$\sin \theta = \frac{\text{opp}}{\text{hyp}} \qquad \csc \theta = \frac{\text{hyp}}{\text{opp}}$$

$$\cos \theta = \frac{\text{adj}}{\text{hyp}} \qquad \sec \theta = \frac{\text{hyp}}{\text{adj}}$$

$$\tan \theta = \frac{\text{opp}}{\text{adj}} \qquad \cot \theta = \frac{\text{adj}}{\text{opp}}$$

The abbreviations opp, adj, and hyp represent the lengths of the three sides of the right triangle, as follows.

opp = the length of the side *opposite* θ
adj = the length of the side *adjacent* to θ
hyp = the length of the *hypotenuse*

EXAMPLE 1 *Evaluating Trigonometric Functions*

Find the values of the six trigonometric functions of θ, as shown in Figure 2.23.

SOLUTION

By the Pythagorean Theorem, it follows that

$$\text{hyp} = \sqrt{3^2 + 4^2} = \sqrt{25} = 5.$$

Thus, we have adj = 3, opp = 4, and hyp = 5, so that the six trigonometric functions of θ have the following values.

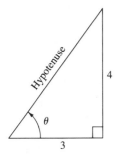

FIGURE 2.23

$$\sin \theta = \frac{\text{opp}}{\text{hyp}} = \frac{4}{5} \qquad \csc \theta = \frac{\text{hyp}}{\text{opp}} = \frac{5}{4}$$

$$\cos \theta = \frac{\text{adj}}{\text{hyp}} = \frac{3}{5} \qquad \sec \theta = \frac{\text{hyp}}{\text{adj}} = \frac{5}{3}$$

$$\tan \theta = \frac{\text{opp}}{\text{adj}} = \frac{4}{3} \qquad \cot \theta = \frac{\text{adj}}{\text{opp}} = \frac{3}{4}$$

In Example 1, we were given the lengths of the sides of the right triangle, but we were not given the angle θ. A much more common problem in trigonometry is to be asked to find the trigonometric functions for a *given* acute angle θ. To do this, we construct a right triangle having θ as one of its angles. This procedure is demonstrated in Examples 2 and 3.

EXAMPLE 2 *Evaluating Trigonometric Functions of 45°*

Find the value of sin 45°, cos 45°, and tan 45°.

SOLUTION

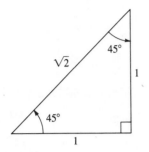

FIGURE 2.24

First we construct a right triangle having 45° as one of its acute angles, as shown in Figure 2.24. We arbitrarily choose the length of the adjacent side to be 1. From geometry, we know that the other acute angle is also 45° and therefore the triangle is isosceles. Hence, the length of the opposite side is also 1. Then, using the Pythagorean Theorem, we find the length of the hypotenuse to be

$$\text{hyp} = \sqrt{1^2 + 1^2} = \sqrt{2}.$$

Finally, we have the following.

$$\sin 45° = \frac{\text{opp}}{\text{hyp}} = \frac{1}{\sqrt{2}} = \frac{\sqrt{2}}{2}$$

$$\cos 45° = \frac{\text{adj}}{\text{hyp}} = \frac{1}{\sqrt{2}} = \frac{\sqrt{2}}{2}$$

$$\tan 45° = \frac{\text{opp}}{\text{adj}} = \frac{1}{1} = 1$$

EXAMPLE 3 *Evaluating Trigonometric Functions of 30° and 60°*

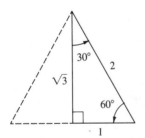

FIGURE 2.25

Use the equilateral triangle shown in Figure 2.25 to find the value of sin 60°, cos 60°, sin 30°, and cos 30°.

SOLUTION

First of all, try using the Pythagorean Theorem and the 60°-60°-60° equilateral triangle to verify the lengths of the sides given in Figure 2.25. For $\theta = 60°$, we have adj = 1, opp = $\sqrt{3}$, and hyp = 2. Therefore,

$$\sin 60° = \frac{\text{opp}}{\text{hyp}} = \frac{\sqrt{3}}{2}$$

and

$$\cos 60° = \frac{\text{adj}}{\text{hyp}} = \frac{1}{2}.$$

For $\theta = 30°$, we have adj $= \sqrt{3}$, opp $= 1$, and hyp $= 2$. Therefore,

$$\sin 30° = \frac{\text{opp}}{\text{hyp}} = \frac{1}{2}$$

and

$$\cos 30° = \frac{\text{adj}}{\text{hyp}} = \frac{\sqrt{3}}{2}.$$

Because the angles 30°, 45°, and 60° occur frequently in trigonometry, we suggest that you learn to construct the triangles shown in Figures 2.24 and 2.25. The values of the sine, cosine, and tangent of 30°, 45°, and 60° are summarized in the following table.

Sine, Cosine, and Tangent of Special Angles

$\sin 30° = \dfrac{1}{2}$	$\cos 30° = \dfrac{\sqrt{3}}{2}$	$\tan 30° = \dfrac{1}{\sqrt{3}}$
$\sin 45° = \dfrac{\sqrt{2}}{2}$	$\cos 45° = \dfrac{\sqrt{2}}{2}$	$\tan 45° = 1$
$\sin 60° = \dfrac{\sqrt{3}}{2}$	$\cos 60° = \dfrac{1}{2}$	$\tan 60° = \sqrt{3}$

Trigonometric Identities

In the preceding table, note that for the angles 30° and 60°, we have $\sin 30° = 1/2 = \cos 60°$. This occurs because 30° and 60° are complementary angles, and, in general, it can be shown that *cofunctions of complementary angles are equal.* That is, if θ is an acute angle, then the following relationships are true.

$$\sin(90° - \theta) = \cos \theta \qquad \cos(90° - \theta) = \sin \theta$$
$$\tan(90° - \theta) = \cot \theta \qquad \cot(90° - \theta) = \tan \theta$$
$$\sec(90° - \theta) = \csc \theta \qquad \csc(90° - \theta) = \sec \theta$$

For instance, since 10° and 80° are complementary angles, it follows that $\sin 10° = \cos 80°$ and $\tan 10° = \cot 80°$.

In trigonometry, a great deal of time is spent studying relationships between trigonometric functions. You will need to memorize many of these relationships (identities). We begin with some basic identities that are easily established from the definitions of the six trigonometric functions.

Fundamental Trigonometric Identities

Reciprocal Identities:

$$\sin \theta = \frac{1}{\csc \theta} \qquad \sec \theta = \frac{1}{\cos \theta} \qquad \tan \theta = \frac{1}{\cot \theta}$$

$$\csc \theta = \frac{1}{\sin \theta} \qquad \cos \theta = \frac{1}{\sec \theta} \qquad \cot \theta = \frac{1}{\tan \theta}$$

Tangent and Cotangent Identities:

$$\tan \theta = \frac{\sin \theta}{\cos \theta} \qquad \cot \theta = \frac{\cos \theta}{\sin \theta}$$

Pythagorean Identities:

$$\sin^2 \theta + \cos^2 \theta = 1 \qquad 1 + \tan^2 \theta = \sec^2 \theta$$
$$1 + \cot^2 \theta = \csc^2 \theta$$

Remark: We use $\sin^2 \theta$ to represent $(\sin \theta)^2$.

In the next two examples, we show how trigonometric identities can be used to find exact values of trigonometric functions.

EXAMPLE 4 Applying Trigonometric Identities

Let θ be the acute angle such that $\sin \theta = 0.6$. Find the values of (a) $\cos \theta$ and (b) $\tan \theta$.

SOLUTION

(a) To find the value of $\cos \theta$, we use the Pythagorean Identity

$$\sin^2 \theta + \cos^2 \theta = 1.$$

Thus, we have

$$(0.6)^2 + \cos^2 \theta = 1$$
$$\cos^2 \theta = 1 - (0.6)^2 = 0.64$$
$$\cos \theta = \sqrt{0.64} = 0.8.$$

Note that we choose the positive square root because θ is given to be an acute angle.

(b) Now, knowing the sine and cosine of θ, we find the tangent of θ to be

$$\tan \theta = \frac{\sin \theta}{\cos \theta} = \frac{0.6}{0.8} = 0.75.$$

Try using the triangle shown in Figure 2.26 to check these results.

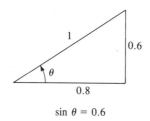

$\sin \theta = 0.6$

FIGURE 2.26

Remark: The triangle in Figure 2.26 was obtained from the fact that

$$\sin \theta = 0.6 = \frac{0.6}{1} = \frac{\text{opp}}{\text{hyp}}.$$

Thus, opp $= 0.6$, hyp $= 1$, and by the Pythagorean Theorem it follows that adj $= \sqrt{1^2 - (0.6)^2} = \sqrt{0.64} = 0.8$.

EXAMPLE 5 Applying Trigonometric Identities

Let θ be an acute angle such that $\tan \theta = 3$. Find the values of (a) $\cot \theta$ and (b) $\sec \theta$.

SOLUTION

(a) Using the reciprocal identity

$$\cot \theta = \frac{1}{\tan \theta}$$

we have

$$\cot \theta = \frac{1}{3}.$$

$\tan \theta = 3$

FIGURE 2.27

(b) Using the Pythagorean Identity $1 + \tan^2 \theta = \sec^2 \theta$, we have

$$\sec^2 \theta = 1 + 3^2 = 10$$
$$\sec \theta = \sqrt{10}.$$

Try using the triangle shown in Figure 2.27 to check these results.

To use a calculator to evaluate trigonometric functions of angles measured in degrees, we first set the calculator in *degree mode* and then proceed as demonstrated in the previous section. For instance, we can find the values of $\cos 28°$ and $\sec 28°$ as follows.

Function	*Mode*	*Keystrokes*	*Display*
$\cos 28°$	Degree	28 **cos**	0.8829476
$\sec 28°$	Degree	28 **cos** **1/x**	1.1325701

Throughout this text, we follow the convention that angles are assumed to be measured in radians unless noted otherwise. For example, when we write $\sin 1$, we will always mean the sine of 1 *radian*. When we want to denote the sine of 1 degree, we will write $\sin 1°$.

EXAMPLE 6 *Using a Calculator to Evaluate Trigonometric Functions*

Use a calculator to evaluate sec(5° 40′ 12″).

SOLUTION

Converting first to decimal form, we have

$$5° \; 40′ \; 12″ = 5° + \left(\frac{40}{60}\right)° + \left(\frac{12}{3600}\right)° = 5.67°.$$

Hence, it follows that

$$\sec(5° \; 40′ \; 12″) = \sec 5.67° = \frac{1}{\cos 5.67°} \approx 1.00492.$$

Applications Involving Right Triangles

Many applications of trigonometry involve a process called **solving right triangles.** In this type of application, we are usually given two sides of a right triangle and asked to find one of its acute angles, *or* we are given one side and one of the acute angles and asked to find one of the other sides.

EXAMPLE 7 *Solving a Right Triangle Given an Angle and One Side*

A surveyor is standing 50 feet from the base of a large tree, as shown in Figure 2.28. The surveyor measures the angle of elevation to the top of the tree as 71.5°. How tall is the tree?

SOLUTION

From Figure 2.28, we see that

$$\tan 71.5° = \frac{\text{opp}}{\text{adj}} = \frac{y}{x}$$

where $x = 50$ and y is the height of the tree. Thus, we can determine the height of the tree to be

$$y = x \tan 71.5° \approx 50(2.98868) \approx 149.4 \text{ feet.}$$

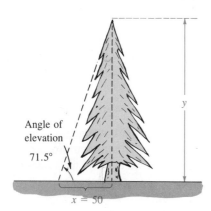

Angle of
elevation

71.5°

y

$x = 50$

FIGURE 2.28

Note in Example 7 that we were given one side and one of the acute angles of a right triangle and were asked to find the opposite side. In Example 8, we are given two sides and are asked to find one of the acute angles.

Trigonometric Functions of an Acute Angle

EXAMPLE 8 *Solving a Right Triangle Given Two Sides*

A person is standing 200 yards from a river. Rather than walk directly to the river, the person walks 400 yards along a straight path to the river's edge. Find the acute angle θ between this path and the river's edge, as indicated in Figure 2.29.

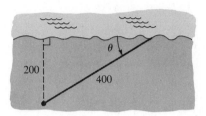

FIGURE 2.29

SOLUTION

From Figure 2.29, we see that the sine of the angle θ is given by

$$\sin \theta = \frac{\text{opp}}{\text{hyp}} = \frac{200}{400} = \frac{1}{2}.$$

Now, we recognize that $\theta = 30°$.

In Example 8, we were able to recognize that the acute angle that satisfies the equation $\sin \theta = 1/2$ is $\theta = 30°$. Suppose, however, that we were given the equation $\sin \theta = 0.6$ and asked to find the acute angle θ. Since

$$\sin 30° = \frac{1}{2} = 0.5000 \qquad \text{and} \qquad \sin 45° = \frac{1}{\sqrt{2}} \approx 0.7071$$

we know that θ lies somewhere between 30° and 45°. A more precise value of θ can be found using the $\boxed{\text{INV}}$ key on a calculator. To do this, we use the following keystroke sequence.

Degree Mode: .6 $\boxed{\text{INV}}$ $\boxed{\text{sin}}$ *Display* 36.8699

Thus, we conclude that if $\sin \theta = 0.6$, then $\theta \approx 36.87°$.

Remark: Instead of an inverse key $\boxed{\text{INV}}$, some calculators have a second function key $\boxed{\text{2nd f}}$. In Section 2.8, we will explain the concepts involved in the use of the $\boxed{\text{INV}}$ key.

Trigonometry

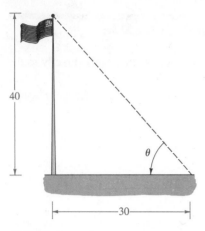

FIGURE 2.30

EXAMPLE 9 Solving a Right Triangle Given Two Sides

A 40-foot flagpole casts a 30-foot shadow, as shown in Figure 2.30. Find θ, the angle of elevation of the sun.

SOLUTION

From Figure 2.30 we see that the *opposite* and *adjacent* sides are known. Thus, we write

$$\tan \theta = \frac{\text{opp}}{\text{adj}}$$
$$= \frac{40}{30}.$$

With a calculator in degree mode we use the keystrokes

$$40 \; \boxed{\div} \; 30 \; \boxed{=} \; \boxed{\text{INV}} \; \boxed{\text{tan}}$$

to obtain $\theta \approx 53.13°$.

WARM UP

Convert the given angles to radian measure.

1. $30°$ **2.** $135°$

Convert the given angles to degree measure.

3. $\dfrac{\pi}{3}$ radians   **4.** $-\dfrac{3\pi}{2}$

In each case, find the length of the hypotenuse of the right triangle whose sides have the given lengths.

5. $a = 5, b = 3$ **6.** $a = 2, b = 2$

Perform the indicated operations. (Round your answers to two decimal places.)

7. 0.300×4.125 **8.** 7.30×43.50

9. $\dfrac{19{,}500}{0.007}$ **10.** $\dfrac{(10.5)(3401)}{1240}$

EXERCISES 2.3

In Exercises 1–8, find the exact value of the six trigonometric functions of the angle θ given in the accompanying figure. (Use the Pythagorean Theorem to find the third side of the triangle.)

1.

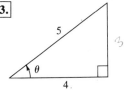

2.

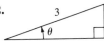

3.

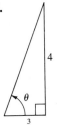

4.

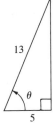

5.

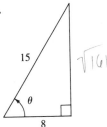

6.

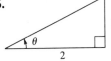

7.

8.

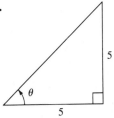

In Exercises 9–16, sketch a right triangle corresponding to the trigonometric function of the acute angle θ, and find the other five trigonometric functions of θ.

9. $\sin \theta = \frac{2}{3}$

10. $\cot \theta = 5$

11. $\sec \theta = 2$

12. $\cos \theta = \frac{5}{7}$

13. $\tan \theta = 3$

14. $\csc \theta = \frac{17}{4}$

15. $\cot \theta = \frac{3}{2}$

16. $\sin \theta = \frac{3}{8}$

In Exercises 17–20, use the given function values to evaluate the required trigonometric functions.

17. $\sin 60° = \frac{\sqrt{3}}{2}$, $\cos 60° = \frac{1}{2}$

(a) $\tan 60°$ (b) $\sin 30°$
(c) $\cos 30°$ (d) $\cot 60°$

18. $\sin 30° = \frac{1}{2}$, $\tan 30° = \frac{\sqrt{3}}{3}$

(a) $\csc 30°$ (b) $\cot 60°$
(c) $\cos 30°$ (d) $\cot 30°$

19. $\csc \theta = 3$, $\sec \theta = \frac{3\sqrt{2}}{4}$

(a) $\sin \theta$ (b) $\cos \theta$
(c) $\tan \theta$ (d) $\sec(90° - \theta)$

20. $\sec \theta = 5$, $\tan \theta = 2\sqrt{6}$

(a) $\cos \theta$ (b) $\cot \theta$
(c) $\cot(90° - \theta)$ (d) $\sin \theta$

In Exercises 21–24, evaluate the given trigonometric function by memory or by constructing an appropriate triangle for the special angles 30°, 45°, and 60°.

21. (a) $\cos 60°$ (b) $\tan 30°$
22. (a) $\csc 30°$ (b) $\sin 45°$
23. (a) $\cot 45°$ (b) $\cos 45°$
24. (a) $\sin 60°$ (b) $\csc 45°$

In Exercises 25–34, use a calculator to evaluate each function. Round your answers to four decimal places. (Be sure the calculator is in the correct mode.)

25. (a) $\sin 10°$ (b) $\cos 80°$
26. (a) $\tan 23.5°$ (b) $\cot 66.5°$
27. (a) $\sec 42° \ 12'$ (b) $\csc 48° \ 7'$
28. (a) $\cos 16° \ 18'$ (b) $\sin 73° \ 56'$
29. (a) $\sin 16.35°$ (b) $\csc 16.35°$
30. (a) $\cos 4° \ 50' \ 15''$ (b) $\sec 4° \ 50' \ 15''$
31. (a) $\cot \frac{\pi}{16}$ (b) $\tan \frac{\pi}{16}$
32. (a) $\sec 0.75$ (b) $\cos 0.75$
33. (a) $\csc 1$ (b) $\sec\left(\frac{\pi}{2} - 1\right)$
34. (a) $\tan \frac{1}{2}$ (b) $\cot\left(\frac{\pi}{2} - \frac{1}{2}\right)$

In Exercises 35–40, find the value of θ in degrees (0° < θ < 90°) and radians (0 < θ < π/2) without a calculator.

35. (a) $\sin \theta = \frac{1}{2}$ (b) $\csc \theta = 2$

36. (a) $\cos \theta = \frac{\sqrt{2}}{2}$ (b) $\tan \theta = 1$

37. (a) $\sec \theta = 2$ (b) $\cot \theta = 1$

38. (a) $\tan \theta = \sqrt{3}$ (b) $\cos \theta = \frac{1}{2}$

39. (a) $\csc \theta = \frac{2\sqrt{3}}{3}$ (b) $\sin \theta = \frac{\sqrt{2}}{2}$

40. (a) $\cot \theta = \frac{\sqrt{3}}{3}$ (b) $\sec \theta = \sqrt{2}$

In Exercises 41–44, find the value of θ in degrees (0° < θ < 90°) and radians (0 < θ < π/2) by using the inverse key on a calculator. Round your answers to two decimal places.

41. (a) $\sin \theta = 0.8191$
(b) $\cos \theta = 0.0175$

42. (a) $\cos \theta = 0.9848$
(b) $\cos \theta = 0.8746$

43. (a) $\tan \theta = 1.1920$
(b) $\tan \theta = 0.4663$

44. (a) $\sin \theta = 0.3746$
(b) $\cos \theta = 0.3746$

45. Solve for y.

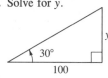

46. Solve for x.

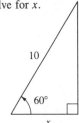

47. Solve for x.

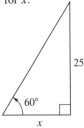

48. Solve for r.

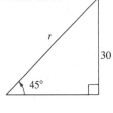

49. Solve for r.

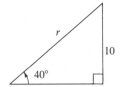

50. Solve for x.

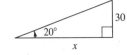

51. Solve for y.

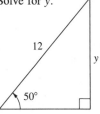

52. Solve for r.

53. A six-foot person standing 12 feet from a streetlight casts an eight-foot shadow (see figure). What is the height of the streetlight?

54. A guy wire is stretched from a broadcasting tower at a point 200 feet above the ground to an anchor 125 feet from the base (see figure). How long is the wire?

FIGURE FOR 53

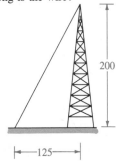

FIGURE FOR 54

55. A 20-foot ladder leaning against the side of a house makes a 75° angle with the ground (see figure). How far up the side of the house does the ladder reach?

56. A biologist wants to know the width w of a river in order to properly set instruments for studying the pollutants in the water. From point A, the biologist walks downstream 100 feet and sights to point C. From this sighting, it is determined that θ = 50° (see figure). How wide is the river?

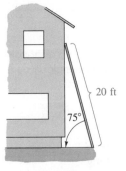

FIGURE FOR 55

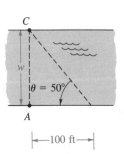

FIGURE FOR 56

57. From a 150-foot observation tower on the coast, a Coast Guard officer sights a boat in difficulty. The angle of depression of the boat is 4° (see figure). How far is the boat from the shoreline?

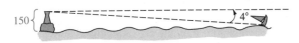

FIGURE FOR 57

58. A ramp $17\frac{1}{2}$ feet in length rises to a loading platform that is $3\frac{1}{3}$ feet off the ground (see figure). Find the angle θ that the ramp makes with the ground.

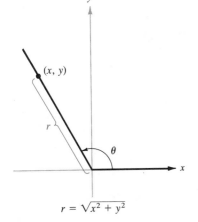

FIGURE FOR 58

In Exercises 59–64, determine whether the statement is true or false, and give reasons.

59. $\sin 60° \csc 60° = 1$ **60.** $\sec 30° = \csc 60°$

61. $\sin 45° + \cos 45° = 1$

62. $\cot^2 10° - \csc^2 10° = -1$

63. $\dfrac{\sin 60°}{\sin 30°} = \sin 2°$

64. $\tan[(0.8)^2] = \tan^2 (0.8)$

2.4 Trigonometric Functions of Any Angle

In Section 2.3, we restricted the evaluation of trigonometric functions to acute angles. In this section, we expand our evaluation techniques to include any angle θ.

Trigonometric Functions of Any Angle

Let θ be an angle in standard position with (x, y) any point (except the origin) on the terminal side of θ and $r = \sqrt{x^2 + y^2}$, as shown in Figure 2.31.

$$\sin \theta = \frac{y}{r} \qquad\qquad \csc \theta = \frac{r}{y}, \quad y \neq 0$$

$$\cos \theta = \frac{x}{r} \qquad\qquad \sec \theta = \frac{r}{x}, \quad x \neq 0$$

$$\tan \theta = \frac{y}{x}, \quad x \neq 0 \qquad \cot \theta = \frac{x}{y}, \quad y \neq 0$$

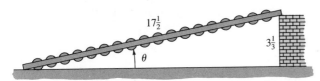

$$r = \sqrt{x^2 + y^2}$$

FIGURE 2.31

Trigonometry

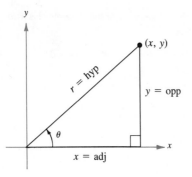

FIGURE 2.32

This result can be easily derived from the unit circle definition of the six trigonometric functions. By referring back to Section 2.2, you will see that the only difference is that in the unit circle definition the point (x, y) lies on the unit circle (which implies that $1 = \sqrt{x^2 + y^2}$), whereas, in the above description, (x, y) can be *any* point in the plane (other than the origin) and the positive number r is given by $r = \sqrt{x^2 + y^2}$.

Moreover, if θ is an *acute* angle, then the six ratios listed above coincide with those given in the previous section. To see this, note in Figure 2.32 that for an acute angle θ, $x = $ adj, $y = $ opp, and $r = $ hyp.

EXAMPLE 1 *Evaluating Trigonometric Functions*

Let $(-3, 4)$ be a point on the terminal side of θ. Find the sine, cosine, and tangent of θ.

SOLUTION

Referring to Figure 2.33, we see that $x = -3$, $y = 4$, and

$$r = \sqrt{x^2 + y^2} = \sqrt{(-3)^2 + 4^2} = \sqrt{25} = 5.$$

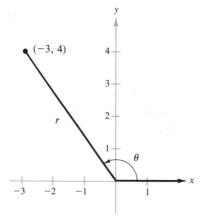

FIGURE 2.33

Thus, we have

$$\sin \theta = \frac{y}{r} = \frac{4}{5}$$

$$\cos \theta = \frac{x}{r} = \frac{-3}{5} = -\frac{3}{5}$$

$$\tan \theta = \frac{y}{x} = \frac{4}{-3} = -\frac{4}{3}.$$

The *signs* of the trigonometric functions in the four quadrants can be easily determined from the definitions of the functions. For instance, since $\cos \theta = x/r$, it follows that $\cos \theta$ is positive wherever $x > 0$, which is in Quadrants I and IV. (Remember, r is always positive.) In a similar manner we can verify the results shown in Figure 2.34.

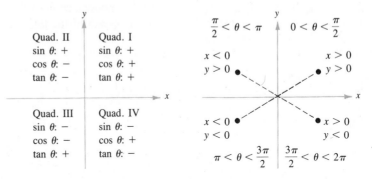

Signs of Trigonometric Functions

FIGURE 2.34

EXAMPLE 2 *Evaluating Trigonometric Functions*

Given $\tan \theta = -5/4$ and $\cos \theta > 0$, find $\sin \theta$ and $\sec \theta$.

SOLUTION

Note that θ lies in Quadrant IV because that is the only quadrant in which the tangent is negative and the cosine is positive. Moreover, using

$$\tan \theta = \frac{y}{x} = -\frac{5}{4}$$

and the fact that y is negative in Quadrant IV, we can conclude that $y = -5$ and $x = 4$. Hence, $r = \sqrt{25 + 16} = \sqrt{41}$ and we have

$$\sin \theta = \frac{y}{r} = \frac{-5}{\sqrt{41}} \approx -0.7809$$

$$\sec \theta = \frac{r}{x} = \frac{\sqrt{41}}{4} \approx 1.6008.$$

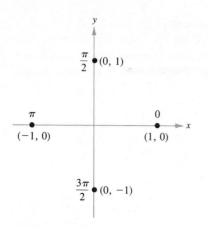

FIGURE 2.35

EXAMPLE 3   *Trigonometric Functions of Quadrant Angles*

Evaluate the sine function at the four quadrant angles 0, $\pi/2$, π, and $3\pi/2$.

SOLUTION

First we choose a point on the terminal side of each angle, as shown in Figure 2.35. For each of the four given points, $r = 1$, and we have

$$\sin 0 = \frac{y}{r} = \frac{0}{1} = 0 \qquad\qquad (x, y) = (1, 0)$$

$$\sin \frac{\pi}{2} = \frac{y}{r} = \frac{1}{1} = 1 \qquad\qquad (x, y) = (0, 1)$$

$$\sin \pi = \frac{y}{r} = \frac{0}{1} = 0 \qquad\qquad (x, y) = (-1, 0)$$

$$\sin \frac{3\pi}{2} = \frac{y}{r} = \frac{-1}{1} = -1. \qquad\qquad (x, y) = (0, -1)$$

Try using Figure 2.35 to evaluate some of the other trigonometric functions at the four quadrant angles.

Reference Angles

The values of the trigonometric functions of angles greater than 90° (or less than 0°) can be determined from their values at corresponding acute angles called **reference angles.**

Definition of Reference Angles

Let θ be an angle in standard position. Its **reference angle** is the acute angle θ' formed by the terminal side of θ and the horizontal axis.

Figure 2.36 illustrates the reference angle for θ in Quadrants II, III, and IV.

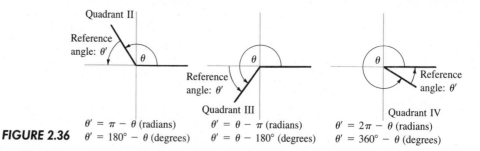

FIGURE 2.36

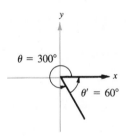

(a) θ in Quadrant IV

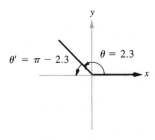

(b) θ in Quadrant II

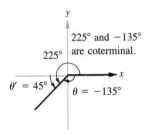

(c) θ in Quadrant III

FIGURE 2.37

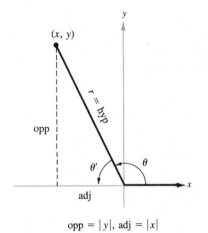

opp = $|y|$, adj = $|x|$

FIGURE 2.38

EXAMPLE 4 Finding Reference Angles

Find the reference angle θ' for each of the following.

(a) $\theta = 300°$

(b) $\theta = 2.3$

(c) $\theta = -135°$

SOLUTION

(a) Since $\theta = 300°$ lies in Quadrant IV, the angle it makes with the x-axis is

$$\theta' = 360° - 300° = 60°. \qquad \textit{Degrees}$$

(b) Since $\theta = 2.3$ lies between $\pi/2 \approx 1.5708$ and $\pi \approx 3.1416$, it follows that θ is in Quadrant II and its reference angle is

$$\theta' = \pi - 2.3 \approx 0.8416. \qquad \textit{Radians}$$

(c) First, we determine that $-135°$ is coterminal with $225°$, which lies in Quadrant III. Hence, the reference angle is

$$\theta' = 225° - 180° = 45°. \qquad \textit{Degrees}$$

Figure 2.37 shows each angle θ and its reference angle θ'.

To see how a reference angle is used to *evaluate* a trigonometric function, consider the point (x, y) on the terminal side of θ, as shown in Figure 2.38. We know that

$$\sin \theta = \frac{y}{r} \qquad \text{and} \qquad \tan \theta = \frac{y}{x}.$$

For the triangle with acute angle θ' and sides of lengths $|x|$ and $|y|$, we have

$$\sin \theta' = \frac{\text{opp}}{\text{hyp}} = \frac{|y|}{r} \qquad \text{and} \qquad \tan \theta' = \frac{\text{opp}}{\text{adj}} = \frac{|y|}{|x|}.$$

Thus, it follows that $\sin \theta$ and $\sin \theta'$ are equal, *except possibly in sign*. The same is true for $\tan \theta$ and $\tan \theta'$ *and* for the other four trigonometric functions. In all cases, the sign of the function value can be determined by the quadrant in which θ lies.

The following steps can be used to evaluate trigonometric functions of any angle.

Trigonometry

Evaluating Trigonometric Functions of Any Angle

To find the value of a trigonometric function of any angle θ, use the following steps.

1. Determine the function value for the associated reference angle θ'.
2. Depending on the quadrant in which θ lies, prefix the appropriate sign to the function value.

By using reference angles together with the special angles discussed in the previous section, we can greatly extend our scope of *exact* trigonometric values. For instance, knowing the function values of 30° means that we know the function values of all angles for which 30° is a reference angle. For convenience, we provide the following table of the exact values of the trigonometric functions of special angles and quadrant angles. You should memorize these values.

TABLE 2.1

θ (degrees)	0°	30°	45°	60°	90°	180°	270°
θ (radians)	0	$\dfrac{\pi}{6}$	$\dfrac{\pi}{4}$	$\dfrac{\pi}{3}$	$\dfrac{\pi}{2}$	π	$\dfrac{3\pi}{2}$
$\sin \theta$	0	$\dfrac{1}{2}$	$\dfrac{\sqrt{2}}{2}$	$\dfrac{\sqrt{3}}{2}$	1	0	-1
$\cos \theta$	1	$\dfrac{\sqrt{3}}{2}$	$\dfrac{\sqrt{2}}{2}$	$\dfrac{1}{2}$	0	-1	0
$\tan \theta$	0	$\dfrac{\sqrt{3}}{3}$	1	$\sqrt{3}$	undef.	0	undef.

EXAMPLE 5 *Trigonometric Functions of Nonacute Angles*

Evaluate the following.

(a) $\cos \dfrac{4\pi}{3}$

(b) $\tan(-210°)$

(c) $\csc \dfrac{11\pi}{4}$

SOLUTION

(a) Since $\theta = 4\pi/3$ lies in Quadrant III, the reference angle is $\theta' = (4\pi/3) - \pi = \pi/3$, as shown in Figure 2.39(a). Moreover, the cosine is negative in Quadrant III, so that

$$\cos \frac{4\pi}{3} = (-)\cos \frac{\pi}{3} = -\frac{1}{2}. \qquad \textit{Special angle, } \pi/3$$

(b) Since $-210° + 360° = 150°$, it follows that $-210°$ is coterminal with the second-quadrant angle $150°$. Therefore, the reference angle is $\theta' = 180° - 150° = 30°$, as shown in Figure 2.39(b). Finally, since the tangent is negative in Quadrant II, we have

$$\tan(-210°) = (-)\tan 30° = -\frac{\sqrt{3}}{3}. \qquad \textit{Special angle, } 30°$$

(c) Since $(11\pi/4) - 2\pi = 3\pi/4$, it follows that $11\pi/4$ is coterminal with the second-quadrant angle $3\pi/4$. Therefore, the reference angle is $\theta' = \pi - (3\pi/4) = \pi/4$, as shown in Figure 2.39(c). Because the cosecant is positive in Quadrant II, we have

$$\csc \frac{11\pi}{4} = (+)\csc \frac{\pi}{4} = \frac{1}{\sin(\pi/4)} = \sqrt{2}. \qquad \textit{Special angle, } \pi/4$$

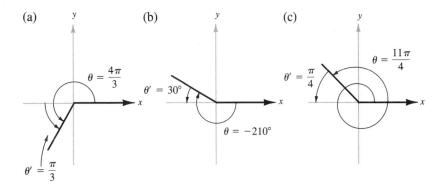

FIGURE 2.39

The fundamental trigonometric identities listed in the previous section (for an acute angle θ) are also valid when θ is any angle. In Example 6, we show how these identities can be used to find the exact value of trigonometric functions.

EXAMPLE 6 *Using Identities to Evaluate Trigonometric Functions*

Let θ be an angle in Quadrant II such that $\sin \theta = 1/3$. Find (a) $\cos \theta$ and (b) $\tan \theta$.

SOLUTION

(a) Since $\sin \theta = 1/3$, we can use the Pythagorean Identity $\sin^2 \theta + \cos^2 \theta = 1$ to obtain

$$\left(\frac{1}{3}\right)^2 + \cos^2 \theta = 1$$

$$\cos^2 \theta = 1 - \frac{1}{9} = \frac{8}{9}.$$

Since $\cos \theta < 0$ in Quadrant II, we use the negative root

$$\cos \theta = -\frac{\sqrt{8}}{\sqrt{9}} = -\frac{2\sqrt{2}}{3}.$$

(b) Using the result from part (a) and the trigonometric identity $\tan \theta = \sin \theta / \cos \theta$, we obtain

$$\tan \theta = \frac{1/3}{-2\sqrt{2}/3} = -\frac{1}{2\sqrt{2}} = -\frac{\sqrt{2}}{4}.$$

Scientific calculators can be used to approximate the values of trigonometric functions of any angle, as demonstrated in Example 7. (Note that some calculators restrict the magnitude of the angles they will accept.)

EXAMPLE 7 *Evaluating Trigonometric Functions with a Calculator*

Use a calculator to approximate the following values.

(a) $\cot 410°$ (b) $\sin(-7)$ (c) $\tan \dfrac{14\pi}{5}$

SOLUTION

Function	Mode	Keystrokes	Display
(a) $\cot 410°$	Degree	410 [tan] [1/x]	0.8390996
(b) $\sin(-7)$	Radian	7 [+/-] [sin]	−0.6569866
(c) $\tan \dfrac{14\pi}{5}$	Radian	14 [×] π [÷] 5 [=] [tan]	−0.7265425

Exercises

At this point, we have completed our introduction to basic trigonometry. We have measured angles in both radians and degrees. We have defined the six trigonometric functions from a right triangle perspective and as functions of real numbers. In our remaining work with trigonometry, we will continue to rely on both perspectives. For instance, in the next three sections on graphing techniques we will think of the trigonometric functions as functions of real numbers. Later, in Chapter 4, we will look at applications involving angles and triangles.

For your convenience, we have included on the inside cover of this text a summary of basic trigonometry.

WARM UP

Evaluate the trigonometric functions from memory.

1. $\sin 30°$ **2.** $\tan 45°$ **3.** $\cos \dfrac{\pi}{4}$ **4.** $\cot \dfrac{\pi}{3}$ **5.** $\sec \dfrac{\pi}{6}$ **6.** $\csc \dfrac{\pi}{4}$

In each case use the given trigonometric function of an *acute* angle θ to find the values of the remaining trigonometric functions.

7. $\tan \theta = \dfrac{3}{2}$ **8.** $\cos \theta = \dfrac{2}{3}$ **9.** $\sin \theta = \dfrac{1}{5}$ **10.** $\sec \theta = 3$

EXERCISES 2.4

In Exercises 1–4, determine the exact value of the six trigonometric functions of the given angle θ.

1. (a)
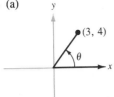
(3, 4)

(b)
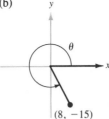
(8, −15)

3. (a)
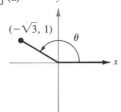
$(-\sqrt{3},\ 1)$

(b)
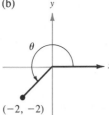
(−2, −2)

2. (a)
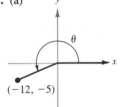
(−12, −5)

(b)
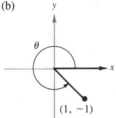
(1, −1)

4. (a)
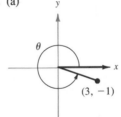
(3, −1)

(b)
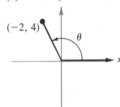
(−2, 4)

In Exercises 5–8, the given point is on the terminal side of an angle in standard position. Determine the exact value of the six trigonometric functions of the angle.

5. (a) $(7, 24)$ (b) $(7, -24)$

6. (a) $(8, 15)$ (b) $(-9, -40)$

7. (a) $(-4, 10)$ (b) $(3, -5)$

8. (a) $(-5, -2)$ (b) $(-\frac{3}{2}, 3)$

In Exercises 9–12, use the two similar triangles in the accompanying figure to find (a) the unknown sides of the triangles and (b) the six trigonometric functions of the angles α_1 and α_2.

9. $a_1 = 3$, $b_1 = 4$, $a_2 = 9$

10. $b_1 = 12$, $c_1 = 13$, $c_2 = 26$

11. $a_1 = 1$, $c_1 = 2$, $b_2 = 5$

12. $b_1 = 4$, $a_2 = 4$, $b_2 = 10$

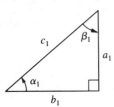 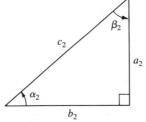

FIGURE FOR 9–12

In Exercises 13–16, determine the quadrant in which θ lies.

13. (a) $\sin \theta < 0$ and $\cos \theta < 0$
     (b) $\sin \theta > 0$ and $\cos \theta < 0$

14. (a) $\sin \theta > 0$ and $\cos \theta > 0$
     (b) $\sin \theta < 0$ and $\cos \theta > 0$

15. (a) $\sin \theta > 0$ and $\tan \theta < 0$
     (b) $\cos \theta > 0$ and $\tan \theta < 0$

16. (a) $\sec \theta > 0$ and $\cot \theta < 0$
     (b) $\csc \theta < 0$ and $\tan \theta > 0$

In Exercises 17–26, find the exact value (if possible) of the six trigonometric functions of θ.

17. θ lies in Quadrant II, $\sin \theta = \frac{3}{5}$

18. θ lies in Quadrant III, $\cos \theta = -\frac{4}{5}$

19. $\sin \theta < 0$, $\tan \theta = -\frac{15}{8}$ **20.** $\tan \theta < 0$, $\cos \theta = \frac{8}{17}$

21. $\sin \theta > 0$, $\sec \theta = -2$

22. $\frac{\pi}{2} \le \theta \le \frac{3\pi}{2}$, $\cot \theta$ is undefined

23. $\sin \theta = 0$, $\sec \theta = -1$

24. $\pi \le \theta \le 2\pi$, $\tan \theta$ is undefined

25. The terminal side of θ is in Quadrant III and lies on the line $y = 2x$.

26. The terminal side of θ is in Quadrant IV and lies on the line $4x + 3y = 0$.

In Exercises 27–34, find the reference angle θ', and draw a sketch.

27. (a) $\theta = 203°$ (b) $\theta = 127°$

28. (a) $\theta = 309°$ (b) $\theta = 226°$

29. (a) $\theta = -245°$ (b) $\theta = -72°$

30. (a) $\theta = -145°$ (b) $\theta = -239°$

31. (a) $\theta = \dfrac{2\pi}{3}$ (b) $\theta = \dfrac{7\pi}{6}$

32. (a) $\theta = \dfrac{7\pi}{4}$ (b) $\theta = \dfrac{8\pi}{9}$

33. (a) $\theta = 3.5$ (b) $\theta = 5.8$

34. (a) $\theta = \dfrac{11\pi}{3}$ (b) $\theta = -\dfrac{7\pi}{10}$

In Exercises 35–44, evaluate the sine, cosine, and tangent of the given angles without using a calculator.

35. (a) $225°$ (b) $-225°$

36. (a) $300°$ (b) $330°$

37. (a) $750°$ (b) $510°$

38. (a) $-405°$ (b) $-120°$

39. (a) $\dfrac{4\pi}{3}$ (b) $\dfrac{2\pi}{3}$

40. (a) $\dfrac{\pi}{4}$ (b) $\dfrac{5\pi}{4}$

41. (a) $-\dfrac{\pi}{6}$ (b) $\dfrac{5\pi}{6}$

42. (a) $-\dfrac{\pi}{2}$ (b) $\dfrac{\pi}{2}$

43. (a) $\dfrac{11\pi}{4}$ (b) $-\dfrac{13\pi}{6}$

44. (a) $\dfrac{10\pi}{3}$ (b) $\dfrac{17\pi}{3}$

In Exercises 45–52, use a calculator to evaluate the given trigonometric functions to four decimal places. (Be sure the calculator is set in the correct mode.)

45. (a) $\sin 10°$ (b) $\csc 10°$

46. (a) $\sec 225°$ (b) $\sec 135°$

47. (a) $\tan \dfrac{\pi}{9}$ (b) $\tan \dfrac{10\pi}{9}$

48. (a) $\cot 1.35$ (b) $\tan 1.35$

49. (a) $\cos(-110°)$ (b) $\cos 250°$

50. (a) $\sin(-0.65)$ (b) $\sin 5.63$

51. (a) $\tan 240°$ (b) $\cot 210°$

52. (a) $\csc 2.62$ (b) $\csc 150°$

In Exercises 53–58, find two values of θ that satisfy the given equation. List your answers in degrees ($0° \le \theta < 360°$) *and* radians ($0 \le \theta < 2\pi$). Do not use a calculator.

53. (a) $\sin \theta = \frac{1}{2}$ (b) $\sin \theta = -\frac{1}{2}$

54. (a) $\cos \theta = \dfrac{\sqrt{2}}{2}$ (b) $\cos \theta = -\dfrac{\sqrt{2}}{2}$

55. (a) $\csc \theta = \dfrac{2\sqrt{3}}{3}$ (b) $\cot \theta = -1$

56. (a) $\sec \theta = 2$ (b) $\sec \theta = -2$

57. (a) $\tan \theta = 1$ (b) $\cot \theta = -\sqrt{3}$

58. (a) $\sin \theta = \dfrac{\sqrt{3}}{2}$ (b) $\sin \theta = -\dfrac{\sqrt{3}}{2}$

In Exercises 59 and 60, use a calculator to approximate two values of θ ($0° \le \theta < 360°$) that satisfy the given equation. Round your answers to two decimal places.

59. (a) $\sin \theta = 0.8191$ (b) $\sin \theta = -0.2589$

60. (a) $\cos \theta = 0.8746$ (b) $\cos \theta = -0.2419$

In Exercises 61–64, use a calculator to approximate *two* values of θ ($0 \le \theta < 2\pi$) that satisfy the given equation. Round your answers to three decimal places.

61. (a) $\cos \theta = 0.9848$ (b) $\cos \theta = -0.5890$

62. (a) $\sin \theta = 0.0175$ (b) $\sin \theta = -0.6691$

63. (a) $\tan \theta = 1.192$ (b) $\tan \theta = -8.144$

64. (a) $\cot \theta = 5.671$ (b) $\cot \theta = -1.280$

In Exercises 65–68, evaluate the expression without using a calculator.

65. $\sin^2 2 + \cos^2 2$

66. $\tan^2 20° - \sec^2 20°$

67. $2 \sin^2 \dfrac{7\pi}{6} - \sin \dfrac{7\pi}{6} - 1$

68. $\sec^2 \dfrac{3\pi}{4} - 2 \tan \dfrac{3\pi}{4} - 2$

69. The average daily temperature (in degrees Fahrenheit) for a certain city is given by

$$T = 45 - 23 \cos\left[\frac{2\pi}{365}(t - 32)\right]$$

where t is the time in days, with $t = 1$ corresponding to January 1. Find the average temperature on (a) January 1, (b) July 4 ($t = 185$), and (c) October 18 ($t = 291$).

70. A company that produces a product with seasonal demands forecasts monthly sales over the next two years to be

$$S = 23.1 + 0.442t + 4.3 \sin\left(\frac{\pi t}{6}\right)$$

where S is measured in thousands of units and t is the time in months, with $t = 1$ representing January, 1988. Predict the sales for the following months.

(a) February, 1988 (b) February, 1989

(c) September, 1988 (d) September, 1989

2.5 Graphs of Sine and Cosine

In this section we look at techniques for sketching the graphs of the sine and cosine functions. To accommodate the familiar *xy*-coordinate system, we use the variable *x* in place of θ or *t*. For example, we will write $y = \sin x$ and $y = \cos x$.

In Section 2.2 we discovered that the *domain* of the sine function is the set of all real numbers, and the *range* is the interval $[-1, 1]$. Moreover, the sine function has a *period* of 2π which implies that

$$\sin(x + 2\pi n) = \sin x$$

for all integers *n*.

Using this information, reference angles, and the function values from Table 2.1 (in the previous section), we can plot several points for $y = \sin x$ as *x* varies over the interval $[0, 2\pi]$. Then, by connecting these points with a smooth curve, we obtain the solid portion of the graph shown in Figure 2.40. This solid portion of the graph is called a **sine wave** or one **cycle** of the sine curve. The gray portion of the graph indicates that the basic sine wave repeats indefinitely to the right and left.

A similar analysis of the cosine function shows that its domain is also the set of all real numbers, its range is the interval $[-1, 1]$, and it has a period of 2π. The graph of the cosine function is shown in Figure 2.41.

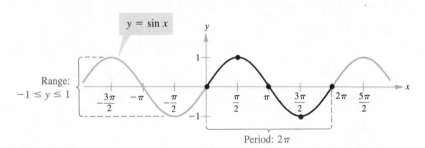

FIGURE 2.40

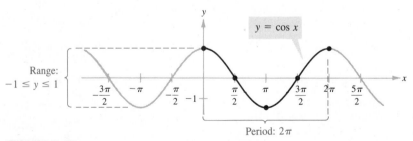

FIGURE 2.41

Note from Figures 2.40 and 2.41 that the sine graph is symmetric with respect to the *origin*, whereas the cosine graph is symmetric with respect to the *y-axis*. These properties of symmetry follow from the fact that the sine function is odd:

$$\sin(-x) = -\sin x \qquad\qquad \textit{Origin symmetry}$$

whereas the cosine function is even:

$$\cos(-x) = \cos x. \qquad\qquad \textit{y-axis symmetry}$$

Key Points on Graphs of Sine and Cosine Functions

To help you construct the graphs of the basic sine and cosine functions, we note five **key points** in one period of each graph: the intercepts, maximum points, and minimum points. For the sine function, the key points are

$$(0, 0), \quad \left(\frac{\pi}{2}, 1\right), \quad (\pi, 0), \quad \left(\frac{3\pi}{2}, -1\right), \quad \text{and} \quad (2\pi, 0).$$

For the cosine function, the key points are

$$(0, 1), \quad \left(\frac{\pi}{2}, 0\right), \quad (\pi, -1), \quad \left(\frac{3\pi}{2}, 0\right), \quad \text{and} \quad (2\pi, 1).$$

Note how the *x*-coordinates of these points divide the period of sin *x* and cos *x* into *four* equal parts, as indicated in Figure 2.42.

In Example 1, note how we use these key points to sketch the graph of $y = 2 \sin x$.

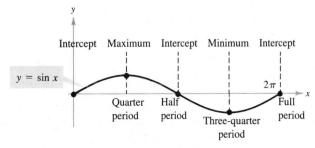

Period: 2π

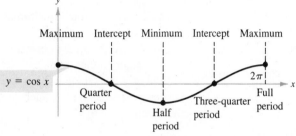

FIGURE 2.42

Period: 2π

EXAMPLE 1   *Using Key Points to Sketch a Sine Curve*

Sketch the graph of $y = 2 \sin x$ on the interval $[-\pi, 4\pi]$.

SOLUTION

In this case, note that $y = 2 \sin x = 2(\sin x)$ indicates that the y-values for the key points will have twice the magnitude of the graph of $y = \sin x$. Thus, the key points for $y = 2 \sin x$ are

$$(0, 0), \quad \left(\frac{\pi}{2}, 2\right), \quad (\pi, 0), \quad \left(\frac{3\pi}{2}, -2\right), \quad \text{and} \quad (2\pi, 0).$$

By connecting these key points with a smooth curve and extending the curve in both directions over the interval $[-\pi, 4\pi]$, we obtain the graph shown in Figure 2.43.

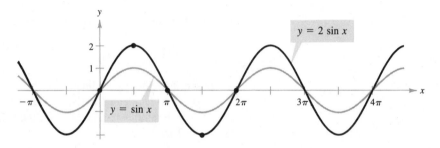

FIGURE 2.43

Transformations of Sine and Cosine Curves

In the rest of this section we look at variations in the graphs of the basic functions $y = \sin x$ and $y = \cos x$. In particular, we want to investigate the graphic effect of each of the constants a, b, and c in equations of the forms

$$y = a \sin(bx - c) \quad \text{and} \quad y = a \cos(bx - c).$$

A quick review of the transformations studied in Section 1.6 should help in this investigation.

In Figure 2.43, the constant factor 2 in $y = 2 \sin x$ acts as a vertical *stretch* factor so that y ranges between -2 and 2 instead of between -1 and 1. Similarly, the factor $1/2$ in $y = \frac{1}{2} \sin x$ *shrinks* the y-values so that y ranges between $-1/2$ and $1/2$. Such factors are referred to as the **amplitudes** of the functions.

Definition of Amplitude of Sine and Cosine Curves

The **amplitude** of $y = a \sin x$ and $y = a \cos x$ is the largest value of y and is given by

amplitude $= |a|$.

EXAMPLE 2 Using Amplitude to Sketch Graphs

On the same coordinate axes, sketch graphs of

$$y = \frac{1}{2} \cos x \quad \text{and} \quad y = 3 \cos x.$$

SOLUTION

Since the amplitude of $y = \frac{1}{2} \cos x$ is $1/2$, the maximum value is $1/2$ and the minimum value is $-1/2$. For one cycle, $0 \le x \le 2\pi$, the key points are

$$\left(0, \frac{1}{2}\right), \quad \left(\frac{\pi}{2}, 0\right), \quad \left(\pi, -\frac{1}{2}\right), \quad \left(\frac{3\pi}{2}, 0\right), \quad \text{and} \quad \left(2\pi, \frac{1}{2}\right).$$

A similar analysis shows that the amplitude of $y = 3 \cos x$ is 3, and the key points are

$$(0, 3), \quad \left(\frac{\pi}{2}, 0\right), \quad (\pi, -3), \quad \left(\frac{3\pi}{2}, 0\right), \quad \text{and} \quad (2\pi, 3).$$

The graphs of these two functions are shown in Figure 2.44.

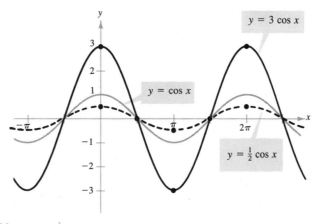

FIGURE 2.44   Amplitude Determines Vertical Stretch or Shrink

Trigonometry

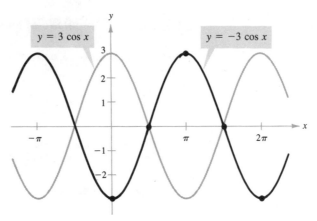

FIGURE 2.45   Reflection in the x-Axis

We know from Section 1.6 that the graph of $y = -f(x)$ is a **reflection** (in the x-axis) of the graph of $y = f(x)$. For instance, as shown in Figure 2.45, the graph of $y = -3 \cos x$ is a reflection of the graph of $y = 3 \cos x$.

Next, we consider the effect of the *positive* constant b on the graphs of

$$y = a \sin bx$$

and

$$y = a \cos bx.$$

Because $y = a \sin x$ completes one cycle from $x = 0$ to $x = 2\pi$, it follows that $y = a \sin bx$ completes one cycle from $bx = 0$ to $bx = 2\pi$, that is,

$$bx = 0 \implies x = 0$$

$$bx = 2\pi \implies x = \frac{2\pi}{b}.$$

Consequently, $y = a \sin bx$ completes one cycle from $x = 0$ to $x = 2\pi/b$, and hence its *period* is $2\pi/b$.

Period of Sine and Cosine Functions

Let b be a positive real number. The **period** of $y = a \sin bx$ and $y = a \cos bx$ is $2\pi/b$.

Note that if $0 < b < 1$, the period of $y = a \sin bx$ is greater than 2π and represents a *horizontal stretching* of the graph of $y = a \sin x$. Similarly, if $b > 1$, the period of $y = a \sin bx$ is less than 2π and represents a *horizontal shrinking* of the graph of $y = a \sin x$.

Graphs of Sine and Cosine

If b is negative, we use the identities $\sin(-x) = -\sin x$ and $\cos(-x) = \cos x$ to rewrite the function. For example, the period of

$$y = \sin(-2x)$$
$$= -\sin 2x$$

is $2\pi/2 = \pi$, and the period of

$$y = \cos(-3x)$$
$$= \cos 3x$$

is $2\pi/3$.

EXAMPLE 3 *Horizontal Stretching*

Sketch the graph of

$$y = \sin \frac{x}{2}.$$

SOLUTION

The amplitude is 1. Moreover, since $b = 1/2$, the period is

$$\frac{2\pi}{b} = \frac{2\pi}{1/2} = 4\pi.$$

Now, dividing the interval $[0, 4\pi]$ into four equal parts, we obtain the following key points on the graph:

$$(0, 0), \quad (\pi, 1), \quad (2\pi, 0), \quad (3\pi, -1), \quad \text{and} \quad (4\pi, 0).$$

The graph is shown in Figure 2.46.

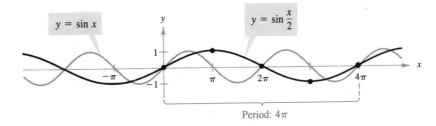

FIGURE 2.46

Trigonometry

EXAMPLE 4   *Horizontal Shrinking*

Sketch the graph of

$$y = \sin 3x.$$

SOLUTION

The amplitude is 1, and since $b = 3$, the period is

$$\frac{2\pi}{b} = \frac{2\pi}{3}.$$

Dividing the interval $[0, 2\pi/3]$ into four equal parts, we obtain the following key points on the graph:

$$(0, 0), \quad \left(\frac{\pi}{6}, 1\right), \quad \left(\frac{\pi}{3}, 0\right), \quad \left(\frac{\pi}{2}, -1\right), \quad \text{and} \quad \left(\frac{2\pi}{3}, 0\right).$$

The graph is shown in Figure 2.47.

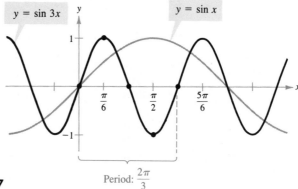

FIGURE 2.47

Our fourth graphic transformation is the horizontal shift caused by the constant c in the general equations

$$y = a \sin(bx - c) \qquad \text{and} \qquad y = a \cos(bx - c).$$

Comparing $y = a \sin bx$ with $y = a \sin(bx - c)$, we find that the graph of $y = a \sin(bx - c)$ completes one cycle from $bx - c = 0$ to $bx - c = 2\pi$. By solving for x, we find the interval for one cycle to be

Left Endpoint   Right Endpoint

$$\frac{c}{b} \le x \le \frac{c}{b} + \frac{2\pi}{b}.$$

Period

Graphs of Sine and Cosine

This implies that the period of $y = a \sin(bx - c)$ is $2\pi/b$, and the graph of $y = a \sin bx$ is shifted by an amount c/b. We call the number c/b the **phase shift.** We summarize these results as follows.

Graphs of the Sine and Cosine Functions

The graphs of $y = a \sin(bx - c)$ and $y = a \cos(bx - c)$ have the following characteristics. (Assume $b > 0$.)

$$\text{amplitude} = |a|$$
$$\text{period} = 2\pi/b$$

The **phase shift** and resulting interval for one cycle are solutions to the equations

$$bx - c = 0 \quad \text{and} \quad bx - c = 2\pi.$$

Note how we use this information to sketch graphs of the sine and cosine functions in Examples 5 and 6.

EXAMPLE 5 *Using Amplitude, Period, and Shift to Sketch Graphs*

Sketch the graph of

$$y = \frac{1}{2} \sin\left(x - \frac{\pi}{3}\right).$$

SOLUTION

The amplitude is $1/2$ and the period is 2π. By solving the equations

$$x - \frac{\pi}{3} = 0 \quad \text{and} \quad x - \frac{\pi}{3} = 2\pi$$
$$x = \frac{\pi}{3} \qquad\qquad x = \frac{7\pi}{3}$$

we see that the interval $[\pi/3, 7\pi/3]$ corresponds to one cycle of the graph. Dividing this interval into four equal parts produces the following key points:

$$\left(\frac{\pi}{3}, 0\right), \quad \left(\frac{5\pi}{6}, \frac{1}{2}\right), \quad \left(\frac{4\pi}{3}, 0\right), \quad \left(\frac{11\pi}{6}, -\frac{1}{2}\right), \quad \text{and} \quad \left(\frac{7\pi}{3}, 0\right).$$

The graph is shown in Figure 2.48.

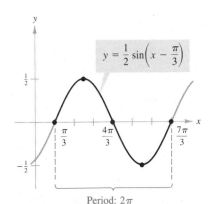

FIGURE 2.48

Trigonometry

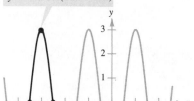

$y = -3 \cos(2\pi x + 4\pi)$

Period: 1

FIGURE 2.49

EXAMPLE 6 *Using Amplitude, Period, and Shift to Sketch Graphs*

Sketch the graph of $y = -3 \cos(2\pi x + 4\pi)$.

SOLUTION

For this function the amplitude is 3 and the period is $2\pi/2\pi = 1$. By solving the equations

$$2\pi x + 4\pi = 0 \qquad \text{and} \qquad 2\pi x + 4\pi = 2\pi$$
$$x = -2 \qquad\qquad\qquad\qquad x = -1$$

we see that one cycle corresponds to the interval $[-2, -1]$. Dividing this interval into four equal parts produces the following key points:

$$(-2, -3), \quad \left(-\frac{7}{4}, 0\right), \quad \left(-\frac{3}{2}, 3\right), \quad \left(-\frac{5}{4}, 0\right), \quad \text{and} \quad (-1, -3).$$

The graph is shown in Figure 2.49. ▬

Our final graphic transformation is the *vertical shift* caused by the constant d in the equations

$$y = d + a \sin(bx - c) \qquad \text{and} \qquad y = d + a \cos(bx - c).$$

The shift is d units upward for $d > 0$ and downward for $d < 0$. In other words, the graph oscillates about the horizontal line $y = d$ instead of the x-axis.

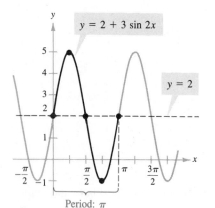

$y = 2 + 3 \sin 2x$

$y = 2$

Period: π

FIGURE 2.50

EXAMPLE 7 *A Vertical Translation*

Sketch the graph of $y = 2 + 3 \sin 2x$.

SOLUTION

The amplitude is 3 and the period is π. The key points over the interval $[0, \pi]$ are

$$(0, 2), \quad \left(\frac{\pi}{4}, 5\right), \quad \left(\frac{\pi}{2}, 2\right), \quad \left(\frac{3\pi}{4}, -1\right), \quad \text{and} \quad (\pi, 2).$$

The graph is shown in Figure 2.50. ▬

For our last example in this section, we reverse the situation and show how to determine an equation for a given graph.

EXAMPLE 8 Finding an Equation for a Given Graph

Find the amplitude, period, and phase shift for the sine function whose graph is shown in Figure 2.51 Write an equation for this graph.

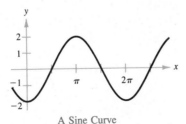

FIGURE 2.51   A Sine Curve

SOLUTION

The amplitude for this sine curve is 2. The period is 2π, and there is a right phase shift of $\pi/2$. Thus, we can write the following equation.

$$y = 2\sin\left(x - \frac{\pi}{2}\right)$$

It is of interest to note that the graph of $y = \cos x$ corresponds to a *left* shift of $\pi/2$ units of the graph of $y = \sin x$, as shown in Figure 2.52. Similarly, the graph of $y = \sin x$ is a *right* shift of the graph of $y = \cos x$. This is consistent with the cofunction identities discussed in Section 2.3. That is,

$$\sin\left(\frac{\pi}{2} - x\right) = \cos x \qquad \text{and} \qquad \cos\left(\frac{\pi}{2} - x\right) = \sin x.$$

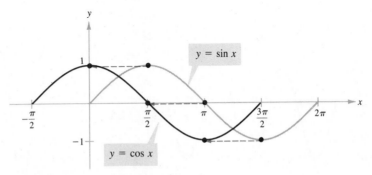

FIGURE 2.52   Shift the graph of $y = \sin x$ to the left $\frac{\pi}{2}$ units to obtain the graph of $y = \cos x$.

WARM UP

Simplify the expressions.

1. $\dfrac{2\pi}{1/3}$

2. $\dfrac{2\pi}{4\pi}$

Solve for *x*.

3. $2x - \dfrac{\pi}{3} = 0$

4. $2x - \dfrac{\pi}{3} = 2\pi$

5. $3\pi x + 6\pi = 0$

6. $3\pi x + 6\pi = 2\pi$

Evaluate the trigonometric functions from memory.

7. $\sin \dfrac{\pi}{2}$

8. $\sin \pi$

9. $\cos 0$

10. $\cos \dfrac{\pi}{2}$

EXERCISES 2.5

In Exercises 1–14, determine the period and amplitude of the given function.

1. $y = 2 \sin 2x$

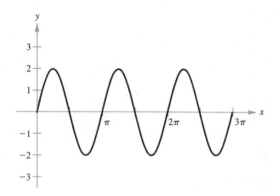

2. $y = 3 \cos 3x$

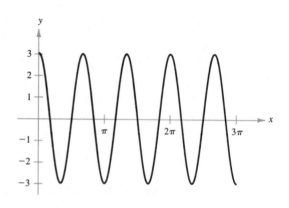

3. $y = \dfrac{3}{2} \cos \dfrac{x}{2}$

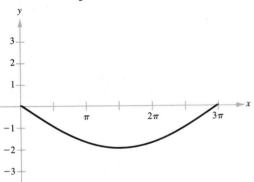

6. $y = \dfrac{5}{2} \cos \dfrac{\pi x}{2}$

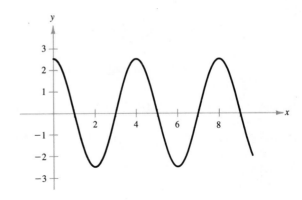

4. $y = -2 \sin \dfrac{x}{3}$

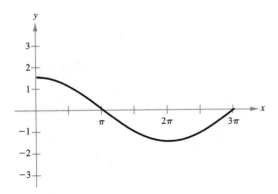

7. $y = 2 \sin x$

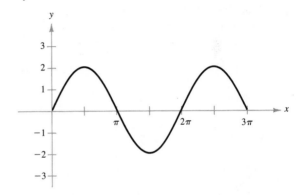

5. $y = \frac{1}{2} \sin \pi x$

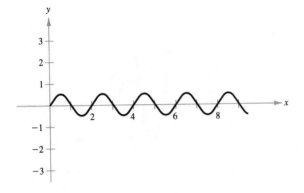

8. $y = -\cos \dfrac{2x}{3}$

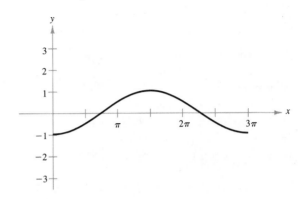

9. $y = -2 \sin 10x$

10. $y = \frac{1}{3} \sin 8x$

11. $y = \frac{1}{2} \cos \frac{2x}{3}$

12. $y = \frac{5}{2} \cos \frac{x}{4}$

13. $y = 3 \sin 4\pi x$

14. $y = \frac{2}{3} \cos \frac{\pi x}{10}$

In Exercises 15–22, describe the relationship between the graphs of f and g.

15. $f(x) = \sin x$
    $g(x) = \sin(x - \pi)$

16. $f(x) = \cos x$
    $g(x) = \cos(x + \pi)$

17. $f(x) = \cos 2x$
    $g(x) = -\cos 2x$

18. $f(x) = \sin 3x$
    $g(x) = \sin(-3x)$

19. $f(x) = \cos x$
    $g(x) = \cos 2x$

20. $f(x) = \sin x$
    $g(x) = \sin 3x$

21. $f(x) = \sin x$
    $g(x) = 2 + \sin x$

22. $f(x) = \cos 4x$
    $g(x) = -2 + \cos 4x$

In Exercises 23–30, sketch the graphs of the two functions on the same coordinate plane. (Include two full periods.)

 23. $f(x) = -2 \sin x$
    $g(x) = 4 \sin x$

24. $f(x) = \sin x$
    $g(x) = \sin \frac{x}{3}$

25. $f(x) = \cos x$
    $g(x) = 1 + \cos x$

26. $f(x) = 2 \cos 2x$
    $g(x) = -\cos 4x$

27. $f(x) = -\frac{1}{2} \sin \frac{x}{2}$
    $g(x) = 3 - \frac{1}{2} \sin \frac{x}{2}$

28. $f(x) = 4 \sin \pi x$
    $g(x) = 4 \sin \pi x - 3$

29. $f(x) = 2 \cos x$
    $g(x) = 2 \cos(x + \pi)$

30. $f(x) = -\cos x$
    $g(x) = -\cos(x - \pi)$

In Exercises 31–56, sketch the graph of the given function. (Include two full periods.)

31. $y = -2 \sin 6x$

32. $y = -3 \cos 4x$

33. $y = \cos 2\pi x$

34. $y = \frac{3}{2} \sin \frac{\pi x}{4}$

35. $y = -\sin \frac{2\pi x}{3}$

36. $y = 10 \cos \frac{\pi x}{6}$

37. $y = 2 - \sin \frac{2\pi x}{3}$

38. $y = 2 \cos x - 3$

39. $y = \sin\left(x - \frac{\pi}{4}\right)$

40. $y = \frac{1}{2} \sin(x - \pi)$

41. $y = 3 \cos(x + \pi)$

42. $y = 4 \cos\left(x + \frac{\pi}{4}\right)$

43. $y = 3 \cos(x + \pi) - 3$

44. $y = 4 \cos\left(x + \frac{\pi}{4}\right) + 4$

45. $y = \frac{2}{3} \cos\left(\frac{x}{2} - \frac{\pi}{4}\right)$

46. $y = -3 \cos(6x + \pi)$

47. $y = -2 \sin(4x + \pi)$

48. $y = -4 \sin\left(\frac{2}{3}x - \frac{\pi}{3}\right)$

49. $y = \cos\left(2\pi x - \frac{\pi}{2}\right) + 1$

50. $y = 3 \cos\left(\frac{\pi x}{2} + \frac{\pi}{2}\right) - 2$

51. $y = -0.1 \sin\left(\frac{\pi x}{10} + \pi\right)$

52. $y = 5 \sin(\pi - 2x) + 10$

53. $y = 5 \cos(\pi - 2x) + 2$

54. $y = \frac{1}{100} \sin 120\pi t$

55. $y = \frac{1}{10} \cos 60\pi x$

56. $y = -3 + 5 \cos \frac{\pi t}{12}$

In Exercises 57–60, find a, b, and c so that the graph of the function matches the graph in the accompanying figure. (Assume $a > 0$.)

57. $y = a \sin(bx - c)$

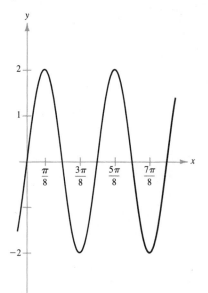

Exercises

58. $y = a \sin(bx - c)$

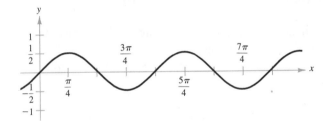

59. $y = a \cos(bx - c)$

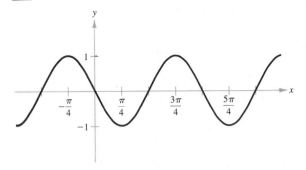

60. $y = a \sin(bx - c)$

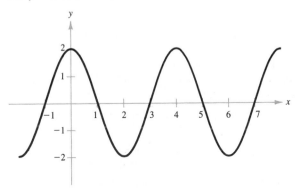

In Exercises 61–64, sketch the graph of f and g on the same coordinate axes and show that $f(x) = g(x)$ for all x. (Include two full periods.)

61. $f(x) = \sin x$

$\quad g(x) = \cos\left(x - \dfrac{\pi}{2}\right)$

62. $f(x) = \sin x$

$\quad g(x) = -\cos\left(x + \dfrac{\pi}{2}\right)$

63. $f(x) = \cos x$

$\quad g(x) = -\sin\left(x - \dfrac{\pi}{2}\right)$

64. $f(x) = \cos x$

$\quad g(x) = -\cos(x - \pi)$

65. For a person at rest, the velocity v (in liters per second) of air flow during a respiratory cycle is

$$v = 0.85 \sin \frac{\pi t}{3}$$

where t is the time in seconds. (Inhalation occurs when $v > 0$, and exhalation occurs when $v < 0$.)
(a) Find the time for one full respiratory cycle.
(b) Find the number of cycles per minute.
(c) Sketch the graph of the velocity function.

66. After exercising for a few minutes, a person has a respiratory cycle for which the velocity of air flow is approximated by

$$v = 1.75 \sin \frac{\pi t}{2}.$$

Use this model to repeat Exercise 65.

67. When tuning a piano, a technician strikes a tuning fork for the A above middle C and sets up wave motion that can be approximated by

$$y = 0.001 \sin 880\pi t$$

where t is the time in seconds.
(a) What is the period p of this function?
(b) The frequency f is given by $f = 1/p$. What is the frequency of this note?
(c) Sketch the graph of this function.

2.6 *Graphs of Other Trigonometric Functions*

In this section we continue our discussion of the graphs of the trigonometric functions, starting with the graph of the tangent function. Recall from Section 2.2 that the tangent function is odd, that is, $\tan(-x) = -\tan x$. Consequently, the graph of

$$y = \tan x$$

is symmetric with respect to the origin. We know also from the identity

$$\tan x = \frac{\sin x}{\cos x}$$

that the tangent is undefined when $\cos x = 0$. Two such values are $x = \pm\pi/2 \approx \pm 1.5708$. We examine this in more detail in the following table.

x	$-\dfrac{\pi}{2}$	-1.57	-1.5	-1	0	1	1.5	1.57	$\dfrac{\pi}{2}$
$\tan x$	undef.	-1255.8	-14.1	-1.56	0	1.56	14.1	1255.8	undef.

> $\tan x$ approaches $-\infty$ as x approaches $-\pi/2$ from the right

> $\tan x$ approaches ∞ as x approaches $\pi/2$ from the left

As indicated in this table, $\tan x$ increases without bound as x approaches $\pi/2$ from the left, and decreases without bound as x approaches $-\pi/2$ from the right. Thus, the graph of $y = \tan x$ has *vertical asymptotes* at $x = \pi/2$ and $-\pi/2$, as shown in Figure 2.53. Note in the graph that the period of the tangent function is π. Consequently, the tangent function has vertical asymptotes when $x = \pi/2 \pm n\pi$. The domain of the tangent function is the set of

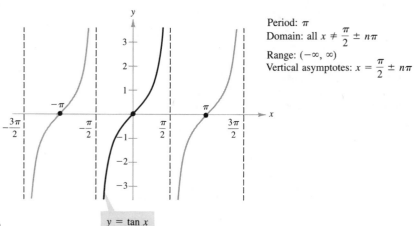

Period: π
Domain: all $x \neq \dfrac{\pi}{2} \pm n\pi$

Range: $(-\infty, \infty)$
Vertical asymptotes: $x = \dfrac{\pi}{2} \pm n\pi$

$y = \tan x$

FIGURE 2.53

all real numbers other than $x = \pi/2 \pm n\pi$, and the range is the set of all real numbers. Note from Figure 2.53 that the tangent function is increasing between each pair of consecutive asymptotes and that an x-intercept occurs at the midpoint of each cycle.

Sketching the graph of a function with the form

$$y = a \tan(bx - c)$$

is similar to sketching the graph of $y = a \sin(bx - c)$ in that we locate key points which identify the intercepts and asymptotes. Two consecutive asymptotes can be found by solving the equations

$$bx - c = -\frac{\pi}{2} \quad \text{and} \quad bx - c = \frac{\pi}{2}.$$

The midpoint between two consecutive asymptotes is an x-intercept of the graph. After plotting the asymptotes and the x-intercept, plot a few additional points between the two asymptotes and sketch one cycle. Additional cycles to the right or left can easily be sketched.

Remark: The period of the function $y = a \tan(bx - c)$ is the distance between two consecutive asymptotes. The amplitude of a tangent function is not defined.

EXAMPLE 1 *Sketching the Graph of a Tangent Function*

Sketch the graph of

$$y = \tan \frac{x}{2}.$$

SOLUTION

From the equations

$$\frac{x}{2} = -\frac{\pi}{2} \quad \text{and} \quad \frac{x}{2} = \frac{\pi}{2}$$

$$x = -\pi \qquad\qquad x = \pi$$

we see that two consecutive asymptotes occur at $x = -\pi$ and $x = \pi$. Between these two asymptotes, we plot a few points, as shown in the table, including the x-intercept, and complete the graph, as shown in Figure 2.54.

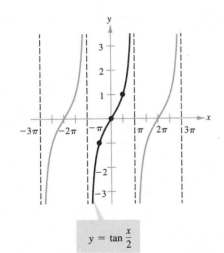

$$y = \tan \frac{x}{2}$$

FIGURE 2.54

x	$-\dfrac{\pi}{2}$	0	$\dfrac{\pi}{2}$
$\tan \dfrac{x}{2}$	-1	0	1

Trigonometry

EXAMPLE 2 *Sketching the Graph of a Tangent Function*

Sketch the graph of $y = -3 \tan 2x$.

SOLUTION

By solving the equations

$$2x = -\frac{\pi}{2} \quad \text{and} \quad 2x = \frac{\pi}{2}$$

$$x = -\frac{\pi}{4} \quad\quad\quad x = \frac{\pi}{4}$$

we see that two consecutive asymptotes occur at $x = -\pi/4$ and $x = \pi/4$. Between these two asymptotes, we plot a few points, as shown in the following table, and complete one cycle.

x	$-\dfrac{\pi}{8}$	0	$\dfrac{\pi}{8}$
$-3 \tan 2x$	3	0	-3

Four cycles of the graph are shown in Figure 2.55.

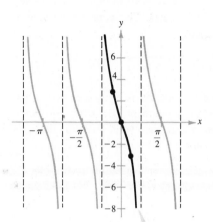

$y = -3 \tan 2x$

FIGURE 2.55

The graph of the cotangent function is similar to the graph of the tangent function. It also has a period of π. However, from the identity

$$y = \cot x = \frac{\cos x}{\sin x}$$

we can see that the cotangent function has vertical asymptotes at $x = n\pi$, because $\sin x$ is zero at these x-values. The graph of the cotangent function is shown in Figure 2.56.

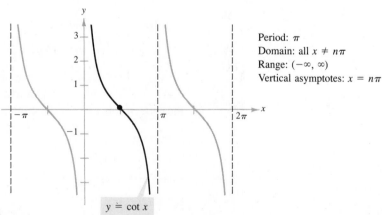

Period: π
Domain: all $x \ne n\pi$
Range: $(-\infty, \infty)$
Vertical asymptotes: $x = n\pi$

$y = \cot x$

FIGURE 2.56

EXAMPLE 3 *Sketching the Graph of a Cotangent Function*

Sketch the graph of

$$y = 2 \cot \frac{x}{3}.$$

SOLUTION

To locate two consecutive vertical asymptotes of the graph, we solve

$$\frac{x}{3} = 0 \qquad \text{and} \qquad \frac{x}{3} = \pi$$

$$x = 0 \qquad\qquad\qquad x = 3\pi.$$

Then, between these two asymptotes we plot the points shown in the following table, and complete one cycle of the graph. (Note that the period is 3π, the distance between consecutive asymptotes.)

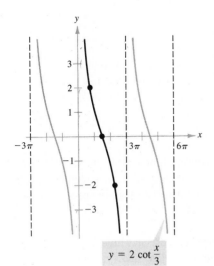

$$y = 2 \cot \frac{x}{3}$$

FIGURE 2.57

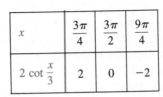

x	$\dfrac{3\pi}{4}$	$\dfrac{3\pi}{2}$	$\dfrac{9\pi}{4}$
$2 \cot \dfrac{x}{3}$	2	0	−2

Three cycles of the graph are shown in Figure 2.57.

Graphs of the Reciprocal Functions

The graphs of the two remaining trigonometric functions can be obtained from the graphs of the sine and cosine functions using the reciprocal identities

$$\csc x = \frac{1}{\sin x} \qquad \text{and} \qquad \sec x = \frac{1}{\cos x}.$$

For instance, at a given value for x, the y-coordinate for $\sec x$ is the reciprocal of the y-coordinate for $\cos x$. Of course, when $\cos x = 0$, the reciprocal does not exist. Near such values for x, the behavior of the secant function is similar to that of the tangent function. In other words, the graphs of

$$\tan x = \frac{\sin x}{\cos x} \qquad \text{and} \qquad \sec x = \frac{1}{\cos x}$$

have vertical asymptotes at $x = (\pi/2) + n\pi$, because the cosine is zero at these x-values. Similarly,

$$\cot x = \frac{\cos x}{\sin x} \qquad \text{and} \qquad \csc x = \frac{1}{\sin x}$$

have vertical asymptotes where $\sin x = 0$, that is, at $x = n\pi$.

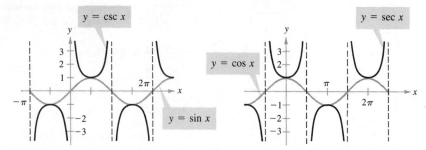

Period: 2π
Domain: all $x \neq n\pi$
Range: all y not in $(-1, 1)$
Vertical asymptotes: $x = n\pi$

Period: 2π
Domain: all $x \neq \dfrac{\pi}{2} + n\pi$
Range: all y not in $(-1, 1)$
Vertical asymptotes: $x = \dfrac{\pi}{2} + n\pi$

FIGURE 2.58

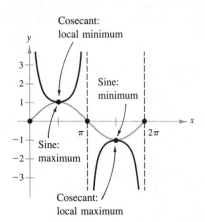

FIGURE 2.59

To sketch the graph of a secant or cosecant function we suggest that you first make a sketch of its reciprocal function. For instance, to sketch the graph of $y = \csc x$, first sketch the graph of $y = \sin x$, then take reciprocals of the y-coordinates to obtain points on the graph of $y = \csc x$. We use this procedure to obtain the graphs shown in Figure 2.58.

In comparing the graphs of the secant and cosecant functions with those of the sine and cosine functions, note that the "hills" and "valleys" are interchanged. For example, a hill (or maximum point) on the sine curve corresponds to a valley (a local minimum) on the cosecant curve. Similarly, a valley (or minimum point) on the sine curve corresponds to a hill (a local maximum) on the cosecant curve, as shown in Figure 2.59.

EXAMPLE 4 *Sketching the Graph of a Cosecant Function*

Sketch the graph of

$$y = 2 \csc\left(x + \frac{\pi}{4}\right).$$

SOLUTION

We begin by sketching the graph of

$$2 \sin\left(x + \frac{\pi}{4}\right).$$

Graphs of Other Trigonometric Functions

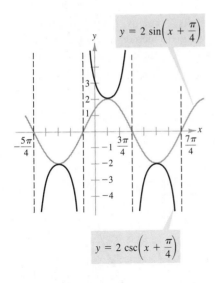

$$y = 2\sin\left(x + \frac{\pi}{4}\right)$$

$$y = 2\csc\left(x + \frac{\pi}{4}\right)$$

FIGURE 2.60

For this function, the amplitude is 2 and the period is 2π. By solving the equations

$$x + \frac{\pi}{4} = 0 \qquad \text{and} \qquad x + \frac{\pi}{4} = 2\pi$$

$$x = -\frac{\pi}{4} \qquad\qquad x = \frac{7\pi}{4}$$

we see that one cycle of the sine function corresponds to the interval from $x = -\pi/4$ to $x = 7\pi/4$. The graph of this sine function is represented by the gray curve in Figure 2.60. Because the sine function is zero at the endpoints of this interval, the corresponding cosecant function

$$y = 2\csc\left(x + \frac{\pi}{4}\right)$$

$$= 2\left(\frac{1}{\sin[x + (\pi/4)]}\right)$$

has vertical asymptotes at $x = -\pi/4$ and $7\pi/4$. The graph of the cosecant function is represented by the solid curve.

EXAMPLE 5   Sketching the Graph of a Secant Function

Sketch the graph of $y = \sec 2x$.

SOLUTION

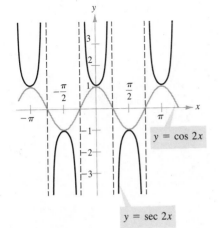

$$y = \cos 2x$$

$$y = \sec 2x$$

FIGURE 2.61

We begin by sketching the graph of $\cos 2x$ as indicated by the gray curve in Figure 2.61. Then, we form the graph of $y = \sec 2x$ as the solid curve in the figure. Note that the x-intercepts of $\cos 2x$

$$\left(\frac{\pi}{4}, 0\right), \quad \left(\frac{3\pi}{4}, 0\right), \quad \left(\frac{5\pi}{4}, 0\right), \quad \cdots$$

correspond to the vertical asymptotes

$$x = \frac{\pi}{4}, \quad x = \frac{3\pi}{4}, \quad x = \frac{5\pi}{4}, \quad \cdots$$

of the graph of $y = \sec 2x$.

Trigonometry

In Figure 2.62, we summarize the graphs, domains, ranges, and periods of the six basic trigonometric functions. Be sure to memorize this information.

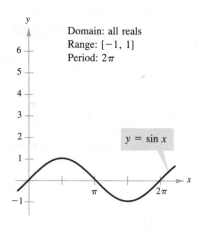

Domain: all reals
Range: $[-1, 1]$
Period: 2π

$y = \sin x$

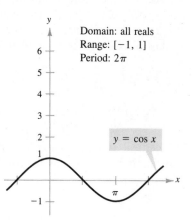

Domain: all reals
Range: $[-1, 1]$
Period: 2π

$y = \cos x$

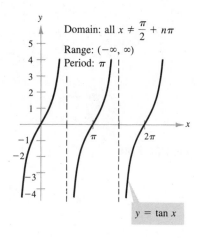

Domain: all $x \neq \dfrac{\pi}{2} + n\pi$
Range: $(-\infty, \infty)$
Period: π

$y = \tan x$

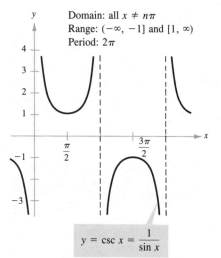

Domain: all $x \neq n\pi$
Range: $(-\infty, -1]$ and $[1, \infty)$
Period: 2π

$y = \csc x = \dfrac{1}{\sin x}$

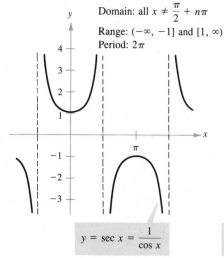

Domain: all $x \neq \dfrac{\pi}{2} + n\pi$
Range: $(-\infty, -1]$ and $[1, \infty)$
Period: 2π

$y = \sec x = \dfrac{1}{\cos x}$

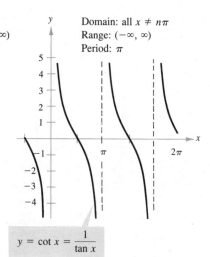

Domain: all $x \neq n\pi$
Range: $(-\infty, \infty)$
Period: π

$y = \cot x = \dfrac{1}{\tan x}$

Graphs of the Six Trigonometric Functions

FIGURE 2.62

WARM UP

Evaluate the trigonometric functions from memory.

1. $\tan 0$

2. $\cos \dfrac{\pi}{4}$

3. $\tan \dfrac{\pi}{4}$

4. $\cot \dfrac{\pi}{2}$

5. $\sin \pi$

6. $\cos \dfrac{\pi}{2}$

Sketch the graph of each function. (Include two full periods.)

7. $y = -2 \cos 2x$

8. $y = 3 \sin \dfrac{x}{4}$

9. $y = \frac{3}{2} \sin 2\pi x$

10. $y = -2 \cos \dfrac{\pi x}{2}$

EXERCISES 2.6

In Exercises 1–8, match the trigonometric function with the correct graph and give the period of the function. [The graphs are labeled (a)–(h).]

1. $y = \sec 2x$

2. $y = \tan 3x$

3. $y = \tan \dfrac{x}{2}$

4. $y = 2 \csc \dfrac{x}{2}$

5. $y = \cot \pi x$

6. $y = \frac{1}{2} \sec \pi x$

7. $y = -\sec x$

8. $y = -2 \csc 2\pi x$

(c)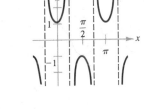

(d)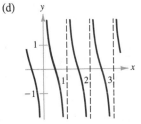

(a)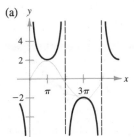

(b)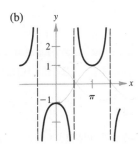

(e)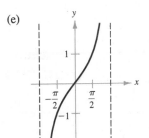

(f)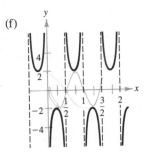

(g)

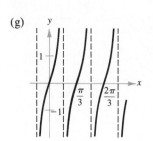

(h)

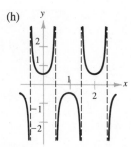

In Exercises 9–36, sketch the graph of the function through two periods.

9. $y = \tan 2x$

10. $y = -\tan 2x$

11. $y = \tan \dfrac{x}{3}$

12. $y = -3 \tan \pi x$

13. $y = -2 \sec 4x$

14. $y = 2 \sec 4x$

15. $y = -\sec \pi x$

16. $y = \sec \pi x$

17. $y = \sec \pi x - 1$

18. $y = -2 \sec 4x + 2$

19. $y = \csc \dfrac{x}{2}$

20. $y = \csc \dfrac{x}{3}$

21. $y = \cot \dfrac{x}{2}$

22. $y = 3 \cot \dfrac{\pi x}{2}$

23. $y = \frac{1}{2} \sec 2x$

24. $y = -\frac{1}{2} \tan x$

25. $y = \tan\left(x - \dfrac{\pi}{4}\right)$

26. $y = \sec(x + \pi)$

27. $y = \dfrac{1}{4} \csc\left(x + \dfrac{\pi}{4}\right)$

28. $y = -\csc(4x - \pi)$

29. $y = \dfrac{1}{4} \cot\left(x - \dfrac{\pi}{2}\right)$

30. $y = 2 \cot\left(x + \dfrac{\pi}{2}\right)$

31. $y = 2 \sec(2x - \pi)$

32. $y = \dfrac{1}{3} \sec\left(\dfrac{\pi x}{2} + \dfrac{\pi}{2}\right)$

33. $y = \tan \dfrac{\pi x}{4}$

34. $y = 0.1 \tan\left(\dfrac{\pi x}{4} + \dfrac{\pi}{4}\right)$

35. $y = \csc(\pi - x)$

36. $y = \sec(\pi - x)$

37. A plane flying at an altitude of 6 miles over level ground will pass directly over a radar antenna (see figure). Let d be the ground distance from the antenna to the point directly under the plane and let x be the angle of elevation to the plane from the antenna. Write d as a function of x, $0 < x < \pi/2$.

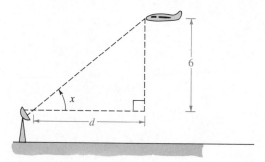

FIGURE FOR 37

38. A television camera is on a reviewing platform 100 feet from the street on which a parade will be passing from left to right (see figure). Express the distance d from the camera to a particular unit in the parade as a function of the angle x and sketch the graph of the function over the appropriate domain. (Consider x as negative when a unit in the parade approaches from the left.)

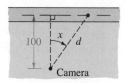

FIGURE FOR 38

39. Use the graph of $f(x) = \sec x$ to verify that the secant is an even function.

40. Use the graph of $f(x) = \tan x$ to verify that the tangent is an odd function.

2.7 Additional Graphing Techniques

Addition of Ordinates

The behavior of some physical phenomena can be represented by more than one trigonometric function, or by a combination of algebraic and trigonometric functions. We can use a technique called **addition of ordinates** (*y*-values)

Additional Graphing Techniques

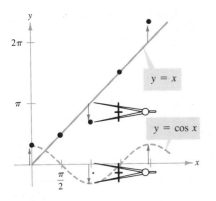

FIGURE 2.63

to sketch the graphs of functions like

$$y = \sin x - \cos 2x \quad \text{or} \quad y = x + \cos x.$$

For example, the graph of $y = x + \cos x$ can be obtained by first making sketches of $y = x$ and $y = \cos x$ on the same set of axes and then geometrically adding the ordinates (y-values) for several representative x-values. This addition of ordinates is aided by the use of a compass or ruler to measure the vertical displacements, as shown in Figure 2.63.

As with previous trigonometric graphs, the *key points* to plot are those for which one or both functions have an intercept, maximum point, or minimum point.

EXAMPLE 1 *Graphing by Addition of Ordinates*

Use addition of ordinates to sketch the graph of

$$y = x + \cos x.$$

SOLUTION

First, we make sketches (shown in gray) of the graphs of

$$y = x \quad \text{and} \quad y = \cos x$$

on the same coordinate plane. For $y = \cos x$, the key points are

$$(0, 1), \quad \left(\frac{\pi}{2}, 0\right), \quad (\pi, -1), \quad \left(\frac{3\pi}{2}, 0\right), \quad \text{and} \quad (2\pi, 1).$$

The points lying on the graph of $y = x$ directly above (or below) these five points are

$$(0, 0), \quad \left(\frac{\pi}{2}, \frac{\pi}{2}\right), \quad (\pi, \pi), \quad \left(\frac{3\pi}{2}, \frac{3\pi}{2}\right), \quad \text{and} \quad (2\pi, 2\pi).$$

We geometrically add the y-values of the two functions at these key points

$$(0, 1), \quad \left(\frac{\pi}{2}, \frac{\pi}{2}\right), \quad (\pi, -1 + \pi), \quad \left(\frac{3\pi}{2}, \frac{3\pi}{2}\right), \quad (2\pi, 1 + 2\pi)$$

and plot the results. Then we connect the resulting points by a smooth curve, obtaining the graph shown in Figure 2.64.

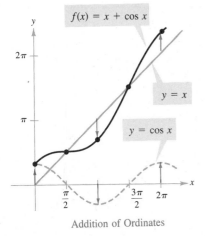

Addition of Ordinates

FIGURE 2.64

Note that the function in Figure 2.64 is not periodic as we defined the concept. However, the next example involves two trigonometric functions whose resulting difference is a periodic function.

EXAMPLE 2 *Graphing by Addition of Ordinates*

Sketch the graph of $y = \sin x - \cos 2x$.

SOLUTION

In this case, we make sketches (shown in gray) of the graphs of

$$y = \sin x \qquad \text{and} \qquad y = -\cos 2x$$

noting that their respective periods are 2π and π. From the shorter period (the one for $-\cos 2x$), we consider the key points given in the following table.

x	0	$\dfrac{\pi}{4}$	$\dfrac{\pi}{2}$	$\dfrac{3\pi}{4}$	π	$\dfrac{5\pi}{4}$	$\dfrac{3\pi}{2}$	$\dfrac{7\pi}{4}$	2π
$-\cos 2x$	-1	0	1	0	-1	0	1	0	-1
$\sin x$	0	$\dfrac{\sqrt{2}}{2}$	1	$\dfrac{\sqrt{2}}{2}$	0	$-\dfrac{\sqrt{2}}{2}$	-1	$-\dfrac{\sqrt{2}}{2}$	0
$\sin x - \cos 2x$	-1	$\dfrac{\sqrt{2}}{2}$	2	$\dfrac{\sqrt{2}}{2}$	-1	$-\dfrac{\sqrt{2}}{2}$	0	$-\dfrac{\sqrt{2}}{2}$	-1

By plotting the points indicated in the fourth row of the table and connecting them with a smooth curve, we obtain the graph shown in Figure 2.65. Note that the function has a period of 2π.

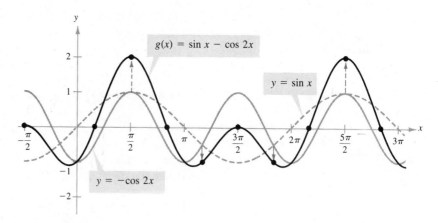

FIGURE 2.65   Addition of Ordinates

EXAMPLE 3   *Graphing by Addition of Ordinates*

Sketch the graph of $y = \sin x + 2 \cos x$.

SOLUTION

In this case, we make sketches (shown in gray) of the graphs of

$$y = \sin x \quad \text{and} \quad y = 2 \cos x$$

noting that both have periods of 2π. Then we consider the key points given in the following table.

x	0	$\dfrac{\pi}{2}$	π	$\dfrac{3\pi}{2}$	2π
$2 \cos x$	2	0	-2	0	2
$\sin x$	0	1	0	-1	0
$\sin x + 2 \cos x$	2	1	-2	-1	2

By plotting the points indicated in the fourth row of the table and connecting them with a smooth curve, we obtain the graph shown in Figure 2.66. Note that this function has a period of 2π.

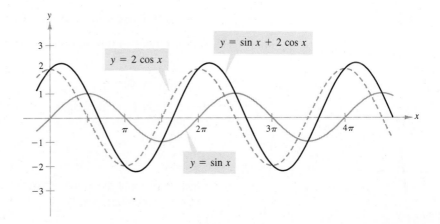

FIGURE 2.66

One of the simplest uses of addition of ordinates occurs in vertical shifts like those in the graphs of

$$y = 2 + \sin 3x \quad \text{and} \quad y = 3 + \cos\left(x - \frac{\pi}{4}\right).$$

We know from Sections 1.6 and 2.5 that adding a constant does not change the shape (or period) of a trigonometric graph—it only changes its vertical location. For example, using the graph of $y = \sin 3x$, we can easily sketch the graph of

$$y = 2 + \sin 3x$$

by adding 2 to each ordinate, as shown in Figure 2.67.

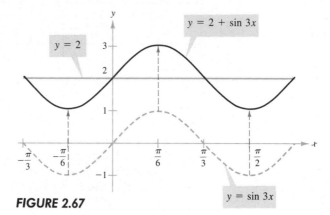

FIGURE 2.67

Damped Trigonometric Graphs

A *product* of two functions can be graphed using properties of the individual functions involved. For instance, consider the function

$$f(x) = x \sin x$$

as the product of the functions $y = x$ and $y = \sin x$. Using properties of absolute value and the fact that $|\sin x| \le 1$, we have $0 \le |x| \, |\sin x| \le |x|$. Consequently,

$$-|x| \le x \sin x \le |x|$$

which means that the graph of $f(x) = x \sin x$ lies between the lines $y = -x$ and $y = x$. Furthermore, since

$$f(x) = x \sin x = \pm x \quad \text{at} \quad x = \frac{\pi}{2} + n\pi$$

$$f(x) = x \sin x = 0 \quad \text{at} \quad x = n\pi$$

the graph of f touches the line $y = -x$ or the line $y = x$ at $x = (\pi/2) + n\pi$ and has x-intercepts at $x = n\pi$. A sketch of f is shown in Figure 2.68.

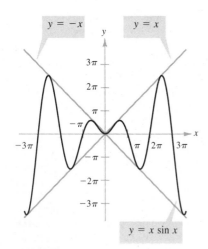

FIGURE 2.68

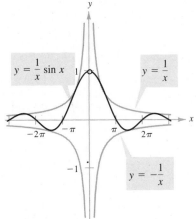

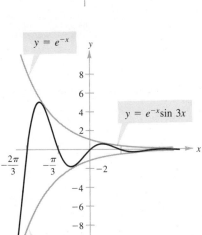

FIGURE 2.69

In the function $f(x) = x \sin x$ the factor x is called the **damping factor.** By changing the damping factor, we can change the graph significantly. For example, look in Figure 2.69 at the graphs of

$$y = \frac{1}{x} \sin x \qquad \text{and} \qquad y = e^{-x} \sin 3x.$$

We show another instance of a damping factor in Example 4.

EXAMPLE 4 *Damped Cosine Wave*

Sketch the graph of $f(x) = 2^{-x/2} \cos x$.

SOLUTION

Consider $f(x)$ as the product of the two functions

$$y = 2^{-x/2} \qquad \text{and} \qquad y = \cos x$$

each of which has the set of real numbers as its domain. For any real number x, we know that $2^{-x/2} \geq 0$ and $|\cos x| \leq 1$. Therefore, $|2^{-x/2}|\,|\cos x| \leq 2^{-x/2}$, which means that

$$-2^{-x/2} \leq 2^{-x/2} \cos x \leq 2^{-x/2}.$$

Furthermore, since

$$f(x) = 2^{-x/2} \cos x = \pm 2^{-x/2} \qquad \text{at} \qquad x = n\pi$$

and

$$f(x) = 2^{-x/2} \cos x = 0 \qquad \text{at} \qquad x = \frac{\pi}{2} + n\pi$$

the graph of f touches the curves $y = -2^{-x/2}$ and $y = 2^{-x/2}$ at $x = n\pi$ and has intercepts at $x = (\pi/2) + n\pi$. A sketch is shown in Figure 2.70.

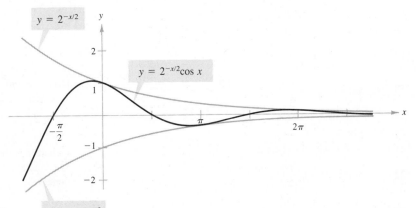

FIGURE 2.70

WARM UP

Find the x-values in the interval $[0, 2\pi]$ for which $f(x)$ is -1, 0, or 1.

1. $f(x) = \sin x$ **2.** $f(x) = \cos x$

3. $f(x) = \sin 2x$ **4.** $f(x) = \cos \dfrac{x}{2}$

Sketch the graph of each function.

5. $y = |x|$ **6.** $y = e^{-x}$

7. $y = \sin \pi x$ **8.** $y = \cos 2x$

Evaluate $f(x)$ when $x = 0$, $\pi/6$, $\pi/4$, $\pi/3$, and $\pi/2$.

9. $f(x) = x \cos x$ **10.** $f(x) = x + \sin x$

EXERCISES 2.7

In Exercises 1–26, use addition of ordinates to sketch the graph of the function.

1. $y = 2 - 2 \sin \dfrac{x}{2}$ **2.** $y = -3 + \cos x$

3. $y = 4 - 2 \cos \pi x$ **4.** $y = 5 - \frac{1}{2} \sin 2\pi x$

5. $y = -1 + \cot x$ **6.** $y = 2 + \tan \pi x$

7. $y = 1 + \csc x$ **8.** $y = 1 - \sec x$

9. $y = x + \sin x$ **10.** $y = x + \cos x$

11. $y = \frac{1}{2}x - 2 \cos x$ **12.** $y = 2x - \sin x$

13. $y = \sin x + \cos x$ **14.** $y = \cos x + \cos 2x$

15. $y = 2 \sin x + \sin 2x$

16. $y = 2 \sin x + \cos 2x$

17. $y = \cos x - \cos \dfrac{x}{2}$

18. $y = \sin x - \dfrac{1}{2} \sin \dfrac{x}{2}$

19. $y = \sin x + \frac{1}{3} \sin 5x$

20. $y = \cos x - \frac{1}{4} \cos 2x$

21. $y = -3 + \cos x + 2 \sin 2x$

22. $y = \sin \pi x + \sin \dfrac{\pi x}{2}$

23. $y = -x + \sin \dfrac{\pi x}{2}$

24. $y = -\dfrac{x}{2} - \dfrac{1}{2} \sin \dfrac{\pi x}{4}$

25. $y = \sin x + \cos\left(x + \dfrac{\pi}{2}\right)$

26. $y = \sin x - \cos\left(x + \dfrac{\pi}{2}\right)$

In Exercises 27–34, sketch the graph of the function.

27. $y = x \cos x$ **28.** $y = |x| \sin x$

29. $y = |x| \cos x$ **30.** $y = 2^{-x/4} \cos \pi x$

31. $y = e^{-x^2/2} \sin x$ **32.** $y = e^{-t} \cos t$

33. $y = \sin^2 x$ **34.** $y = \cos^2 \dfrac{\pi x}{2}$

35. The monthly sales S (in thousands of units) of a seasonal product is approximated by

$$S = 74.50 + 43.75 \sin \dfrac{\pi t}{6}$$

where t is the time in months, with $t = 1$ corresponding to January. Sketch the graph of this sales function over one year.

36. The function

$$P = 100 - 20 \cos \frac{5\pi t}{3}$$

approximates the blood pressure P (in millimeters of mercury) for a person at rest, where the time t is measured in seconds. Sketch the graph of this function over a 10-second interval of time.

37. Suppose that the population of a certain predator at time t (in months) in a given region is estimated to be

$$P = 10,000 + 3000 \sin \frac{2\pi t}{24}$$

and the population of its primary food source (its prey) is estimated to be

$$p = 15,000 + 5000 \cos \frac{2\pi t}{24}.$$

Sketch both of these functions on the same graph and explain the oscillations in the size of each population.

38. An object weighing W pounds is suspended from the ceiling by a steel spring (see figure). The weight is pulled downward (positive direction) from its equilibrium position and released. The resulting motion of the weight is described by the function

$$y = \frac{1}{2}e^{-t/4} \cos 4t$$

where y is distance in feet and t is the time in seconds. Sketch the graph of the function.

39. Use a calculator to evaluate the function

$$f(x) = \frac{1 - \cos x}{x}$$

at several points in the interval $[-1, 1]$, and then use these points to sketch the graph of f. This function is undefined when $x = 0$. From your graph, estimate the value that $f(x)$ is approaching as x approaches 0.

x	-0.5	-0.4	-0.3	-0.2	-0.1
$\dfrac{1 - \cos x}{x}$					

x	0.1	0.2	0.3	0.4	0.5
$\dfrac{1 - \cos x}{x}$					

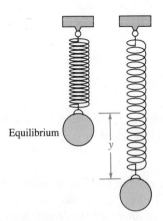

Equilibrium

y

FIGURE FOR 38

2.8 *Inverse Trigonometric Functions*

Up to this point, much of our time has been spent evaluating trigonometric functions at specified angles or real numbers. However, in Section 2.3 we introduced the *inverse* problem: *Given the value of sin x, find x.* There we used the calculator key INV or 2nd f and promised to later explain the functions involved. We now investigate these **inverse trigonometric functions.**

Recall from Section 1.7 that, in order for a function to have an inverse, it must be one-to-one. From Figure 2.71 it is obvious that $y = \sin x$ is not one-to-one because different values of x yield the same y-value. However, if we restrict the domain to the interval $-\pi/2 \le x \le \pi/2$ (corresponding to the solid portion of the graph in Figure 2.71), the following properties hold.

1. On the interval $[-\pi/2, \pi/2]$, the function $y = \sin x$ is increasing.
2. On the interval $[-\pi/2, \pi/2]$, $y = \sin x$ takes on its full range of values, $-1 \le \sin x \le 1$.
3. On the interval $[-\pi/2, \pi/2]$, $y = \sin x$ is a one-to-one function.

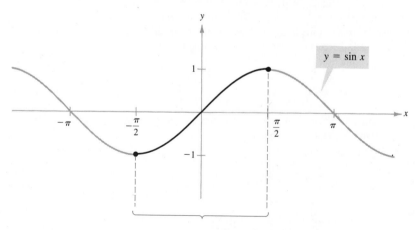

FIGURE 2.71   sin x is one-to-one on this interval.

Thus, on the restricted domain $-\pi/2 \le x \le \pi/2$, $y = \sin x$ has a unique inverse called the **inverse sine function.** It is denoted by

$$y = \arcsin x \qquad \text{or} \qquad y = \sin^{-1} x.$$

The notation $\sin^{-1} x$ is consistent with the inverse function notation $f^{-1}(x)$ used in Section 1.7. The arcsin x notation (read as "the arc sine of x") comes from the association of a central angle with its subtended *arc length* on a unit circle. Thus, arcsin x means the angle (or arc) whose sine is x. The values of arcsin x lie in the interval $-\pi/2 \le \arcsin x \le \pi/2$. The graph of $y = \arcsin x$ is shown in Example 2.

Inverse Trigonometric Functions

Definition of Inverse Sine Function

The **inverse sine function** is defined by

$$y = \arcsin x \quad \text{if and only if} \quad \sin y = x$$

where $-1 \le x \le 1$ and $-\pi/2 \le y \le \pi/2$. The domain of $y = \arcsin x$ is $[-1, 1]$ and the range is $[-\pi/2, \pi/2]$.

Remark: When evaluating the inverse sine function, it helps to remember the phrase "the arcsine of x is the angle (or number) whose sine is x."

Both notations, $\arcsin x$ and $\sin^{-1} x$, are commonly used in mathematics, so remember that $\sin^{-1} x$ denotes the *inverse* sine function rather than $1/\sin x$.

EXAMPLE 1 *Evaluating the Inverse Sine Function*

Find the values of the following (if possible).

(a) $\arcsin\left(-\dfrac{1}{2}\right)$ (b) $\sin^{-1}\dfrac{\sqrt{3}}{2}$ (c) $\sin^{-1} 2$

SOLUTION

(a) By definition, $y = \arcsin(-1/2)$ implies that

$$\sin y = -\frac{1}{2}, \quad \text{for } -\frac{\pi}{2} \le y \le \frac{\pi}{2}.$$

Because $\sin(-\pi/6) = -1/2$, we conclude that $y = -\pi/6$ and

$$\arcsin\left(-\frac{1}{2}\right) = -\frac{\pi}{6}.$$

(b) By definition, $y = \sin^{-1}(\sqrt{3}/2)$ implies that

$$\sin y = \frac{\sqrt{3}}{2}, \quad \text{for } -\frac{\pi}{2} \le y \le \frac{\pi}{2}.$$

Because $\sin(\pi/3) = \sqrt{3}/2$, we conclude that $y = \pi/3$ and

$$\sin^{-1}\frac{\sqrt{3}}{2} = \frac{\pi}{3}.$$

(c) It is not possible to evaluate $y = \sin^{-1} x$ when $x = 2$ because there is no angle whose sine is 2. Remember that the domain of the inverse sine function is $[-1, 1]$.

Trigonometry

From Section 1.7 we know that graphs of inverse functions are reflections of each other in the line $y = x$. In Example 2 we demonstrate this result for the inverse sine function.

EXAMPLE 2 *Graphing the Arcsine Function*

Sketch a graph of $y = \arcsin x$.

SOLUTION

By definition, the equations

$$y = \arcsin x \qquad \text{and} \qquad \sin y = x$$

are equivalent. Hence, their graphs are the same. By assigning values to y in the second equation, we can make the following table of values.

y	$-\dfrac{\pi}{2}$	$-\dfrac{\pi}{4}$	$-\dfrac{\pi}{6}$	0	$\dfrac{\pi}{6}$	$\dfrac{\pi}{4}$	$\dfrac{\pi}{2}$
$x = \sin y$	-1	$-\dfrac{\sqrt{2}}{2}$	$-\dfrac{1}{2}$	0	$\dfrac{1}{2}$	$\dfrac{\sqrt{2}}{2}$	1

The resulting graph for $y = \arcsin x$ is shown in Figure 2.72. Note that it is the reflection (in line $y = x$) of the solid part of Figure 2.71. Be sure you see that Figure 2.72 shows the *entire* graph of the inverse sine function. Remember that the range of $y = \arcsin x$ is the closed interval $[-\pi/2, \pi/2]$.

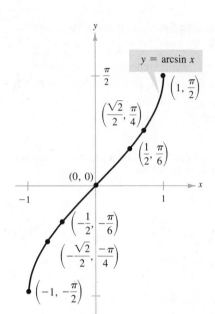

FIGURE 2.72

The cosine function is decreasing on the interval $0 \le x \le \pi$, as shown in Figure 2.73. Consequently, on this interval the cosine has an inverse

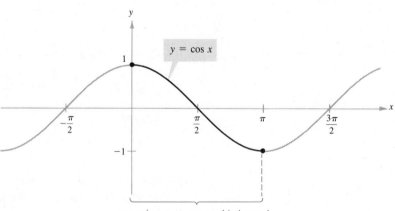

FIGURE 2.73

cos x is one-to-one on this interval.

function, which we call the **inverse cosine function** and denote by

$$y = \arccos x$$

or

$$y = \cos^{-1} x.$$

Similarly, we can define an **inverse tangent function** by restricting the domain of $y = \tan x$ to the interval $(-\pi/2, \pi/2)$. In the following list, we summarize the definitions of the three most common inverse trigonometric functions. The remaining three are discussed in the exercise set. (We summarize the graphs, domains, and ranges of *all six* inverse trigonometric functions in the Appendix.)

Definition of the Inverse Trigonometric Functions

Function	Domain	Range
$y = \textbf{arcsin } x$ if and only if $\sin y = x$	$-i \le x \le 1$	$-\dfrac{\pi}{2} \le y \le \dfrac{\pi}{2}$
$y = \textbf{arccos } x$ if and only if $\cos y = x$	$-1 \le x \le 1$	$0 \le y \le \pi$
$y = \textbf{arctan } x$ if and only if $\tan y = x$	$-\infty < x < \infty$	$-\dfrac{\pi}{2} < y < \dfrac{\pi}{2}$

The graphs of these three inverse trigonometric functions are shown in Figure 2.74.

Domain: $[-1, 1]$
Range: $[-\pi/2, \pi/2]$

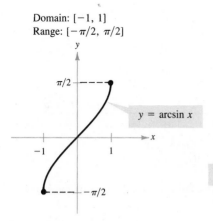

Domain: $[-1, 1]$
Range: $[0, \pi]$

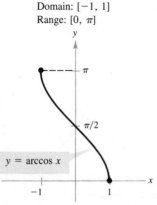

Domain: $(-\infty, \infty)$
Range: $(-\pi/2, \pi/2)$

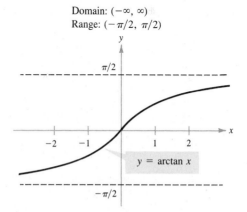

FIGURE 2.74

EXAMPLE 3 *Evaluating Inverse Trigonometric Functions*

Evaluate the following.

(a) $\arccos \dfrac{\sqrt{2}}{2}$ (b) $\arccos(-1)$ (c) $\arctan 0$

SOLUTION

(a) Because $\cos(\pi/4) = \sqrt{2}/2$, it follows that

$$\arccos \frac{\sqrt{2}}{2} = \frac{\pi}{4}.$$

(b) Because $\cos \pi = -1$, it follows that

$$\arccos(-1) = \pi.$$

(c) Because $\tan 0 = 0$, it follows that

$$\arctan 0 = 0.$$

In Example 3, we were able to evaluate the given inverse trigonometric function without a calculator. In the next example, a calculator is necessary to evaluate the functions.

EXAMPLE 4 *Evaluating Inverse Trigonometric Functions*

If possible, find the value of each of the following.

(a) $\arctan(-8.45)$ (b) $\arcsin 0.2447$ (c) $\arccos 2$

SOLUTION

Function	Mode	Keystrokes	Display
(a) $\arctan(-8.45)$	Radian	8.45 `+/-` `INV` `tan`	-1.453001
(b) $\arcsin 0.2447$	Radian	0.2447 `INV` `sin`	0.2472103
(c) $\arccos 2$	Radian	2 `INV` `cos`	ERROR

Note that the *error* in part (c) occurs because the domain of the inverse cosine function is $[-1, 1]$.

Remark: In Example 4, had we set the calculator to degree mode, the display would have been in degrees rather than radians. This convention is peculiar to calculators. By definition, the values of inverse trigonometric functions are always *in radians*.

Compositions of Trigonometric and Inverse Trigonometric Functions

Recall from Section 1.7 that inverse functions possess the properties

$$f(f^{-1}(x)) = x$$

and

$$f^{-1}(f(x)) = x.$$

The inverse trigonometric versions of these properties are given in the following list.

Inverse Properties

If $-1 \leq x \leq 1$ and $-\pi/2 \leq y \leq \pi/2$, then

$\sin(\arcsin x) = x$ and $\arcsin(\sin y) = y.$

If $-1 \leq x \leq 1$ and $0 \leq y \leq \pi$, then

$\cos(\arccos x) = x$ and $\arccos(\cos y) = y.$

If $-\pi/2 < y < \pi/2$, then

$\tan(\arctan x) = x$ and $\arctan(\tan y) = y.$

Remark: Keep in mind that these inverse properties do not apply for arbitrary values of x and y. For instance,

$$\arcsin\left(\sin \frac{3\pi}{2}\right) = \arcsin(-1)$$

$$= -\frac{\pi}{2} \neq \frac{3\pi}{2}.$$

In other words, the property $\arcsin(\sin y) = y$ is not valid for values of y outside the interval $[-\pi/2, \pi/2]$.

EXAMPLE 5 *Using Inverse Properties*

If possible, find the value of the following.

(a) $\tan[\arctan(-5)]$ 　　　(b) $\arcsin\left(\sin\dfrac{5\pi}{3}\right)$ 　　　(c) $\cos(\cos^{-1}\pi)$

SOLUTION

(a) Since -5 lies in the domain of the arctan x, the inverse property applies, and we have

$$\tan[\arctan(-5)] = -5.$$

(b) In this case, $5\pi/3$ does not lie within the range of the arcsine function, $-\pi/2 \le x \le \pi/2$. However, $5\pi/3$ is coterminal with

$$\frac{5\pi}{3} - 2\pi = -\frac{\pi}{3}$$

which does lie in the range of the arcsine function, and we have

$$\arcsin\left(\sin\frac{5\pi}{3}\right) = \arcsin\left[\sin\left(-\frac{\pi}{3}\right)\right] = -\frac{\pi}{3}.$$

(c) The expression $\cos(\cos^{-1}\pi)$ is not defined because $\cos^{-1}\pi$ is not defined. Remember that the domain of the inverse cosine function is $[-1, 1]$.

As with the trigonometric functions, much of the work with the inverse trigonometric functions can be done by *exact* calculations rather than by calculator approximations. Exact calculations help to increase our understanding of the inverse functions by relating them to the triangle definitions of the trigonometric functions.

In Example 6, we show how to use right triangles to find exact values of functions of inverse functions. Then, in Example 7, we show how to use triangles to convert a trigonometric expression into an algebraic one. This conversion technique is used frequently in calculus.

EXAMPLE 6 *Evaluating Functions of Inverse Trigonometric Functions*

Find the exact value of the following.

(a) $\tan\left(\arccos\dfrac{2}{3}\right)$ 　　　(b) $\cos\left[\arcsin\left(-\dfrac{3}{5}\right)\right]$

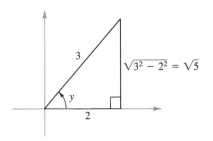

FIGURE 2.75

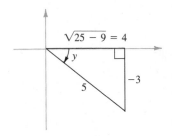

FIGURE 2.76

SOLUTION

(a) If we let $y = \arccos(2/3)$, then $\cos y = 2/3$. Because $\cos y$ is positive, y is a *first* quadrant angle. We can sketch and label y as shown in Figure 2.75. Consequently,

$$\tan\left(\arccos\frac{2}{3}\right) = \tan y$$

$$= \frac{\text{opp}}{\text{adj}} = \frac{\sqrt{5}}{2}.$$

(b) Let $y = \arcsin(-3/5)$. Then $\sin y = -3/5$. Because $\sin y$ is negative, y is a *fourth* quadrant angle. We can sketch and label y as shown in Figure 2.76. Consequently,

$$\cos\left[\arcsin\left(-\frac{3}{5}\right)\right] = \cos y$$

$$= \frac{\text{adj}}{\text{hyp}} = \frac{4}{5}.$$

EXAMPLE 7 *Some Problems from Calculus*

Write each of the following as an algebraic expression in x.

(a) $\sin(\arccos 3x), \quad 0 \le x \le \dfrac{1}{3}$ (b) $\cot(\arccos 3x), \quad 0 \le x \le \dfrac{1}{3}$

SOLUTION

Let $y = \arccos 3x$. Then we have $\cos y = 3x$. Since

$$\cos y = \frac{3x}{1} = \frac{\text{adj}}{\text{hyp}}$$

we can sketch a right triangle with acute angle y, as shown in Figure 2.77. From this triangle, we can easily convert each expression to algebraic form.

(a) $\sin(\arccos 3x) = \sin y = \dfrac{\text{opp}}{\text{hyp}} = \sqrt{1 - 9x^2}, \quad 0 \le x \le \dfrac{1}{3}$

(b) $\cot(\arccos 3x) = \cot y = \dfrac{\text{adj}}{\text{opp}} = \dfrac{3x}{\sqrt{1 - 9x^2}}, \quad 0 \le x \le \dfrac{1}{3}$

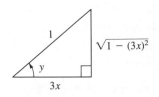

FIGURE 2.77

Remark: In Example 7, a similar argument can be made for x-values lying in the interval $[-1/3, 0]$.

WARM UP

Evaluate the trigonometric functions from memory.

1. $\sin\left(-\dfrac{\pi}{2}\right)$

2. $\cos \pi$

3. $\tan\left(-\dfrac{\pi}{4}\right)$

4. $\sin \dfrac{\pi}{4}$

Find an angle x in the interval $[-\pi/2, \pi/2]$ that has the same sine value as the given value.

5. $\sin 2\pi$

6. $\sin \dfrac{5\pi}{6}$

Find an angle x in the interval $[0, \pi]$ that has the same cosine value as the given value.

7. $\cos 3\pi$

8. $\cos\left(-\dfrac{\pi}{4}\right)$

Find an angle x in the interval $(-\pi/2, \pi/2)$ that has the same tangent value as the given value.

9. $\tan 4\pi$

10. $\tan \dfrac{3\pi}{4}$

EXERCISES 2.8

In Exercises 1–16, evaluate the given expression without using a calculator.

1. $\arcsin \frac{1}{2}$

2. $\arcsin 0$

3. $\arccos \frac{1}{2}$

4. $\arccos 0$

5. $\arctan \dfrac{\sqrt{3}}{3}$

6. $\arctan(-1)$

7. $\arccos\left(-\dfrac{\sqrt{3}}{2}\right)$

8. $\arcsin\left(-\dfrac{\sqrt{2}}{2}\right)$

9. $\arctan(-\sqrt{3})$

10. $\text{arccot}(-\sqrt{3})$

11. $\arccos(-\frac{1}{2})$

12. $\arcsin \dfrac{\sqrt{2}}{2}$

13. $\arcsin \dfrac{\sqrt{3}}{2}$

14. $\arccos 1$

15. $\arctan 0$

16. $\arctan\left(-\dfrac{\sqrt{3}}{3}\right)$

In Exercises 17–28, use a calculator to approximate the given value. [Round your answers to two decimal places.]

17. $\arccos 0.28$

18. $\arcsin 0.45$

19. $\arcsin(-0.75)$

20. $\arccos(-0.8)$

21. $\arctan(-2)$

22. $\arctan 15$

23. $\arcsin 0.31$

24. $\arccos 0.26$

25. $\arccos(-0.41)$

26. $\arcsin(-0.125)$

27. $\arctan 0.92$

28. $\arctan 2.8$

In Exercises 29–34, use the properties of inverse trigonometric functions to evaluate the given expression.

29. $\sin(\arcsin 0.3)$

30. $\tan(\arctan 25)$

31. $\cos[\arccos(-0.1)]$

32. $\sin[\arcsin(-0.2)]$

33. $\arcsin(\sin 3\pi)$

34. $\arccos\left(\cos \dfrac{7\pi}{2}\right)$

In Exercises 35–42, find the exact value of the given expression without using a calculator. [*Hint:* Make a sketch of a right triangle, as illustrated in Example 6.]

35. $\sin\left(\arctan\frac{3}{4}\right)$ **36.** $\sec\left(\arcsin\frac{4}{5}\right)$

37. $\cos(\arctan 2)$ **38.** $\sin\left(\arccos\frac{\sqrt{5}}{5}\right)$

39. $\cos\left(\arcsin\frac{5}{13}\right)$ **40.** $\csc\left[\arctan\left(-\frac{5}{12}\right)\right]$

41. $\sec\left[\arctan\left(-\frac{3}{5}\right)\right]$ **42.** $\tan\left[\arcsin\left(-\frac{5}{6}\right)\right]$

In Exercises 43–52, write an algebraic expression that is equivalent to the given expression. [*Hint:* Sketch a right triangle, as demonstrated in Example 7.]

43. $\cot(\arctan x)$ **44.** $\sin(\arctan x)$

45. $\cos(\arcsin 2x)$ **46.** $\sec(\arctan 3x)$

47. $\sin(\arccos x)$ **48.** $\cot\left(\arctan\frac{1}{x}\right)$

49. $\tan\left(\arccos\frac{x}{3}\right)$ **50.** $\sec[\arcsin(x-1)]$

51. $\csc\left(\arctan\frac{x}{\sqrt{2}}\right)$ **52.** $\cos\left(\arcsin\frac{x-h}{r}\right)$

In Exercises 53–56, fill in the blanks.

53. $\arctan\dfrac{9}{x} = \arcsin(\quad)$

54. $\arcsin\dfrac{\sqrt{36-x^2}}{x} = \arccos(\quad), \quad |x| \le 6$

55. $\arccos\dfrac{3}{\sqrt{x^2-2x+10}} = \arcsin(\quad)$

56. $\arccos\dfrac{x-2}{2} = \arctan(\quad), \quad |x-2| \le 2$

In Exercises 57–60, sketch the graph of the function.

57. $f(x) = \arcsin(x-1)$ **58.** $f(x) = \arctan x + \dfrac{\pi}{2}$

59. $f(x) = \arccos 2x$ **60.** $f(x) = \arccos\dfrac{x}{4}$

61. A photographer is taking a picture of a four-foot painting hung in an art gallery. The camera lens is one foot below the lower edge of the painting (see figure). The angle β subtended by the camera lens x feet from the painting is given by

$$\beta = \arctan\frac{4x}{x^2+5}.$$

Find β when (a) $x = 3$ feet and (b) $x = 6$ feet.

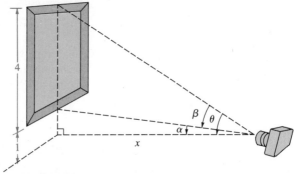

FIGURE FOR 61

62. In calculus, it is shown that the area of the region bounded by the graphs of $y = 0$, $y = 1/(x^2 + 1)$, $x = a$, and $x = b$ is given by

$$\text{Area} = \arctan b - \arctan a$$

(see figure). Find the area for the following values of a and b.

(a) $a = 0$, $b = 1$
(b) $a = -1$, $b = 1$
(c) $a = 0$, $b = 3$
(d) $a = -1$, $b = 3$

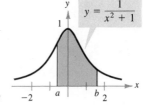

FIGURE FOR 62

63. Define the inverse cotangent function by restricting the domain of the cotangent to the interval $(0, \pi)$.

64. Define the inverse secant function by restricting the domain of the secant to the intervals $[0, \pi/2)$ and $(\pi/2, \pi]$.

65. Use the result of Exercise 64 to evaluate the following.
(a) $\operatorname{arcsec} \sqrt{2}$ (b) $\operatorname{arcsec} 1$

66. Define the inverse cosecant function by restricting the domain of the cosecant to the intervals $[-\pi/2, 0)$ and $(0, \pi/2]$.

In Exercises 67–72, prove the identity.

67. $\arcsin(-x) = -\arcsin x$

68. $\arctan(-x) = -\arctan x$

69. $\arccos(-x) = \pi - \arccos x$

70. $\arctan x + \arctan\dfrac{1}{x} = \dfrac{\pi}{2}, \quad x > 0$

71. $\arcsin x + \arccos x = \dfrac{\pi}{2}$

72. $\arcsin x = \arctan\dfrac{x}{\sqrt{1-x^2}}$

2.9 Applications of Trigonometry

In keeping with our twofold perspective of trigonometry, this section includes both right triangle applications and applications that emphasize the periodic nature of the trigonometric functions.

Applications Involving Right Triangles

In this section we will denote the three angles of a right triangle by the letters A, B, and C (where C is the right angle), and the lengths of the sides opposite these angles by the letters a, b, and c (where c is the hypotenuse).

In Section 2.3, we introduced the problem of solving a right triangle. We review that procedure in the first four examples.

EXAMPLE 1 *Solving a Right Triangle, Given One Acute Angle and One Side*

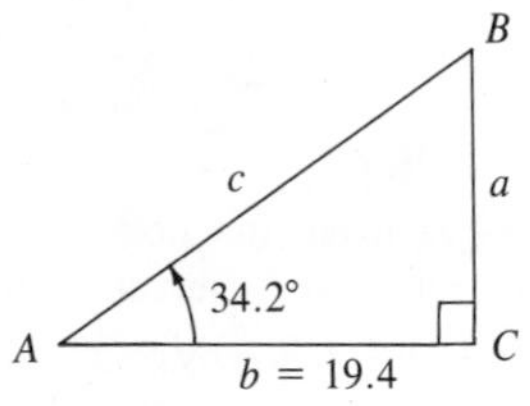

Solve the right triangle having $A = 34.2°$ and $b = 19.4$, as shown in Figure 2.78.

SOLUTION

Since $C = 90°$, it follows that $A + B = 90°$ and

$$B = 90° - 34.2° = 55.8°.$$

To solve for a we use the fact that

$$\tan A = \frac{\text{opp}}{\text{adj}} = \frac{a}{b} \implies a = b \tan A.$$

Thus, we have

$$a = 19.4 \tan 34.2° \approx 13.18.$$

Similarly, to solve for c we use the fact that

$$\cos A = \frac{\text{adj}}{\text{hyp}} = \frac{b}{c} \implies c = \frac{b}{\cos A}.$$

Thus, we have

$$c = \frac{19.4}{\cos 34.2°} \approx 23.46.$$

FIGURE 2.78

Applications of Trigonometry

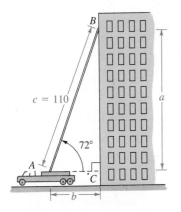

FIGURE 2.79

EXAMPLE 2 *Finding a Side of a Right Triangle*

A safety regulation states that the maximum angle of elevation for a rescue ladder is 72°. If a fire department's longest ladder is 110 feet, what is the maximum safe rescue height?

SOLUTION

A sketch is shown in Figure 2.79. From the equation

$$\sin A = \frac{a}{c}$$

it follows that

$$a = c \sin A = 110(\sin 72°) \approx 104.6 \text{ feet.}$$

In Example 2, we used the term **angle of elevation** to represent the angle from the horizontal upward to an object. For objects that lie below the horizontal, it is common to use the term **angle of depression,** as shown in Figure 2.80.

FIGURE 2.80

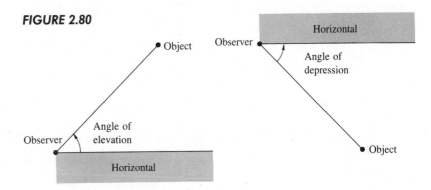

EXAMPLE 3 *Finding a Side of a Right Triangle*

From a point 200 feet from the base of a building, the angle of elevation to the *bottom* of a smokestack is 35°, while the angle of elevation to the *top* is 53°, as shown in Figure 2.81. Find the height, *s*, of the smokestack alone.

SOLUTION

Note from Figure 2.81 that this problem involves two right triangles. In the smaller right triangle, we use the fact that tan 35° = *a*/200 to conclude that the height of the building is

$$a = 200 \tan 35°.$$

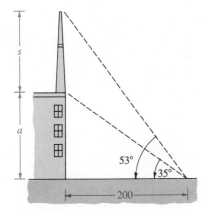

FIGURE 2.81

Now, from the larger right triangle, we use the equation

$$\tan 53° = \frac{a + s}{200}$$

to conclude that $a + s = 200 \tan 53°$. Hence, the height of the smokestack is

$$s = 200 \tan 53° - a = 200 \tan 53° - 200 \tan 35° \approx 125.4 \text{ feet.}$$

In Examples 1 through 3 we found the lengths of the sides of a right triangle, given an acute angle and the length of one of the sides. We can also find the angles of a right triangle given only the lengths of two sides, as demonstrated in Example 4.

EXAMPLE 4   *Finding an Acute Angle of a Right Triangle*

A swimming pool is 20 meters long and 12 meters wide. The bottom of the pool is slanted so that the water depth is 1.3 meters at the shallow end and 4 meters at the deep end, as shown in Figure 2.82. Find the angle of depression of the bottom of the pool.

SOLUTION

Using the tangent function, we see that

$$\tan A = \frac{\text{opp}}{\text{adj}} = \frac{2.7}{20} = 0.135.$$

Thus, the angle of depression is given by

$$A = \arctan 0.135 \approx 0.13419 \text{ (radians)} \approx 7.69°.$$

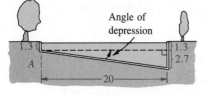

FIGURE 2.82

In surveying and navigation, directions are generally given in terms of **bearings.** A bearing measures the acute angle a path or line of sight makes with a fixed north-south line. For instance, in Figure 2.83(a), the bearing is S 35° E, meaning *35 degrees east of south*. Similarly, the bearings in parts (b) and (c) are N 80° W and N 45° E, respectively.

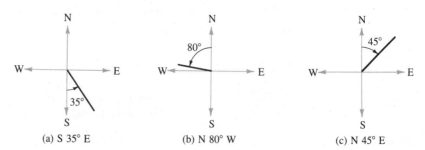

FIGURE 2.83   (a) S 35° E   (b) N 80° W   (c) N 45° E

EXAMPLE 5 *Finding Directions in Terms of Bearings*

A ship leaves port at noon and heads due west at 20 knots (nautical miles per hour). At 2 P.M., to avoid a storm, it changes course to N 54° W, as shown in Figure 2.84. Find the ship's bearing and distance from the port of departure at 3 P.M.

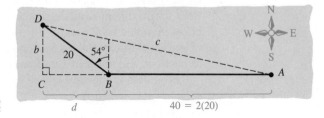

FIGURE 2.84

SOLUTION

In triangle BCD, we have $B = 90° - 54° = 36°$. The two sides of this triangle are determined as follows.

$$\sin B = \frac{b}{20}$$

$$b = 20 \sin 36°$$

$$\cos B = \frac{d}{20}$$

$$d = 20 \cos 36°$$

Now, in triangle ACD, we can determine angle A as follows.

$$\tan A = \frac{b}{d + 40}$$

$$= \frac{20 \sin 36°}{20 \cos 36° + 40} \approx 0.2092494$$

$$A \approx \arctan 0.2092494 \approx 0.2062732 \text{ radians} \approx 11.82°$$

The angle with the north-south line is $90° - 11.82° = 78.18°$. Therefore, the bearing of the ship is

  N 78.18° W.   *Bearing*

Finally, from triangle ACD, we have $\sin A = b/c$ which yields

$$c = \frac{b}{\sin A} = \frac{20 \sin 36°}{\sin(\arctan 0.2092494)}$$

$$c \approx 57.4 \text{ nautical miles.}$$ *Distance from port*

Harmonic Motion

The periodic nature of the trigonometric functions is useful for describing the motion of a point on an object that vibrates, oscillates, rotates, or is moved by wave motion.

For example, consider a ball that is bobbing up and down on the end of a spring, as shown in Figure 2.85. Suppose that 10 centimeters is the maximum distance the ball moves vertically upward or downward from its equilibrium (at rest) position. Suppose further that the time it takes for the ball to move from its maximum displacement above zero to its maximum displacement below zero and back again is $t = 4$ seconds. Assuming the ideal conditions of perfect elasticity and no friction or air resistance, the ball would continue to move up and down in a uniform and regular manner.

From this spring we can conclude that the period (time for one complete cycle) of the motion is

period = 4 seconds

and that its amplitude (maximum displacement from equilibrium) is

amplitude = 10 centimeters.

Motion of this nature can be described by a sine or cosine function, and is called **simple harmonic motion.**

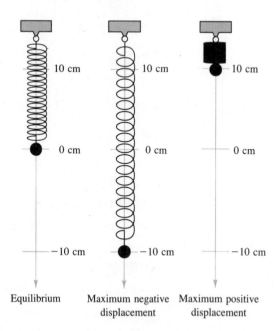

| Equilibrium | Maximum negative displacement | Maximum positive displacement |

FIGURE 2.85   Simple Harmonic Motion

Applications of Trigonometry

Definition of Simple Harmonic Motion

A point that moves on a coordinate line is said to be in **simple harmonic motion** if its distance d from the origin at time t is given by either

$$d = a \sin \omega t \quad \text{or} \quad d = a \cos \omega t$$

where a and ω are real numbers such that $\omega > 0$. The motion has **amplitude** $|a|$, **period** $2\pi/\omega$, and **frequency** $\omega/2\pi$.

EXAMPLE 6 *Simple Harmonic Motion*

Write the equation for the simple harmonic motion of the ball described in Figure 2.85, where the period is 4 seconds. What is the frequency for this motion?

SOLUTION

Since the spring is at equilibrium ($d = 0$) when $t = 0$, we use the equation

$$d = a \sin \omega t.$$

Moreover, since the maximum displacement from zero is 10 and the period is 4, we have

$$\text{amplitude} = |a| = 10$$

$$\text{period} = \frac{2\pi}{\omega} = 4 \implies \omega = \frac{\pi}{2}.$$

Consequently, the equation of motion is

$$d = 10 \sin \frac{\pi}{2}t.$$

Note that the choice of $a = 10$ or $a = -10$ depends on whether the ball initially moves up or down. The frequency is given by

$$\text{frequency} = \frac{\omega}{2\pi}$$

$$= \frac{\pi/2}{2\pi}$$

$$= \frac{1}{4} \text{ cycle per second.}$$

Trigonometry

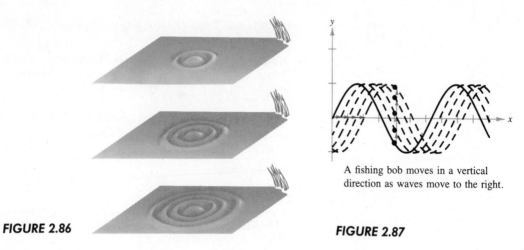

FIGURE 2.86

A fishing bob moves in a vertical direction as waves move to the right.

FIGURE 2.87

One clear illustration of the relation between sine waves and harmonic motion is seen in the wave motion resulting from dropping a stone into a calm pool of water. The waves move outward in roughly the shape of sine (or cosine) waves, as shown in Figure 2.86. As an example, suppose you are fishing and your fishing bob is attached so that it does not move horizontally. As the waves move outward from the dropped stone, your fishing bob will move up and down in simple harmonic motion, as shown in Figure 2.87.

EXAMPLE 7 *Simple Harmonic Motion*

Given the equation for simple harmonic motion

$$d = 6 \cos \frac{3\pi}{4} t$$

find (a) the maximum displacement, (b) the frequency, (c) the value of d when $t = 4$, and (d) the least positive value of t for which $d = 0$.

SOLUTION

The given equation has the form $d = a \cos \omega t$, with $a = 6$ and $\omega = 3\pi/4$.

(a) The maximum displacement (from the point of equilibrium) is given by the amplitude. Thus, the maximum displacement is 6.

(b) The frequency is

$$\text{frequency} = \frac{\omega}{2\pi} = \frac{3\pi/4}{2\pi} = \frac{3}{8} \text{ cycle per unit of time.}$$

(c) When $t = 4$, the position is given by

$$d = 6 \cos\left[\frac{3\pi}{4}(4)\right] = 6 \cos 3\pi = 6(-1) = -6.$$

(d) To find the least positive value of t for which $d = 0$, we solve the equation

$$d = 6 \cos \frac{3\pi}{4}t = 0$$

to obtain

$$\frac{3\pi}{4}t = \frac{\pi}{2}, \frac{3\pi}{2}, \frac{5\pi}{2}, \cdots$$

$$t = \frac{2}{3}, 2, \frac{10}{3}, \cdots$$

Thus, the least positive value of t is $t = 2/3$.

Many other physical phenomena can be characterized by wave motion. These include electromagnetic waves such as radio waves, television waves, and microwaves. Radio waves transmit sound in two different ways. For an AM station, the *amplitude* of the wave is modified to carry sound (AM stands for **amplitude modulation**). See Figure 2.88(a). An FM radio signal has its *frequency* modified in order to carry sound, hence the term **frequency modulation**. See Figure 2.88(b).

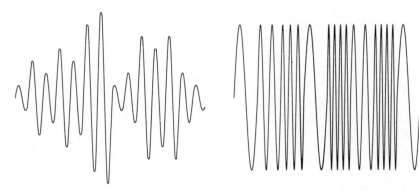

(a) AM: Amplitude Modulation   (b) FM: Frequency Modulation

Radio Waves

FIGURE 2.88

WARM UP

Evaluate the expressions. (Round your answers to two decimal places.)

1. $20 \sin 25°$

2. $42 \tan 62°$

3. $\arcsin 0.8723$

4. $\arctan 2.8703$

Solve for x. [Round your answers to two decimal places.]

5. $\cos 22° = \dfrac{x + 13 \sin 22°}{13 \sin 54°}$

6. $\tan 36° = \dfrac{x + 85 \tan 18°}{85}$

Find the amplitude and period of each function.

7. $f(x) = -4 \sin 2x$

8. $f(x) = \frac{1}{2} \sin \pi x$

9. $f(x) = 3 \cos 3\pi x$

10. $f(x) = 0.2 \cos \dfrac{x}{4}$

EXERCISES 2.9

In Exercises 1–10, solve the right triangle shown in the figure. (Round your answers to two decimal places.)

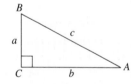

FIGURE FOR 1–10

1. $A = 20°, \quad b = 10$

2. $B = 54°, \quad c = 15$

3. $B = 71°, \quad b = 24$

4. $A = 8.4°, \quad a = 40.5$

5. $A = 12° \ 15', \quad c = 430.5$

6. $B = 65° \ 12', \quad a = 14.2$

7. $a = 6, \quad b = 10$

8. $a = 25, \quad c = 35$

9. $b = 16, \quad c = 52$

10. $b = 1.32, \quad c = 9.45$

11. A ladder of length 16 feet leans against the side of a house (see figure). Find the height h of the top of the ladder if the angle of elevation of the ladder is 74°.

12. The length of the shadow of a tree is 125 feet when the angle of elevation of the sun is 33° (see figure). Approximate the height h of the tree.

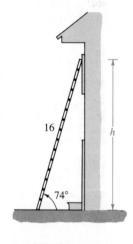

FIGURE FOR 11

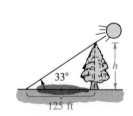

FIGURE FOR 12

13. An isosceles triangle has two angles of 52° (see figure). The base of the triangle is 4 inches. Find the altitude of the triangle.

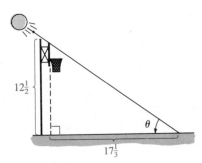

FIGURE FOR 13

14. The height of the top of an outdoor basketball backboard is $12\frac{1}{2}$ feet, and the backboard casts a shadow $17\frac{1}{3}$ feet long (see figure). Find the angle of elevation of the sun.

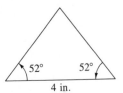

FIGURE FOR 14

15. An amateur radio operator erects a 75-foot vertical tower for his antenna. Find the angle of elevation to the top of the tower at a point on level ground 50 feet from the base.

16. Find the angle of depression from the top of a lighthouse 250 feet above water level to the water line of a ship 2 miles offshore.

17. A spacecraft is traveling in a circular orbit 100 miles above the surface of the earth (see figure). Find the angle of depression from the spacecraft to the horizon. Assume that the radius of the earth is 4000 miles.

FIGURE FOR 17

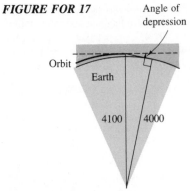

18. A sign on the roadway at the top of a mountain indicates that for the next 4 miles the grade is 12.5° (see figure). Find the change in elevation for a car descending the mountain.

FIGURE FOR 18

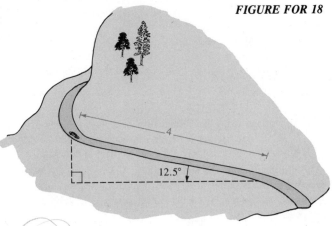

19. From a point 50 feet in front of a church, the angles of elevation to the base of the steeple and the top of the steeple are 35° and 47° 40′, respectively (see figure). Find the height of the steeple.

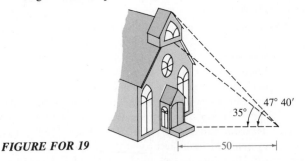

FIGURE FOR 19

20. From a point 100 feet in front of a public library, the angles of elevation to the base of a flagpole and to the top of the pole are 28° and 39° 45′, respectively. The flagpole is mounted on the front of the library's roof (see figure). Find the height of the pole.

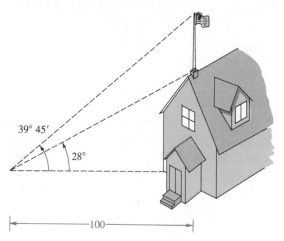

39° 45′

28°

100

FIGURE FOR 20

21. An airplane flying at 550 miles per hour has a bearing of N 52° E. After flying 1.5 hours, how far north and how far east has the plane traveled from its point of departure?

22. A ship leaves port at noon and has a bearing of S 27° W. If the ship is sailing at 20 knots, how many nautical miles south and how many nautical miles west has the ship traveled by 6 P.M.?

23. A ship is 45 miles east and 30 miles south of port. If the captain wants to travel directly to port, what bearing should be taken?

24. A plane is 120 miles north and 85 miles east of an airport. If the pilot wants to fly directly to the airport, what bearing should be taken?

25. A surveyor wishes to find the distance across a swamp (see figure). The bearing from A to B is N 32° W. The surveyor walks 50 yards from A, and at the point C the bearing to B is N 68° W. (a) Find the bearing from A to C. (b) Find the distance from A to B.

26. Two fire towers are 20 miles apart, tower A being due north of tower B. A fire is spotted from the towers, and its bearings from A and B are S 14° E and N 34° E (see figure). Find the distance d of the fire from the line segment AB. [*Hint:* Use the fact that d = 20/(cot 14° + cot 34°).]

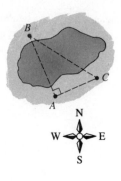

B

C

A

N

W   E

S

FIGURE FOR 25

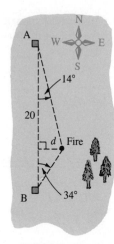

N

A   W   E

S

14°

20

d Fire

B   34°

FIGURE FOR 26

27. An observer in a lighthouse 300 feet above sea level spots two ships directly offshore. The angles of depression to the ships are 4° and 6.5° (see figure). How far apart are the ships?

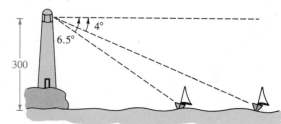

4°

6.5°

300

(Note: Angles are not drawn to scale.)

FIGURE FOR 27

28. A passenger in an airplane flying at 30,000 feet sees two towns directly to the left of the plane. The angles of depression to the towns are 28° and 55° (see figure). How far apart are the towns?

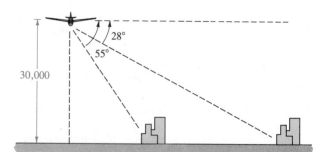

28°

55°

30,000

FIGURE FOR 28

29. A plane is observed approaching your home, and you assume that it is traveling at approximately 550 miles per hour. If the angle of elevation of the plane is 16° at one time and one minute later the angle is 57°, approximate the altitude.

30. In traveling across flat land, you notice a mountain directly in front of you. Its angle of elevation (to the peak) is 3.5°. After you drive 13 miles closer to the mountain, the angle of elevation is 9°. Approximate the height of the mountain.

31. A regular pentagon is inscribed in a circle of radius 25 inches. Find the length of the sides of the pentagon.

32. Repeat Exercise 31 using a hexagon.

33. Use the accompanying figure to find the distance y across the flat sides of the hexagonal nut as a function of r.

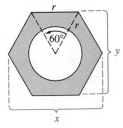

FIGURE FOR 33

34. The accompanying figure shows a circular sheet of diameter 25 cm, containing 12 equally spaced bolt holes. Determine the straight-line distance between the centers of the bolt heads.

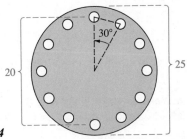

FIGURE FOR 34

In Exercises 35 and 36, find the lengths of all the unknown members of the given truss.

35.

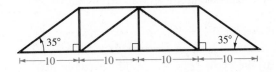

36.

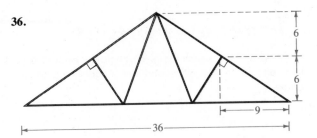

In Exercises 37–40, for the simple harmonic motion described by the given trigonometric function, find (a) the maximum displacement, (b) the frequency, and (c) the least positive value of t for which $d = 0$.

37. $d = 4 \cos 8\pi t$

38. $d = \frac{1}{2} \cos 20\pi t$

39. $d = \frac{1}{16} \sin 120\pi t$

40. $d = \frac{1}{64} \sin 792\pi t$

41. A point on the end of a tuning fork moves in simple harmonic motion described by $d = a \sin \omega t$. Find ω given that the tuning fork for middle C has a frequency of 264 vibrations per second.

42. A buoy oscillates in simple harmonic motion as waves move past. At a given time it is noted that the buoy moves a total of 3.5 feet from its low point to its high point, and that it returns to the high point every 10 seconds. Write an equation describing the motion of the buoy if, at $t = 0$, it is at its high point.

CHAPTER 2 REVIEW EXERCISES

In Exercises 1–4, sketch the given angle in standard position, and list one positive and one negative coterminal angle.

1. $\dfrac{11\pi}{4}$

2. $-405°$

3. $-110°$

4. $\dfrac{2\pi}{9}$

In Exercises 5–8, convert the angle measurement to decimal form. Round each answer to two decimal places.

5. $135°\ 16'\ 45''$

6. $-234°\ 50''$

7. $5°\ 22'\ 53''$

8. $280°\ 8'\ 50''$

In Exercises 9–12, convert the angle measurement to D° M′ S″ form.

9. $135.27°$

10. $25.1°$

11. $-85.15°$

12. $-327.85°$

In Exercises 13–16, convert the angle measurement from radians to degrees. Round each answer to two decimal places.

13. $\dfrac{5\pi}{7}$

14. $-\dfrac{3\pi}{5}$

15. -3.5

16. 1.75

In Exercises 17–20, convert the angle measurement from degrees to radians. Round each answer to four decimal places.

17. $480°$

18. $-16.5°$

19. $-33°\ 45'$

20. $84°\ 15'$

In Exercises 21–24, find the reference angle for the given angle.

21. $-\dfrac{6\pi}{5}$

22. $640°$

23. $252°$

24. $\dfrac{17\pi}{3}$

In Exercises 25–28, find the six trigonometric functions of the angle θ (in standard position) whose terminal side passes through the given point.

25. $(-7, 2)$

26. $(4, -8)$

27. $(-4, -6)$

28. $\left(\tfrac{2}{3}, \tfrac{5}{2}\right)$

In Exercises 29–32, use a right triangle to find the remaining five trigonometric functions of θ.

29. $\sec\theta = \tfrac{6}{5}$, $\tan\theta < 0$

30. $\tan\theta = -\tfrac{12}{5}$, $\sin\theta > 0$

31. $\sin\theta = \tfrac{3}{8}$, $\cos\theta < 0$

32. $\cos\theta = -\tfrac{2}{5}$, $\sin\theta > 0$

In Exercises 33–36, evaluate the given trigonometric functions without the use of a calculator.

33. $\sin\dfrac{5\pi}{3}$

34. $\cot\left(-\dfrac{5\pi}{6}\right)$

35. $\cos 495°$

36. $\csc\dfrac{3\pi}{2}$

In Exercises 37–40, use a calculator to evaluate the given trigonometric functions. Round each answer to two decimal places.

37. $\tan 33°$

38. $\csc 105°$

39. $\sec\dfrac{12\pi}{5}$

40. $\sin\left(-\dfrac{\pi}{9}\right)$

In Exercises 41–44, find two values of θ in degrees ($0° \le \theta < 360°$) and in radians ($0 \le \theta < 2\pi$) without using a calculator.

41. $\cos\theta = -\dfrac{\sqrt{2}}{2}$

42. $\sec\theta$ is undefined

43. $\csc\theta = -2$

44. $\tan\theta = \dfrac{\sqrt{3}}{3}$

In Exercises 45–48, find two values of θ in degrees ($0° \le \theta < 360°$) and in radians ($0 \le \theta < 2\pi$) by using a calculator.

45. $\sin\theta = 0.8387$

46. $\cot\theta = -1.5399$

47. $\sec\theta = -1.0353$

48. $\csc\theta = 11.4737$

In Exercises 49–52, use a right triangle to write an algebraic expression for the given expression.

49. $\sec[\arcsin(x - 1)]$

50. $\tan\left(\arccos\dfrac{x}{2}\right)$

51. $\sin\left(\arccos\dfrac{x^2}{4 - x^2}\right)$

52. $\csc(\arcsin 10x)$

In Exercises 53–70, sketch the graph of the given function.

53. $f(x) = 5 \sin \dfrac{2x}{5}$

54. $f(x) = 8 \cos\left(-\dfrac{x}{4}\right)$

55. $f(x) = -\dfrac{1}{4} \cos \dfrac{\pi x}{4}$

56. $f(x) = -\tan \dfrac{\pi x}{4}$

57. $g(t) = \dfrac{5}{2} \sin(\pi - \pi t)$

58. $g(t) = \sec\left(t - \dfrac{\pi}{4}\right)$

59. $h(t) = \csc\left(3t - \dfrac{\pi}{2}\right)$

60. $h(t) = 3 \csc\left(2t + \dfrac{\pi}{4}\right)$

61. $f(\theta) = \cot \dfrac{\pi \theta}{8}$

62. $E(t) = 110 \cos\left(120\pi t - \dfrac{\pi}{3}\right)$

63. $f(x) = \dfrac{x}{4} - \sin x$

64. $g(x) = 3\left(\sin \dfrac{\pi x}{3} + 1\right)$

65. $h(\theta) = \theta \sin \pi \theta$

66. $f(t) = 2.5 e^{-t/4} \sin 2\pi t$

67. $f(x) = \arcsin \dfrac{x}{2}$

68. $f(x) = \arccos(x - \pi)$

69. $f(x) = \dfrac{\pi}{2} + \arctan x$

70. $f(x) = 2 \arccos x$

71. An observer 2.5 miles from the launch pad of a space shuttle measures the angle of elevation to the base of the vehicle to be $28°$ soon after liftoff (see figure). How high is the shuttle at that instant? Assume that the shuttle is still moving vertically.

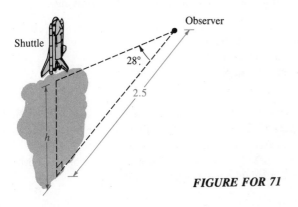

Observer

Shuttle

$28°$

2.5

FIGURE FOR 71

72. From city A to city B, a plane flies 650 miles at a bearing that is N 48° E. From city B to city C the plane flies 810 miles at a bearing of S 65° E. Find the distance from A to C and the bearing from A to C.

73. A train travels 2.5 miles (horizontally) on a straight track with a grade of 1° 10′ (see figure). What is the vertical rise of the train in that distance?

1° 10′

2.5

FIGURE FOR 73

74. In calculus it is shown that the sine and cosine functions can be represented by the infinite sums shown below, where x is in radians. Use the first four terms of the sum to approximate the given functional values and compare the result with that given by a calculator.

$$\sin x = x - \frac{x^3}{3!} + \frac{x^5}{5!} - \frac{x^7}{7!} + \cdots + \frac{(-1)^{n-1}x^{2n-1}}{(2n-1)!} + \cdots$$

$$\cos x = 1 - \frac{x^2}{2!} + \frac{x^4}{4!} - \frac{x^6}{6!} + \cdots + \frac{(-1)^{n-1}x^{2n-2}}{(2n-2)!} + \cdots$$

(a) $\sin 1$

(b) $\cos 1$

(c) $\sin \dfrac{1}{2}$

(d) $\cos(-1)$

(e) $\cos \dfrac{\pi}{4}$

(f) $\sin \dfrac{\pi}{6}$

C H A P T E R   3

Analytic Trigonometry

3.1 Applications of Fundamental Identities

In Chapter 2, we studied the basic definitions, properties, graphs, and applications of the individual trigonometric functions. In this chapter, we will study algebraic combinations of these functions. In particular, we will show how to use the fundamental identities to perform the following.

1. Simplify trigonometric expressions.
2. Develop additional trigonometric identities.
3. Solve trigonometric equations.

In our study, we will make use of many algebraic skills such as finding special products, factoring, performing operations with fractional expressions, rationalizing denominators, and solving equations.

If you have not already done so, we suggest that you memorize the following list of fundamental trigonometric identities.

Fundamental Identities

Reciprocal Identities

$$\sin x = \frac{1}{\csc x} \qquad \sec x = \frac{1}{\cos x} \qquad \tan x = \frac{1}{\cot x}$$

$$\csc x = \frac{1}{\sin x} \qquad \cos x = \frac{1}{\sec x} \qquad \cot x = \frac{1}{\tan x}$$

Tangent and Cotangent Identities

$$\tan x = \frac{\sin x}{\cos x} \qquad \cot x = \frac{\cos x}{\sin x}$$

Pythagorean Identities

$$\sin^2 x + \cos^2 x = 1 \qquad 1 + \tan^2 x = \sec^2 x \qquad 1 + \cot^2 x = \csc^2 x$$

Cofunction Identities

$$\sin\left(\frac{\pi}{2} - x\right) = \cos x \qquad \sec\left(\frac{\pi}{2} - x\right) = \csc x \qquad \tan\left(\frac{\pi}{2} - x\right) = \cot x$$

$$\csc\left(\frac{\pi}{2} - x\right) = \sec x \qquad \cos\left(\frac{\pi}{2} - x\right) = \sin x \qquad \cot\left(\frac{\pi}{2} - x\right) = \tan x$$

Negative Angle Identities

$$\sin(-x) = -\sin x \qquad \sec(-x) = \sec x \qquad \tan(-x) = -\tan x$$
$$\csc(-x) = -\csc x \qquad \cos(-x) = \cos x \qquad \cot(-x) = -\cot x$$

Remark: Pythagorean identities are sometimes used in radical form such as

$$\sin x = \pm\sqrt{1 - \cos^2 x} \qquad \text{or} \qquad \tan x = \pm\sqrt{\sec^2 x - 1}$$

where the sign depends on the choice of x.

EXAMPLE 1   *Using Identities to Evaluate a Function*

Use the fundamental identities to find $\csc x$, given that

$$\tan x = -6 \qquad \text{and} \qquad \sin x > 0.$$

SOLUTION

Since $\tan x < 0$ and $\sin x > 0$, it follows that x lies in Quadrant II. Using a reciprocal identity, we have

$$\cot x = \frac{1}{\tan x} = \frac{1}{-6}.$$

Analytic Trigonometry

From a Pythagorean identity, we obtain

$$\csc^2 x = 1 + \cot^2 x = 1 + \left(\frac{-1}{6}\right)^2 = \frac{37}{36}.$$

Finally, because $\csc x$ is positive when x is in Quadrant II, we choose the positive root and conclude that

$$\csc x = \frac{\sqrt{37}}{6}.$$

EXAMPLE 2   *Using Identities to Evaluate a Function*

Use the fundamental identities to find $\sin x$ and $\tan x$, given that

$$\sec x = -\frac{3}{2} \quad \text{and} \quad \tan x > 0.$$

SOLUTION

Because $\sec x < 0$ and $\tan x > 0$, it follows that x lies in Quadrant III. Moreover, by a reciprocal identity, we have

$$\cos x = \frac{1}{\sec x} = \frac{1}{-3/2} = -\frac{2}{3}.$$

Thus, by a Pythagorean identity, we obtain

$$\sin^2 x = 1 - \cos^2 x = 1 - \left(-\frac{2}{3}\right)^2 = 1 - \frac{4}{9} = \frac{5}{9}.$$

Because $\sin x$ is negative when x is in Quadrant III, we choose the negative root and obtain

$$\sin x = -\frac{\sqrt{5}}{3}.$$

Furthermore,

$$\tan x = \frac{\sin x}{\cos x} = \frac{-\sqrt{5}/3}{-2/3} = \frac{\sqrt{5}}{2}.$$

In the next five examples, we use algebraic techniques and the fundamental identities to factor and/or simplify trigonometric *expressions* such as

$$\cot x - \cos x \sin x, \quad \frac{1 - \sin x}{\cos^2 x}, \quad \text{and} \quad \frac{\tan x}{\sec x - 1}.$$

EXAMPLE 3 *Simplifying a Trigonometric Expression*

Simplify the expression $\sin x \cos^2 x - \sin x$.

SOLUTION

In this case, we first factor out a common monomial factor and then use a fundamental identity.

$$
\begin{aligned}
\sin x \cos^2 x - \sin x &= \sin x(\cos^2 x - 1) &&\textit{Monomial factor}\\
&= -\sin x(1 - \cos^2 x)\\
&= -\sin x(\sin^2 x) &&\textit{Identity}\\
&= -\sin^3 x
\end{aligned}
$$

EXAMPLE 4 *Factoring Trigonometric Expressions*

Factor the following expressions.

(a) $\sec^2 \theta - 1$

(b) $4 \tan^2 \theta + \tan \theta - 3$

SOLUTION

(a) Here we have the difference of two squares, which factors as

$$\sec^2 \theta - 1 = (\sec \theta - 1)(\sec \theta + 1).$$

(b) This expression has the polynomial form, $ax^2 + bx + c$, and it factors as

$$4 \tan^2 \theta + \tan \theta - 3 = (4 \tan \theta - 3)(\tan \theta + 1).$$

On occasion, factoring or simplifying can best be done by first rewriting the expression in terms of just *one* trigonometric function or in terms of *sine and cosine alone*. This is demonstrated in the next three examples.

EXAMPLE 5 *Factoring a Trigonometric Expression*

Factor the expression $\csc^2 x - \cot x - 3$.

SOLUTION

As given, this expression cannot be factored, so we use the identity $\csc^2 x = 1 + \cot^2 x$ to rewrite the expression in terms of the cotangent alone. We then factor to obtain

$$
\begin{aligned}
\csc^2 x - \cot x - 3 &= (1 + \cot^2 x) - \cot x - 3 && \text{\textit{Identity}} \\
&= \cot^2 x - \cot x - 2 && \text{\textit{Combine terms}} \\
&= (\cot x - 2)(\cot x + 1). && \text{\textit{Factor}} \quad \blacksquare
\end{aligned}
$$

EXAMPLE 6 *Simplifying a Trigonometric Expression*

Simplify the expression $\sin t + \cot t \cos t$.

SOLUTION

Since this expression is not factorable as given, we convert all terms to sines and cosines to see what can be done.

$$
\begin{aligned}
\sin t + \cot t \cos t &= \sin t + \left(\frac{\cos t}{\sin t}\right) \cos t \\
&= \frac{\sin^2 t + \cos^2 t}{\sin t} && \text{\textit{Add fractions}} \\
&= \frac{1}{\sin t} && \text{\textit{Identity}} \\
&= \csc t. && \text{\textit{Identity}} \quad \blacksquare
\end{aligned}
$$

EXAMPLE 7 *Simplifying a Trigonometric Expression*

Simplify the function $f(x) = (\sin x + \cos x - 1)(\sin x + \cos x + 1)$.

SOLUTION

$$
\begin{aligned}
f(x) &= (\sin x + \cos x - 1)(\sin x + \cos x + 1) \\
&= [(\sin x + \cos x) - 1][(\sin x + \cos x) + 1] \\
&= (\sin x + \cos x)^2 - 1 = \sin^2 x + 2 \sin x \cos x + \cos^2 x - 1 \\
&= 2 \sin x \cos x + 1 - 1 = 2 \sin x \cos x
\end{aligned}
$$

$\blacksquare$

Applications of Fundamental Identities

EXAMPLE 8 *Combining Fractional Expressions*

Perform the indicated addition and simplify the result.

$$\frac{\sin\theta}{1+\cos\theta} + \frac{\cos\theta}{\sin\theta}$$

SOLUTION

Using the definition of addition of fractions

$$\frac{a}{b} + \frac{c}{d} = \frac{ad+bc}{bd}$$

we can write

$$\frac{\sin\theta}{1+\cos\theta} + \frac{\cos\theta}{\sin\theta} = \frac{\sin\theta(\sin\theta)+\cos\theta(1+\cos\theta)}{(1+\cos\theta)\sin\theta}$$

$$= \frac{\sin^2\theta + \cos^2\theta + \cos\theta}{(1+\cos\theta)\sin\theta}$$

$$= \frac{1+\cos\theta}{(1+\cos\theta)\sin\theta} \qquad \textit{Identity}$$

$$= \frac{1}{\sin\theta} \qquad\qquad \textit{Reduce}$$

$$= \csc\theta. \qquad\qquad \textit{Identity}$$

The last two examples of this section involve techniques for rewriting expressions into forms that are useful in calculus.

EXAMPLE 9 *Rewriting a Trigonometric Expression*

Rewrite the following expression so that it is *not* in fractional form.

$$\frac{1}{1+\sin x}$$

SOLUTION

From the Pythagorean identity

$$\cos^2 x = 1 - \sin^2 x = (1 - \sin x)(1 + \sin x)$$

we can see that by multiplying both the numerator and the denominator by $(1 - \sin x)$ we will obtain a monomial denominator. That is,

$$
\begin{aligned}
\frac{1}{1 + \sin x} &= \frac{1}{1 + \sin x} \cdot \frac{1 - \sin x}{1 - \sin x} \\[6pt]
&= \frac{1 - \sin x}{1 - \sin^2 x} && \textit{Multiply} \\[6pt]
&= \frac{1 - \sin x}{\cos^2 x} && \textit{Identity} \\[6pt]
&= \frac{1}{\cos^2 x} - \frac{\sin x}{\cos^2 x} \\[6pt]
&= \frac{1}{\cos^2 x} - \frac{\sin x}{\cos x} \cdot \frac{1}{\cos x} \\[6pt]
&= \sec^2 x - \tan x \sec x. && \textit{Identity}
\end{aligned}
$$

EXAMPLE 10 *Trigonometric Substitution*

Use the substitution $x = 2 \tan \theta$, $0 < \theta < \pi/2$ to express $\sqrt{4 + x^2}$ as a trigonometric function of θ.

SOLUTION

Letting $x = 2 \tan \theta$, we have

$$
\begin{aligned}
\sqrt{4 + x^2} &= \sqrt{4 + (2 \tan \theta)^2} \\
&= \sqrt{4(1 + \tan^2 \theta)} \\
&= \sqrt{4 \sec^2 \theta} && \textit{Pythagorean identity} \\
&= 2 \sec \theta. && \textit{sec } \theta > 0 \text{ for } 0 < \theta < \frac{\pi}{2}
\end{aligned}
$$

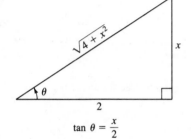

$$\tan \theta = \frac{x}{2}$$

FIGURE 3.1

Figure 3.1 shows the right triangle illustration of the trigonometric substitution in Example 10. For $0 < \theta < \pi/2$, we have opp $= x$, adj $= 2$, and hyp $= \sqrt{4 + x^2}$. Thus, we can write

$$\sec \theta = \frac{\sqrt{4 + x^2}}{2} \quad \Longrightarrow \quad \sqrt{4 + x^2} = 2 \sec \theta.$$

Exercises

WARM UP

Use a right triangle to find the other five trigonometric functions of the acute angle θ.

1. $\tan \theta = \frac{3}{2}$

2. $\sec \theta = 3$

Determine the exact value of all six trigonometric functions of θ. Assume that the given point is on the terminal side of an angle θ, in standard position.

3. $(7, -3)$

4. $(-10, 5)$

Simplify the expressions.

5. $\sqrt{1 - \left(\dfrac{\sqrt{3}}{2}\right)^2}$

6. $\sqrt{\left(\dfrac{3}{4}\right)^2 + 1}$

7. $\sqrt{1 + \left(\dfrac{3}{8}\right)^2}$

8. $\sqrt{1 - \left(\dfrac{\sqrt{5}}{3}\right)^2}$

Perform the indicated operations and simplify.

9. $\dfrac{4}{1 + x} + \dfrac{x}{4}$

10. $\dfrac{3}{1 - x} - \dfrac{5}{1 + x}$

EXERCISES 3.1

In Exercises 1–14, use the fundamental identities to find the values of the other four trigonometric functions.

1. $\sin x = \dfrac{1}{2}, \quad \cos x = \dfrac{\sqrt{3}}{2}$

2. $\tan x = \dfrac{\sqrt{3}}{3}, \quad \cos x = -\dfrac{\sqrt{3}}{2}$

3. $\sec \theta = \sqrt{2}, \quad \sin \theta = -\dfrac{\sqrt{2}}{2}$

4. $\csc \theta = \dfrac{5}{3}, \quad \tan \theta = \dfrac{3}{4}$

5. $\tan x = \dfrac{5}{12}, \quad \sec x = -\dfrac{13}{12}$

6. $\cot \phi = -3, \quad \sin \phi = \dfrac{\sqrt{10}}{10}$

7. $\sec \phi = -1, \quad \sin \phi = 0$

8. $\cos\left(\dfrac{\pi}{2} - x\right) = \dfrac{3}{5}, \quad \cos x = \dfrac{4}{5}$

9. $\sin(-x) = -\dfrac{2}{3}, \quad \tan x = -\dfrac{2\sqrt{5}}{5}$

10. $\csc x = 5, \quad \cos x > 0$

11. $\tan \theta = 2, \quad \sin \theta < 0$

12. $\sec \theta = -3, \quad \tan \theta < 0$

13. $\sin \theta = -1, \quad \cot \theta = 0$

14. $\tan \theta$ is undefined, $\quad \sin \theta > 0$

In Exercises 15–20, match the trigonometric expression with one of the following.

(a) -1 (b) $\cos x$ (c) $\cot x$

(d) 1 (e) $-\tan x$ (f) $\sin x$

15. $\sec x \cos x$

16. $\dfrac{\sin(-x)}{\cos(-x)}$

17. $\tan^2 x - \sec^2 x$

18. $\dfrac{1 - \cos^2 x}{\sin x}$

19. $\cot x \sin x$

20. $\dfrac{\sin[(\pi/2) - x]}{\cos[(\pi/2) - x]}$

Analytic Trigonometry

In Exercises 21–26, match the trigonometric expression with one of the following.

(a) $\csc x$ (b) $\tan x$ (c) $\sin^2 x$
(d) $\sin x \tan x$ (e) $\sec^2 x$ (f) $\sec^2 x + \tan^2 x$

21. $\sin x \sec x$ **22.** $\cos^2 x(\sec^2 x - 1)$

23. $\dfrac{\sec^2 x - 1}{\sin^2 x}$ **24.** $\cot x \sec x$

25. $\sec^4 x - \tan^4 x$ **26.** $\dfrac{\cos^2[(\pi/2) - x]}{\cos x}$

In Exercises 27–40, use the fundamental identities to simplify the given expression.

27. $\tan \phi \csc \phi$ **28.** $\sin \phi(\csc \phi - \sin \phi)$

29. $\cos \beta \tan \beta$ **30.** $\sec \alpha \dfrac{\sin \alpha}{\tan \alpha}$

31. $\dfrac{\cot x}{\csc x}$ **32.** $\dfrac{\csc \theta}{\sec \theta}$

33. $\sec^2 x(1 - \sin^2 x)$ **34.** $\dfrac{1}{\tan^2 x + 1}$

35. $\dfrac{\sin(-x)}{\cos x}$ **36.** $\dfrac{\tan^2 \theta}{\sec^2 \theta}$

37. $\cos\left(\dfrac{\pi}{2} - x\right)\sec x$ **38.** $\cot\left(\dfrac{\pi}{2} - x\right)\cos x$

39. $\dfrac{\cos^2 y}{1 - \sin y}$ **40.** $\cos t(1 + \tan^2 t)$

In Exercises 41–48, factor each expression and use the fundamental identities to simplify the result.

41. $\tan^2 x - \tan^2 x \sin^2 x$ **42.** $\sec^2 x \tan^2 x + \sec^2 x$

43. $\sin^2 x \sec^2 x - \sin^2 x$ **44.** $\dfrac{\sec^2 x - 1}{\sec x - 1}$

45. $\tan^4 x + 2 \tan^2 x + 1$ **46.** $1 - 2 \cos^2 x + \cos^4 x$

47. $\sin^4 x - \cos^4 x$ **48.** $\csc^3 x - \csc^2 x - \csc x + 1$

In Exercises 49–52, perform the multiplication and use the fundamental identities to simplify the result.

49. $(\sin x + \cos x)^2$
50. $(\cot x + \csc x)(\cot x - \csc x)$
51. $(\sec x + 1)(\sec x - 1)$
52. $(3 - 3 \sin x)(3 + 3 \sin x)$

In Exercises 53–56, perform the addition and use the fundamental identities to simplify the result.

53. $\dfrac{1}{1 + \cos x} + \dfrac{1}{1 - \cos x}$ **54.** $\dfrac{1}{\sec x + 1} - \dfrac{1}{\sec x - 1}$

55. $\dfrac{\cos x}{1 + \sin x} + \dfrac{1 + \sin x}{\cos x}$ **56.** $\tan x - \dfrac{\sec^2 x}{\tan x}$

In Exercises 57–60, rewrite the given expression so that it is *not* in fractional form.

57. $\dfrac{\sin^2 y}{1 - \cos y}$ **58.** $\dfrac{5}{\tan x + \sec x}$

59. $\dfrac{3}{\sec x - \tan x}$ **60.** $\dfrac{\tan^2 x}{\csc x + 1}$

In Exercises 61–70, use the given trigonometric substitution to write the algebraic expression as a trigonometric expression involving θ, where $0 < \theta < \pi/2$.

61. $\sqrt{25 - x^2}$, $x = 5 \sin \theta$
62. $\sqrt{16 - 4x^2}$, $x = 2 \sin \theta$
63. $\sqrt{x^2 - 9}$, $x = 3 \sec \theta$
64. $\sqrt{x^2 - 4}$, $x = 2 \sec \theta$
65. $\sqrt{x^2 + 25}$, $x = 5 \tan \theta$
66. $\sqrt{x^2 + 100}$, $x = 10 \tan \theta$
67. $\sqrt{1 - (x - 1)^2}$, $x - 1 = \sin \theta$
68. $\sqrt{1 - e^{2x}}$, $e^x = \sin \theta$
69. $\sqrt{(9 + x^2)^3}$, $x = 3 \tan \theta$
70. $\sqrt{(x^2 - 16)^3}$, $x = 4 \sec \theta$

In Exercises 71 and 72, determine the values of θ, $0 \leq \theta < 2\pi$, for which the equation is true.

71. $\sec \theta = \sqrt{1 + \tan^2 \theta}$ **72.** $\cos \theta = -\sqrt{1 - \sin^2 \theta}$

In Exercises 73 and 74, rewrite the given expression as a single logarithm and simplify.

73. $\ln|\cos \theta| - \ln|\sin \theta|$ **74.** $\ln|\cot t| + \ln(1 + \tan^2 t)$

In Exercises 75–78, determine whether the statement is true or false, and give a reason for your answer.

75. $\dfrac{\sin k\theta}{\cos k\theta} = \tan \theta$, k is constant

76. $5 \sec \theta = \dfrac{1}{5 \cos \theta}$

77. $\sin \theta \csc \theta = 1$ **78.** $\sin \theta \csc \phi = 1$

In Exercises 79–82, use a calculator to demonstrate that the identity is true for the given values of θ.

79. $\csc^2 \theta - \cot^2 \theta = 1$

   (a) $\theta = 132°$

   (b) $\theta = \dfrac{2\pi}{7}$

80. $\tan^2 \theta + 1 = \sec^2 \theta$

   (a) $\theta = 346°$

   (b) $\theta = 3.1$

81. $\cos\left(\dfrac{\pi}{2} - \theta\right) = \sin \theta$

   (a) $\theta = 80°$

   (b) $\theta = 0.8$

82. $\sin(-\theta) = -\sin \theta$

   (a) $\theta = 250°$

   (b) $\theta = \frac{1}{2}$

83. Express each of the other trigonometric functions of θ in terms of $\sin \theta$.

84. Express each of the other trigonometric functions of θ in terms of $\cos \theta$.

3.2 *Verifying Trigonometric Identities*

In the previous section, we showed how to rewrite trigonometric expressions in equivalent forms. In this section, we will demonstrate methods for proving (or verifying) trigonometric identities. And in Section 3.3, we will show how to solve trigonometric equations. The key to verifying identities and solving equations is the ability to use the fundamental identities and the rules of algebra to rewrite trigonometric expressions.

Before going on, let's review some distinctions among trigonometric expressions, equations, and identities. An *expression* has no equal sign. It is merely a combination of functions. When simplifying expressions, we use an equal sign only to indicate the equivalence of the original expression and the new form. An *equation* is a statement containing an equal sign that is true for a specific set of values. In this sense, it is really a *conditional* equation. For example, the equation

$$\sin x = -1$$

is true only for $x = (3\pi/2) \pm 2n\pi$. Hence, it is a conditional equation. On the other hand, an equation that is true for all real values in the domain of the variable is called an *identity*. For example, the familiar equation

$$\sin^2 x = 1 - \cos^2 x$$

is true for all real x; hence, it is an identity.

Though there are similarities, proving that a trigonometric equation is an identity is quite different from solving an equation. There is no well-defined set of rules to follow in verifying trigonometric identities, and the process is best learned by practice. However, the following guidelines should be helpful.

Guidelines for Verifying Trigonometric Identities

1. Work with one side of the equation at a time. It is often better to work with the more complicated side.
2. Look for opportunities to factor an expression, add fractions, square a binomial, or create a monomial denominator.
3. Look for opportunities to use the fundamental identities. Note which functions are in the final expression you want. Sines and cosines pair up well, as do secants and tangents, and cosecants and cotangents.
4. If the preceding guidelines do not help, try converting all terms to sines and cosines.
5. Do not just sit and stare at the problem. Try something! Even paths that lead to dead ends can give you insights.

Note how we use these guidelines in the examples in this section.

EXAMPLE 1 *Verifying a Trigonometric Identity*

Verify the identity

$$\frac{\sec^2 \theta - 1}{\sec^2 \theta} = \sin^2 \theta.$$

SOLUTION

Because the left side is more complicated, we will work with it.

$$\frac{\sec^2 \theta - 1}{\sec^2 \theta} = \frac{(\tan^2 \theta + 1) - 1}{\sec^2 \theta} \qquad \textit{Identity}$$

$$= \frac{\tan^2 \theta}{\sec^2 \theta} \qquad \textit{Simplify}$$

$$= \tan^2 \theta (\cos^2 \theta) \qquad \textit{Identity}$$

$$= \frac{\sin^2 \theta}{\cos^2 \theta} (\cos^2 \theta) \qquad \textit{Identity}$$

$$= \sin^2 \theta \qquad \textit{Simplify}$$

ALTERNATIVE SOLUTION

Sometimes it is helpful to separate a fraction into two parts. In this case, we have

$$\frac{\sec^2 \theta - 1}{\sec^2 \theta} = \frac{\sec^2 \theta}{\sec^2 \theta} - \frac{1}{\sec^2 \theta} \qquad \textit{Separate fractions}$$

$$= 1 - \cos^2 \theta \qquad \textit{Identity}$$

$$= \sin^2 \theta. \qquad \textit{Identity}$$

Remark: As you can see from Example 1, there can be more than one way to verify an identity. Your method may differ from that used by your instructor or fellow students. Here is a good chance to be creative and establish your own style, but try to be as efficient as possible.

EXAMPLE 2 Combining Fractions before Using Identities

Verify the identity

$$\frac{1}{1 - \sin \alpha} + \frac{1}{1 + \sin \alpha} = 2 \sec^2 \alpha.$$

SOLUTION

Let's add the two fractions and see where we can go from there.

$$\frac{1}{1 - \sin \alpha} + \frac{1}{1 + \sin \alpha} = \frac{1 + \sin \alpha + 1 - \sin \alpha}{(1 - \sin \alpha)(1 + \sin \alpha)}$$

$$= \frac{2}{1 - \sin^2 \alpha} = \frac{2}{\cos^2 \alpha} = 2 \sec^2 \alpha$$

EXAMPLE 3 Verifying a Trigonometric Identity

Verify the identity

$$(\tan^2 x + 1)(\cos^2 x - 1) = -\tan^2 x.$$

SOLUTION

By multiplying the factors on the left side, we obtain

$$(\tan^2 x + 1)(\cos^2 x - 1) = \tan^2 x(\cos^2 x) - \tan^2 x + \cos^2 x - 1$$

$$= \frac{\sin^2 x}{\cos^2 x}(\cos^2 x) + \cos^2 x - \tan^2 x - 1$$

$$= (\sin^2 x + \cos^2 x) - \tan^2 x - 1$$

$$= 1 - \tan^2 x - 1 = -\tan^2 x.$$

ALTERNATIVE SOLUTION

By applying identities before multiplying, we obtain

$$(\tan^2 x + 1)(\cos^2 x - 1) = \sec^2 x(-\sin^2 x)$$

$$= -\frac{\sin^2 x}{\cos^2 x} = -\tan^2 x.$$

EXAMPLE 4 *Converting to Sines and Cosines*

Verify the identity

$$\tan x + \cot x = \sec x \csc x.$$

SOLUTION

In this case there appear to be no fractions to add, no products to find, and no opportunity to use one of the Pythagorean identities. Hence, we try converting the left side into sines and cosines to see what happens.

$$\tan x + \cot x = \frac{\sin x}{\cos x} + \frac{\cos x}{\sin x}$$

$$= \frac{\sin^2 x + \cos^2 x}{\cos x \sin x} \qquad \textit{Add fractions}$$

$$= \frac{1}{\cos x \sin x} \qquad \textit{sin}^2\ x + \cos^2\ x = 1$$

$$= \frac{1}{\cos x} \cdot \frac{1}{\sin x} \qquad \textit{Product of fractions}$$

$$= \sec x \csc x \qquad \textit{Reciprocal identities}$$

Recall from algebra that *rationalizing the denominator* is, on occasion, a powerful simplification technique. A related form of this technique works for simplifying trigonometric expressions as well.

EXAMPLE 5 *Verifying a Trigonometric Identity*

Verify the identity

$$\sec y + \tan y = \frac{\cos y}{1 - \sin y}.$$

SOLUTION

Let's work with the *right* side. Note that we can create a monomial denominator by multiplying the numerator and denominator by $(1 + \sin y)$.

$$\frac{\cos y}{1 - \sin y} = \frac{\cos y}{1 - \sin y}\left(\frac{1 + \sin y}{1 + \sin y}\right)$$

$$= \frac{\cos y + \cos y \sin y}{1 - \sin^2 y}$$

$$= \frac{\cos y + \cos y \sin y}{\cos^2 y} = \frac{\cos y}{\cos^2 y} + \frac{\cos y \sin y}{\cos^2 y}$$

$$= \frac{1}{\cos y} + \frac{\sin y}{\cos y} = \sec y + \tan y$$

So far in this section, we have been verifying trigonometric identities by working with one side of the equation and converting to the form given on the other side. On occasion it is practical to work with each side *separately*, to obtain one common form equivalent to both sides.

EXAMPLE 6 Working with Each Side Separately

Verify the identity

$$\frac{\cot^2 \theta}{1 + \csc \theta} = \frac{1 - \sin \theta}{\sin \theta}.$$

SOLUTION

Working with the left side, we have

$$\frac{\cot^2 \theta}{1 + \csc \theta} = \frac{\csc^2 \theta - 1}{1 + \csc \theta} \qquad cot^2\ \theta = csc^2\ \theta - 1$$

$$= \frac{(\csc \theta - 1)(\cancel{\csc \theta + 1})}{\cancel{1 + \csc \theta}} \qquad Factor$$

$$= \csc \theta - 1. \qquad Reduce$$

Now, simplifying the right side, we have

$$\frac{1 - \sin \theta}{\sin \theta} = \frac{1}{\sin \theta} - \frac{\sin \theta}{\sin \theta} = \csc \theta - 1.$$

The identity is verified since both sides are equal to $\csc \theta - 1$.

Analytic Trigonometry

In the last example, we rewrite powers of trigonometric functions as more complicated sums of products of trigonometric functions. This is a common procedure used to simplify the operations of calculus.

EXAMPLE 7 An Example from Calculus

Verify the following identities.

(a) $\tan^4 x = \tan^2 x \sec^2 x - \tan^2 x$

(b) $\sin^3 x \cos^4 x = (\cos^4 x - \cos^6 x)\sin x$

SOLUTION

Note the use of the Pythagorean identities in the verifications.

(a) $\tan^4 x = (\tan^2 x)(\tan^2 x)$

$\qquad = \tan^2 x(\sec^2 x - 1)$

$\qquad = \tan^2 x \sec^2 x - \tan^2 x$

(b) $\sin^3 x \cos^4 x = \sin^2 x \cos^4 x \sin x$

$\qquad = (1 - \cos^2 x)\cos^4 x \sin x$

$\qquad = (\cos^4 x - \cos^6 x)\sin x$

WARM UP

Factor the expressions and, if possible, simplify the results.

1. (a) $x^2 - x^2 y^2$
(b) $\sin^2 x - \sin^2 x \cos^2 x$

2. (a) $x^2 + x^2 y^2$
(b) $\cos^2 x + \cos^2 x \tan^2 x$

3. (a) $x^4 - 1$
(b) $\tan^4 x - 1$

4. (a) $z^3 + 1$
(b) $\tan^3 x + 1$

5. (a) $x^3 - x^2 + x - 1$
(b) $\cot^3 x - \cot^2 x + \cot x - 1$

6. (a) $x^4 - 2x^2 + 1$
(b) $\sin^4 x - 2 \sin^2 x + 1$

Perform the indicated additions and subtractions and, if possible, simplify the results.

7. (a) $\dfrac{y^2}{x} - x$

(b) $\dfrac{\csc^2 x}{\cot x} - \cot x$

8. (a) $1 - \dfrac{1}{x^2}$.

(b) $1 - \dfrac{1}{\sec^2 x}$

9. (a) $\dfrac{y}{1 + z} + \dfrac{1 + z}{y}$

(b) $\dfrac{\sin x}{1 + \cos x} + \dfrac{1 + \cos x}{\sin x}$

10. (a) $\dfrac{y}{z} - \dfrac{z}{1 + y}$

(b) $\dfrac{\tan x}{\sec x} - \dfrac{\sec x}{1 + \tan x}$

EXERCISES 3.2

In Exercises 1–60, verify the given identity.

1. $\sin t \csc t = 1$

2. $\tan y \cot y = 1$

3. $(1 + \sin \alpha)(1 - \sin \alpha) = \cos^2 \alpha$

4. $\cot^2 y(\sec^2 y - 1) = 1$

5. $\cos^2 \beta - \sin^2 \beta = 1 - 2 \sin^2 \beta$

6. $\cos^2 \beta - \sin^2 \beta = 2 \cos^2 \beta - 1$

7. $\tan^2 \theta + 4 = \sec^2 \theta + 3$

8. $2 - \sec^2 z = 1 - \tan^2 z$

9. $\sin^2 \alpha - \sin^4 \alpha = \cos^2 \alpha - \cos^4 \alpha$

10. $\cos x + \sin x \tan x = \sec x$

11. $\dfrac{\sec^2 x}{\tan x} = \sec x \csc x$

12. $\dfrac{\cot^3 t}{\csc t} = \cos t(\csc^2 t - 1)$

13. $\dfrac{\cot^2 t}{\csc t} = \csc t - \sin t$

14. $\dfrac{1}{\sin x} - \sin x = \dfrac{\cos^2 x}{\sin x}$

15. $\sin^{1/2} x \cos x - \sin^{5/2} x \cos x = \cos^3 x \sqrt{\sin x}$

16. $\sec^6 x(\sec x \tan x) - \sec^4 x(\sec x \tan x) = \sec^5 x \tan^3 x$

17. $\dfrac{1}{\sec x \tan x} = \csc x - \sin x$

18. $\dfrac{\sec \theta - 1}{1 - \cos \theta} = \sec \theta$

19. $\cos x + \sin x \tan x = \sec x$

20. $\sec x - \cos x = \sin x \tan x$

21. $\csc x - \sin x = \cos x \cot x$

22. $\dfrac{\sec x + \tan x}{\sec x - \tan x} = (\sec x + \tan x)^2$

23. $\dfrac{1}{\tan x} + \dfrac{1}{\cot x} = \tan x + \cot x$

24. $\dfrac{1}{\sin x} - \dfrac{1}{\csc x} = \csc x - \sin x$

25. $\dfrac{\cos \theta \cot \theta}{1 - \sin \theta} - 1 = \csc \theta$

26. $\dfrac{1 + \sin \theta}{\cos \theta} + \dfrac{\cos \theta}{1 + \sin \theta} = 2 \sec \theta$

27. $\dfrac{1}{\cot x + 1} + \dfrac{1}{\tan x + 1} = 1$

28. $\cos x - \dfrac{\cos x}{1 - \tan x} = \dfrac{\sin x \cos x}{\sin x - \cos x}$

29. $2 \sec^2 x - 2 \sec^2 x \sin^2 x - \sin^2 x - \cos^2 x = 1$

30. $\csc x(\csc x - \sin x) + \dfrac{\sin x - \cos x}{\sin x} + \cot x = \csc^2 x$

31. $2 + \cos^2 x - 3 \cos^4 x = \sin^2 x(2 + 3 \cos^2 x)$

32. $4 \tan^4 x + \tan^2 x - 3 = \sec^2 x(4 \tan^2 x - 3)$

33. $\csc^4 x - 2 \csc^2 x + 1 = \cot^4 x$

34. $\sin x(1 - 2 \cos^2 x + \cos^4 x) = \sin^5 x$

35. $\sec^4 \theta - \tan^4 \theta = 1 + 2 \tan^2 \theta$

36. $\csc^4 \theta - \cot^4 \theta = 2 \csc^2 \theta - 1$

37. $\dfrac{\sin \beta}{1 - \cos \beta} = \dfrac{1 + \cos \beta}{\sin \beta}$

38. $\dfrac{\cot \alpha}{\csc \alpha - 1} = \dfrac{\csc \alpha + 1}{\cot \alpha}$

39. $\dfrac{\tan^3 \alpha - 1}{\tan \alpha - 1} = \tan^2 \alpha + \tan \alpha + 1$

40. $\dfrac{\sin^3 \beta + \cos^3 \beta}{\sin \beta + \cos \beta} = 1 - \sin \beta \cos \beta$

41. $\cos\left(\dfrac{\pi}{2} - x\right)\csc x = 1$

42. $\dfrac{\cos[(\pi/2) - x]}{\sin[(\pi/2) - x]} = \tan x$

43. $\dfrac{\csc(-x)}{\sec(-x)} = -\cot x$

44. $(1 + \sin y)[1 + \sin(-y)] = \cos^2 y$

45. $\dfrac{\cos(-\theta)}{1 + \sin(-\theta)} = \sec \theta + \tan \theta$

46. $\dfrac{1 + \sec(-\theta)}{\sin(-\theta) + \tan(-\theta)} = -\csc \theta$

47. $\dfrac{\sin x \cos y + \cos x \sin y}{\cos x \cos y - \sin x \sin y} = \dfrac{\tan x + \tan y}{1 - \tan x \tan y}$

48. $\dfrac{\tan x + \tan y}{1 - \tan x \tan y} = \dfrac{\cot x + \cot y}{\cot x \cot y - 1}$

49. $\dfrac{\tan x + \cot y}{\tan x \cot y} = \tan y + \cot x$

50. $\dfrac{\cos x - \cos y}{\sin x + \sin y} + \dfrac{\sin x - \sin y}{\cos x + \cos y} = 0$

51. $\sqrt{\dfrac{1 + \sin \theta}{1 - \sin \theta}} = \dfrac{1 + \sin \theta}{|\cos \theta|}$

52. $\sqrt{\dfrac{1 - \cos \theta}{1 + \cos \theta}} = \dfrac{1 - \cos \theta}{|\sin \theta|}$

53. $\ln|\tan \theta| = \ln|\sin \theta| - \ln|\cos \theta|$

54. $\ln|\sec \theta| = -\ln|\cos \theta|$

55. $-\ln(1 + \cos \theta) = \ln(1 - \cos \theta) - 2 \ln|\sin \theta|$

56. $-\ln|\sec \theta + \tan \theta| = \ln|\sec \theta - \tan \theta|$

Analytic Trigonometry

57. $\sin^2 x + \sin^2\left(\dfrac{\pi}{2} - x\right) = 1$

58. $\sec^2 y - \cot^2\left(\dfrac{\pi}{2} - y\right) = 1$

59. $\csc x \cos\left(\dfrac{\pi}{2} - x\right) = 1$

60. $\sec^2\left(\dfrac{\pi}{2} - x\right) - 1 = \cot^2 x$

In Exercises 61–64, explain why the equation is *not* an identity and find one value of the variable for which the equation is not true.

61. $\sin\theta = \sqrt{1 - \cos^2\theta}$ **62.** $\tan\theta = \sqrt{\sec^2\theta - 1}$

63. $\sqrt{\tan^2 x} = \tan x$

64. $\sqrt{\sin^2 x + \cos^2 x} = \sin x + \cos x$

3.3 *Solving Trigonometric Equations*

We now switch from *verifying* trigonometric identities to *solving* trigonometric equations. To see the difference, consider the following two equations.

$$\sin^2 x + \cos^2 x = 1 \qquad \text{and} \qquad \sin x = 1$$

The first equation is an identity because it is true for *all* real values of x. The second equation, however, is true only for *some* values of x. When we find these values, we say we are solving the equation.

To solve a trigonometric equation, we use standard algebraic techniques such as collecting like terms and factoring to isolate the trigonometric function involved in the equation. For example, in the equation

$$2 \sin x - 1 = 0$$

we isolate $\sin x$ as follows.

$$2 \sin x = 1$$
$$\sin x = \frac{1}{2}$$

Now, to solve for x, we note that the equation $\sin x = 1/2$ has solutions $x = \pi/6$ and $x = 5\pi/6$ in the interval $[0, 2\pi)$. Moreover, because $\sin x$ has a period of 2π, there are infinitely many other solutions, which can be written as

$$x = \frac{\pi}{6} + 2n\pi \qquad \text{and} \qquad x = \frac{5\pi}{6} + 2n\pi$$

where n is an integer, as shown in Figure 3.2. We call this the **general form** of the solution.

Another way to see that the equation $\sin x = 1/2$ has infinitely many solutions is indicated in Figure 3.3. For $0 \le x < 2\pi$, the solutions are $x = \pi/6$ and $x = 5\pi/6$. Any angles that are coterminal with $\pi/6$ or $5\pi/6$ will also be solutions of the equation.

Solving Trigonometric Equations

FIGURE 3.2

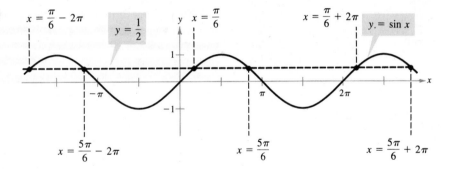

FIGURE 3.3

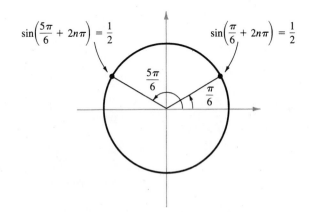

EXAMPLE 1 Collecting Like Terms

Solve the equation $\sin x + \sqrt{2} = -\sin x$.

SOLUTION

Collecting like terms and isolating $\sin x$, we obtain

$$\sin x + \sin x = -\sqrt{2}$$
$$2 \sin x = -\sqrt{2}$$
$$\sin x = -\frac{\sqrt{2}}{2}.$$

Since $\sin x$ has a period of 2π, we first find all solutions in the interval $[0, 2\pi)$. These are $x = 5\pi/4$ and $x = 7\pi/4$. Now we add $2n\pi$ to each of these solutions to get the general form

$$x = \frac{5\pi}{4} + 2n\pi \quad \text{and} \quad x = \frac{7\pi}{4} + 2n\pi$$

where n is an integer.

The general form of the solution depends upon the period of the function involved. For equations involving sin x, cos x, sec x, or csc x, we add $2n\pi$ to each solution in the interval $[0, 2\pi)$. For equations involving tan x or cot x, we add $n\pi$. Unless stated otherwise, "solve the equation" means that we are to find the *general form* of the solution.

EXAMPLE 2 *Taking Square Roots*

Solve the equation $3 \tan^2 x - 1 = 0$.

SOLUTION

First we isolate $\tan^2 x$, and then we take the square root of both sides.

$$3 \tan^2 x = 1$$

$$\tan^2 x = \frac{1}{3}$$

$$\tan x = \pm \frac{1}{\sqrt{3}}$$

Since tan x has a period of π, we find all solutions in the interval $[0, \pi)$. These are $x = \pi/6$ and $x = 5\pi/6$. Finally, we add $n\pi$ to each and obtain the general form

$$x = \frac{\pi}{6} + n\pi \qquad \text{and} \qquad x = \frac{5\pi}{6} + n\pi$$

where n is an integer.

The equations in Examples 1 and 2 involved only one trigonometric function. When two or more functions occur in the same equation, we collect all terms to one side and try to separate the functions by factoring. This may produce factors that yield no solutions, as illustrated in Example 3.

EXAMPLE 3 *Factoring*

Solve the equation $\cot x \cos^2 x = 2 \cot x$.

SOLUTION

Collecting terms to one side and factoring produces

$$\cot x \cos^2 x - 2 \cot x = 0$$

$$\cot x(\cos^2 x - 2) = 0.$$

By setting each of these factors to zero, we obtain the following.

$$\cot x = 0, \qquad\qquad \cos^2 x - 2 = 0$$
$$x = \frac{\pi}{2} \qquad\qquad\quad \cos^2 x = 2$$
$$\cos x = \pm\sqrt{2}$$

No solution is obtained from $\cos x = \pm\sqrt{2}$ because $\pm\sqrt{2}$ are outside the range of the cosine function. Therefore, the general form of the solution is obtained by adding multiples of π to $x = \pi/2$, to get

$$x = \frac{\pi}{2} + n\pi$$

where n is an integer.

Equations of Quadratic Type

Many trigonometric equations are of quadratic type. Here are a couple of examples.

Quadratic in sin x		*Quadratic in sec x*
$2 \sin^2 x - \sin x - 1 = 0$	and	$\sec^2 x - 3 \sec x - 2 = 0$

To solve equations of this type, we factor the quadratic or, if that is not possible, we use the quadratic formula.

When working with an equation of quadratic type, be sure that the equation involves a *single* trigonometric function. For instance, we can rewrite the equation

$$2 \sin^2 x + 3 \cos x - 3 = 0$$

in terms of a single trigonometric function as follows.

$$2(1 - \cos^2 x) + 3 \cos x - 3 = 0 \qquad \textit{Identity}$$
$$2 \cos^2 x - 3 \cos x + 1 = 0 \qquad \textit{Multiply by } -1$$

At this point, we can factor the equation to obtain

$$(2 \cos x - 1)(\cos x - 1) = 0.$$

Finally, by setting each factor equal to zero, the solutions in the interval $[0, 2\pi)$ are determined to be $x = 0$, $x = \pi/3$, and $x = 5\pi/3$. The general solution is therefore

$$x = 2n\pi, \qquad x = \frac{\pi}{3} + 2n\pi, \qquad x = \frac{5\pi}{3} + 2n\pi$$

where n is an integer.

Analytic Trigonometry

EXAMPLE 4 *Factoring an Equation of Quadratic Type*

Find all solutions of $2 \sin^2 x - \sin x - 1 = 0$ in the interval $[0, 2\pi)$.

SOLUTION

Treating the equation as a quadratic in $\sin x$ and factoring, we obtain

$$2 \sin^2 x - \sin x - 1 = 0$$
$$(2 \sin x + 1)(\sin x - 1) = 0.$$

Setting each factor to zero, we obtain the following solutions.

$$2 \sin x + 1 = 0, \qquad\qquad \sin x - 1 = 0$$
$$\sin x = -\frac{1}{2} \qquad\qquad \sin x = 1$$
$$x = \frac{7\pi}{6}, \frac{11\pi}{6} \qquad\qquad x = \frac{\pi}{2}$$

Remark: In Example 4, the general solution would be

$$x = \frac{7\pi}{6} + 2n\pi, \qquad x = \frac{11\pi}{6} + 2n\pi, \qquad x = \frac{\pi}{2} + 2n\pi$$

where n is an integer.

Sometimes we must square both sides of an equation to obtain a quadratic, as demonstrated in the next example. Because this procedure can introduce extraneous solutions, you should check any solutions in the original equation to see if they are valid or extraneous.

EXAMPLE 5 *Squaring and Converting to Quadratic Type*

Find all solutions of $\cos x + 1 = \sin x$ in the interval $[0, 2\pi)$.

SOLUTION

It is not immediately clear how to rewrite this equation in terms of a single trigonometric function. Let's see what happens when we square both sides of the equation.

$$\cos x + 1 = \sin x \qquad\qquad \textit{Given}$$
$$\cos^2 x + 2 \cos x + 1 = \sin^2 x \qquad\qquad \textit{Square both sides}$$
$$\cos^2 x + 2 \cos x + 1 = 1 - \cos^2 x \qquad\qquad \textit{Identity}$$
$$2 \cos^2 x + 2 \cos x = 0 \qquad\qquad \textit{Collect terms}$$
$$2 \cos x(\cos x + 1) = 0 \qquad\qquad \textit{Factor}$$

Solving Trigonometric Equations

Setting each factor to zero produces the following.

$$2 \cos x = 0, \qquad\qquad \cos x + 1 = 0$$
$$\cos x = 0 \qquad\qquad\qquad \cos x = -1$$
$$x = \frac{\pi}{2}, \frac{3\pi}{2} \qquad\qquad\qquad x = \pi$$

Because we squared the original equation, we check for extraneous solutions. Of the three possible solutions, $x = 3\pi/2$ turns out to be extraneous. (Try checking this.) Thus, in the interval $[0, 2\pi)$, the only two solutions are $x = \pi/2$ and $x = \pi$.

Remark: In Example 5, the general solution would be

$$x = \frac{\pi}{2} + 2n\pi, \qquad x = \pi + 2n\pi$$

where n is an integer.

For trigonometric functions of *multiple angles*, extra care is needed to determine all possible solutions. We show how this is done in Example 6.

EXAMPLE 6 Functions of Multiple Angles

Find all solutions of $2 \cos 3t = 1$.

SOLUTION

Solving for $\cos 3t$, we have

$$2 \cos 3t = 1$$
$$\cos 3t = \frac{1}{2}.$$

In the interval $[0, 2\pi)$, we know that $3t = \pi/3$ and $3t = 5\pi/3$ so that, in general,

$$3t = \frac{\pi}{3} + 2n\pi \qquad \text{and} \qquad 3t = \frac{5\pi}{3} + 2n\pi.$$

Dividing this result by 3, we obtain the general form

$$t = \frac{\pi}{9} + \frac{2n\pi}{3} \qquad \text{and} \qquad t = \frac{5\pi}{9} + \frac{2n\pi}{3}$$

where n is an integer.

EXAMPLE 7 *Functions of Multiple Angles*

Find all solutions of $3 \tan(x/2) + 3 = 0$.

SOLUTION

First we isolate $\tan(x/2)$, as follows.

$$3 \tan \frac{x}{2} = -3$$

$$\tan \frac{x}{2} = \frac{-3}{3} = -1$$

In the interval $[0, \pi)$, we know that $x/2 = 3\pi/4$ so that, in general, we have

$$\frac{x}{2} = \frac{3\pi}{4} + n\pi.$$

Multiplication by 2 yields the general form

$$x = \frac{3\pi}{2} + 2n\pi$$

where n is an integer.

Using Inverse Functions and a Calculator

So far in this section, we have chosen examples for which the solutions are special values like

$$0, \ \pm\frac{\pi}{6}, \ \pm\frac{\pi}{4}, \ \pm\frac{\pi}{3}, \ \pm\frac{\pi}{2}$$

and so on. In the remaining examples, we solve more general trigonometric equations by using inverse trigonometric functions or a calculator.

EXAMPLE 8 *Using Inverse Functions*

Find all solutions of $\sec^2 x - 2 \tan x = 4$.

SOLUTION

Collecting terms and substituting $(1 + \tan^2 x)$ for $\sec^2 x$, we have

$$1 + \tan^2 x - 2 \tan x - 4 = 0$$
$$\tan^2 x - 2 \tan x - 3 = 0$$
$$(\tan x - 3)(\tan x + 1) = 0.$$

Setting each factor equal to zero, we obtain two solutions in the interval $(-\pi/2, \ \pi/2)$. [Recall that the range of the inverse tangent function is $(-\pi/2, \ \pi/2)$.]

$$\tan x = 3, \qquad\qquad \tan x = -1$$

$$x = \arctan 3 \qquad\qquad x = \arctan(-1) = -\frac{\pi}{4}$$

Finally, by adding multiples of π (the period of the tangent), we obtain the general solution

$$x = \arctan 3 + n\pi$$

and

$$x = -\frac{\pi}{4} + n\pi$$

where n is an integer.

Remark: If you have access to a computer or calculator with plotting capabilities, you might want to check your solutions graphically. For instance, the graph of the function

$$f(x) = \tan^2 x - 2 \tan x - 3$$

shown in Figure 3.4, has the two x-intercepts

$$x = \arctan 3 \approx 1.2490$$

and

$$x = -\frac{\pi}{4} \approx -0.7854$$

in the interval $(-\pi/2, \ \pi/2)$.

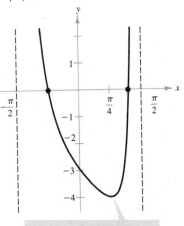

FIGURE 3.4

$f(x) = \tan^2 x - 2 \tan x - 3$

Analytic Trigonometry

When using a calculator for arcsin x, arccos x, and arctan x, the displayed solution may need to be adjusted to obtain solutions in the desired interval, as shown in the next example.

EXAMPLE 9 Using the Quadratic Formula

Find all solutions of

$$\sin^2 t - 3 \sin t - 2 = 0$$

in the interval $[0, 2\pi)$.

SOLUTION

Since the left side of this equation is not factorable, we use the quadratic formula.

$$\sin t = \frac{-(-3) \pm \sqrt{(-3)^2 - 4(1)(-2)}}{2(1)} = \frac{3 \pm \sqrt{17}}{2}$$

$$\sin t \approx 3.561553 \quad \text{or} \quad -0.5615528$$

Because the range of the sine function is $[-1, 1]$, the equation $\sin t = 3.561553$ has no solution. To solve the equation $\sin t = -0.5615528$, we use a calculator and the inverse sine function as follows.

$$t \approx \arcsin(-0.5615528) \approx -0.5962613$$

Note that this solution is not in the interval $[0, 2\pi)$. To find the solutions in $[0, 2\pi)$, it is helpful to make a sketch of the graph of the sine function. From Figure 3.5, we see that the two solutions that lie in the interval $[0, 2\pi)$ are

$$t \approx \pi + 0.5962613 \approx 3.737854 \qquad \textit{Quadrant III}$$

and

$$t \approx 2\pi - 0.5962613 \approx 5.686924. \qquad \textit{Quadrant IV}$$

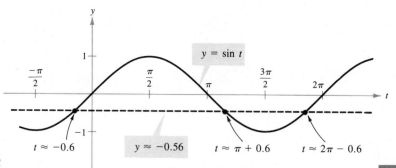

FIGURE 3.5

Exercises

WARM UP

Find two values of θ ($0 \leq \theta < 2\pi$) that satisfy each equation.

1. $\cos \theta = -\frac{1}{2}$

2. $\sin \theta = \frac{\sqrt{3}}{2}$

3. $\cos \theta = \frac{\sqrt{2}}{2}$

4. $\sin \theta = -\frac{\sqrt{2}}{2}$

Find one value of θ ($0 \leq \theta \leq \pi$) that satisfies each equation.

5. $\tan \theta = \sqrt{3}$

6. $\tan \theta = -1$

Solve for x.

7. $\frac{x}{3} + \frac{x}{5} = 1$

8. $2x(x + 3) - 5(x + 3) = 0$

9. $2x^2 - 4x - 5 = 0$

10. $\frac{1}{x} = \frac{x}{2x + 3}$

EXERCISES 3.3

In Exercises 1–6, verify that the given values of x are solutions of the equation.

1. $2 \cos x - 1 = 0$

    (a) $x = \frac{\pi}{3}$ (b) $x = \frac{5\pi}{3}$

2. $\csc x - 2 = 0$

    (a) $x = \frac{\pi}{6}$ (b) $x = \frac{5\pi}{6}$

3. $3 \tan^2 2x - 1 = 0$

    (a) $x = \frac{\pi}{12}$ (b) $x = \frac{5\pi}{12}$

4. $2 \cos^2 4x - 1 = 0$

    (a) $x = \frac{\pi}{16}$ (b) $x = \frac{3\pi}{16}$

5. $2 \sin^2 x - \sin x - 1 = 0$

    (a) $x = \frac{\pi}{2}$ (b) $x = \frac{7\pi}{6}$

6. $\sec^4 x - 4 \sec^2 x = 0$

    (a) $x = \frac{2\pi}{3}$ (b) $x = \frac{5\pi}{3}$

In Exercises 7–20, find all solutions of the equation. (Do not use a calculator.)

7. $2 \cos x + 1 = 0$

8. $2 \sin x - 1 = 0$

9. $\sqrt{3} \csc x - 2 = 0$

10. $\tan x + 1 = 0$

11. $2 \sin^2 x = 1$

12. $\tan^2 x = 3$

13. $3 \sec^2 x - 4 = 0$

14. $\csc^2 x - 2 = 0$

15. $\tan x(\tan x - 1) = 0$

16. $\sin^2 x = 3 \cos^2 x$

17. $\sin x(\sin x + 1) = 0$

18. $4 \sin^2 x - 3 = 0$

19. $\cos x(2 \cos x + 1) = 0$

20. $(3 \tan^2 x - 1)(\tan^2 x - 3) = 0$

In Exercises 21–40, find all solutions in the interval $[0, 2\pi)$. (Do not use a calculator.)

21. $\sec x \csc x - 2 \csc x = 0$

22. $\sec^2 x - \sec x - 2 = 0$

23. $2 \sin^2 x + 3 \sin x + 1 = 0$

24. $3 \tan^3 x - \tan x = 0$

25. $\cos^3 x = \cos x$

26. $4 \sin^3 x + 2 \sin^2 x - 2 \sin x - 1 = 0$

27. $2 \sin^2 x = 2 + \cos x$

28. $\csc^2 x = (1 + \sqrt{3}) - (1 - \sqrt{3})\cot x$

29. $2 \sec^2 x + \tan^2 x - 3 = 0$

30. $\sec^2 x = (1 + \sqrt{3}) - (1 - \sqrt{3})\tan x$

31. $2 \sin x + \csc x = 0$

32. $\csc x + \cot x = 1$

33. $\sin 2x = -\dfrac{\sqrt{3}}{2}$

34. $\tan 3x = 1$

35. $\cos \dfrac{x}{2} = \dfrac{\sqrt{2}}{2}$

36. $\sec 4x = 2$

37. $\dfrac{1 + \cos x}{1 - \cos x} = 0$

38. $\cos x + \sin x \tan x = 2$

39. $\dfrac{1 + \sin x}{\cos x} + \dfrac{\cos x}{1 + \sin x} = 4$

40. $\dfrac{\cos x \cot x}{1 - \sin x} = 3$

In Exercises 41–50, use a calculator to find all solutions in the interval $[0, 2\pi)$.

41. $2 \tan^2 x + 7 \tan x - 15 = 0$

42. $12 \cos^2 x + 5 \cos x - 3 = 0$

43. $12 \sin^2 x - 13 \sin x + 3 = 0$

44. $3 \tan^2 x + 4 \tan x - 4 = 0$

45. $6 \cos^2 x - 13 \cos x + 6 = 0$

46. $\sin^2 x + \sin x - 20 = 0$

47. $\tan^2 x - 8 \tan x + 13 = 0$

48. $2 \cos^2 x + 6 \cos x - 1 = 0$

49. $\sin^2 x + 2 \sin x - 1 = 0$

50. $4 \cos^2 x - 4 \cos x - 1 = 0$

51. The function $f(x) = \sin x + \cos x$ has maximum or minimum values when

$$\cos x - \sin x = 0.$$

Find all solutions of this equation in the interval $[0, 2\pi)$ and sketch a graph of the function f.

52. The function $f(x) = 2 \sin x + \cos 2x$ has maximum or minimum values when

$$2 \cos x - 4 \sin x \cos x = 0.$$

Find all solutions of this equation in the interval $[0, 2\pi)$ and sketch a graph of the function f.

53. A 5-pound weight is oscillating on the end of a spring, and the position of the weight relative to the point of equilibrium is given by

$$h(t) = \tfrac{1}{4}(\cos 8t - 3 \sin 8t)$$

where t is the time in seconds. Find the times when the weight is at the point of equilibrium $[h(t) = 0]$ for $0 \le t \le 1$.

54. The monthly sales (in thousands of units) of a seasonal product is approximated by

$$S = 74.50 + 43.75 \sin \dfrac{\pi t}{6}$$

where t is the time in months, with $t = 1$ corresponding to January. Determine the months when sales exceed 100,000 units.

55. A batted baseball leaves the bat at an angle of θ with the horizontal, with a velocity of $v_0 = 100$ feet per second, and is caught by an outfielder 300 feet from home plate (see figure). Find θ if the range of a projectile is given by

$$r = \tfrac{1}{32}v_0^2 \sin 2\theta.$$

56. A gun with a muzzle velocity of 1200 feet per second is pointed at a target 1000 yards away (see figure). Neglecting air resistance, what should be the minimum angle of elevation of the gun? (Use the formula for range given in Exercise 55.)

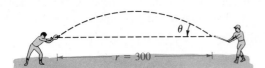

FIGURE FOR 55

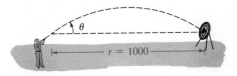

FIGURE FOR 56

3.4 Sum and Difference Formulas

In this and the following section, we show the derivations and uses of several trigonometric identities (or formulas) that are useful in scientific applications.

We begin with six sum and difference formulas that express trigonometric functions of $(u \pm v)$ as functions of u and v alone.

Sum and Difference Formulas

Sine
$$\sin(u + v) = \sin u \cos v + \cos u \sin v$$
$$\sin(u - v) = \sin u \cos v - \cos u \sin v$$

Cosine
$$\cos(u + v) = \cos u \cos v - \sin u \sin v$$
$$\cos(u - v) = \cos u \cos v + \sin u \sin v$$

Tangent
$$\tan(u + v) = \frac{\tan u + \tan v}{1 - \tan u \tan v}$$

$$\tan(u - v) = \frac{\tan u - \tan v}{1 + \tan u \tan v}$$

PROOF

We prove only the formulas for $\cos(u \pm v)$. In Figure 3.6 we let A be the point $(1, 0)$ and then use u and v to locate the points $B = (x_1, y_1)$, $C = (x_2, y_2)$, and $D = (x_3, y_3)$ on the unit circle. Thus, $x_i^2 + y_i^2 = 1$ for $i = 1, 2, 3$. For convenience, we assume that $0 < v < u < 2\pi$.

From Figure 3.7, note that arcs AC and BD have the same length. Hence, *line segments AC and BD* are also equal in length, which implies that

$$\sqrt{(x_2 - 1)^2 + (y_2 - 0)^2} = \sqrt{(x_3 - x_1)^2 + (y_3 - y_1)^2}$$
$$x_2^2 - 2x_2 + 1 + y_2^2 = x_3^2 - 2x_1x_3 + x_1^2 + y_3^2 - 2y_1y_3 + y_1^2$$
$$(x_2^2 + y_2^2) + 1 - 2x_2 = (x_3^2 + y_3^2) + (x_1^2 + y_1^2) - 2x_1x_3 - 2y_1y_3$$
$$1 + 1 - 2x_2 = 1 + 1 - 2x_1x_3 - 2y_1y_3$$
$$x_2 = x_3x_1 + y_3y_1.$$

Finally, by substituting the values $x_2 = \cos(u - v)$, $x_3 = \cos u$, $x_1 = \cos v$, $y_3 = \sin u$, and $y_1 = \sin v$, we obtain

$$\cos(u - v) = \cos u \cos v + \sin u \sin v.$$

The formula for $\cos(u + v)$ can be established by considering $u + v = u - (-v)$ and using the formula just derived to obtain

$$\cos(u + v) = \cos[u - (-v)]$$
$$= \cos u \cos(-v) + \sin u \sin(-v)$$
$$= \cos u \cos v - \sin u \sin v.$$

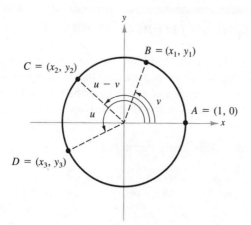

FIGURE 3.6

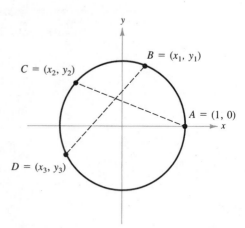

FIGURE 3.7

Remark: Note that $\sin(u + v) \neq \sin u + \sin v$. Similar statements can be made for $\cos(u + v)$ and $\tan(u + v)$.

Applications of Sum and Difference Formulas

In the remainder of this section, we show a variety of uses of sum and difference formulas. First, we show how sum and difference formulas can be used to find exact values of trigonometric functions involving sums or differences of special angles.

EXAMPLE 1 *Evaluating a Trigonometric Function*

Find the exact value of cos 75°.

SOLUTION

To find the *exact* value of cos 75°, we use the fact that

$$75° = 30° + 45°.$$

Consequently, the formula for $\cos(u + v)$ yields

$$\cos 75° = \cos(30° + 45°)$$
$$= \cos 30° \cos 45° - \sin 30° \sin 45°$$
$$= \frac{\sqrt{3}}{2}\left(\frac{\sqrt{2}}{2}\right) - \frac{1}{2}\left(\frac{\sqrt{2}}{2}\right)$$
$$= \frac{\sqrt{6} - \sqrt{2}}{4}.$$

Remark: Try checking the result obtained in Example 1 on your calculator. You will find that

$$\cos 75° = \frac{\sqrt{6} - \sqrt{2}}{4} \approx 0.259.$$

EXAMPLE 2 *Evaluating a Trigonometric Function*

Find the exact value of

$$\cos \frac{\pi}{12}.$$

Analytic Trigonometry

SOLUTION

Using the fact that

$$\frac{\pi}{12} = \frac{\pi}{3} - \frac{\pi}{4}$$

together with the formula for $\cos(u - v)$, we obtain

$$\cos\frac{\pi}{12} = \cos\left(\frac{\pi}{3} - \frac{\pi}{4}\right)$$

$$= \cos\frac{\pi}{3}\cos\frac{\pi}{4} + \sin\frac{\pi}{3}\sin\frac{\pi}{4}$$

$$= \frac{1}{2}\left(\frac{\sqrt{2}}{2}\right) + \frac{\sqrt{3}}{2}\left(\frac{\sqrt{2}}{2}\right)$$

$$= \frac{\sqrt{2} + \sqrt{6}}{4}.$$

EXAMPLE 3 *Evaluating a Trigonometric Expression*

Find the exact value of $\sin 42° \cos 12° - \cos 42° \sin 12°$.

SOLUTION

Recognizing that this expression fits the formula for $\sin(u - v)$, we can write

$$\sin 42° \cos 12° - \cos 42° \sin 12° = \sin(42° - 12°)$$
$$= \sin 30°$$
$$= \frac{1}{2}.$$

EXAMPLE 4 *Evaluating a Trigonometric Expression*

Find the exact value of

$$\frac{\tan 80° + \tan 55°}{1 - \tan 80° \tan 55°}.$$

SOLUTION

From the formula for $\tan(u + v)$, we have

$$\frac{\tan 80° + \tan 55°}{1 - \tan 80° \tan 55°} = \tan(80° + 55°) = \tan 135° = -\tan 45° = -1.$$

EXAMPLE 5 An Application of a Difference Formula

Find $\cos(u - v)$ given that

$$\cos u = -\frac{15}{17}, \quad \pi < u < \frac{3\pi}{2}$$

and

$$\sin v = \frac{4}{5}, \quad 0 < v < \frac{\pi}{2}.$$

SOLUTION

Using the given values for $\cos u$ and $\sin v$, we can sketch angles u and v as shown in Figure 3.8. This implies that

$$\cos v = \frac{3}{5} \quad \text{and} \quad \sin u = -\frac{8}{17}.$$

Therefore,

$$\cos(u - v) = \cos u \cos v + \sin u \sin v$$
$$= \left(-\frac{15}{17}\right)\left(\frac{3}{5}\right) + \left(-\frac{8}{17}\right)\left(\frac{4}{5}\right)$$
$$= -\frac{77}{85}.$$

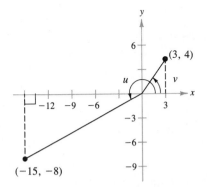

FIGURE 3.8

EXAMPLE 6 Proving a Cofunction Identity

Use the formula for $\cos(u - v)$ to prove the cofunction identity

$$\cos\left(\frac{\pi}{2} - x\right) = \sin x.$$

SOLUTION

Using the difference formula

$$\cos(u - v) = \cos u \cos v + \sin u \sin v$$

we have

$$\cos\left(\frac{\pi}{2} - x\right) = \cos\frac{\pi}{2} \cos x + \sin\frac{\pi}{2} \sin x$$
$$= (0)\cos x + (1)\sin x$$
$$= \sin x.$$

Sum and difference formulas can be used to derive **reduction formulas** involving expressions like

$$\sin\left(\theta + \frac{n\pi}{2}\right) \qquad \text{and} \qquad \cos\left(\theta + \frac{n\pi}{2}\right)$$

where n is an integer. We show two such formulas in Example 7.

EXAMPLE 7 *Deriving Reduction Formulas*

Simplify the following expressions.

(a) $\cos\left(\theta - \frac{3\pi}{2}\right)$ 　　　　　　　　(b) $\sin(\theta + 3\pi)$

SOLUTION

(a) Using the formula

$$\cos(u - v) = \cos u \cos v + \sin u \sin v$$

we have

$$\cos\left(\theta - \frac{3\pi}{2}\right) = \cos\theta\cos\frac{3\pi}{2} + \sin\theta\sin\frac{3\pi}{2}$$
$$= (\cos\theta)(0) + (\sin\theta)(-1)$$
$$= -\sin\theta.$$

(b) Using the formula

$$\sin(u + v) = \sin u \cos v + \cos u \sin v$$

we have

$$\sin(\theta + 3\pi) = \sin\theta\cos 3\pi + \cos\theta\sin 3\pi$$
$$= (\sin\theta)(-1) + (\cos\theta)(0)$$
$$= -\sin\theta.$$

EXAMPLE 8 *Solving a Trigonometric Equation*

Find all solutions of

$$\sin\left(x + \frac{\pi}{4}\right) + \sin\left(x - \frac{\pi}{4}\right) = -1$$

in the interval $[0, 2\pi)$.

SOLUTION

Using sum and difference formulas, we rewrite the given equation as

$$\sin x \cos \frac{\pi}{4} + \cos x \sin \frac{\pi}{4} + \sin x \cos \frac{\pi}{4} - \cos x \sin \frac{\pi}{4} = -1$$

$$2 \sin x \cos \frac{\pi}{4} = -1$$

$$2(\sin x)\left(\frac{\sqrt{2}}{2}\right) = -1$$

$$\sin x = -\frac{1}{\sqrt{2}}$$

$$\sin x = -\frac{\sqrt{2}}{2}.$$

Therefore, the only solutions in the interval $[0, 2\pi)$ are

$$x = \frac{5\pi}{4} \quad \text{and} \quad x = \frac{7\pi}{4}.$$

Our last example shows how a sum formula allows us to rewrite a trigonometric expression in a form that is useful in calculus.

EXAMPLE 9 *An Application from Calculus*

Verify that

$$\frac{\sin(x + h) - \sin x}{h} = (\cos x)\left(\frac{\sin h}{h}\right) - (\sin x)\left(\frac{1 - \cos h}{h}\right)$$

where $h \neq 0$.

SOLUTION

Using the formula for $\sin(u + v)$, we have

$$\frac{\sin(x + h) - \sin x}{h} = \frac{\sin x \cos h + \cos x \sin h - \sin x}{h}$$

$$= \frac{\cos x \sin h - \sin x(1 - \cos h)}{h}$$

$$= (\cos x)\left(\frac{\sin h}{h}\right) - (\sin x)\left(\frac{1 - \cos h}{h}\right).$$

Analytic Trigonometry

WARM UP

Find $\sin \theta$.

1. $\tan \theta = \frac{1}{3}$ (θ in Quadrant I)

2. $\cot \theta = \frac{3}{5}$ (θ in Quadrant III)

3. $\cos \theta = \frac{3}{4}$ (θ in Quadrant IV)

4. $\sec \theta = -3$ (θ in Quadrant II)

Solve for x.

5. $\sin x = \dfrac{\sqrt{2}}{2}$

6. $\cos x = 0$

Simplify the expressions.

7. $\tan x \sec^2 x - \tan x$

8. $\dfrac{\cos x \csc x}{\tan x}$

9. $\dfrac{\cos x}{1 - \sin x} - \tan x$

10. $1 - 3 \cos^2 x + 2 \cos^4 x$

EXERCISES 3.4

In Exercises 1–10, determine the exact values of the sine, cosine, and tangent of the given angle.

1. $75° = 30° + 45°$

2. $15° = 45° - 30°$

3. $105° = 60° + 45°$

4. $165° = 135° + 30°$

5. $195° = 225° - 30°$

6. $255° = 300° - 45°$

7. $\dfrac{11\pi}{12} = \dfrac{3\pi}{4} + \dfrac{\pi}{6}$

8. $\dfrac{7\pi}{12} = \dfrac{\pi}{3} + \dfrac{\pi}{4}$

9. $\dfrac{17\pi}{12} = \dfrac{9\pi}{4} - \dfrac{5\pi}{6}$

10. $-\dfrac{\pi}{12} = \dfrac{\pi}{6} - \dfrac{\pi}{4}$

In Exercises 11–20, simplify the given expression.

11. $\cos 25° \cos 15° - \sin 25° \sin 15°$

12. $\sin 140° \cos 50° + \cos 140° \sin 50°$

13. $\sin 230° \cos 30° - \cos 230° \sin 30°$

14. $\cos 20° \cos 30° + \sin 20° \sin 30°$

15. $\dfrac{\tan 325° - \tan 86°}{1 + \tan 325° \tan 86°}$

16. $\dfrac{\tan 140° - \tan 60°}{1 + \tan 140° \tan 60°}$

17. $\sin 3 \cos 1.2 - \cos 3 \sin 1.2$

18. $\cos \dfrac{\pi}{7} \cos \dfrac{\pi}{5} - \sin \dfrac{\pi}{7} \sin \dfrac{\pi}{5}$

19. $\dfrac{\tan 2x + \tan x}{1 - \tan 2x \tan x}$

20. $\cos 3x \cos 2y + \sin 3x \sin 2y$

In Exercises 21–24, find the exact value of the trigonometric function given that

$$\sin u = \frac{5}{13}, \; 0 < u < \frac{\pi}{2} \quad \text{and} \quad \cos v = -\frac{3}{5}, \frac{\pi}{2} < v < \pi.$$

21. $\sin(u + v)$

22. $\cos(v - u)$

23. $\cos(v + u)$

24. $\sin(u - v)$

In Exercises 25–28, find the exact value of the trigonometric function given that

$$\sin u = \frac{7}{25}, \frac{\pi}{2} < u < \pi \quad \text{and} \quad \cos v = \frac{4}{5}, \frac{3\pi}{2} < v < 2\pi.$$

25. $\cos(u + v)$

26. $\sin(u + v)$

27. $\sin(v - u)$

28. $\cos(u - v)$

In Exercises 29–48, verify the given identity.

29. $\sin\left(\dfrac{\pi}{2} + x\right) = \cos x$ EXPAND

30. $\sin(3\pi - x) = \sin x$

31. $\cos\left(\dfrac{3\pi}{2} - x\right) = -\sin x$

32. $\cos(\pi + x) = -\cos x$

33. $\sin\left(\dfrac{\pi}{6} + x\right) = \dfrac{1}{2}(\cos x + \sqrt{3} \sin x)$

34. $\cos\left(\dfrac{5\pi}{4} - x\right) = -\dfrac{\sqrt{2}}{2}(\cos x + \sin x)$

35. $\cos(\pi - \theta) + \sin\left(\dfrac{\pi}{2} + \theta\right) = 0$

36. $\sin\left(\dfrac{3\pi}{2} + \theta\right) + \sin(\pi - \theta) = \sin\theta - \cos\theta$

37. $\tan(\pi + \theta) = \tan\theta$

38. $\tan\left(\dfrac{\pi}{4} - \theta\right) = \dfrac{1 - \tan\theta}{1 + \tan\theta}$

39. $\cos(x + y)\cos(x - y) = \cos^2 x - \sin^2 y$

40. $\sin(x + y)\sin(x - y) = \sin^2 x - \sin^2 y$

41. $\sin(x + y) + \sin(x - y) = 2\sin x \cos y$

42. $\cos(x + y) + \cos(x - y) = 2\cos x \cos y$

43. $\sin(x + y + z) = \sin x \cos y \cos z + \sin y \cos x \cos z + \sin z \cos x \cos y - \sin x \sin y \sin z$

44. $\cos(x + y + z) = \cos x \cos y \cos z - \cos x \sin y \sin z - \cos y \sin x \sin z - \cos z \sin x \sin y$

45. $\cos(n\pi + \theta) = (-1)^n \cos\theta, \quad n$ is an integer

46. $\sin(n\pi + \theta) = (-1)^n \sin\theta, \quad n$ is an integer

47. $a\sin B\theta + b\cos B\theta = \sqrt{a^2 + b^2}\,\sin(B\theta + C)$,

where $C = \arctan \dfrac{b}{a}$

48. $a\sin B\theta + b\cos B\theta = \sqrt{a^2 + b^2}\,\cos(B\theta - C)$,

where $C = \arctan \dfrac{a}{b}$

In Exercises 49–52, use the formulas given in Exercises 47 and 48 to write the given trigonometric expression in the following forms.

(a) $\sqrt{a^2 + b^2}\,\sin(B\theta + C)$ (b) $\sqrt{a^2 + b^2}\,\cos(B\theta - C)$

49. $\sin\theta + \cos\theta$

50. $3\sin 2\theta + 4\cos 2\theta$

51. $12\sin 3\theta + 5\cos 3\theta$

52. $\sin 2\theta - \cos 2\theta$

In Exercises 53 and 54, use the formulas given in Exercises 47 and 48 to write the given trigonometric expression in the form $a\sin B\theta + b\cos B\theta$.

53. $2\sin\left(\theta + \dfrac{\pi}{4}\right)$ **54.** $5\cos\left(\theta + \dfrac{3\pi}{4}\right)$

In Exercises 55 and 56, write the trigonometric expression as an algebraic expression in x. [*Hint:* See Examples 6 and 7 in Section 2.8.]

55. $\sin(\arcsin x + \arccos x)$

56. $\sin(\arctan 2x - \arccos x)$

In Exercises 57–62, find all solutions in the interval $[0, 2\pi)$.

57. $\sin\left(x + \dfrac{\pi}{3}\right) + \sin\left(x - \dfrac{\pi}{3}\right) = 1$

58. $\sin\left(x + \dfrac{\pi}{6}\right) - \sin\left(x - \dfrac{\pi}{6}\right) = \dfrac{1}{2}$

59. $\cos\left(x + \dfrac{\pi}{4}\right) + \cos\left(x - \dfrac{\pi}{4}\right) = 1$

60. $\cos\left(x + \dfrac{\pi}{4}\right) - \cos\left(x - \dfrac{\pi}{4}\right) = 1$

61. $\tan(x + \pi) + 2\sin(x + \pi) = 0$

62. $\tan(x + \pi) - \cos\left(x + \dfrac{\pi}{2}\right) = 0$

63. Show that

$$\frac{\cos(x + h) - \cos x}{h} = \cos x\left(\frac{\cos h - 1}{h}\right) - \sin x\left(\frac{\sin h}{h}\right).$$

64. A weight is attached to a spring suspended vertically from the ceiling. When a certain driving force is applied to the system, the weight moves vertically from its equilibrium position and this motion is described by the model

$$y = \tfrac{1}{3}\sin 2t + \tfrac{1}{4}\cos 2t$$

where y is the distance from equilibrium measured in feet and t is time in seconds.

(a) Write the model in the form

$$y = \sqrt{a^2 + b^2}\,\sin(Bt + C).$$

   (See Exercise 47.)

(b) Find the amplitude of the oscillations of the weight.

(c) Find the frequency of the oscillations of the weight.

65. Use the difference formula

$$\cos(u - v) = \cos u \cos v + \sin u \sin v$$

together with the cofunction identity

$$\sin(u + v) = \cos\left[\frac{\pi}{2} - (u + v)\right] = \cos\left[\left(\frac{\pi}{2} - u\right) - v\right]$$

to prove the sum formula

$$\sin(u + v) = \sin u \cos v + \cos u \sin v.$$

66. Use the sum formula for $\sin(u + v)$ and $\cos(u + v)$ to prove the sum formula

$$\tan(u + v) = \frac{\tan u + \tan v}{1 - \tan u \tan v}.$$

Analytic Trigonometry

3.5 Multiple-Angle Formulas and Product-Sum Formulas

In this section we look at two other categories of trigonometric identities. The first category involves functions of multiple angles such as $\sin ku$ or $\cos ku$. The second category involves products of trigonometric functions such as $\sin u \cos v$.

Multiple-Angle Formulas

The most commonly used multiple-angle formulas are the double-angle formulas. They are used often enough so that you should memorize them.

Double-Angle Formulas

$$\sin 2u = 2 \sin u \cos u$$
$$\cos 2u = \cos^2 u - \sin^2 u = 2 \cos^2 u - 1 = 1 - 2 \sin^2 u$$
$$\tan 2u = \frac{2 \tan u}{1 - \tan^2 u}$$

PROOF

To prove the first formula, we let $v = u$ in the formula for $\sin(u + v)$, and obtain

$$\sin 2u = \sin(u + u)$$
$$= \sin u \cos u + \cos u \sin u$$
$$= 2 \sin u \cos u.$$

The other double-angle formulas can be proved in a similar way, and we leave their proofs for you to do.

Remark: Note that $\sin 2u \neq 2 \sin u$. Similar statements can be made for $\cos 2u$ and $\tan 2u$.

EXAMPLE 1 *Solving a Trigonometric Equation*

Find all solutions of $2 \cos x + \sin 2x = 0$.

Multiple-Angle Formulas and Product-Sum Formulas

SOLUTION

We begin by rewriting the equation so that it involves functions of x (rather than $2x$). Then we can factor and solve as usual.

$2 \cos x + \sin 2x = 0$	*Given*
$2 \cos x + 2 \sin x \cos x = 0$	*Double-angle formula*
$2 \cos x(1 + \sin x) = 0$	*Factor*
$\cos x = 0, \qquad\qquad 1 + \sin x = 0$	*Set factors to zero*
$x = \dfrac{\pi}{2}, \dfrac{3\pi}{2} \qquad\qquad x = \dfrac{3\pi}{2}$	*Solutions in* $[0, 2\pi)$

Therefore, the general solution is

$$x = \frac{\pi}{2} + 2n\pi \qquad \text{and} \qquad x = \frac{3\pi}{2} + 2n\pi$$

where n is an integer.

EXAMPLE 2 Evaluating Functions Involving Double Angles

Use the fact that

$$\cos \theta = \frac{5}{13}, \quad \frac{3\pi}{2} < \theta < 2\pi$$

to find $\sin 2\theta$, $\cos 2\theta$, and $\tan 2\theta$.

SOLUTION

From Figure 3.9, we can see that

$$\sin \theta = \frac{y}{r} = -\frac{12}{13}.$$

Consequently,

$$\sin 2\theta = 2 \sin \theta \cos \theta = 2\left(\frac{-12}{13}\right)\left(\frac{5}{13}\right) = -\frac{120}{169}$$

$$\cos 2\theta = 2 \cos^2 \theta - 1 = 2\left(\frac{25}{169}\right) - 1 = -\frac{119}{169}$$

$$\tan 2\theta = \frac{\sin 2\theta}{\cos 2\theta} = \frac{120}{119}.$$

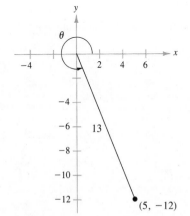

FIGURE 3.9

The double-angle formulas are not restricted to angles 2θ and θ. Other *double* combinations like 4θ and 2θ or 6θ and 3θ are also valid. Here are a couple of examples.

$$\sin 4\theta = 2 \sin 2\theta \cos 2\theta$$
$$\cos 6\theta = \cos^2 3\theta - \sin^2 3\theta$$

By using double-angle formulas together with the sum formulas derived in the previous section, we can form other multiple-angle formulas. For instance, in the following example, we derive a *triple*-angle formula for $\sin 3x$.

EXAMPLE 3 *Deriving a Triple-Angle Formula*

Express $\sin 3x$ in terms of $\sin x$.

SOLUTION

Considering $3x = 2x + x$, we have

$$
\begin{aligned}
\sin 3x &= \sin(2x + x) \\
&= \sin 2x \cos x + \cos 2x \sin x \\
&= 2 \sin x \cos x \cos x + (1 - 2 \sin^2 x)\sin x \\
&= 2 \sin x \cos^2 x + \sin x - 2 \sin^3 x \\
&= 2 \sin x(1 - \sin^2 x) + \sin x - 2 \sin^3 x \\
&= 2 \sin x - 2 \sin^3 x + \sin x - 2 \sin^3 x \\
&= 3 \sin x - 4 \sin^3 x.
\end{aligned}
$$

The double-angle formulas can be used to obtain the following **power-reducing formulas.**

Power-Reducing Formulas

$$\sin^2 u = \frac{1 - \cos 2u}{2}$$

$$\cos^2 u = \frac{1 + \cos 2u}{2}$$

$$\tan^2 u = \frac{1 - \cos 2u}{1 + \cos 2u}$$

PROOF

The first two formulas can be verified by solving for $\sin^2 u$ and $\cos^2 u$, respectively, in the double-angle formulas

$$\cos 2u = 1 - 2 \sin^2 u \quad \text{and} \quad \cos 2u = 2 \cos^2 u - 1.$$

The third formula can be verified using the fact that

$$\tan^2 u = \frac{\sin^2 u}{\cos^2 u}.$$

Example 4 shows a typical power reduction that is used in calculus.

EXAMPLE 4 *Reducing the Power of a Trigonometric Function*

Rewrite $\sin^4 x$ as a sum involving first powers of the cosine of multiple angles.

SOLUTION

Note the repeated use of power-reducing formulas in the following procedure.

$$\begin{aligned}
\sin^4 x = (\sin^2 x)^2 &= \left(\frac{1 - \cos 2x}{2} \right)^2 \\
&= \frac{1}{4}(1 - 2 \cos 2x + \cos^2 2x) \\
&= \frac{1}{4}\left(1 - 2 \cos 2x + \frac{1 + \cos 4x}{2} \right) \\
&= \frac{1}{4} - \frac{1}{2} \cos 2x + \frac{1}{8} + \frac{1}{8} \cos 4x \\
&= \frac{3}{8} - \frac{1}{2} \cos 2x + \frac{1}{8} \cos 4x \\
&= \frac{1}{8}(3 - 4 \cos 2x + \cos 4x)
\end{aligned}$$

We can derive some useful alternative forms of the power-reducing formulas by replacing u with $u/2$. The results are called **half-angle formulas.**

$$\sin \frac{u}{2} = \pm \sqrt{\frac{1 - \cos u}{2}} \qquad \cos \frac{u}{2} = \pm \sqrt{\frac{1 + \cos u}{2}}$$

The signs depend upon the quadrant in which $u/2$ lies.

$$\tan \frac{u}{2} = \frac{1 - \cos u}{\sin u} = \frac{\sin u}{1 + \cos u}$$

EXAMPLE 5 *Using a Half-Angle Formula*

Find the exact value of sin 105°.

SOLUTION

We begin by noting that 105° is half of 210°. Then, using the half-angle formula for $\sin(u/2)$ and the fact that 105° lies in Quadrant II, we have

$$\sin 105° = \sqrt{\frac{1 - \cos 210°}{2}} = \sqrt{\frac{1 - (-\cos 30°)}{2}}$$

$$= \sqrt{\frac{1 + (\sqrt{3}/2)}{2}} = \frac{\sqrt{2 + \sqrt{3}}}{2}.$$

Note that we chose the positive square root because sin θ is positive in Quadrant II.

Remark: Try using your calculator to verify the result obtained in Example 5. That is, evaluate

$$\sin 105° \quad \text{and} \quad \frac{\sqrt{2 + \sqrt{3}}}{2}$$

and you will see that both values are approximately 0.9659258.

EXAMPLE 6 *Solving a Trigonometric Equation*

Find all solutions of

$$2 - \sin^2 x = 2 \cos^2 \frac{x}{2}.$$

SOLUTION

$$2 - \sin^2 x = 2\left(\frac{1 + \cos x}{2}\right) \qquad \text{\textit{Half-angle formula}}$$

$$2 - \sin^2 x = 1 + \cos x$$

$$2 - (1 - \cos^2 x) = 1 + \cos x \qquad \text{\textit{Pythagorean identity}}$$

$$\cos^2 x - \cos x = 0 \qquad \text{\textit{Simplify}}$$

$$\cos x \,(\cos x - 1) = 0 \qquad \text{\textit{Factor}}$$

$$\cos x = 0, \qquad\qquad \cos x = 1 \qquad \text{\textit{Set factors to zero}}$$

$$x = \frac{\pi}{2}, \frac{3\pi}{2} \qquad\qquad x = 0 \qquad \text{\textit{Solutions in }} [0, 2\pi)$$

Therefore, the general solution is

$$x = 2n\pi, \qquad x = \frac{\pi}{2} + 2n\pi, \qquad \text{and} \qquad x = \frac{3\pi}{2} + 2n\pi$$

where n is an integer.

Product-Sum Formulas

Each of the following **product-to-sum formulas** are easily verified using the sum and difference formulas discussed in the preceding section.

Product-to-Sum Formulas

$$\sin u \sin v = \frac{1}{2}[\cos(u - v) - \cos(u + v)]$$

$$\cos u \cos v = \frac{1}{2}[\cos(u - v) + \cos(u + v)]$$

$$\sin u \cos v = \frac{1}{2}[\sin(u + v) + \sin(u - v)]$$

$$\cos u \sin v = \frac{1}{2}[\sin(u + v) - \sin(u - v)]$$

EXAMPLE 7   Writing Products as Sums

Rewrite $\cos 5x \sin 4x$ as a sum or difference.

SOLUTION

$$\cos 5x \sin 4x = \frac{1}{2}[\sin(5x + 4x) - \sin(5x - 4x)]$$

$$= \frac{1}{2}\sin 9x - \frac{1}{2}\sin x$$

Occasionally, it is useful to reverse the procedure and write a sum of trigonometric functions as a product. This can be accomplished with the following **sum-to-product formulas.**

Analytic Trigonometry

Sum-to-Product Formulas

$$\sin x + \sin y = 2 \sin\left(\frac{x+y}{2}\right)\cos\left(\frac{x-y}{2}\right)$$

$$\sin x - \sin y = 2 \cos\left(\frac{x+y}{2}\right)\sin\left(\frac{x-y}{2}\right)$$

$$\cos x + \cos y = 2 \cos\left(\frac{x+y}{2}\right)\cos\left(\frac{x-y}{2}\right)$$

$$\cos x - \cos y = -2 \sin\left(\frac{x+y}{2}\right)\sin\left(\frac{x-y}{2}\right)$$

PROOF

To prove the first formula, we let $x = u + v$ and $y = u - v$. Then, substituting $u = (x + y)/2$ and $v = (x - y)/2$ in the product-to-sum formula

$$\sin u \cos v = \frac{1}{2}[\sin(u + v) + \sin(u - v)]$$

we get

$$\sin\left(\frac{x+y}{2}\right)\cos\left(\frac{x-y}{2}\right) = \frac{1}{2}(\sin x + \sin y)$$

or equivalently,

$$\sin x + \sin y = 2 \sin\left(\frac{x+y}{2}\right)\cos\left(\frac{x-y}{2}\right).$$

EXAMPLE 8 *Using a Sum-to-Product Formula*

Find the exact value of $\cos 195° + \cos 105°$

SOLUTION

Using the appropriate sum-to-product formula, we get

$$\cos 195° + \cos 105° = 2 \cos\left(\frac{195° + 105°}{2}\right)\cos\left(\frac{195° - 105°}{2}\right)$$

$$= 2 \cos 150° \cos 45°$$

$$= 2\left(-\frac{\sqrt{3}}{2}\right)\left(\frac{\sqrt{2}}{2}\right)$$

$$= -\frac{\sqrt{6}}{2}.$$

EXAMPLE 9 *Solving a Trigonometric Equation*

Find all solutions of $\sin 5x + \sin 3x = 0$.

SOLUTION

$$\sin 5x + \sin 3x = 0 \qquad\qquad \textit{Given}$$

$$2 \sin\left(\frac{5x + 3x}{2}\right)\cos\left(\frac{5x - 3x}{2}\right) = 0 \qquad \textit{Sum-to-product formula}$$

$$2 \sin 4x \cos x = 0 \qquad\qquad \textit{Simplify}$$

$$\sin 4x = 0, \qquad \cos x = 0 \qquad\qquad \textit{Set factors to zero}$$

In the interval $[0, 2\pi)$, the solutions of $\sin 4x = 0$ are

$$x = 0, \quad \frac{\pi}{4}, \quad \frac{\pi}{2}, \quad \frac{3\pi}{4}, \quad \pi, \quad \frac{5\pi}{4}, \quad \frac{3\pi}{2}, \quad \frac{7\pi}{4}.$$

Moreover, the equation $\cos x = 0$ yields no additional solutions, and we can conclude that the solutions are of the form

$$x = \frac{n\pi}{4}$$

where n is an integer.

EXAMPLE 10 *Verifying a Trigonometric Identity*

Verify the identity

$$\frac{\sin t + \sin 3t}{\cos t + \cos 3t} = \tan 2t.$$

SOLUTION

Using appropriate sum-to-product formulas, we have

$$\frac{\sin t + \sin 3t}{\cos t + \cos 3t} = \frac{2 \sin 2t \cos(-t)}{2 \cos 2t \cos(-t)}$$

$$= \frac{\sin 2t}{\cos 2t}$$

$$= \tan 2t.$$

WARM UP

Factor the expressions.

1. $2 \sin x + \sin x \cos x$

2. $\cos^2 x - \cos x - 2$

Find all solutions in the interval $[0, 2\pi)$.

3. $\sin 2x = 0$

4. $\cos 2x = 0$

5. $\cos \dfrac{x}{2} = 0$

6. $\sin \dfrac{x}{2} = 0$

Simplify the expressions.

7. $\dfrac{1 - \cos(\pi/4)}{2}$

8. $\dfrac{1 + \cos(\pi/3)}{2}$

9. $\dfrac{2 \sin 3x \cos x}{2 \cos 3x \cos x}$

10. $(1 - 2 \sin^2 x)\cos x - 2 \sin x \cos x \sin x$

EXERCISES 3.5

In Exercises 1–6, use a double-angle formula to determine the exact values of the sine, cosine, and tangent of the given angle.

1. $90° = 2(45°)$

2. $180° = 2(90°)$

3. $60° = 2(30°)$

4. $120° = 2(60°)$

5. $\dfrac{2\pi}{3} = 2\left(\dfrac{\pi}{3}\right)$

6. $\dfrac{3\pi}{2} = 2\left(\dfrac{3\pi}{4}\right)$

In Exercises 7–12, find the exact values of $\sin 2u$, $\cos 2u$, and $\tan 2u$.

7. $\sin u = \dfrac{3}{5}, \quad 0 < u < \dfrac{\pi}{2}$

8. $\cos u = -\dfrac{2}{3}, \quad \dfrac{\pi}{2} < u < \pi$

9. $\tan u = \dfrac{1}{2}, \quad \pi < u < \dfrac{3\pi}{2}$

10. $\cot u = -4, \quad \dfrac{3\pi}{2} < u < 2\pi$

11. $\sec u = -\dfrac{5}{2}, \quad \dfrac{\pi}{2} < u < \pi$

12. $\csc u = 3, \quad \dfrac{\pi}{2} < u < \pi$

In Exercises 13–18, use half-angle formulas to determine the exact values of the sine, cosine, and tangent of the given angle.

13. $105° = \frac{1}{2}(210°)$

14. $165° = \frac{1}{2}(330°)$

15. $112° \, 30' = \frac{1}{2}(225°)$

16. $67° \, 30' = \frac{1}{2}(135°)$

17. $\dfrac{\pi}{8} = \dfrac{1}{2}\left(\dfrac{\pi}{4}\right)$

18. $\dfrac{\pi}{12} = \dfrac{1}{2}\left(\dfrac{\pi}{6}\right)$

In Exercises 19–24, find the exact values of $\sin(u/2)$, $\cos(u/2)$, and $\tan(u/2)$ by using the half-angle formulas and the given information.

19. $\sin u = \dfrac{5}{13}, \quad \dfrac{\pi}{2} < u < \pi$

20. $\cos u = \dfrac{3}{5}, \quad 0 < u < \dfrac{\pi}{2}$

21. $\tan u = -\dfrac{5}{8}, \quad \dfrac{3\pi}{2} < u < 2\pi$

22. $\cot u = 3, \quad \pi < u < \dfrac{3\pi}{2}$

23. $\csc u = -\dfrac{5}{3}, \quad \pi < u < \dfrac{3\pi}{2}$

24. $\sec u = -\dfrac{7}{2}, \quad \dfrac{\pi}{2} < u < \pi$

In Exercises 25–28, use the half-angle formulas to simplify the given expression.

25. $\sqrt{\dfrac{1 - \cos 6x}{2}}$ **26.** $\sqrt{\dfrac{1 + \cos 4x}{2}}$

27. $-\sqrt{\dfrac{1 - \cos 8x}{1 + \cos 8x}}$ **28.** $-\sqrt{\dfrac{1 - \cos(x - 1)}{2}}$

In Exercises 29 and 30, use the power-reducing formulas to write each expression in terms of the first power of the cosine.

29. (a) $\cos^4 x$
    (b) $\sin^2 x \cos^4 x$

30. (a) $\cos^6 x$
    (b) $\sin^2 x \cos^2 x$

In Exercises 31–40, rewrite the given product as a sum.

31. $6 \sin \dfrac{\pi}{4} \cos \dfrac{\pi}{4}$ **32.** $4 \sin \dfrac{\pi}{3} \cos \dfrac{5\pi}{6}$

33. $\sin 5\theta \cos 3\theta$ **34.** $3 \sin 2\alpha \sin 3\alpha$

35. $5 \cos(-5\beta) \cos 3\beta$ **36.** $\cos 2\theta \cos 4\theta$

37. $\sin(x + y) \sin(x - y)$ **38.** $\sin(x + y) \cos(x - y)$

39. $\sin(\theta + \pi) \cos(\theta - \pi)$

40. $10 \cos 75° \cos 15°$

In Exercises 41–50, express the given sum (or difference) as a product.

41. $\sin 60° + \sin 30°$ **42.** $\cos 120° + \cos 30°$

43. $\cos \dfrac{3\pi}{4} - \cos \dfrac{\pi}{4}$ **44.** $\sin 5\theta - \sin 3\theta$

45. $\cos 6x + \cos 2x$ **46.** $\sin x + \sin 5x$

47. $\sin(\alpha + \beta) - \sin(\alpha - \beta)$

48. $\cos\left(\theta + \dfrac{\pi}{2}\right) - \cos\left(\theta - \dfrac{\pi}{2}\right)$

49. $\cos(\phi + 2\pi) + \cos \phi$

50. $\sin\left(x + \dfrac{\pi}{2}\right) + \sin\left(x - \dfrac{\pi}{2}\right)$

In Exercises 51–72, verify the given identity.

51. $\csc 2\theta = \dfrac{\csc \theta}{2 \cos \theta}$

52. $\sec 2\theta = \dfrac{\sec^2 \theta}{2 - \sec^2 \theta}$

53. $\cos^2 2\alpha - \sin^2 2\alpha = \cos 4\alpha$

54. $\sin \dfrac{\alpha}{3} \cos \dfrac{\alpha}{3} = \dfrac{1}{2} \sin \dfrac{2\alpha}{3}$

55. $(\sin x + \cos x)^2 = 1 + \sin 2x$

56. $\cos^4 x - \sin^4 x = \cos 2x$

57. $\cos 3\beta = \cos^3 \beta - 3 \sin^2 \beta \cos \beta$

58. $\sin 4\beta = 4 \sin \beta \cos \beta(1 - 2 \sin^2 \beta)$

59. $1 + \cos 10y = 2 \cos^2 5y$

60. $\dfrac{\cos 3\beta}{\cos \beta} = 1 - 4 \sin^2 \beta$

61. $\sec \dfrac{u}{2} = \pm\sqrt{\dfrac{2 \tan u}{\tan u + \sin u}}$

62. $\tan \dfrac{u}{2} = \csc u - \cot u$

63. $\dfrac{\cos 4x + \cos 2x}{\sin 4x + \sin 2x} = \cot 3x$

64. $\dfrac{\cos 3x - \cos x}{\sin 3x - \sin x} = -\tan 2x$

65. $\dfrac{\cos 4x - \cos 2x}{2 \sin 3x} = -\sin x$

66. $\dfrac{\sin x \pm \sin y}{\cos x + \cos y} = \tan \dfrac{x \pm y}{2}$

67. $\dfrac{\sin x \pm \sin y}{\cos x - \cos y} = -\cot \dfrac{x \mp y}{2}$

68. $\dfrac{\sin x + \sin y}{\sin x - \sin y} = \dfrac{\tan[(x + y)/2]}{\tan[(x - y)/2]}$

69. $\dfrac{\cos t + \cos 3t}{\sin 3t - \sin t} = \cot t$

70. $\dfrac{\sin 6t - \sin 2t}{\cos 2t + \cos 6t} = \tan 2t$

71. $\sin^2 4x - \sin^2 2x = \sin 2x \sin 6x$

72. $\sin\left(\dfrac{\pi}{6} + x\right) + \sin\left(\dfrac{\pi}{6} - x\right) = \cos x$

In Exercises 73–84, find all solutions in the interval $[0, 2\pi)$.

73. $4 \sin x \cos x = 1$ **74.** $\sin 2x \sin x = \cos x$

75. $\cos 2x = \cos x$ **76.** $\cos 3x - \cos x = 0$

77. $\sin 4x + 2 \sin 2x = 0$

78. $\tan\left(x + \dfrac{\pi}{4}\right) - \tan\left(x - \dfrac{\pi}{4}\right) = 4$

79. $(\sin 2x + \cos 2x)^2 = 1$

80. $\cos 2x - \cos 6x = 0$

81. $\sin 6x + \sin 2x = 0$ **82.** $\sin^2 3x - \sin^2 x = 0$

83. $\dfrac{\cos 2x}{\sin 3x - \sin x} = 1$

84. $\sin 2x + \sin 4x + \sin 6x = 0$

In Exercises 85 and 86, sketch the graph by using the power-reducing formulas.

85. $f(x) = \sin^2 x$ **86.** $f(x) = \cos^2 x$

In Exercises 87 and 88, write the trigonometric expression as an algebraic expression in x.

87. $\sin(2 \arcsin x)$ **88.** $\cos(2 \arccos x)$

In Exercises 89 and 90, verify the identity for the complementary angles ϕ and θ.

89. $\sin(\phi - \theta) = \cos 2\theta$ **90.** $\cos(\phi - \theta) = \sin 2\theta$

In Exercises 91–93, prove the product-to-sum formulas.

91. $\cos u \cos v = \frac{1}{2}[\cos(u - v) + \cos(u + v)]$

92. $\sin u \cos v = \frac{1}{2}[\sin(u + v) + \sin(u - v)]$

93. $\cos u \sin v = \frac{1}{2}[\sin(u + v) - \sin(u - v)]$

CHAPTER 3 REVIEW EXERCISES

In Exercises 1–10, simplify the given expression.

1. $\dfrac{1}{\cot^2 x + 1}$

2. $\dfrac{\sin 2\alpha}{\cos^2 \alpha - \sin^2 \alpha}$

3. $\dfrac{\sin^2 \alpha - \cos^2 \alpha}{\sin^2 \alpha - \sin \alpha \cos \alpha}$

4. $\dfrac{\sin^3 \beta + \cos^3 \beta}{\sin \beta + \cos \beta}$

5. $\cos^2 \beta + \cos^2 \beta \tan^2 \beta$

6. $\dfrac{\sin \theta}{1 + \cos \theta} + \dfrac{1 + \cos \theta}{\sin \theta}$

7. $\tan^2 \theta(\csc^2 \theta - 1)$

8. $\dfrac{2 \tan(x + 1)}{1 - \tan^2(x + 1)}$

9. $1 - 4 \sin^2 x \cos^2 x$

10. $\sqrt{\dfrac{1 - \cos^2 x}{1 + \cos x}}$

In Exercises 11–40, verify the given identity.

11. $\tan x(1 - \sin^2 x) = \frac{1}{2} \sin 2x$

12. $\cos^3 x \sin^2 x = (\sin^2 x - \sin^4 x) \cos x$

13. $\sin^5 x \cos^2 x = (\cos^2 x - 2 \cos^4 x + \cos^6 x) \sin x$

14. $\sin^4 2x = \frac{1}{8}(\cos 8x - 4 \cos 4x + 3)$

15. $\sin^2 x \cos^4 x = \frac{1}{16}(1 - \cos 4x + 2 \sin^2 2x \cos 2x)$

16. $\sin 3x \cos 2x = \frac{1}{2}(\sin 5x + \sin x)$

17. $\sin 3\theta \sin \theta = \frac{1}{2}(\cos 2\theta - \cos 4\theta)$

18. $\sqrt{1 - \cos x} = \dfrac{|\sin x|}{\sqrt{1 + \cos x}}$

19. $\sqrt{\dfrac{1 - \sin \theta}{1 + \sin \theta}} = \dfrac{1 - \sin \theta}{|\cos \theta|}$

20. $\sin 4x = 8 \cos^3 x \sin x - 4 \cos x \sin x$

21. $\cos 3x = 4 \cos^3 x - 3 \cos x$

22. $\cos 4x = 8 \cos^4 x - 8 \cos^2 x + 1$

23. $\sin\left(x - \dfrac{3\pi}{2}\right) = \cos x$

24. $\cos\left(x + \dfrac{\pi}{2}\right) = -\sin x$

25. $\dfrac{\sec x - 1}{\tan x} = \tan \dfrac{x}{2}$

26. $\dfrac{2 \cos 3x}{\sin 4x - \sin 2x} = \csc x$

27. $\dfrac{\cos 3x - \cos x}{\sin 3x - \sin x} = -\tan 2x$

28. $1 - \cos 2x = 2 \sin^2 x$

29. $\sin(\pi - x) = \sin x$

30. $\cot\left(\dfrac{\pi}{2} - x\right) = \tan x$

31. $2 \sin y \cos y \sec 2y = \tan 2y$

32. $\cot \dfrac{u}{2} = \pm\sqrt{\dfrac{\sec u + 1}{\sec u - 1}}$

33. $\sin \dfrac{\theta}{2} + \cos \dfrac{\theta}{2} = \pm\sqrt{1 + \sin \theta}$

34. $\cos^2 5x - \cos^2 x = -\sin 4x \sin 6x$

35. $\tan^2 x = \dfrac{1 - \cos 2x}{1 + \cos 2x}$

36. $\dfrac{\sin(\alpha + \beta)}{\cos \alpha \cos \beta} = \tan \alpha + \tan \beta$

37. $\sin 2x + \sin 4x - \sin 6x = 4 \sin x \sin 2x \sin 3x$

38. $\sin 2x + \sin 4x + \sin 6x = 4 \cos x \cos 2x \sin 3x$

39. $1 + \cos 2x + \cos 4x + \cos 6x = 4 \cos x \cos 2x \cos 3x$

40. $1 - \cos 2x + \cos 4x - \cos 6x = 4 \sin x \cos 2x \sin 3x$

In Exercises 41–44, find the exact value of the trigonometric function by using the sum, difference, or half-angle formulas.

41. $\sin \dfrac{5\pi}{12} = \sin\left(\dfrac{2\pi}{3} - \dfrac{\pi}{4}\right)$

42. $\cos 285° = \cos(225° + 60°)$

43. $\cos(157° \; 30') = \cos \dfrac{315°}{2}$

44. $\sin \dfrac{3\pi}{8} = \sin\left[\dfrac{1}{2}\left(\dfrac{3\pi}{4}\right)\right]$

In Exercises 45–50, find the exact value of the trigonometric function given that

$\sin u = \frac{3}{4}$ and $\cos v = -\frac{5}{13}$ (u and v in Quadrant II).

45. $\sin(u + v)$ **46.** $\tan(u + v)$

47. $\cos(u - v)$ **48.** $\sin 2v$

49. $\cos \dfrac{u}{2}$ **50.** $\tan 2v$

In Exercises 51–60, find all solutions of the given equation in the interval $[0, 2\pi)$.

51. $\sin x - \tan x = 0$ **52.** $\csc x - 2 \cot x = 0$

53. $\dfrac{1 + \sin x}{\cos x} + \dfrac{\cos x}{1 + \sin x} = 4$

54. $\cos x = \cos \dfrac{x}{2}$

55. $\sin 2x + \sqrt{2} \sin x = 0$

56. $\cos 4x - 7 \cos 2x = 8$

57. $\cos^2 x + \sin x = 1$ **58.** $\sin 4x - \sin 2x = 0$

59. $\tan^3 x - \tan^2 x + 3 \tan x - 3 = 0$

60. $\sin x + \sin 3x + \sin 5x = 0$

In Exercises 61 and 62, write the trigonometric expression as a product.

61. $\cos 3\theta + \cos 2\theta$

62. $\sin\left(x + \dfrac{\pi}{4}\right) - \sin\left(x - \dfrac{\pi}{4}\right)$

In Exercises 63 and 64, write the trigonometric product as a sum or difference.

63. $\sin 3\alpha \sin 2\alpha$ **64.** $\cos \dfrac{x}{2} \cos \dfrac{x}{4}$

65. A standing wave on a string of given length is modeled by the equation

$$y = A\left(\cos\left[2\pi\left(\dfrac{t}{T} - \dfrac{x}{\lambda}\right)\right] + \cos\left[2\pi\left(\dfrac{t}{T} + \dfrac{x}{\lambda}\right)\right]\right).$$

Use the trigonometric identities for the cosine of the sum and difference of two angles to verify that the following equation is an equivalent model for the standing wave

$$y = 2A \cos \dfrac{2\pi t}{T} \cos \dfrac{2\pi x}{\lambda}.$$

66. Write $\cos(2 \arccos 2x)$ as an algebraic expression in x.

C H A P T E R 4

Additional Applications of Trigonometry

4.1 *Law of Sines*

In Chapter 2 we looked at techniques for solving right triangles. In this section and the next, we will solve triangles that have no right angles. Such triangles are called **oblique.** As standard notation, we label the vertices of a triangle as A, B, and C, and their opposite sides as a, b, and c, as shown in Figure 4.1.

To solve an oblique triangle, we need to know the measure of at least one side and any two other parts of the triangle—either two sides, two angles, or one angle and one side. This breaks down into the following four possible cases.

1. Two angles and any side (AAS or ASA).
2. Two sides and an angle opposite one of them (SSA).
3. Three sides (SSS).
4. Two sides and their included angle (SAS).

The first two cases can be solved using what is called the **Law of Sines,** while the last two cases require the **Law of Cosines** (to be discussed in Section 4.2).

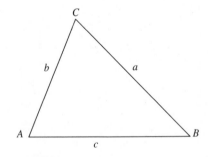

FIGURE 4.1

230

Law of Sines

If ABC is a triangle with sides a, b, and c, then

$$\frac{a}{\sin A} = \frac{b}{\sin B} = \frac{c}{\sin C}.$$

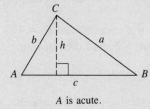 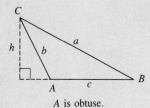

A is acute.   A is obtuse.

Oblique Triangles

FIGURE 4.2

PROOF

Let h be the altitude of either triangle shown in Figure 4.2. Then, we have

$$\sin A = \frac{h}{b} \qquad \text{or} \qquad h = b \sin A$$

$$\sin B = \frac{h}{a} \qquad \text{or} \qquad h = a \sin B.$$

Equating these two values of h, we have

$$a \sin B = b \sin A \qquad \text{or} \qquad \frac{a}{\sin A} = \frac{b}{\sin B}.$$

Note that $\sin A \neq 0$ and $\sin B \neq 0$ because no angle of a triangle can have a measure of $0°$ or $180°$. By constructing an altitude from vertex B to extended side AC, we can show that

$$\frac{a}{\sin A} = \frac{c}{\sin C}.$$

Hence, the Law of Sines is established.

Remark: The Law of Sines can also be written in the form

$$\frac{\sin A}{a} = \frac{\sin B}{b} = \frac{\sin C}{c}.$$

Additional Applications of Trigonometry

When using a calculator with the Law of Sines, remember to store all intermediate calculations. By not rounding until the final result, you minimize the round-off error.

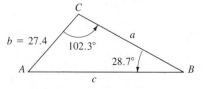

FIGURE 4.3

EXAMPLE 1 *Given Two Angles and One Side—AAS*

Given a triangle with $C = 102.3°$, $B = 28.7°$, and $b = 27.4$, as shown in Figure 4.3, find the remaining angle and sides.

SOLUTION

The third angle of the triangle is

$$A = 180° - B - C = 180° - 28.7° - 102.3° = 49.0°.$$

By the Law of Sines, we have

$$\frac{a}{\sin 49°} = \frac{b}{\sin 28.7°} = \frac{c}{\sin 102.3°}.$$

Because $b = 27.4$, we obtain

$$a = \frac{27.4}{\sin 28.7°}(\sin 49°) \approx 43.06$$

and

$$c = \frac{27.4}{\sin 28.7°}(\sin 102.3°) \approx 55.75.$$

Note that the ratio $(27.4)/(\sin 28.7°)$ occurs in both solutions, and you can save time by storing this result for repeated use.

When solving triangles, a careful sketch is useful as a quick test for the feasibility of an answer. Remember that the longest side lies opposite the largest angle, and the shortest side lies opposite the smallest angle of a triangle.

EXAMPLE 2 *Given Two Angles and One Side—ASA*

A pole tilts *toward* the sun at an 8° angle from vertical, and it casts a 22-foot shadow. The angle of elevation from the tip of the shadow to the top of the pole is 43°. How tall is the pole?

Law of Sines

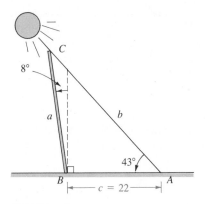

FIGURE 4.4

SOLUTION

From Figure 4.4, note that $A = 43°$, and $B = 90° + 8° = 98°$. Thus, the third angle is

$$C = 180° - A - B = 180° - 43° - 98° = 39°.$$

By the Law of Sines, we have

$$\frac{a}{\sin 43°} = \frac{c}{\sin 39°}.$$

Because $c = 22$ feet, the length of the pole is

$$a = \frac{22}{\sin 39°}(\sin 43°) \approx 23.84 \text{ feet.}$$

Remark: For practice, try reworking Example 2 for a pole that tilts *away from* the sun under the same conditions.

In Examples 1 and 2 we saw that two angles (whose sum is less than 180°) and one side determine a unique triangle. However, if two sides and one opposite angle are given, three possible situations can occur: (1) no such triangle exists, (2) one such triangle exists, or (3) two distinct triangles may satisfy the conditions. The possibilities in this *ambiguous* case (SSA) are summarized in the following table.

The Ambiguous Case (SSA) (Given: a, b, and A)

	A Is Acute				A Is Obtuse	
Sketch ($h = b \sin A$)						
Necessary Condition	$a < h$	$a = h$	$a > b$	$h < a < b$	$a \le b$	$a > b$
Triangles Possible	None	One	One	Two	None	One

In order to determine which of the possibilities hold for a given pair of sides and opposite angle, we suggest that you make a sketch, as demonstrated in the next three examples.

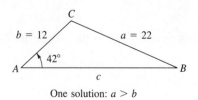

One solution: $a > b$

FIGURE 4.5

EXAMPLE 3   *Single Solution Case—SSA*

Given a triangle with $a = 22$, $b = 12$, and $A = 42°$, find the remaining side and angles.

SOLUTION

Since A is acute and $a > b$, we know that there is only one triangle that satisfies the given conditions, as shown in Figure 4.5. Thus, by the Law of Sines, we have

$$\frac{22}{\sin 42°} = \frac{12}{\sin B}$$

which implies that

$$\sin B = 12\left(\frac{\sin 42°}{22}\right) \approx 0.3649803$$
$$B \approx 21.41°.$$

Now, we can determine that $C \approx 180° - 42° - 21.41° = 116.59°$, and the remaining side is given by

$$\frac{c}{\sin 116.59°} = \frac{22}{\sin 42°}$$

$$c = \sin 116.59°\left(\frac{22}{\sin 42°}\right) \approx 29.40.$$

EXAMPLE 4   *No Solution Case—SSA*

Show that there is no triangle that satisfies either of the following conditions.

(a) $a = 15$, $b = 25$, $A = 85°$
(b) $a = 15.2$, $b = 20$, $A = 110°$

SOLUTION

(a) We begin by making the sketch shown in Figure 4.6. From this figure it appears that no triangle is formed. We can verify this using the Law of Sines, as follows.

$$\frac{a}{\sin A} = \frac{b}{\sin B}$$

$$\frac{15}{\sin 85°} = \frac{25}{\sin B}$$

$$\sin B = 25\left(\frac{\sin 85°}{15}\right) \approx 1.660 > 1$$

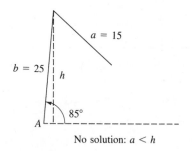

No solution: $a < h$

FIGURE 4.6

Law of Sines

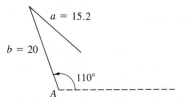

No solution: $a < b$ and $A > 90°$

FIGURE 4.7

This contradicts the fact that $|\sin B| \le 1$. Hence, no triangle can be formed having sides $a = 15$ and $b = 25$ and an angle of $A = 85°$.

(b) Because A is obtuse and $a = 15.2$ is less than $b = 20$, we conclude that there is *no solution*, as shown in Figure 4.7. Try using the Law of Sines to verify this. ▬

EXAMPLE 5 *Two-Solution Case—SSA*

Find two triangles for which $a = 12$, $b = 31$, and $A = 20.5°$.

SOLUTION

To begin, we note that

$$h = b \sin A = 31(\sin 20.5°) \approx 10.86.$$

Hence, $h < a < b$ and we conclude that there are two possible triangles. By the Law of Sines, we obtain

$$\frac{a}{\sin A} = \frac{b}{\sin B}$$

which implies that

$$\sin B = b\left(\frac{\sin A}{a}\right) = 31\left(\frac{\sin 20.5°}{12}\right) \approx 0.9047.$$

There are two angles $B_1 \approx 64.8°$ and $B_2 \approx 115.2°$ between $0°$ and $180°$ whose sine is 0.9047. For $B_1 \approx 64.8°$, we obtain

$$C \approx 180° - 20.5° - 64.8° = 94.7°$$

$$c = \frac{a}{\sin A}(\sin C) = \frac{12}{\sin 20.5°}(\sin 94.7°) \approx 34.15.$$

For $B_2 \approx 115.2°$, we obtain

$$C \approx 180° - 20.5° - 115.2° = 44.3°$$

$$c = \frac{a}{\sin A}(\sin C) = \frac{12}{\sin 20.5°}(\sin 44.3°) \approx 23.9.$$

The resulting triangles are shown in Figures 4.8 and 4.9.

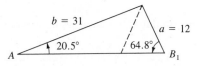

FIGURE 4.8

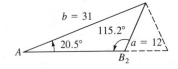

FIGURE 4.9

▬

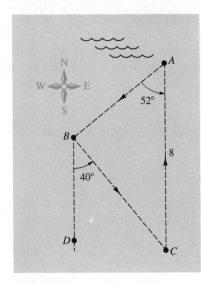

FIGURE 4.10

EXAMPLE 6 An Application of the Law of Sines

The course for a boat race starts at point A and proceeds in the direction S 52° W to point B, then in the direction S 40° E to point C, and finally back to A, as shown in Figure 4.10. The point C lies 8 kilometers directly south of point A. Approximate the total distance of the race course.

SOLUTION

Because lines BD and AC are parallel, it follows that $\angle BCA = \angle DBC$. Consequently, triangle ABC has the measures shown in Figure 4.11. For angle B, we have

$$B = 180° - 52° - 40° = 88°.$$

Thus, using the Law of Sines,

$$\frac{a}{\sin 52°} = \frac{b}{\sin 88°} = \frac{c}{\sin 40°},$$

we let $b = 8$ and obtain

$$a = \frac{8}{\sin 88°}(\sin 52°) \approx 6.308$$

$$c = \frac{8}{\sin 88°}(\sin 40°) \approx 5.145.$$

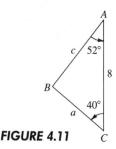

FIGURE 4.11

Finally, the total length of the course is approximately

Length $\approx 8 + 6.308 + 5.145 = 19.453$ kilometers.

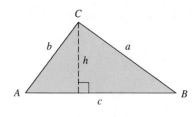

A is acute.

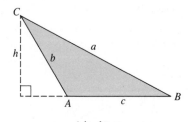

A is obtuse.

Oblique Triangles

FIGURE 4.12

The procedure used to prove the Law of Sines leads to a simple formula for the area of an oblique triangle. Referring to Figure 4.12, we note that each triangle has a height of

$$h = b \sin A.$$

Consequently, the area of each triangle is given by

$$\text{Area} = \frac{1}{2}(\text{base})(\text{height}) = \frac{1}{2}(c)(b \sin A) = \frac{1}{2}bc \sin A.$$

By similar arguments, we can develop the formulas

$$\text{Area} = \frac{1}{2}ab \sin C = \frac{1}{2}ac \sin B.$$

This leads to the following theorem.

Law of Sines

Area of an Oblique Triangle

The area of any triangle is given by one-half the product of the lengths of two sides times the sine of their included angle. That is,

$$\text{Area} = \frac{1}{2}bc \sin A = \frac{1}{2}ab \sin C = \frac{1}{2}ac \sin B.$$

EXAMPLE 7   *Finding the Area of an Oblique Triangle*

Find the area of a triangular lot having two sides of lengths 90 meters and 52 meters and an included angle of 102°.

SOLUTION

Consider $a = 90$ m, $b = 52$ m, and angle $C = 102°$, as shown in Figure 4.13. Then the area of the triangle is

$$\text{Area} = \frac{1}{2}ab \sin C = \frac{1}{2}(90)(52)(\sin 102°)$$

$$\approx 2289 \text{ square meters.}$$

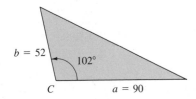

FIGURE 4.13

WARM UP

Solve the *right* triangle shown in the figure.

1. $a = 3$, $c = 6$ **2.** $a = 5$, $b = 5$

3. $b = 15$, $c = 17$ **4.** $A = 42°$, $a = 7.5$

5. $B = 10°$, $b = 4$ **6.** $B = 72° \, 15'$, $c = 150$

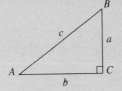

Find the altitude h of each triangle.

7.

8.

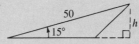

Solve for x.

9. $\dfrac{2}{\sin 30°} = \dfrac{9}{x}$

10. $\dfrac{100}{\sin 72°} = \dfrac{x}{\sin 60°}$

EXERCISES 4.1

In Exercises 1–16, find the remaining sides and angles of the triangle.

1.

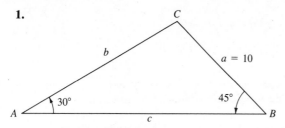

2.

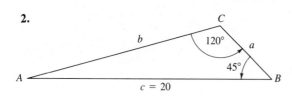

3.

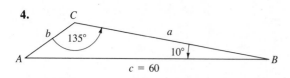

4.

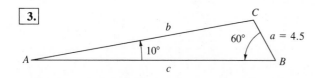

5. $A = 36°$, $a = 8$, $b = 5$ **6.** $A = 60°$, $a = 9$, $c = 10$

7. $A = 150°$, $C = 20°$, $a = 200$

8. $A = 24.3°$, $C = 54.6°$, $c = 2.68$

9. $A = 83° 20'$, $C = 54.6°$, $c = 18.1$

10. $A = 5° 40'$, $B = 8° 15'$, $b = 4.8$

11. $B = 15° 30'$, $a = 4.5$, $b = 6.8$

12. $C = 85° 20'$, $a = 35$, $c = 50$

13. $C = 145°$, $b = 4$, $c = 14$

14. $A = 100°$, $a = 125$, $c = 10$

15. $A = 110° 15'$, $a = 48$, $b = 16$

16. $B = 2° 45'$, $b = 6.2$, $c = 5.8$

In Exercises 17–22, solve the triangle, if possible. If two solutions exist, find both.

17. $A = 58°$, $a = 4.5$, $b = 12.8$

18. $A = 58°$, $a = 11.4$, $b = 12.8$

19. $A = 58°$, $a = 4.5$, $b = 5$

20. $A = 58°$, $a = 42.4$, $b = 50$

21. $A = 110°$, $a = 125$, $b = 200$

22. $A = 110°$, $a = 125$, $b = 100$

In Exercises 23 and 24, find a value for b such that the triangle has (a) one solution, (b) two solutions, and (c) no solution.

23. $A = 36°$, $a = 5$ **24.** $A = 60°$, $a = 10$

In Exercises 25–30, find the area of the triangle having the indicated sides and angles.

25. $C = 120°$, $a = 4$, $b = 6$

26. $B = 72° 30'$, $a = 105$, $c = 64$

27. $A = 43° 45'$, $b = 57$, $c = 85$

28. $A = 5° 15'$, $b = 4.5$, $c = 22$

29. $B = 130°$, $a = 62$, $c = 20$

30. $C = 84° 30'$, $a = 16$, $b = 20$

31. Find the length d of the brace required to support the streetlight shown in the figure.

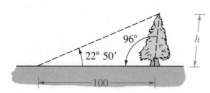

FIGURE FOR 31

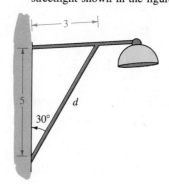

FIGURE FOR 32

32. Because of prevailing winds, a tree grew so that it was leaning 6° from the vertical. At a point 100 feet from the tree, the angle of elevation to the top of the tree is 22° 50′ (see figure). Find the height h of the tree.

33. A bridge is to be built across a small lake from B to C (see figure). The bearing from B to C is S 41° W. From a point A, 100 yards from B, the bearings to B and C are S 74° E and S 28° E, respectively. Find the distance from B to C.

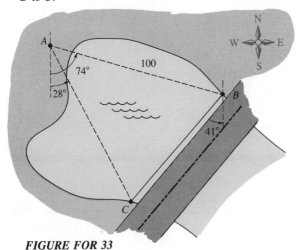

FIGURE FOR 33

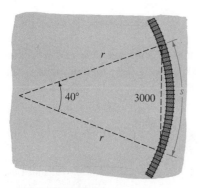

FIGURE FOR 34

34. The circular arc of a railroad curve has a chord of length 3000 feet, and a central angle of 40° (see figure). Find (a) the radius r of the circular arc and (b) the length s of the circular arc.

35. The angles of elevation to an airplane from two points A and B on level ground are 51° and 68°, respectively. A and B are 6 miles apart, and the airplane is between A and B in the same vertical plane. Find the altitude of the airplane.

36. The angles of elevation to an airplane from two points A and B on level ground are 51° and 68°, respectively. A and B are 2.5 miles apart, and the airplane is to the east of A and B in the same vertical plane. Find the altitude of the airplane.

37. Two fire towers A and B are 18.5 miles apart. The bearing from A to B is N 65° E. A fire is spotted by the ranger in each tower, and its bearings from A and B are N 28° E and N 16.5° W, respectively (see figure). Find the distance of the fire from each tower.

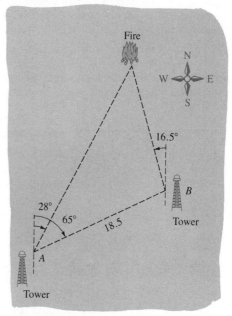

FIGURE FOR 37

38. A boat is sailing due east parallel to the shoreline at a speed of 10 miles per hour. At a given time the bearing to a lighthouse is S 72° E, and 15 minutes later the bearing is S 66° E (see figure). Find the distance from the boat to the shoreline if the lighthouse is at the shoreline.

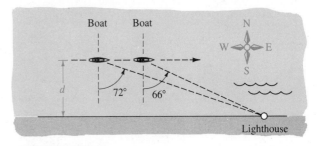

FIGURE FOR 38

39. A family is traveling due west on a road that passes a famous landmark. At a given time the bearing to the landmark is N 62° W, and after the family travels 5 miles farther the bearing is N 38° W. What is the closest the family will come to the landmark while on the road?

40. A rescue vehicle is located near an apartment complex, and its emergency light is turning at the rate of 30 revolutions per minute. One-eighth of a second after illuminating the nearest point on the apartment complex, the lightbeam reaches a point 50 feet along the apartments. One-eighth of a second later, it reaches a point 70 feet further along the apartment wall. How far is the rescue truck from the apartment complex?

41. The following information about a triangular parcel of land is given at a zoning board meeting: "One side is 450 feet long, and another is 120 feet long. The angle opposite the shorter side is 30°." Could this information be correct?

42. The angles of elevation, θ and ϕ, to an airplane are being continuously monitored at two observation points A and B which are two miles apart (see figure). Write an equation giving the distance d between the plane and point B in terms of θ and ϕ.

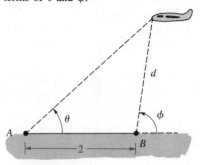

FIGURE FOR 42

4.2 Law of Cosines

Two cases remain in our list of conditions needed to solve an oblique triangle—SSS and SAS. In Section 4.1, we mentioned that the Law of Sines would not work in either of these cases. To see why, consider the three ratios given in the Law of Sines:

$$\frac{a}{\sin A} = \frac{b}{\sin B} = \frac{c}{\sin C}.$$

To use the Law of Sines we must know at least one side and its opposite angle. If we are given three sides (SSS), or two sides and their included angle (SAS), none of the above ratios would be complete. In such cases we rely on the **Law of Cosines.**

Law of Cosines

If ABC is a triangle with sides a, b, and c, then the following equations are valid.

Standard Form	*Alternative Form*
$a^2 = b^2 + c^2 - 2bc \cos A$	$\cos A = \dfrac{b^2 + c^2 - a^2}{2bc}$
$b^2 = a^2 + c^2 - 2ac \cos B$	$\cos B = \dfrac{a^2 + c^2 - b^2}{2ac}$
$c^2 = a^2 + b^2 - 2ab \cos C$	$\cos C = \dfrac{a^2 + b^2 - c^2}{2ab}$

Law of Cosines

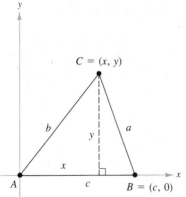

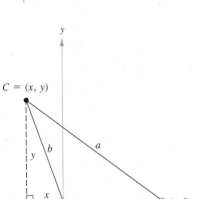

FIGURE 4.14

PROOF

In Figure 4.14, vertex B has coordinates $(c, 0)$. Furthermore, C has coordinates (x, y), where $x = b \cos A$ and $y = b \sin A$. Since a is the distance from vertex C to vertex B, it follows that

$$a = \sqrt{(x - c)^2 + (y - 0)^2}$$
$$a^2 = (b \cos A - c)^2 + (b \sin A)^2$$
$$= b^2 \cos^2 A - 2bc \cos A + c^2 + b^2 \sin^2 A$$
$$= b^2(\sin^2 A + \cos^2 A) + c^2 - 2bc \cos A.$$

Using the identity $\sin^2 A + \cos^2 A = 1$, we then obtain

$$a^2 = b^2 + c^2 - 2bc \cos A.$$

Similar arguments with angles B and C in standard position establish the other two equations.

Note that if $A = 90°$ in the right-hand triangle in Figure 4.14, then $\cos A = 0$ and the first form of the Law of Cosines becomes the Pythagorean Theorem:

$$a^2 = b^2 + c^2.$$

Thus, the Pythagorean Theorem is actually just a special case of the more general Law of Cosines.

EXAMPLE 1 *Given Three Sides of a Triangle—SSS*

Find the three angles of the triangle whose sides have lengths $a = 8.65$, $b = 19.2$, and $c = 13.7$.

SOLUTION

It is a good idea first to find the angle opposite the longest side—side b in this case (see Figure 4.15). Using the Law of Cosines, we find that

$$\cos B = \frac{a^2 + c^2 - b^2}{2ac}$$

$$= \frac{(8.65)^2 + (13.7)^2 - (19.2)^2}{2(8.65)(13.7)}$$

$$\approx -0.4477765.$$

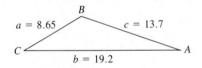

FIGURE 4.15

Since cos B is negative, we know B is an *obtuse* angle given by

$$B \approx 116.60°.$$

At this point we could use the Law of Cosines to find cos A and cos C. However, knowing that $B \approx 116.60°$, it is simpler to use the Law of Sines to obtain

$$\frac{b}{\sin B} = \frac{a}{\sin A}$$

$$\sin A = a\left(\frac{\sin B}{b}\right)$$

$$\approx 8.65\left(\frac{\sin 116.60°}{19.2}\right)$$

$$\approx 0.4028351.$$

Since B is obtuse, we know that A must be acute, because a triangle can have, at most, one obtuse angle. Thus, $A \approx 23.76°$ and

$$C \approx 180° - 23.76° - 116.60° = 39.64°.$$

Do you see why it was wise to find the largest angle *first* in Example 1? Knowing the cosine of an angle, we can determine whether the angle is acute or obtuse. That is,

$$\cos \theta > 0 \quad \text{for} \quad 0° < \theta < 90° \qquad \textit{(Acute)}$$
$$\cos \theta < 0 \quad \text{for} \quad 90° < \theta < 180°. \qquad \textit{(Obtuse)}$$

So, in Example 1, once we found that B was obtuse, we subsequently knew that angles A and C were both acute. If the largest angle is acute, then the remaining two angles will be acute also.

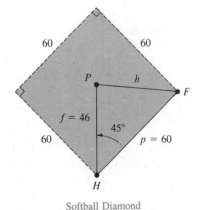

Softball Diamond

FIGURE 4.16

EXAMPLE 2 *Given Two Sides and the Included Angle—SAS*

The pitcher's mound on a softball field is 46 feet from home plate, and the distance between the bases is 60 feet, as shown in Figure 4.16. How far is the pitcher's mound from first base? (Note that the pitcher's mound is *not* halfway between home plate and second base.)

SOLUTION

In triangle HPF, we have $H = 45°$ (line HP bisects the right angle at H), $f = 46$, and $p = 60$. Using the Law of Cosines for this SAS case, we have

$$h^2 = f^2 + p^2 - 2fp \cos H$$
$$= 46^2 + 60^2 - 2(46)(60) \cos 45°$$
$$\approx 1812.8.$$

Therefore, the approximate distance from the pitcher's mound to first base is

$$h \approx \sqrt{1812.8} \approx 42.58 \text{ feet.}$$

EXAMPLE 3 Given Two Sides and the Included Angle—SAS

A ship travels 60 miles due east, then adjusts its course 15° northward, as shown in Figure 4.17. After traveling 80 miles in that direction, how far is the ship from its point of departure?

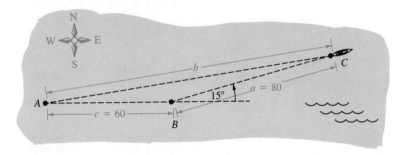

FIGURE 4.17

SOLUTION

We have $c = 60$, $B = 180° - 15° = 165°$, and $a = 80$. Consequently, by the Law of Cosines, we find that

$$b^2 = a^2 + c^2 - 2ac \cos B$$
$$= 80^2 + 60^2 - 2(80)(60) \cos 165°$$
$$\approx 19,273.$$

Therefore, the distance b is

$$b \approx \sqrt{19,273} \approx 138.8 \text{ miles.}$$

The Law of Cosines can be used to establish the following formula for the area of a triangle. This formula is credited to the Greek mathematician Heron (c. 100 B.C.).

Additional Applications of Trigonometry

Heron's Area Formula

Given any triangle with sides of length a, b, and c, the area of the triangle is

$$\text{Area} = \sqrt{s(s-a)(s-b)(s-c)}$$

where $s = (a + b + c)/2$.

PROOF

From the previous section, we know that

$$\text{Area} = \frac{1}{2}bc \sin A = \sqrt{\frac{1}{4}b^2c^2 \sin^2 A} = \sqrt{\frac{1}{4}b^2c^2(1 - \cos^2 A)}$$

$$= \sqrt{\left[\frac{1}{2}bc(1 + \cos A)\right]\left[\frac{1}{2}bc(1 - \cos A)\right]}.$$

Using the Law of Cosines, we can show that

$$\frac{1}{2}bc(1 + \cos A) = \frac{a + b + c}{2} \cdot \frac{-a + b + c}{2}$$

and

$$\frac{1}{2}bc(1 - \cos A) = \frac{a - b + c}{2} \cdot \frac{a + b - c}{2}.$$

(See Exercises 37 and 38.) Letting $s = (a + b + c)/2$, these two equations can be rewritten as

$$\frac{1}{2}bc(1 + \cos A) = s(s - a) \quad \text{and} \quad \frac{1}{2}bc(1 - \cos A) = (s - b)(s - c).$$

Thus, we conclude that

$$\text{Area} = \sqrt{s(s - a)(s - b)(s - c)}.$$

EXAMPLE 4 *Using Heron's Area Formula*

Find the area of the triangular region having sides of lengths $a = 47$ yards, $b = 58$ yards, and $c = 78.6$ yards.

SOLUTION

Since

$$s = \frac{1}{2}(a + b + c) = \frac{183.6}{2} = 91.8$$

Heron's Formula yields

$$\text{Area} = \sqrt{s(s-a)(s-b)(s-c)}$$
$$= \sqrt{91.8(44.8)(33.8)(13.2)}$$
$$\approx 1{,}354.58 \text{ square yards.}$$

WARM UP

Simplify the expressions.

1. $\sqrt{(7-3)^2 + [1-(-5)]^2}$

2. $\sqrt{[-2-(-5)]^2 + (12-6)^2}$

Find the distance between the two points.

3. $(4, -2), (8, 10)$

4. $(1, 3), (7, 12)$

Find the area of each triangle.

5.

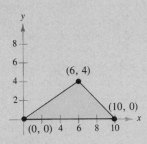

6.

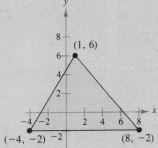

Find the remaining sides and angles of each triangle (if possible).

7. $A = 10°, C = 100°, b = 25$

8. $A = 20°, C = 90°, c = 100$

9. $B = 30°, b = 6.5, c = 15$

10. $A = 30°, b = 6.5, a = 10$

EXERCISES 4.2

In Exercises 1–14, use the Law of Cosines to solve the given triangle.

1.

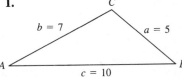

2.

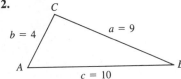

3.

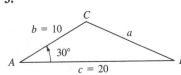

4.

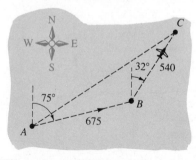

$b = 4.5$
$a = 10$
$110°$

5. $a = 9, b = 12, c = 15$ **6.** $a = 55, b = 25, c = 72$
7. $a = 75.4, b = 52, c = 52$
8. $a = 1.42, b = 0.75, c = 1.25$
9. $A = 120°, b = 3, c = 10$
10. $A = 55°, b = 3, c = 10$
11. $B = 8° 45', a = 25, c = 15$
12. $B = 75° 20', a = 6.2, c = 9.5$
13. $C = 125° 40', a = 32, b = 32$
14. $C = 15°, a = 6.25, b = 2.15$

In Exercises 15–20, use Heron's Formula to find the area of the triangle.

15. $a = 5, b = 7, c = 10$ **16.** $a = 2.5, b = 10.2, c = 9$
17. $a = 12, b = 15, c = 9$
18. $a = 75.4, b = 52, c = 52$
19. $a = 20, b = 20, c = 10$
20. $a = 4.25, b = 1.55, c = 3.00$

21. A boat race occurs along a triangular course marked by buoys A, B, and C. The race starts with the boats going 8000 feet in a northerly direction. The other two sides of the course lie to the east of the first side, and their lengths are 3500 feet and 6500 feet (see figure). Find the bearings for the last two legs of the race.

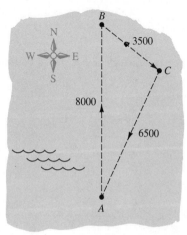

FIGURE FOR 21

22. A plane flies 675 miles from A to B with a bearing of N 75° E. Then it flies 540 miles from B to C with a bearing of N 32° E (see figure). Find the straight-line distance and bearing for the flight from C to A.

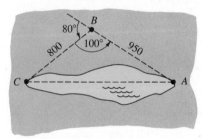

FIGURE FOR 22

23. Two ships leave a port at 9 A.M. One travels at a bearing of N 53° W at 12 miles per hour and the other at a bearing of S 67° W at 16 miles per hour. Approximately how far apart are they at noon that day?

24. A triangular parcel of land has 375 feet of frontage, and the two other boundaries have lengths of 250 feet and 300 feet. What angles does the frontage make with the two other boundaries?

25. A 100-foot vertical tower is to be erected on the side of a hill that makes an 8° angle with the horizontal (see figure). Find the lengths of each of the two guy wires that will be anchored 75 feet uphill and downhill from the base of the tower.

FIGURE FOR 25

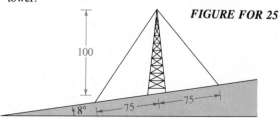

26. To approximate the length of a marsh, a surveyor walks 950 feet from point A to point B, then turns 80° and walks 800 feet to point C (see figure). Approximate the length $\overline{AC}$ of the marsh.

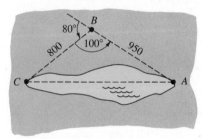

FIGURE FOR 26

27. Determine the angle θ as shown on the streetlight in the figure.

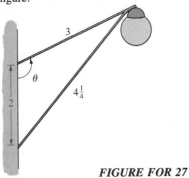

FIGURE FOR 27

28. In order to determine the distance between two aircraft, a tracking station continuously determines the distance to each aircraft and the angle α between them (see figure). Determine the distance a between the planes when $\alpha = 42°$, $b = 35$ miles, and $c = 20$ miles.

FIGURE FOR 28

29. If Q is the midpoint of the line segment $\overline{PR}$, find the lengths of the line segments $\overline{PQ}$, $\overline{QS}$, and $\overline{RS}$ on the truss rafter shown in the figure.

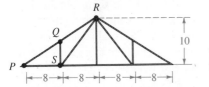

FIGURE FOR 29

30. In a certain process with continuous paper, the paper passes across three rollers of radii 3 inches, 4 inches, and 6 inches (see figure). The centers of the 3-inch and 6-inch rollers are d inches apart, and the length of the arc in contact with the paper on the 4-inch roller is s inches. Complete the following table.

d (inches)	9	10	11	12	13	14	15	16
θ (degrees)								
s (inches)								

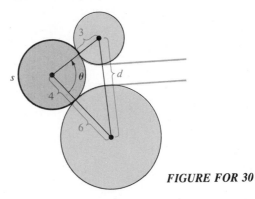

FIGURE FOR 30

31. In a (square) baseball diamond with 90-foot sides the pitcher's mound is 60 feet from home plate.
   (a) How far is it from the pitcher's mound to third base?
   (b) When a runner is halfway from second to third, how far is the runner from the pitcher's mound?

32. On a certain map, Minneapolis is 6.5 inches due west of Albany, Phoenix is 8.5 inches from Minneapolis, and Phoenix is 14.5 inches from Albany.
   (a) Find the bearing of Minneapolis from Phoenix.
   (b) Find the bearing of Albany from Phoenix.

33. On a certain map, Orlando is 7 inches due south of Niagara Falls, Denver is 10.75 inches from Orlando, and Denver is 9.25 inches from Niagara Falls (see figure).
   (a) Find the bearing of Denver from Orlando.
   (b) Find the bearing of Denver from Niagara Falls.

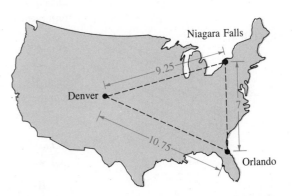

FIGURE FOR 33

34. Let R and r be the radii of the circumscribed and inscribed circles of triangle ABC, respectively, and let $s = (a + b + c)/2$ (see figure). Prove the following.

(a) $2R = \dfrac{a}{\sin A} = \dfrac{b}{\sin B} = \dfrac{c}{\sin C}$

(b) $r = \sqrt{\dfrac{(s - a)(s - b)(s - c)}{s}}$

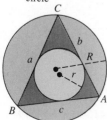

R = radius of the large circle

r = radius of the small circle

FIGURE FOR 34

In Exercises 35 and 36, use the results of Exercise 34.

35. Given the triangle with $a = 25$, $b = 55$, and $c = 72$, find the area of (a) the triangle, (b) the circumscribed circle, and (c) the inscribed circle.

36. Find the length of the largest circular track that can be built on a triangular piece of property whose sides are 200 feet, 250 feet, and 325 feet.

37. Use the Law of Cosines to prove that

$$\frac{1}{2}bc(1 + \cos A) = \frac{a + b + c}{2} \cdot \frac{-a + b + c}{2}.$$

38. Use the Law of Cosines to prove that

$$\frac{1}{2}bc(1 - \cos A) = \frac{a - b + c}{2} \cdot \frac{a + b - c}{2}.$$

4.3 Vectors

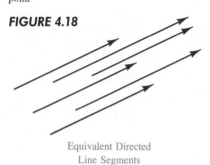

FIGURE 4.18

Equivalent Directed
Line Segments

FIGURE 4.19

Many quantities in geometry and physics, such as area, time, and temperature, can be represented by a single real number.

Other quantities, such as force and velocity, involve both *magnitude* and *direction* and cannot be completely characterized by a single real number. To represent such a quantity, we use a **directed line segment,** as shown in Figure 4.18. The directed line segment $\overrightarrow{PQ}$ has **initial point** P and **terminal point** Q and we denote its **length** by $\|\overrightarrow{PQ}\|$. Two directed line segments that have the same length (or magnitude) and direction are called **equivalent.** For example, the directed line segments in Figure 4.19 are all equivalent. We call the set of all directed line segments that are equivalent to a given directed line segment $\overrightarrow{PQ}$, a **vector v in the plane,** and write $\mathbf{v} = \overrightarrow{PQ}$.

Remark: We denote vectors by lower-case, boldface letters such as $\mathbf{u}$, $\mathbf{v}$, and $\mathbf{w}$.

Be sure you see that a vector in the plane can be represented by many different directed line segments. This is illustrated in the following example.

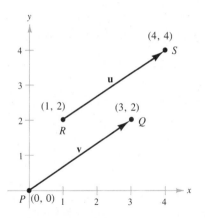

FIGURE 4.20

EXAMPLE 1 Vector Representation by Directed Line Segments

Let **u** be represented by the directed line segment from $P = (0, 0)$ to $Q = (3, 2)$, and let **v** be represented by the directed line segment from $R = (1, 2)$ to $S = (4, 4)$, as shown in Figure 4.20. Show that $\mathbf{u} = \mathbf{v}$.

SOLUTION

From the distance formula, we see that $\overrightarrow{PQ}$ and $\overrightarrow{RS}$ have the *same length*.

$$\|\overrightarrow{PQ}\| = \sqrt{(3 - 0)^2 + (2 - 0)^2} = \sqrt{13}$$

$$\|\overrightarrow{RS}\| = \sqrt{(4 - 1)^2 + (4 - 2)^2} = \sqrt{13}$$

Moreover, both line segments have the *same direction* since they are both directed toward the upper right on lines having a slope of $2/3$. Thus, $\overrightarrow{PQ}$ and $\overrightarrow{RS}$ have the same length and direction, and we conclude that $\mathbf{u} = \mathbf{v}$.

The directed line segment whose initial point is the origin is often the most convenient representative of a set of equivalent directed line segments. We say that this representative of the vector **v** is in **standard position.**

A vector whose initial point is at the origin $(0, 0)$ can be uniquely represented by the coordinates of its terminal point (v_1, v_2). We call this the **component form of a vector v** and write

$$\mathbf{v} = \langle v_1, v_2 \rangle.$$

The coordinates v_1 and v_2 are called the **components** of **v**. If both the initial point and the terminal point lie at the origin, then **v** is called the **zero vector** and is denoted by $\mathbf{0} = \langle 0, 0 \rangle$.

To convert directed line segments to component form we use the following procedure.

Component Form of a Vector

The component form of the vector with initial point $P = (p_1, p_2)$ and terminal point $Q = (q_1, q_2)$ is

$$\overrightarrow{PQ} = \langle q_1 - p_1, q_2 - p_2 \rangle = \langle v_1, v_2 \rangle = \mathbf{v}.$$

The *length* (or magnitude) of **v** is given by

$$\|\mathbf{v}\| = \sqrt{(q_1 - p_1)^2 + (q_2 - p_2)^2} = \sqrt{v_1^2 + v_2^2}.$$

If $\|\mathbf{v}\| = 1$, then **v** is called a **unit vector.** Moreover, $\|\mathbf{v}\| = 0$ if and only if **v** is the zero vector **0**.

Two vectors $\mathbf{u} = \langle u_1, u_2 \rangle$ and $\mathbf{v} = \langle v_1, v_2 \rangle$ are **equal** if and only if $u_1 = v_1$ and $u_2 = v_2$. For instance, in Example 1, the vector $\mathbf{u}$ from $P = (0, 0)$ to $Q = (3, 2)$ is

$$\mathbf{u} = \overrightarrow{PQ} = \langle 3 - 0, 2 - 0 \rangle = \langle 3, 2 \rangle$$

and the vector $\mathbf{v}$ from $R = (1, 2)$ to $S = (4, 4)$ is

$$\mathbf{v} = \overrightarrow{RS} = \langle 4 - 1, 4 - 2 \rangle = \langle 3, 2 \rangle.$$

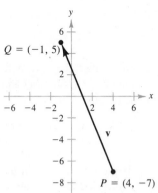

Component form of $\mathbf{v}$:

$$\mathbf{v} = \langle -5, 12 \rangle$$

FIGURE 4.21

EXAMPLE 2 *Finding the Component Form and Length of a Vector*

Find the component form and length of the vector $\mathbf{v}$ that has initial point $(4, -7)$ and terminal point $(-1, 5)$.

SOLUTION

We let $P = (4, -7) = (p_1, p_2)$ and $Q = (-1, 5) = (q_1, q_2)$. Then, the components of $\mathbf{v} = \langle v_1, v_2 \rangle$ are given by

$$v_1 = q_1 - p_1 = -1 - 4 = -5$$
$$v_2 = q_2 - p_2 = 5 - (-7) = 12.$$

Thus, $\mathbf{v} = \langle -5, 12 \rangle$ and the length of $\mathbf{v}$ is

$$\|\mathbf{v}\| = \sqrt{(-5)^2 + 12^2} = \sqrt{169} = 13$$

as shown in Figure 4.21.

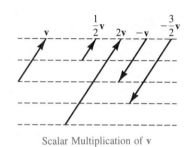

Scalar Multiplication of $\mathbf{v}$

FIGURE 4.22

Vector Operations

The two basic vector operations are **scalar multiplication** and **vector addition.** (In this text, we use the term **scalar** to mean a real number.) Geometrically, the product of a vector $\mathbf{v}$ and a scalar k is the vector that is k times as long as $\mathbf{v}$. If k is positive, then $k\mathbf{v}$ has the same direction as $\mathbf{v}$, and if k is negative, then $k\mathbf{v}$ has the opposite direction of $\mathbf{v}$, as shown in Figure 4.22.

To add two vectors geometrically, we position them (without changing length or direction) so that the initial point of one coincides with the terminal point of the other. The sum $\mathbf{u} + \mathbf{v}$ is formed by joining the initial point of

Vectors

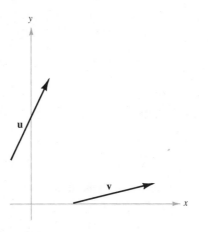

FIGURE 4.23   To find $\mathbf{u} + \mathbf{v}$,   move the initial point of $\mathbf{v}$ to the terminal point of $\mathbf{u}$.

the second vector $\mathbf{v}$ with the terminal point of the first vector $\mathbf{u}$, as shown in Figure 4.23. Since the vector $\mathbf{u} + \mathbf{v}$ is the diagonal of a parallelogram having $\mathbf{u}$ and $\mathbf{v}$ as its adjacent sides, we call this the **parallelogram law** for vector addition.

Vector addition and scalar multiplication can also be defined using components of vectors.

Definition of Vector Addition and Scalar Multiplication

Let $\mathbf{u} = \langle u_1, u_2 \rangle$ and $\mathbf{v} = \langle v_1, v_2 \rangle$ be vectors and let k be a scalar (a real number). Then the **sum** of $\mathbf{u}$ and $\mathbf{v}$ is the vector

$$\mathbf{u} + \mathbf{v} = \langle u_1 + v_1, u_2 + v_2 \rangle \qquad \textit{Sum}$$

and the **scalar multiple** of k times $\mathbf{u}$ is the vector

$$k\mathbf{u} = \langle ku_1, ku_2 \rangle. \qquad \textit{Scalar multiple}$$

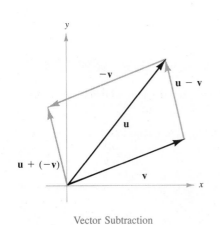

Vector Subtraction

FIGURE 4.24

The **negative** of $\mathbf{v} = \langle v_1, v_2 \rangle$ is

$$-\mathbf{v} = (-1)\mathbf{v} = \langle -v_1, -v_2 \rangle \qquad \textit{Negative}$$

and the **difference** of $\mathbf{u}$ and $\mathbf{v}$ is

$$\mathbf{u} - \mathbf{v} = \mathbf{u} + (-\mathbf{v}) = \langle u_1 - v_1, u_2 - v_2 \rangle. \qquad \textit{Difference}$$

To represent $\mathbf{u} - \mathbf{v}$ graphically, we use directed line segments with the *same* initial points. The difference $\mathbf{u} - \mathbf{v}$ is the vector from the terminal point of $\mathbf{v}$ to the terminal point of $\mathbf{u}$, as shown in Figure 4.24.

(a)

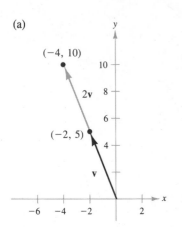

(b)

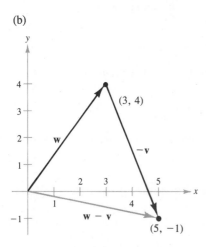

(c)

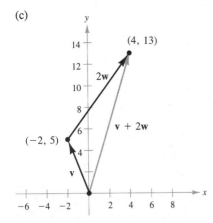

FIGURE 4.25

EXAMPLE 3 *Vector Operations*

Let $\mathbf{v} = \langle -2, 5 \rangle$ and $\mathbf{w} = \langle 3, 4 \rangle$, and find the following vectors.

(a) $2\mathbf{v}$ (b) $\mathbf{w} - \mathbf{v}$ (c) $\mathbf{v} + 2\mathbf{w}$

SOLUTION

(a) Since $\mathbf{v} = \langle -2, 5 \rangle$, we have

$$2\mathbf{v} = \langle 2(-2), 2(5) \rangle = \langle -4, 10 \rangle.$$

A sketch of $2\mathbf{v}$ is shown in Figure 4.25(a).

(b) The difference of $\mathbf{w}$ and $\mathbf{v}$ is given by

$$\mathbf{w} - \mathbf{v} = \langle 3 - (-2), 4 - 5 \rangle = \langle 5, -1 \rangle.$$

A sketch of $\mathbf{w} - \mathbf{v}$ is shown in Figure 4.25(b).

(c) Since $2\mathbf{w} = \langle 6, 8 \rangle$, it follows that

$$\mathbf{v} + 2\mathbf{w} = \langle -2, 5 \rangle + \langle 6, 8 \rangle$$
$$= \langle -2 + 6, 5 + 8 \rangle$$
$$= \langle 4, 13 \rangle.$$

A sketch of $\mathbf{v} + 2\mathbf{w}$ is shown in Figure 4.25(c).

Vector addition and scalar multiplication share many of the properties of ordinary arithmetic.

Properties of Vector Addition and Scalar Multiplication

Let $\mathbf{u}$, $\mathbf{v}$, and $\mathbf{w}$ be vectors and let c and d be scalars. Then the following properties are true.

1. $\mathbf{u} + \mathbf{v} = \mathbf{v} + \mathbf{u}$
2. $(\mathbf{u} + \mathbf{v}) + \mathbf{w} = \mathbf{u} + (\mathbf{v} + \mathbf{w})$
3. $\mathbf{u} + \mathbf{0} = \mathbf{u}$
4. $\mathbf{u} + (-\mathbf{u}) = \mathbf{0}$
5. $c(d\mathbf{u}) = (cd)\mathbf{u}$
6. $(c + d)\mathbf{u} = c\mathbf{u} + d\mathbf{u}$
7. $c(\mathbf{u} + \mathbf{v}) = c\mathbf{u} + c\mathbf{v}$
8. $1(\mathbf{u}) = \mathbf{u}, \quad 0(\mathbf{u}) = \mathbf{0}$
9. $\|c\mathbf{v}\| = |c| \, \|\mathbf{v}\|$

Remark: Property 9 can be stated as follows: The length of the vector $c\mathbf{v}$ is the absolute value of c times the length of $\mathbf{v}$.

In many applications of vectors it is useful to find a unit vector which has the same direction as a given vector **v**. To do this, we divide **v** by its length to obtain

$$\mathbf{u} = \left(\frac{1}{\|\mathbf{v}\|}\right)\mathbf{v} = \frac{\mathbf{v}}{\|\mathbf{v}\|}.$$

Note that **u** is a scalar multiple of **v**. The vector **u** has length 1 and the same direction as **v**. We call **u** a **unit vector in the direction of v.**

EXAMPLE 4 Finding a Unit Vector

Find a unit vector in the direction of $\mathbf{v} = \langle -2, 5 \rangle$ and verify that the result has length 1.

SOLUTION

The unit vector in the direction of **v** is

$$\frac{\mathbf{v}}{\|\mathbf{v}\|} = \frac{\langle -2, 5 \rangle}{\sqrt{(-2)^2 + (5)^2}}$$

$$= \frac{1}{\sqrt{29}}\langle -2, 5 \rangle = \left\langle \frac{-2}{\sqrt{29}}, \frac{5}{\sqrt{29}} \right\rangle.$$

This vector has length 1 since

$$\sqrt{\left(\frac{-2}{\sqrt{29}}\right)^2 + \left(\frac{5}{\sqrt{29}}\right)^2} = \sqrt{\frac{4}{29} + \frac{25}{29}} = \sqrt{\frac{29}{29}} = 1.$$

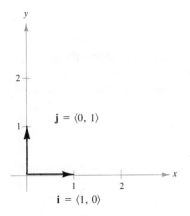

Standard Unit Vectors **i** and **j**

FIGURE 4.26

Standard Unit Vectors

The unit vectors $\langle 1, 0 \rangle$ and $\langle 0, 1 \rangle$ are called the **standard unit vectors** and are denoted by

$$\mathbf{i} = \langle 1, 0 \rangle \qquad \text{and} \qquad \mathbf{j} = \langle 0, 1 \rangle$$

as shown in Figure 4.26. (Note that the lower-case letter **i** is written in boldface to distinguish it from the imaginary number $i = \sqrt{-1}$.) These vectors can be used to represent any vector $\mathbf{v} = \langle v_1, v_2 \rangle$ as follows.

$$\mathbf{v} = \langle v_1, v_2 \rangle = v_1\langle 1, 0 \rangle + v_2\langle 0, 1 \rangle = v_1\mathbf{i} + v_2\mathbf{j}$$

The scalars v_1 and v_2 are called the **horizontal** and **vertical components of v,** respectively. The vector sum $v_1\mathbf{i} + v_2\mathbf{j}$ is called a **linear combination** of the vectors **i** and **j**. Any vector in the plane can be expressed as a linear combination of the standard unit vectors **i** and **j**.

Additional Applications of Trigonometry

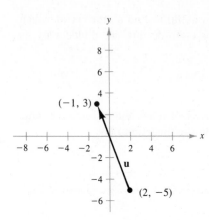

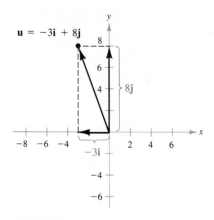

FIGURE 4.27

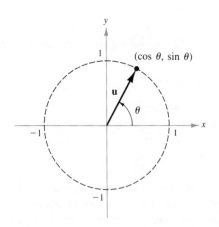

FIGURE 4.28

EXAMPLE 5 Representing a Vector as a Linear Combination of Unit Vectors

Let **u** be the vector with initial point $(2, -5)$ and terminal point $(-1, 3)$. Write **u** as a linear combination of the standard unit vectors **i** and **j**.

SOLUTION

$$
\begin{aligned}
\mathbf{u} &= \langle -1 - 2, 3 + 5 \rangle \\
&= \langle -3, 8 \rangle \\
&= -3\mathbf{i} + 8\mathbf{j}
\end{aligned}
$$

This result is shown graphically in Figure 4.27.

EXAMPLE 6 Vector Operations

Let $\mathbf{u} = -3\mathbf{i} + 8\mathbf{j}$ and $\mathbf{v} = 2\mathbf{i} - \mathbf{j}$. Find $2\mathbf{u} - 3\mathbf{v}$.

SOLUTION

$$
\begin{aligned}
2\mathbf{u} - 3\mathbf{v} &= 2(-3\mathbf{i} + 8\mathbf{j}) - 3(2\mathbf{i} - \mathbf{j}) \\
&= -6\mathbf{i} + 16\mathbf{j} - 6\mathbf{i} + 3\mathbf{j} \\
&= -12\mathbf{i} + 19\mathbf{j}
\end{aligned}
$$

If **u** is a *unit vector* such that θ is the angle (measured counterclockwise) from the positive x-axis to **u**, then the terminal point of **u** lies on the unit circle and we have

$$
\begin{aligned}
\mathbf{u} &= \langle \cos \theta, \sin \theta \rangle \\
&= (\cos \theta)\mathbf{i} + (\sin \theta)\mathbf{j}
\end{aligned}
$$

as shown in Figure 4.28. We call θ the **direction angle** of the vector **u**.

Suppose that **u** is a unit vector with direction angle θ. If **v** is any vector that makes an angle θ with the positive x-axis, then it has the same direction as **u** and we can write

$$
\begin{aligned}
\mathbf{v} &= \|\mathbf{v}\|\langle \cos \theta, \sin \theta \rangle \\
&= \|\mathbf{v}\|(\cos \theta)\mathbf{i} + \|\mathbf{v}\|(\sin \theta)\mathbf{j}.
\end{aligned}
$$

For instance, the vector **v** of length 3 making an angle of 30° with the positive x-axis is given by

$$
\mathbf{v} = 3(\cos 30°)\mathbf{i} + 3(\sin 30°)\mathbf{j} = \frac{3\sqrt{3}}{2}\mathbf{i} + \frac{3}{2}\mathbf{j}
$$

where $\|\mathbf{v}\| = 3$.

Vectors

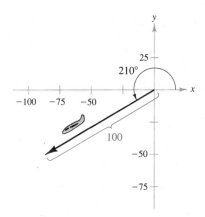

FIGURE 4.29

EXAMPLE 7 *Finding Component Form, Given Magnitude and Direction*

Find the component form of the vector that represents the velocity of an airplane descending at a speed of 100 miles per hour at an angle 30° below horizontal, as shown in Figure 4.29.

SOLUTION

The velocity vector **v** has magnitude of 100 and direction angle of $\theta = 210°$. Hence, the component form of **v** is

$$\mathbf{v} = \|\mathbf{v}\|(\cos \theta)\mathbf{i} + \|\mathbf{v}\|(\sin \theta)\mathbf{j}$$
$$= 100(\cos 210°)\mathbf{i} + 100(\sin 210°)\mathbf{j}$$
$$= 100\left(\frac{-\sqrt{3}}{2}\right)\mathbf{i} + 100\left(\frac{-1}{2}\right)\mathbf{j}$$
$$= -50\sqrt{3}\,\mathbf{i} - 50\mathbf{j}$$
$$= \langle -50\sqrt{3}, -50 \rangle.$$

(You should check to see that $\|\mathbf{v}\| = 100$.)

Since $\mathbf{v} = a\mathbf{i} + b\mathbf{j} = \|\mathbf{v}\| \cos \theta\mathbf{i} + \|\mathbf{v}\| \sin \theta\mathbf{j}$, it follows that the direction angle θ for **v** is determined from

$$\tan \theta = \frac{\sin \theta}{\cos \theta} = \frac{\|\mathbf{v}\| \sin \theta}{\|\mathbf{v}\| \cos \theta} = \frac{b}{a}.$$

We use this result in the next example.

EXAMPLE 8 *Finding Direction Angles of Vectors*

Find the direction angles of the following vectors.

(a) $\mathbf{u} = 3\mathbf{i} + 3\mathbf{j}$ (b) $\mathbf{v} = 3\mathbf{i} - 4\mathbf{j}$

SOLUTION

(a) The direction angle is given by

$$\tan \theta = \frac{b}{a} = \frac{3}{3} = 1.$$

Therefore, $\theta = 45°$, as shown in Figure 4.30.

(b) The direction angle is given by

$$\tan \theta = \frac{b}{a} = \frac{-4}{3}.$$

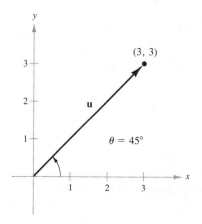

FIGURE 4.30

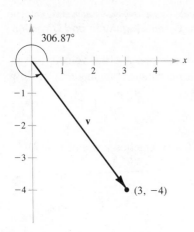

FIGURE 4.31

Moreover, since $\mathbf{v} = 3\mathbf{i} - 4\mathbf{j}$ lies in Quadrant IV, θ lies in Quadrant IV and its reference angle is

$$\theta' = \left| \arctan\left(-\frac{4}{3}\right) \right| \approx |-53.13°| = 53.13°.$$

Therefore, it follows that

$$\theta \approx 360° - 53.13° = 306.87°$$

as shown in Figure 4.31.

Applications of Vectors

Many applications of vectors involve the use of triangles and trigonometry in their solutions.

EXAMPLE 9 An Application

A force of 600 pounds is required to pull a boat and trailer up a ramp inclined at 15° from horizontal. Find the combined weight of the boat and trailer. (Assume no friction is involved.)

SOLUTION

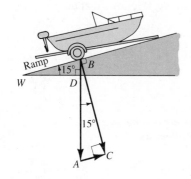

FIGURE 4.32

Based on Figure 4.32, we make the following observations.

$\|\overrightarrow{BA}\|$ = force of gravity = combined weight of boat and trailer

$\|\overrightarrow{BC}\|$ = force against ramp

$\|\overrightarrow{AC}\|$ = force required to move boat up ramp = 600 pounds
(Note that AC is parallel to the ramp.)

By construction, triangles WBD and ABC are similar. Hence, angle ABC is 15°. Therefore, in triangle ABC we have

$$\sin 15° = \frac{\|\overrightarrow{AC}\|}{\|\overrightarrow{BA}\|} = \frac{600}{\|\overrightarrow{BA}\|}$$

$$\|\overrightarrow{BA}\| = \frac{600}{\sin 15°} \approx 2318.$$

Consequently, the combined weight is approximately 2318 pounds.

Vectors

EXAMPLE 10   An Application

An airplane is traveling at a fixed altitude with a negligible wind factor. The airplane is headed N 30° W at a speed of 500 miles per hour, as shown in Figure 4.33. As the airplane reaches a certain point, it encounters a wind with a velocity of 70 miles per hour in the direction N 45° E. What are the resultant speed and direction of the airplane?

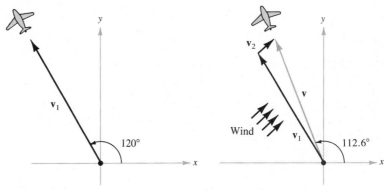

FIGURE 4.33

SOLUTION

Using Figure 4.33, we can represent the velocity of the airplane by the vector

$$\mathbf{v}_1 = 500 \langle \cos 120°, \sin 120° \rangle = \langle -250, 250\sqrt{3} \rangle$$

and the velocity of the wind by the vector

$$\mathbf{v}_2 = 70 \langle \cos 45°, \sin 45° \rangle = \langle 35\sqrt{2}, 35\sqrt{2} \rangle.$$

Thus, the velocity of the airplane is given by the vector

$$\mathbf{v} = \mathbf{v}_1 + \mathbf{v}_2 = \langle -250 + 35\sqrt{2}, 250\sqrt{3} + 35\sqrt{2} \rangle$$
$$\approx \langle -200.5, 482.5 \rangle$$

and the speed of the airplane is

$$\|\mathbf{v}\| = \sqrt{(-200.5)^2 + (482.5)^2} \approx 522.5 \text{ miles per hour.}$$

Finally, if θ is the direction angle of the flight path, we have

$$\tan \theta = \frac{482.5}{-200.5} \approx -2.4065$$

which implies that

$$\theta \approx 180° + \arctan(-2.4065) \approx 180° - 67.4° = 112.6°.$$

WARM UP

Find the distance between the two points.

1. (−2, 6), (5, −15)

2. (0, 0), (−3, −7)

Find an equation of the line passing through the two points.

3. (3, 1), (−2, 4)

4. (−2, −3), (4, 5)

Find the angle θ ($0 \le \theta < 360°$) whose terminal side passes through the given point.

5. (−2, 5)

6. (4, −3)

Find the sine and cosine of θ.

7. $\theta = 30°$

8. $\theta = 120°$

9. $\theta = 300°$

10. $\theta = 210°$

EXERCISES 4.3

In Exercises 1–6, use the figure to sketch a graph of the indicated vector.

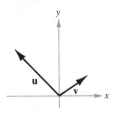

FIGURE FOR 1–6

1. −u

2. 3v

 3. u + v

4. u + 2v

5. u − v

6. v − $\frac{1}{2}$u

In Exercises 7–16, find the component form and the magnitude of the vector **v**.

7.

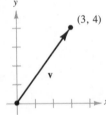

8.

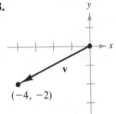

9.

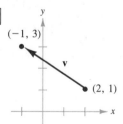

10.

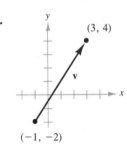

11.

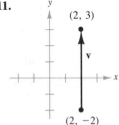

12.

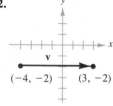

13. Initial Point: (−1, 5)
Terminal Point: (15, 2)

14. Initial Point: (1, 11)
Terminal Point: (9, 3)

15. Initial Point: $(-3, -5)$
   Terminal Point: $(5, -1)$

16. Initial Point: $(-3, 11)$
   Terminal Point: $(9, 40)$

In Exercises 17–26, find (a) $\mathbf{u} + \mathbf{v}$, (b) $\mathbf{u} - \mathbf{v}$, and (c) $2\mathbf{u} - 3\mathbf{v}$.

17. $\mathbf{u} = \langle 1, 2 \rangle$, $\mathbf{v} = \langle 3, 1 \rangle$ **18.** $\mathbf{u} = \langle 2, 3 \rangle$, $\mathbf{v} = \langle 4, 0 \rangle$

19. $\mathbf{u} = \langle -2, 3 \rangle$, $\mathbf{v} = \langle -2, 1 \rangle$

20. $\mathbf{u} = \langle 0, 1 \rangle$, $\mathbf{v} = \langle 0, -1 \rangle$

21. $\mathbf{u} = \langle 4, -2 \rangle$, $\mathbf{v} = \langle 0, 0 \rangle$

22. $\mathbf{u} = \langle 0, 0 \rangle$, $\mathbf{v} = \langle 2, 1 \rangle$

23. $\mathbf{u} = \mathbf{i} + \mathbf{j}$, $\mathbf{v} = 2\mathbf{i} - 3\mathbf{j}$

24. $\mathbf{u} = 2\mathbf{i} - \mathbf{j}$, $\mathbf{v} = -\mathbf{i} + \mathbf{j}$

25. $\mathbf{u} = 2\mathbf{i}$, $\mathbf{v} = \mathbf{j}$ **26.** $\mathbf{u} = 3\mathbf{j}$, $\mathbf{v} = 2\mathbf{i}$

In Exercises 27–34, sketch $\mathbf{v}$ and find its component form. (Assume θ is measured counterclockwise from the x-axis to the vector.)

27. $\|\mathbf{v}\| = 3$, $\theta = 0°$ **28.** $\|\mathbf{v}\| = 1$, $\theta = 45°$

29. $\|\mathbf{v}\| = 1$, $\theta = 150°$ **30.** $\|\mathbf{v}\| = \frac{5}{2}$, $\theta = 45°$

31. $\|\mathbf{v}\| = 3\sqrt{2}$, $\theta = 150°$ **32.** $\|\mathbf{v}\| = 8$, $\theta = 90°$

33. $\|\mathbf{v}\| = 2$, $\mathbf{v}$ in the direction $\mathbf{i} + 3\mathbf{j}$

34. $\|\mathbf{v}\| = 3$, $\mathbf{v}$ in the direction $3\mathbf{i} + 4\mathbf{j}$

In Exercises 35–40, find the component form of $\mathbf{v}$, and sketch the indicated vector operations geometrically, where $\mathbf{u} = 2\mathbf{i} - \mathbf{j}$ and $\mathbf{w} = \mathbf{i} + 2\mathbf{j}$.

35. $\mathbf{v} = \frac{3}{2}\mathbf{u}$ **36.** $\mathbf{v} = \mathbf{u} + \mathbf{w}$

37. $\mathbf{v} = \mathbf{u} + 2\mathbf{w}$ **38.** $\mathbf{v} = -\mathbf{u} + \mathbf{w}$

39. $\mathbf{v} = \frac{1}{2}(3\mathbf{u} + \mathbf{w})$ **40.** $\mathbf{v} = \mathbf{u} - 2\mathbf{w}$

In Exercises 41–44, find the component form of the sum of the vectors $\mathbf{u}$ and $\mathbf{v}$ with direction angles θ_u and θ_v respectively.

41. $\|\mathbf{u}\| = 5$, $\theta_u = 0°$ **42.** $\|\mathbf{u}\| = 2$, $\theta_u = 30°$
   $\|\mathbf{v}\| = 5$, $\theta_v = 90°$ $\|\mathbf{v}\| = 2$, $\theta_v = 90°$

43. $\|\mathbf{u}\| = 20$, $\theta_u = 45°$ **44.** $\|\mathbf{u}\| = 35$, $\theta_u = 25°$
   $\|\mathbf{v}\| = 50$, $\theta_v = 180°$ $\|\mathbf{v}\| = 50$, $\theta_v = 120°$

In Exercises 45–48, find a unit vector in the direction of the given vector.

45. $\mathbf{v} = 4\mathbf{i} - 3\mathbf{j}$ **46.** $\mathbf{v} = \mathbf{i} + \mathbf{j}$

47. $\mathbf{v} = 2\mathbf{j}$ **48.** $\mathbf{v} = \mathbf{i} - 2\mathbf{j}$

In Exercises 49–52, use the Law of Cosines to find the angle α between the given vectors. (Assume $0° \le \alpha \le 180°$.)

49. $\mathbf{v} = \mathbf{i} + \mathbf{j}$, $\mathbf{w} = 2(\mathbf{i} - \mathbf{j})$

50. $\mathbf{v} = 3\mathbf{i} + \mathbf{j}$, $\mathbf{w} = 2\mathbf{i} - \mathbf{j}$

51. $\mathbf{v} = \mathbf{i} + \mathbf{j}$, $\mathbf{w} = 3\mathbf{i} - \mathbf{j}$

52. $\mathbf{v} = \mathbf{i} + 2\mathbf{j}$, $\mathbf{w} = 2\mathbf{i} - \mathbf{j}$

53. Two forces, one of 35 pounds and the other of 50 pounds, act on the same object. The angle between the forces is 30°. Find the magnitude of the resultant (vector sum) of these forces.

54. Two forces, one of 100 pounds and the other of 150 pounds, act on the same object, at angles of 20° and 60°, respectively, with the positive x-axis. Find the direction and magnitude of the resultant (vector sum) of these forces.

55. Three forces of 75 pounds, 100 pounds, and 125 pounds act on the same object, at angles of 30°, 45°, and 120°, respectively, with the positive x-axis. Find the direction and magnitude of the resultant of these forces.

56. Three forces of 70 pounds, 40 pounds, and 60 pounds act on the same object, at angles of −30°, 45°, and 135°, respectively, with the positive x-axis. Find the direction and magnitude of the resultant of these forces.

57. A heavy implement is dragged 10 feet across the floor, using a force of 85 pounds. Find the work done if the direction of the force is 60° above the horizontal (see figure). (Use the formula for work, $W = FD$, where F is the horizontal component of force and D is the horizontal distance.)

58. To carry a 100-pound cylindrical weight, two people lift on the ends of short ropes that are tied to an eyelet on the top center of the cylinder. One of the ropes makes a 20° angle away from the vertical and the other a 30° angle (see figure). Find the vertical component of each person's force.

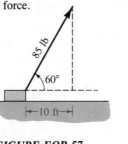

FIGURE FOR 57

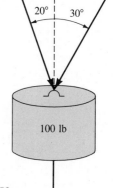

100 lb

FIGURE FOR 58

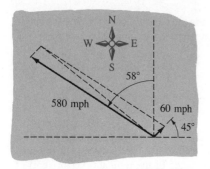

FIGURE FOR 59

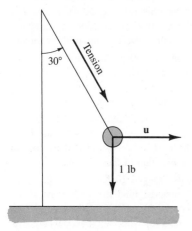

FIGURE FOR 60

59. An airplane's velocity with respect to the air is 580 miles per hour, and it is headed N 58° W. The wind, at the altitude of the plane, is from the southwest and has a velocity of 60 miles per hour (see figure). What is the true direction of the plane, and what is its speed with respect to the ground?

60. A tether ball weighing 1 pound is pulled outward from the pole by a horizontal force **u** until the rope makes a 30° angle with the pole (see figure). Determine the resulting tension in the rope and the magnitude of **u**.

61. An airplane is flying in the direction S 32° E, with an airspeed of 540 miles per hour. Because of the wind, its groundspeed and direction are 500 miles per hour and S 40° E, respectively. Find the direction and speed of the wind.

62. A ball is thrown into the air with an initial velocity of 80 feet per second, at an angle of 50° with the horizontal. Find the vertical and horizontal components of the velocity.

In Exercises 63 and 64, find the angle between the two forces for the given magnitude of their sum.

63. Force One: 45 lb
   Force Two: 60 lb
   Resultant Force: 90 lb

64. Force One: 3000 lb
   Force Two: 1000 lb
   Resultant Force: 3750 lb

4.4 The Dot Product

So far we have studied two operations with vectors—vector addition and multiplication by a scalar—each of which yields another vector. In this section we introduce a third vector operation, called the **dot product.** This product yields a real number, rather than a vector.

Definition of Dot Product

The **dot product** of $\mathbf{u} = \langle u_1, u_2 \rangle$ and $\mathbf{v} = \langle v_1, v_2 \rangle$ is

$$\mathbf{u} \cdot \mathbf{v} = u_1 v_1 + u_2 v_2.$$

Remark: Be sure you see that the dot product of two vectors is a real number and *not* another vector.

EXAMPLE 1 Finding the Dot Product of Two Vectors

Find the dot product of $\mathbf{u} = 5\mathbf{i} - 2\mathbf{j}$ and $\mathbf{v} = 3\mathbf{i} + 2\mathbf{j}$.

SOLUTION

Since $\mathbf{u} = 5\mathbf{i} - 2\mathbf{j} = \langle 5, -2 \rangle$ and $\mathbf{v} = 3\mathbf{i} + 2\mathbf{j} = \langle 3, 2 \rangle$, we have

$$
\begin{aligned}
\mathbf{u} \cdot \mathbf{v} &= \langle 5, -2 \rangle \cdot \langle 3, 2 \rangle \\
&= (5)(3) + (-2)(2) \\
&= 15 - 4 \\
&= 11.
\end{aligned}
$$

The properties listed next follow readily from the definition of dot product.

Properties of the Dot Product

If $\mathbf{u}$, $\mathbf{v}$, and $\mathbf{w}$ are vectors in the plane and c is a scalar, then the following properties are true.

1. $\mathbf{u} \cdot \mathbf{v} = \mathbf{v} \cdot \mathbf{u}$
2. $\mathbf{u} \cdot (\mathbf{v} + \mathbf{w}) = \mathbf{u} \cdot \mathbf{v} + \mathbf{u} \cdot \mathbf{w}$
3. $c(\mathbf{u} \cdot \mathbf{v}) = (c\mathbf{u}) \cdot \mathbf{v} = \mathbf{u} \cdot (c\mathbf{v})$
4. $\mathbf{0} \cdot \mathbf{v} = 0$
5. $\mathbf{v} \cdot \mathbf{v} = \|\mathbf{v}\|^2$

PROOF

We prove the first and fifth properties and leave the remaining proofs as exercises. To prove the first property, let $\mathbf{u} = \langle u_1, u_2 \rangle$ and $\mathbf{v} = \langle v_1, v_2 \rangle$. Then

$$
\mathbf{u} \cdot \mathbf{v} = u_1 v_1 + u_2 v_2 = v_1 u_1 + v_2 u_2 = \mathbf{v} \cdot \mathbf{u}.
$$

For the fifth property, we let $\mathbf{v} = \langle v_1, v_2 \rangle$. Then

$$
\mathbf{v} \cdot \mathbf{v} = v_1{}^2 + v_2{}^2 = (\sqrt{v_1{}^2 + v_2{}^2})^2 = \|\mathbf{v}\|^2.
$$

EXAMPLE 2 *Using Properties of the Dot Product*

For $\mathbf{u} = \langle 2, -2 \rangle$, $\mathbf{v} = \langle 5, 8 \rangle$, and $\mathbf{w} = \langle -4, 3 \rangle$, find the following.

(a) $(\mathbf{u} \cdot \mathbf{v})\mathbf{w}$ (b) $\mathbf{u} \cdot (2\mathbf{v})$ (c) $\|\mathbf{w}\|^2$

SOLUTION

(a) $(\mathbf{u} \cdot \mathbf{v})\mathbf{w} = [2(5) + (-2)(8)]\langle -4, 3 \rangle = -6\langle -4, 3 \rangle = \langle 24, -18 \rangle$

(b) $\mathbf{u} \cdot (2\mathbf{v}) = 2(\mathbf{u} \cdot \mathbf{v}) = 2(-6) = -12$

(c) $\|\mathbf{w}\|^2 = \mathbf{w} \cdot \mathbf{w} = \langle -4, 3 \rangle \cdot \langle -4, 3 \rangle = (-4)(-4) + (3)(3) = 25$

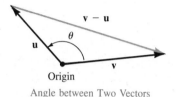

Origin

Angle between Two Vectors

FIGURE 4.34

The dot product of two vectors proves to be a very useful concept, and we devote the remainder of this section to some of its applications. First, we show how the dot product can be used to determine the angle between two vectors.

The **angle between two nonzero vectors** $\mathbf{u}$ and $\mathbf{v}$ is the angle θ, $0 \leq \theta \leq \pi$, between their respective standard position vectors, as shown in Figure 4.34. This means that if $\mathbf{u}$ and $\mathbf{v}$ have the same direction, then $\theta = 0$, and if $\mathbf{u}$ and $\mathbf{v}$ have opposite directions, then $\theta = \pi$. (Note that we do not define the angle between the zero vector and another vector.)

Angle Between Two Vectors

If θ is the angle between two nonzero vectors $\mathbf{u}$ and $\mathbf{v}$, then

$$\cos \theta = \frac{\mathbf{u} \cdot \mathbf{v}}{\|\mathbf{u}\| \, \|\mathbf{v}\|}.$$

PROOF

Consider the triangle determined by vectors $\mathbf{u}$, $\mathbf{v}$, and $\mathbf{v} - \mathbf{u}$, as shown in Figure 4.34. By the Law of Cosines, we have

$$\|\mathbf{v} - \mathbf{u}\|^2 = \|\mathbf{u}\|^2 + \|\mathbf{v}\|^2 - 2\|\mathbf{u}\| \, \|\mathbf{v}\| \cos \theta.$$

Using the properties of the dot product, we can rewrite the left side as follows.

$$\begin{aligned}
\|\mathbf{v} - \mathbf{u}\|^2 &= (\mathbf{v} - \mathbf{u}) \cdot (\mathbf{v} - \mathbf{u}) \\
&= (\mathbf{v} - \mathbf{u}) \cdot \mathbf{v} - (\mathbf{v} - \mathbf{u}) \cdot \mathbf{u} \\
&= \mathbf{v} \cdot \mathbf{v} - \mathbf{u} \cdot \mathbf{v} - \mathbf{v} \cdot \mathbf{u} + \mathbf{u} \cdot \mathbf{u} \\
&= \|\mathbf{v}\|^2 - 2\mathbf{u} \cdot \mathbf{v} + \|\mathbf{u}\|^2
\end{aligned}$$

The Dot Product

Substitution back into the Law of Cosines yields the following.

$$\|\mathbf{v}\|^2 - 2\mathbf{u} \cdot \mathbf{v} + \|\mathbf{u}\|^2 = \|\mathbf{u}\|^2 + \|\mathbf{v}\|^2 - 2\|\mathbf{u}\|\,\|\mathbf{v}\|\cos\theta$$
$$-2\mathbf{u} \cdot \mathbf{v} = -2\|\mathbf{u}\|\,\|\mathbf{v}\|\cos\theta$$
$$\cos\theta = \frac{\mathbf{u} \cdot \mathbf{v}}{\|\mathbf{u}\|\,\|\mathbf{v}\|}$$

If the angle between two vectors is known, then the equation for $\cos\theta$ can be rewritten in the form

$$\mathbf{u} \cdot \mathbf{v} = \|\mathbf{u}\|\,\|\mathbf{v}\|\cos\theta$$

which gives an alternative way of calculating the dot product. This form also shows us that because $\|\mathbf{u}\|$ and $\|\mathbf{v}\|$ are always positive, $\mathbf{u} \cdot \mathbf{v}$ and $\cos\theta$ will always have the same sign.

Figure 4.35 shows the five possible orientations of two vectors.

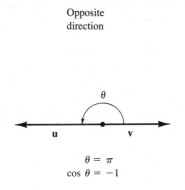

Opposite
direction

$\theta = \pi$
$\cos\theta = -1$

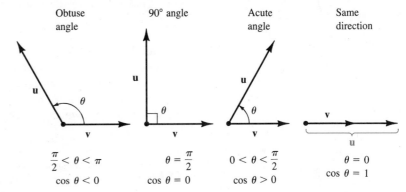

Obtuse
angle

$\dfrac{\pi}{2} < \theta < \pi$

$\cos\theta < 0$

90° angle

$\theta = \dfrac{\pi}{2}$

$\cos\theta = 0$

Acute
angle

$0 < \theta < \dfrac{\pi}{2}$

$\cos\theta > 0$

Same
direction

$\theta = 0$
$\cos\theta = 1$

FIGURE 4.35

EXAMPLE 3   *Finding the Angle Between Two Vectors*

Find the angle θ between the vectors $\mathbf{u} = \langle 2, 4 \rangle$ and $\mathbf{v} = \langle 0, -5 \rangle$.

SOLUTION

Using the formula for the angle between two vectors, we have

$$\cos\theta = \frac{\mathbf{u} \cdot \mathbf{v}}{\|\mathbf{u}\|\,\|\mathbf{v}\|}$$
$$= \frac{0 - 20}{\sqrt{20}\,\sqrt{25}} = \frac{-20}{10\sqrt{5}} = -\frac{2}{\sqrt{5}}.$$

Additional Applications of Trigonometry

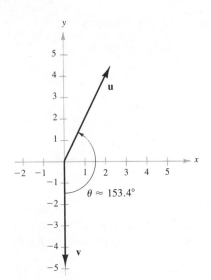

FIGURE 4.36

Therefore, the angle between **u** and **v** is

$$\theta = \arccos\left(-\frac{2}{\sqrt{5}}\right) \approx 2.68 \text{ radians} \approx 153.4°.$$

See Figure 4.36.

If the angle between two vectors is $\theta = \pi/2$, then we call the two vectors **orthogonal** (perpendicular). Since $\cos(\pi/2) = 0$, we have the following definition.

Definition of Orthogonal Vectors

The vectors **u** and **v** are **orthogonal** if

$$\mathbf{u} \cdot \mathbf{v} = 0.$$

We consider the zero vector to be orthogonal to every vector **u** since $\mathbf{0} \cdot \mathbf{u} = 0$.

Informally, we say that two vectors are *parallel* if they have the same or opposite directions. From their coordinate representations, it is easy to determine whether two vectors are parallel. Specifically, two nonzero vectors **u** and **v** have the *same direction* if and only if $\mathbf{u} = k\mathbf{v}$, where $k > 0$. Similarly, **u** and **v** have *opposite directions* if and only if $\mathbf{u} = k\mathbf{v}$, where $k < 0$.

EXAMPLE 4 *Orthogonal and Parallel Vectors*

(a) Determine whether $\mathbf{u} = \langle 3, -1 \rangle$ and $\mathbf{v} = \langle 2, 6 \rangle$ are orthogonal, parallel, or neither.

(b) Determine whether $\mathbf{u} = \langle -1, 2 \rangle$ and $\mathbf{v} = \langle 2, -4 \rangle$ are orthogonal, parallel, or neither.

SOLUTION

(a) Since

$$\cos \theta = \frac{\mathbf{u} \cdot \mathbf{v}}{\|\mathbf{u}\| \, \|\mathbf{v}\|}$$

$$= \frac{6 - 6}{\sqrt{10} \, \sqrt{40}} = \frac{0}{20} = 0$$

it follows that $\mathbf{u} \cdot \mathbf{v} = 0$, which means that **u** and **v** are *orthogonal*. See Figure 4.37(a).

(a)

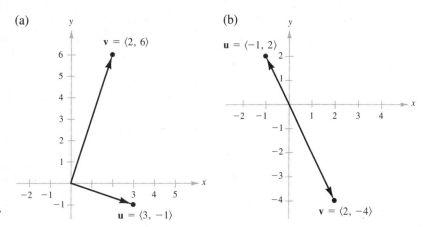

v = $\langle 2, 6 \rangle$

u = $\langle 3, -1 \rangle$

(b)

u = $\langle -1, 2 \rangle$

v = $\langle 2, -4 \rangle$

FIGURE 4.37

(b) The cosine of the angle between **u** and **v** is

$$\cos \theta = \frac{\mathbf{u} \cdot \mathbf{v}}{\|\mathbf{u}\| \, \|\mathbf{v}\|}$$

$$= \frac{-2 - 8}{\sqrt{5} \, \sqrt{20}}$$

$$= \frac{-10}{\sqrt{100}}$$

$$= -1.$$

Therefore, the angle between **u** and **v** is $\theta = \pi$, and **u** and **v** are *parallel*. See Figure 4.37(b).

Vector Components

We have already seen applications in which two vectors are added to produce a resultant vector. There are many applications in physics and engineering in which we want to look at the reverse problem—that is, to decompose a given vector into the sum of two **vector components.** To see the usefulness of this procedure, we look at a physical example.

Consider a boat on an inclined ramp, as shown in Figure 4.38. The force **F** due to gravity pulls the boat *onto* the ramp and *down* the ramp. These two forces, $\mathbf{w}_1$ and $\mathbf{w}_2$, are orthogonal and we call them the vector components of **F** and write

$$\mathbf{F} = \mathbf{w}_1 + \mathbf{w}_2.$$

The forces $\mathbf{w}_1$ and $\mathbf{w}_2$ help us analyze the effect of gravity on the boat. For example, $\mathbf{w}_1$ indicates the force necessary to keep the boat from rolling down the ramp. On the other hand, $\mathbf{w}_2$ indicates the force that the tires must withstand.

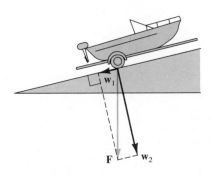

FIGURE 4.38

Additional Applications of Trigonometry

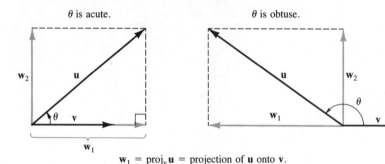

θ is acute. $\qquad$ θ is obtuse.

$\mathbf{w}_1 = \text{proj}_{\mathbf{v}}\,\mathbf{u} = $ projection of $\mathbf{u}$ onto $\mathbf{v}$.
$\mathbf{w}_2 = $ vector component of $\mathbf{u}$ orthogonal to $\mathbf{v}$.

FIGURE 4.39

With this discussion in mind, we make the following definitions. Let $\mathbf{u}$ and $\mathbf{v}$ be nonzero vectors such that

$$\mathbf{u} = \mathbf{w}_1 + \mathbf{w}_2$$

where $\mathbf{w}_1$ is parallel to $\mathbf{v}$ and $\mathbf{w}_2$ is orthogonal to $\mathbf{v}$, as shown in Figure 4.39. The vector $\mathbf{w}_1$ is called the **projection** of $\mathbf{u}$ onto $\mathbf{v}$ and is denoted by

$$\mathbf{w}_1 = \text{proj}_{\mathbf{v}}(\mathbf{u}).$$

The vector $\mathbf{w}_2 = \mathbf{u} - \mathbf{w}_1$ is called the **vector component of u orthogonal to v**.

From this definition, we see that it is easy to find the vector component $\mathbf{w}_2$ once we have found the projection of $\mathbf{u}$ onto $\mathbf{v}$. To find this projection, we use the dot product as indicated in the next theorem.

Orthogonal Projection Using the Dot Product

Let $\mathbf{u}$ and $\mathbf{v}$ be nonzero vectors. Then the projection of $\mathbf{u}$ onto $\mathbf{v}$ is given by

$$\text{proj}_{\mathbf{v}}(\mathbf{u}) = \left(\frac{\mathbf{u} \cdot \mathbf{v}}{\|\mathbf{v}\|^2}\right)\mathbf{v}.$$

PROOF

From Figure 4.39, we let $\mathbf{w}_1 = \text{proj}_{\mathbf{v}}(\mathbf{u})$. Since $\mathbf{w}_1$ is a scalar multiple of $\mathbf{v}$, we can write

$$\mathbf{u} = \mathbf{w}_1 + \mathbf{w}_2 = c\mathbf{v} + \mathbf{w}_2.$$

Now, by taking the dot product of both sides with $\mathbf{v}$, we have

$$\mathbf{u} \cdot \mathbf{v} = (c\mathbf{v} + \mathbf{w}_2) \cdot \mathbf{v} = c\mathbf{v} \cdot \mathbf{v} + \mathbf{w}_2 \cdot \mathbf{v} = c\|\mathbf{v}\|^2 + \mathbf{w}_2 \cdot \mathbf{v}.$$

Since $\mathbf{w}_2$ and $\mathbf{v}$ are orthogonal, $\mathbf{w}_2 \cdot \mathbf{v} = 0$ and we have

$$\mathbf{u} \cdot \mathbf{v} = c\|\mathbf{v}\|^2$$

which implies that

$$c = \frac{\mathbf{u} \cdot \mathbf{v}}{\|\mathbf{v}\|^2}.$$

Therefore, we can write $\mathbf{w}_1$ as

$$\mathbf{w}_1 = \left(\frac{\mathbf{u} \cdot \mathbf{v}}{\|\mathbf{v}\|^2}\right)\mathbf{v}.$$

The projection of $\mathbf{u}$ onto $\mathbf{v}$ can be written as a scalar multiple of a unit vector in the direction of $\mathbf{v}$. That is,

$$\left(\frac{\mathbf{u} \cdot \mathbf{v}}{\|\mathbf{v}\|^2}\right)\mathbf{v} = \left(\frac{\mathbf{u} \cdot \mathbf{v}}{\|\mathbf{v}\|}\right)\frac{\mathbf{v}}{\|\mathbf{v}\|} = (k)\frac{\mathbf{v}}{\|\mathbf{v}\|}.$$

We call k the **component of u in the direction of v.** Thus,

$$k = \frac{\mathbf{u} \cdot \mathbf{v}}{\|\mathbf{v}\|} = \|\mathbf{u}\| \cos \theta.$$

EXAMPLE 5   *Decomposing a Vector into Vector Components*

Find the projection of $\mathbf{u}$ onto $\mathbf{v}$ and the vector component of $\mathbf{u}$ orthogonal to $\mathbf{v}$ for the vectors $\mathbf{u} = 3\mathbf{i} - 5\mathbf{j}$ and $\mathbf{v} = 6\mathbf{i} + 2\mathbf{j}$.

SOLUTION

The projection of $\mathbf{u}$ onto $\mathbf{v}$ is

$$\mathbf{w}_1 = \left(\frac{\mathbf{u} \cdot \mathbf{v}}{\|\mathbf{v}\|^2}\right)\mathbf{v} = \left(\frac{8}{40}\right)(6\mathbf{i} + 2\mathbf{j})$$

$$= \frac{6}{5}\mathbf{i} + \frac{2}{5}\mathbf{j}.$$

See Figure 4.40. The vector component of $\mathbf{u}$ orthogonal to $\mathbf{v}$ is the vector

$$\mathbf{w}_2 = \mathbf{u} - \mathbf{w}_1 = (3\mathbf{i} - 5\mathbf{j}) - \left(\frac{6}{5}\mathbf{i} + \frac{2}{5}\mathbf{j}\right)$$

$$= \frac{9}{5}\mathbf{i} - \frac{27}{5}\mathbf{j}.$$

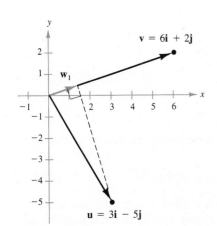

FIGURE 4.40

Now, let's return to the problem involving the boat and the ramp to see how we can use vector components in a physical problem.

EXAMPLE 6 *An Application Using a Vector Component*

A 600-pound boat sets on a ramp inclined at 30°, as shown in Figure 4.41. What force is required to keep the boat from rolling down the ramp?

SOLUTION

Since the force due to gravity is vertical and downward, we represent the gravitational force by the vector

$$\mathbf{F} = -600\mathbf{j}. \qquad \text{\textit{Force due to gravity}}$$

To find the force required to keep the boat from rolling down the ramp, we project $\mathbf{F}$ onto a unit vector $\mathbf{v}$ in the direction of the ramp as follows.

$$\mathbf{v} = \cos 30°\, \mathbf{i} + \sin 30°\, \mathbf{j}$$
$$= \frac{\sqrt{3}}{2}\mathbf{i} + \frac{1}{2}\mathbf{j} \qquad \text{\textit{Unit vector along ramp}}$$

Therefore, the projection of $\mathbf{F}$ onto $\mathbf{v}$ is given by

$$\mathbf{w}_1 = \text{proj}_\mathbf{v}(\mathbf{F}) = \left(\frac{\mathbf{F} \cdot \mathbf{v}}{\|\mathbf{v}\|^2}\right)\mathbf{v}$$
$$= (\mathbf{F} \cdot \mathbf{v})\mathbf{v}$$
$$= (-600)\left(\frac{1}{2}\right)\mathbf{v}$$
$$= -300\left(\frac{\sqrt{3}}{2}\mathbf{i} + \frac{1}{2}\mathbf{j}\right).$$

The magnitude of this force is 300 and therefore, a force of 300 pounds is required to keep the boat from rolling down the ramp.

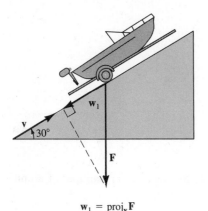

$$\mathbf{w}_1 = \text{proj}_\mathbf{v}\mathbf{F}$$

FIGURE 4.41

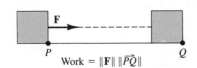

$$\text{Work} = \|\mathbf{F}\|\,\|\overrightarrow{PQ}\|$$

(a) Force acts along the line of motion.

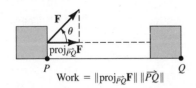

$$\text{Work} = \|\text{proj}_{\overrightarrow{PQ}}\mathbf{F}\|\,\|\overrightarrow{PQ}\|$$

(b) Force acts at angle θ with the line of motion.

FIGURE 4.42

Work

The work W done by constant force $\mathbf{F}$ acting along the line of motion of an object is given by

$$W = (\text{magnitude of force})(\text{distance}) = \|\mathbf{F}\|\,\|\overrightarrow{PQ}\|$$

as shown in Figure 4.42(a). If the constant force $\mathbf{F}$ is not directed along the line of motion, then we can see from Figure 4.42(b) that the work W done by the force is

$$W = \|\text{proj}_{\overrightarrow{PQ}}(\mathbf{F})\|\,\|\overrightarrow{PQ}\| = (\cos\theta)\,\|\mathbf{F}\|\,\|\overrightarrow{PQ}\| = \mathbf{F} \cdot \overrightarrow{PQ}.$$

This notion of work is summarized in the following definition.

Definition of Work

The work W done by a constant force **F** as its point of application moves along the vector $\overrightarrow{PQ}$ is given by either of the following.

1. $W = \|\text{proj}_{\overrightarrow{PQ}}(\mathbf{F})\| \, \|\overrightarrow{PQ}\|$ *Projection form*
2. $W = \mathbf{F} \cdot \overrightarrow{PQ}$ *Dot product form*

EXAMPLE 7 *An Application Involving Work*

To close a sliding door, a man pulls on a rope with a constant force of 50 pounds at a constant angle of 60°, as shown in Figure 4.43. Find the work done in moving the door 12 feet to its closed position.

SOLUTION

Using a projection, we calculate the work as follows.

$$
\begin{aligned}
W &= \|\text{proj}_{\overrightarrow{PQ}}(\mathbf{F})\| \, \|\overrightarrow{PQ}\| \\
&= \cos(60°) \, \|\mathbf{F}\| \, \|\overrightarrow{PQ}\| \\
&= \frac{1}{2}(50)(12) = 300 \text{ ft-lb}
\end{aligned}
$$

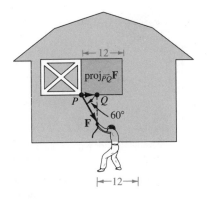

FIGURE 4.43

WARM UP

Find the length of each vector.

1. $\mathbf{v} = \langle 1, 4 \rangle$

2. $\mathbf{v} = \langle -1, 3 \rangle$

Perform the following operations.

3. $3\langle 2, 5 \rangle$

4. $-\frac{1}{2}\langle 4, 6 \rangle$

5. $\langle 8, 5 \rangle - \langle 4, -3 \rangle$

6. $\langle 5, -2 \rangle + \langle -4, 2 \rangle$

7. $3\langle 5, 2 \rangle - 4\langle 2, 3 \rangle$

8. $-\langle 4, 5 \rangle + 3\langle 5, 6 \rangle$

9. $3(2\mathbf{i} - 4\mathbf{j}) + 5(-\mathbf{i} + 2\mathbf{j})$

10. $-2(\mathbf{i} + 4\mathbf{j}) + 4(\mathbf{i} - 4\mathbf{j})$

EXERCISES 4.4

In Exercises 1–6, find (a) $\mathbf{u} \cdot \mathbf{v}$, (b) $\mathbf{u} \cdot \mathbf{u}$, (c) $\|\mathbf{u}\|^2$, (d) $(\mathbf{u} \cdot \mathbf{v})\mathbf{v}$, (e) $\mathbf{u} \cdot (2\mathbf{v})$.

1. $\mathbf{u} = \langle 3, 4 \rangle$, $\mathbf{v} = \langle 2, -3 \rangle$
2. $\mathbf{u} = \langle 5, 12 \rangle$, $\mathbf{v} = \langle -3, 2 \rangle$
3. $\mathbf{u} = \langle 2, -3 \rangle$, $\mathbf{v} = \langle 0, 6 \rangle$
4. $\mathbf{u} = \mathbf{i}$, $\mathbf{v} = \mathbf{i}$
5. $\mathbf{u} = 2\mathbf{i} - \mathbf{j}$, $\mathbf{v} = \mathbf{i} - \mathbf{j}$
6. $\mathbf{u} = 2\mathbf{i} + \mathbf{j}$, $\mathbf{v} = \mathbf{i} - 3\mathbf{j}$

In Exercises 7–14, find the angle θ between the given vectors.

7. $\mathbf{u} = \langle 1, 1 \rangle$, $\mathbf{v} = \langle 2, -2 \rangle$
8. $\mathbf{u} = \langle 3, 1 \rangle$, $\mathbf{v} = \langle 2, -1 \rangle$
9. $\mathbf{u} = 3\mathbf{i} + \mathbf{j}$, $\mathbf{v} = -2\mathbf{i} + 4\mathbf{j}$
10. $\mathbf{u} = 2\mathbf{i} + 3\mathbf{j}$, $\mathbf{v} = -3\mathbf{i} + 2\mathbf{j}$
11. $\mathbf{u} = 3\mathbf{i} + 4\mathbf{j}$, $\mathbf{v} = -2\mathbf{j}$
12. $\mathbf{u} = 2\mathbf{i} - 3\mathbf{j}$, $\mathbf{v} = \mathbf{i} - 2\mathbf{j}$
13. $\mathbf{u} = \cos\left(\dfrac{\pi}{4}\right)\mathbf{i} + \sin\left(\dfrac{\pi}{4}\right)\mathbf{j}$, $\mathbf{v} = \cos\left(\dfrac{\pi}{2}\right)\mathbf{i} + \sin\left(\dfrac{\pi}{2}\right)\mathbf{j}$
14. $\mathbf{u} = \cos\left(\dfrac{\pi}{6}\right)\mathbf{i} + \sin\left(\dfrac{\pi}{6}\right)\mathbf{j}$, $\mathbf{v} = \cos\left(\dfrac{3\pi}{4}\right)\mathbf{i} + \sin\left(\dfrac{3\pi}{4}\right)\mathbf{j}$

In Exercises 15–22, determine whether $\mathbf{u}$ and $\mathbf{v}$ are orthogonal, parallel, or neither.

15. $\mathbf{u} = \langle 4, 0 \rangle$, $\mathbf{v} = \langle 1, 1 \rangle$
16. $\mathbf{u} = \langle 2, 18 \rangle$, $\mathbf{v} = \langle \frac{3}{2}, -\frac{1}{6} \rangle$
17. $\mathbf{u} = \langle 4, 3 \rangle$, $\mathbf{v} = \langle \frac{1}{2}, -\frac{2}{3} \rangle$
18. $\mathbf{u} = -\frac{1}{3}(\mathbf{i} - 2\mathbf{j})$, $\mathbf{v} = 2\mathbf{i} - 4\mathbf{j}$
19. $\mathbf{u} = \mathbf{j}$, $\mathbf{v} = \mathbf{i} - 2\mathbf{j}$
20. $\mathbf{u} = -2\mathbf{i} + 3\mathbf{j}$, $\mathbf{v} = 2\mathbf{i} + \mathbf{j}$
21. $\mathbf{u} = \langle 2, -2 \rangle$, $\mathbf{v} = \langle -1, -1 \rangle$
22. $\mathbf{u} = \langle \cos\theta, \sin\theta \rangle$, $\mathbf{v} = \langle \sin\theta, -\cos\theta \rangle$

In Exercises 23–30, (a) find the projection of $\mathbf{u}$ onto $\mathbf{v}$, and (b) the vector component of $\mathbf{u}$ orthogonal to $\mathbf{v}$.

23. $\mathbf{u} = \langle 2, 3 \rangle$, $\mathbf{v} = \langle 5, 1 \rangle$
24. $\mathbf{u} = \langle 2, -3 \rangle$, $\mathbf{v} = \langle 3, 2 \rangle$
25. $\mathbf{u} = \langle 2, 1 \rangle$, $\mathbf{v} = \langle 0, 3 \rangle$
26. $\mathbf{u} = \langle 0, 4 \rangle$, $\mathbf{v} = \langle 0, 2 \rangle$
27. $\mathbf{u} = \langle 1, 1 \rangle$, $\mathbf{v} = \langle -2, -1 \rangle$

28. $\mathbf{u} = \langle -2, -1 \rangle$, $\mathbf{v} = \langle 1, 1 \rangle$
29. $\mathbf{u} = \langle 5, -4 \rangle$, $\mathbf{v} = \langle 1, 0 \rangle$
30. $\mathbf{u} = \langle 5, -4 \rangle$, $\mathbf{v} = \langle 0, 1 \rangle$

31. A truck with a gross weight of 32,000 pounds is parked on a 15° slope (see figure). Assume the only force to overcome is that due to gravity.
    (a) Find the force required to keep the truck from rolling down the hill.
    (b) Find the force perpendicular to the hill.

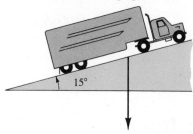

Weight = 32,000 lb   **FIGURE FOR 31**

32. Rework Exercise 31 for a truck that is parked on a 16° slope.

33. An implement is dragged 10 feet across a floor, using a force of 85 pounds. Find the work done if the direction of the force is 60° above the horizontal (see figure).

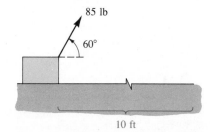

85 lb

60°

10 ft   **FIGURE FOR 33**

34. A toy wagon is pulled by exerting a force of 15 pounds on a handle that makes a 30° angle with the horizontal. Find the work done in pulling the wagon 50 feet (see figure).

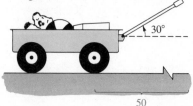

30°

50   **FIGURE FOR 34**

In Exercises 35 and 36, find the work done in moving a particle from P to Q if the magnitude and direction of the force are given by **v**.

35. $P = (0, 0)$, $Q = (4, 7)$, $\mathbf{v} = \langle 1, 4 \rangle$

36. $P = (1, 3)$, $Q = (-3, 5)$, $\mathbf{v} = -2\mathbf{i} + 3\mathbf{j}$

37. The vector $\mathbf{u} = \langle 3240, 1450 \rangle$ gives the number of bushels of corn and wheat raised by a farmer in a certain year. The vector $\mathbf{v} = \langle 2.22, 3.25 \rangle$ gives the price (in dollars) per bushel of each of the crops. Find the dot product, $\mathbf{u} \cdot \mathbf{v}$, and explain what information it gives.

38. The vertices of a triangle are $(1, 2)$, $(3, 4)$, and $(2, 5)$. Find the three angles of the triangle.

39. What is known about θ, the angle between two nonzero vectors $\mathbf{u}$ and $\mathbf{v}$, if the following are true?
(a) $\mathbf{u} \cdot \mathbf{v} = 0$ (b) $\mathbf{u} \cdot \mathbf{v} > 0$ (c) $\mathbf{u} \cdot \mathbf{v} < 0$

40. What can be said about the vectors **u** and **v** if the following are true?
(a) The projection of **u** onto **v** equals **u**.
(b) The projection of **u** onto **v** equals **0**.

41. Use vectors to prove that the diagonals of a rhombus are perpendicular.

42. Prove the following.
$$\|\mathbf{u} - \mathbf{v}\|^2 = \|\mathbf{u}\|^2 + \|\mathbf{v}\|^2 - 2\mathbf{u} \cdot \mathbf{v}$$

43. Prove properties 2, 3, and 4 of dot products.
(a) $\mathbf{u} \cdot (\mathbf{v} + \mathbf{w}) = \mathbf{u} \cdot \mathbf{v} + \mathbf{u} \cdot \mathbf{w}$ Property 2
(b) $c(\mathbf{u} \cdot \mathbf{v}) = (c\mathbf{u}) \cdot \mathbf{v} = \mathbf{u} \cdot (c\mathbf{v})$ Property 3
(c) $\mathbf{0} \cdot \mathbf{v} = 0$ Property 4

44. Prove that if **u** is orthogonal to **v** and **w**, then **u** is orthogonal to $c\mathbf{v} + d\mathbf{w}$ for any scalars c and d.

CHAPTER 4 REVIEW EXERCISES

In Exercises 1–16, solve the triangle (if possible) using the three given parts. If two solutions are possible, list both.

1. $a = 5$, $b = 8$, $c = 10$

2. $a = 6$, $b = 9$, $C = 45°$

3. $A = 12°$, $B = 58°$, $a = 5$

4. $B = 110°$, $C = 30°$, $c = 10.5$

5. $B = 110°$, $a = 4$, $c = 4$

6. $a = 80$, $b = 60$, $c = 100$

7. $A = 75°$, $a = 2.5$, $b = 16.5$

8. $A = 130°$, $a = 50$, $b = 30$

9. $B = 115°$, $a = 7$, $b = 14.5$

10. $c = 50°$, $a = 25$, $c = 22$

11. $A = 15°$, $a = 5$, $b = 10$

12. $B = 150°$, $a = 64$, $b = 10$

13. $B = 150°$, $a = 10$, $c = 20$

14. $a = 2.5$, $b = 15.0$, $c = 4.5$

15. $B = 25°$, $a = 6.2$, $b = 4$

16. $B = 90°$, $a = 5$, $c = 12$

In Exercises 17–20, find the area of the triangle having the given parts.

17. $a = 4$, $b = 5$, $c = 7$ **18.** $a = 15$, $b = 8$, $c = 10$

19. $A = 27°$, $b = 5$, $c = 8$ **20.** $B = 80°$, $a = 4$, $c = 8$

21. Find the height of a tree that stands on a hillside of slope 32° (from the horizontal) if, from a point 75 feet downhill, the angle of elevation to the top of the tree is 48° (see figure).

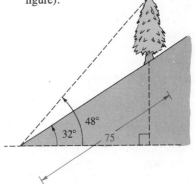

48°
32° 75

FIGURE FOR 21

22. To approximate the length of a marsh, a surveyor walks 450 meters from point A to point B. Then the surveyor turns $65°$ and walks 325 meters to point C. Approximate the length AC of the marsh (see figure).

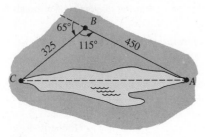

FIGURE FOR 22

23. From a certain distance, the angle of elevation of the top of a building is $17°$. At a point 50 meters closer to the building, the angle of elevation is $31°$. Approximate the height of the building.

24. Determine the width of a river that flows due east, if a tree on the opposite bank has a bearing of N $22°$ $30'$ E and after walking 400 feet downstream, a surveyor finds that the tree has a bearing of N $15°$ W.

25. Two planes leave an airport at approximately the same time. One is flying at 425 miles per hour at a bearing of N $5°$ W, and the other is flying at 530 miles per hour at a bearing of N $67°$ E (see figure). How far apart are the two planes after flying for 2 hours?

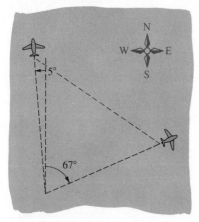

FIGURE FOR 25

26. The lengths of the diagonals of a parallelogram are 10 feet and 16 feet. Find the lengths of the sides of the parallelogram if the diagonals intersect at an angle of $28°$.

In Exercises 27–30, find the component form of the vector **v** satisfying the given conditions.

27. Initial point: $(0, 10)$ **28.** Initial point: $(1, 5)$
Terminal point: $(7, 3)$ Terminal point: $(15, 9)$

29. $\|\mathbf{v}\| = 8$, $\theta = 120°$ **30.** $\|\mathbf{v}\| = \frac{1}{2}$, $\theta = 225°$

In Exercises 31–34, find the component form of the required vector and sketch its graph given that $\mathbf{u} = 6\mathbf{i} - 5\mathbf{j}$ and $\mathbf{v} = 10\mathbf{i} + 3\mathbf{j}$.

31. $\dfrac{1}{\|\mathbf{u}\|}\mathbf{u}$ **32.** $3\mathbf{v}$

33. $4\mathbf{u} - 5\mathbf{v}$ **34.** $-\frac{1}{2}\mathbf{v}$

35. If two forces of 85 pounds and 50 pounds act on a single point, and the angle between them is $15°$, find the magnitude of the resultant of the two forces.

36. A 100-pound weight is being supported by two ropes (see figure). Find the tension in each rope.

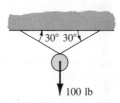

FIGURE FOR 36

37. Find the direction and magnitude of the resultant of the three forces shown in the figure.

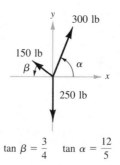

$\tan \beta = \dfrac{3}{4}$ $\tan \alpha = \dfrac{12}{5}$

FIGURE FOR 37

38. A 500-pound motorcycle is on a hill inclined at $12°$. What force is required to keep the motorcycle from rolling back down the hill?

39. An airplane has an airspeed of 450 miles per hour at a bearing of N 30° E. If the wind velocity is 20 miles per hour from the west, find the ground speed and the direction of the plane.

40. Two forces of 60 pounds and 100 pounds have a resultant force of magnitude 125 pounds. Find the angle between the two given forces.

41. Find the vector projection $\mathbf{w}$ of $\mathbf{u} = 3\mathbf{i} + 2\mathbf{j}$ onto $\mathbf{v} = \frac{1}{2}\mathbf{i} - \frac{3}{2}\mathbf{j}$.

42. Use vectors to prove that the midpoints of the sides of *any* quadrilateral are the vertices of a parallelogram.

43. Use vectors to prove that the medians of a triangle pass through a point that is two-thirds of the distance from any vertex to the midpoint of the opposite side.

44. Prove that $\mathbf{w} = (\mathbf{w} \cdot \mathbf{u})\mathbf{u} + (\mathbf{w} \cdot \mathbf{v})\mathbf{v}$ for any vector $\mathbf{w}$ if $\mathbf{u}$ and $\mathbf{v}$ are orthogonal unit vectors.

45. If $\mathbf{u} \neq \mathbf{0}$ and $\mathbf{u} \cdot \mathbf{v} = \mathbf{u} \cdot \mathbf{w}$, does $\mathbf{v} = \mathbf{w}$? Explain your answer.

46. Give a geometrical argument to show that the reflection of the vector $\mathbf{u}$ through the vector $\mathbf{v}$ is given by

$$\mathbf{w} = \left(\frac{2\mathbf{u} \cdot \mathbf{v}}{\mathbf{v} \cdot \mathbf{v}}\right)\mathbf{v} - \mathbf{u}.$$

47. Use the result of Exercise 46 to reflect the vector $\mathbf{u} = \mathbf{i} + 3\mathbf{j}$ through the following vectors.
(a) $\mathbf{v} = \mathbf{i}$ (b) $\mathbf{v} = \mathbf{j}$ (c) $\mathbf{v} = \mathbf{i} + \mathbf{j}$

48. Find the work done in moving an object along the vector $\mathbf{v} = 3\mathbf{i} + 2\mathbf{j}$ if the applied force is $\mathbf{F} = 2\mathbf{i} - \mathbf{j}$.

49. Prove the triangle inequality. That is, prove that if $\mathbf{u}$ and $\mathbf{v}$ are two vectors, then $\|\mathbf{u} + \mathbf{v}\| \leq \|\mathbf{u}\| + \|\mathbf{v}\|$. [*Hint:* Square both sides and use the dot product.]

50. Let $\mathbf{u} = (\cos \alpha)\mathbf{i} + (\sin \alpha)\mathbf{j}$ and $\mathbf{v} = (\cos \beta)\mathbf{i} + (\sin \beta)\mathbf{j}$. Draw these vectors, and by interpreting the dot product $\mathbf{u} \cdot \mathbf{v}$ geometrically, prove that

$$\cos(\alpha - \beta) = \cos \alpha \cos \beta + \sin \alpha \sin \beta.$$

CHAPTER 5

Complex Numbers

5.1 Complex Numbers

In Section 1.2 we noted that if the discriminant $b^2 - 4ac$ of a quadratic is less than zero, then there are no real solutions to the equation $ax^2 + bx + c = 0$. For instance, the set of real numbers is not large enough to accommodate the solutions to an equation as simple as

$$x^2 = -1$$

because $x = \sqrt{-1}$ is not a real number. To overcome this deficiency, mathematicians created an expanded system of numbers using the **imaginary unit i,** defined as

$$i = \sqrt{-1}$$

where $i^2 = -1$. By adding real numbers to real multiples of this imaginary unit, we obtain the set of **complex numbers.** Each complex number can be written in the **standard form,** $a + bi$.

Definition of a Complex Number

For real numbers a and b, the number

$a + bi$

is called a **complex number.** If $a = 0$ and $b \neq 0$, then the complex number bi is called a **pure imaginary number.**

The set of real numbers is a subset of the set of complex numbers because every real number a can be written as a complex number using $b = 0$. That is, for every real number a, we can write $a = a + 0i$.

Two complex numbers $a + bi$ and $c + di$, written in standard form, are **equal** to each other

$$a + bi = c + di$$

Equality of two complex numbers

if and only if $a = c$ and $b = d$.

Operations with Complex Numbers

To add (or subtract) two complex numbers, we add (or subtract) the real and imaginary parts of the numbers separately.

Addition and Subtraction of Complex Numbers

If $a + bi$ and $c + di$ are two complex numbers written in standard form, then their sum and difference are defined as follows.

Sum: $(a + bi) + (c + di) = (a + c) + (b + d)i$

Difference: $(a + bi) - (c + di) = (a - c) + (b - d)i$

The **additive identity** in the complex number system is zero (the same as in the real number system). Furthermore, the **additive inverse** of the complex number $a + bi$ is

$-(a + bi) = -a - bi.$

Additive inverse

Thus, we have

$$(a + bi) + (-a - bi) = 0 + 0i = 0.$$

EXAMPLE 1 *Adding and Subtracting Complex Numbers*

Write the following sums and differences in standard form.

(a) $(3 - i) + (2 + 3i)$ (b) $2i + (-4 - 2i)$

(c) $3 - (-2 + 3i) + (-5 + i)$

SOLUTION

(a) $(3 - i) + (2 + 3i) = 3 - i + 2 + 3i$
$$= 3 + 2 - i + 3i$$
$$= (3 + 2) + (-1 + 3)i$$
$$= 5 + 2i$$

(b) $2i + (-4 - 2i) = 2i - 4 - 2i$
$$= -4 + 2i - 2i$$
$$= -4$$

(c) $3 - (-2 + 3i) + (-5 + i) = 3 + 2 - 3i - 5 + i$
$$= 3 + 2 - 5 - 3i + i$$
$$= 0 - 2i$$
$$= -2i$$

Remark: Note in Example 1(b) that the sum of two complex numbers can be a real number.

Many of the properties of real numbers are valid for complex numbers as well. Here are some.

> Associative Property of Addition and Multiplication
> Commutative Property of Addition and Multiplication
> Distributive Property of Multiplication over Addition

Notice how these properties are used when two complex numbers are multiplied.

$$(a + bi)(c + di) = a(c + di) + bi(c + di) \qquad \text{\textit{Distributive Law}}$$
$$= ac + (ad)i + (bc)i + (bd)i^2 \qquad \text{\textit{Distributive Law}}$$
$$= ac + (ad)i + (bc)i + (bd)(-1) \qquad \text{\textit{Definition of i}}$$
$$= ac - bd + (ad)i + (bc)i \qquad \text{\textit{Commutative Law}}$$
$$= (ac - bd) + (ad + bc)i \qquad \text{\textit{Associative Law}}$$

Rather than trying to memorize this multiplication rule, you may just want to remember how the distributive property is used to derive it. The procedure is similar to multiplying two polynomials and combining like terms.

EXAMPLE 2 *Multiplying Complex Numbers*

Find the following products.

(a) $(i)(-3i)$
(b) $(2 - i)(4 + 3i)$
(c) $(3 + 2i)(3 - 2i)$

SOLUTION

(a) $(i)(-3i) = -3i^2 = -3(-1) = 3$

(b) $(2 - i)(4 + 3i) = 8 + 6i - 4i - 3i^2$ *Binomial product*
$$= 8 + 6i - 4i - 3(-1)   \quad i^2 = -1$$
$$= 8 + 3 + 6i - 4i   \quad \textit{Collect terms}$$
$$= 11 + 2i   \quad \textit{Combine like terms}$$

(c) $(3 + 2i)(3 - 2i) = 9 - 6i + 6i - 4i^2$
$$= 9 - 4(-1) = 9 + 4$$
$$= 13$$

Complex Conjugates

We see from Example 2(c) that the product of complex numbers can be a real number. This occurs with pairs of complex numbers of the forms $a + bi$ and $a - bi$, called **complex conjugates.** In general, we have the following.

$$(a + bi)(a - bi) = a^2 - abi + abi - b^2i^2$$
$$= a^2 - b^2(-1)$$
$$= a^2 + b^2$$

Complex conjugates can be used to divide two complex numbers. That is, to find the quotient

$$\frac{a + bi}{c + di}, \quad c \text{ and } d \text{ not both zero}$$

we multiply the numerator and denominator by the conjugate of the denominator to obtain

$$\frac{a + bi}{c + di} = \frac{a + bi}{c + di}\left(\frac{c - di}{c - di}\right) = \frac{(ac + bd) + (bc - ad)i}{c^2 + d^2}.$$

This procedure is demonstrated in Example 3.

EXAMPLE 3 *Dividing Complex Numbers*

Write the following in standard form.

(a) $\dfrac{1}{1 + i}$ 　　　　　　　　　　　　　(b) $\dfrac{2 + 3i}{4 - 2i}$

SOLUTION

(a) $\dfrac{1}{1 + i} = \dfrac{1}{1 + i}\left(\dfrac{1 - i}{1 - i}\right) = \dfrac{1 - i}{1^2 - i^2}$

$$= \dfrac{1 - i}{1 - (-1)} = \dfrac{1 - i}{2} = \dfrac{1}{2} - \dfrac{1}{2}i$$

(b) $\dfrac{2 + 3i}{4 - 2i} = \dfrac{2 + 3i}{4 - 2i}\left(\dfrac{4 + 2i}{4 + 2i}\right) = \dfrac{8 + 4i + 12i + 6i^2}{16 - 4i^2}$

$$= \dfrac{8 - 6 + 16i}{16 + 4} = \dfrac{1}{20}(2 + 16i)$$

$$= \dfrac{1}{10} + \dfrac{4}{5}i$$

When using the quadratic formula or the square root method to solve quadratic equations, we often obtain a result like $\sqrt{-3}$, which we know is not a real number. By factoring out $i = \sqrt{-1}$, we can write this number in the standard form, as follows.

$$\sqrt{-3} = \sqrt{3(-1)} = \sqrt{3}\sqrt{-1} = \sqrt{3}\,i$$

We call $\sqrt{3}\,i$ the principal square root of -3.

Principal Square Root of a Negative Number

For $a > 0$, the **principal square root** of $-a$ is defined as

$$\sqrt{-a} = \sqrt{a}\,i.$$

In this definition we are using the rule $\sqrt{ab} = \sqrt{a}\sqrt{b}$, for $a > 0$ and $b < 0$. This rule is not valid if *both* a and b are negative. For example,

$$\sqrt{-5}\sqrt{-5} = (\sqrt{5}\,i)(\sqrt{5}\,i) = 5i^2 = -5$$

whereas

$$\sqrt{(-5)(-5)} = \sqrt{25} = 5.$$

Consequently, $\sqrt{(-5)(-5)} \neq \sqrt{-5}\sqrt{-5}$.

Complex Numbers

Remark: When working with square roots of negative numbers, be sure to convert to standard form *before* multiplying.

EXAMPLE 4 *Writing Complex Numbers in Standard Form*

(a) $\sqrt{-3}\sqrt{-12} = \sqrt{3}\,i\sqrt{12}\,i = \sqrt{36}\,i^2 = 6(-1) = -6$

(b) $\sqrt{-48} - \sqrt{-27} = \sqrt{48}\,i - \sqrt{27}\,i = 4\sqrt{3}\,i - 3\sqrt{3}\,i = \sqrt{3}\,i$

(c) $(-1 + \sqrt{-3})^2 = (-1 + \sqrt{3}\,i)^2$

$$= (-1)^2 - 2\sqrt{3}\,i + (\sqrt{3})^2(i^2)$$

$$= 1 - 2\sqrt{3}\,i + 3(-1)$$

$$= -2 - 2\sqrt{3}\,i$$

Example 5 shows how the principal square root of a negative number is used to represent complex solutions of a quadratic equation.

EXAMPLE 5 *Complex Solutions of a Quadratic Equation*

Solve the equation $3x^2 - 2x + 5 = 0$.

SOLUTION

By the quadratic formula, we have

$$x = \frac{-(-2) \pm \sqrt{(-2)^2 - 4(3)(5)}}{2(3)}$$

$$= \frac{2 \pm \sqrt{-56}}{6}$$

$$= \frac{2 \pm 2\sqrt{14}\,i}{6}$$

$$= \frac{1}{3} \pm \frac{\sqrt{14}}{3}\,i.$$

When working with polynomials of degree higher than 2, we occasionally need to raise i to the third or higher power. The pattern looks like this:

$$i^1 = i$$
$$i^2 = -1$$
$$i^3 = i^2 \cdot i = -i$$
$$i^4 = i^2 \cdot i^2 = (-1)(-1) = 1$$
$$i^5 = i^4 \cdot i = i.$$

Complex Numbers

Since the pattern begins to repeat after the fourth power, we can compute the value of i^n for any natural number n. We simply factor out the multiples of 4 in the exponent and compute the remaining portion. For example,

$$i^{38} = i^{36} \cdot i^2$$
$$= (i^4)^9 \cdot i^2$$
$$= (1)^9(-1) = -1.$$

EXAMPLE 6 *Finding Powers of i*

Find the following powers of i.

(a) i^{10}

(b) i^{73}

SOLUTION

(a) Since 8 is the largest multiple of 4 that is less than 10, we write

$$i^{10} = i^8 \cdot i^2 = (i^4)^2 \cdot i^2 = (1)^2 \cdot (-1) = -1.$$

(b) Since 72 is the largest multiple of 4 that is less than 73, we write

$$i^{73} = i^{72} \cdot i^1 = (i^4)^{18} \cdot i = (1)i = i.$$

WARM UP

Simplify the given expressions.

1. $\sqrt{12}$

2. $\sqrt{500}$

3. $\sqrt{20} - \sqrt{5}$

4. $\sqrt{27} - \sqrt{243}$

5. $\sqrt{24}\sqrt{6}$

6. $2\sqrt{18}\sqrt{32}$

7. $\dfrac{1}{\sqrt{3}}$

8. $\dfrac{2}{\sqrt{2}}$

Solve the given quadratic equations.

9. $x^2 + x - 1 = 0$

10. $x^2 + 2x - 1 = 0$

EXERCISES 5.1

1. Write out the first 16 positive integer powers of i (that is, $i, i^2, i^3, ..., i^{16}$), and express each as $i, -i, 1,$ or -1.

2. Express each of the following powers of i as $i, -i, 1,$ or -1.
(a) i^{40} (b) i^{25}
(c) i^{50} (d) i^{67}

In Exercises 3–6, find real numbers a and b so that the equation is true.

3. $a + bi = -10 + 6i$ **4.** $a + bi = 13 + 4i$

5. $(a - 1) + (b + 3)i = 5 + 8i$

6. $(a + 6) + 2bi = 6 - 5i$

In Exercises 7–18, write the complex number in standard form and find its complex conjugate.

7. $4 + \sqrt{-9}$ **8.** $3 + \sqrt{-16}$

9. $2 - \sqrt{-27}$ **10.** $1 + \sqrt{-8}$

11. $\sqrt{-75}$ **12.** 45

13. $-6i + i^2$ **14.** $4i^2 - 2i^3$

15. $-5i^5$ **16.** $(-i)^3$

17. 8 **18.** $(\sqrt{-4})^2 - 5$

In Exercises 19–56, perform the indicated operation and write the result in standard form.

19. $(5 + i) + (6 - 2i)$

20. $(13 - 2i) + (-5 + 6i)$

21. $(8 - i) - (4 - i)$

22. $(3 + 2i) - (6 + 13i)$

23. $(-2 + \sqrt{-8}) + (5 - \sqrt{-50})$

24. $(8 + \sqrt{-18}) - (4 + 3\sqrt{2}\,i)$

25. $-\left(\frac{3}{2} + \frac{5}{2}i\right) + \left(\frac{5}{3} + \frac{11}{3}i\right)$

26. $(1.6 + 3.2i) + (-5.8 + 4.3i)$

27. $\sqrt{-6} \cdot \sqrt{-2}$

28. $\sqrt{-5} \cdot \sqrt{-10}$

29. $(\sqrt{-10})^2$

30. $(\sqrt{-75})^3$

31. $(1 + i)(3 - 2i)$

32. $(6 - 2i)(2 - 3i)$

33. $(4 + 5i)(4 - 5i)$

34. $(6 + 7i)(6 - 7i)$

35. $6i(5 - 2i)$

36. $-8i(9 + 4i)$

37. $-8(5 - 2i)$

38. $(\sqrt{5} - \sqrt{3}\,i)(\sqrt{5} + \sqrt{3}\,i)$

39. $(\sqrt{14} + \sqrt{10}\,i)(\sqrt{14} - \sqrt{10}\,i)$

40. $(3 + \sqrt{-5})(7 - \sqrt{-10})$

41. $(4 + 5i)^2$

42. $(2 - 3i)^3$

43. $\dfrac{4}{4 - 5i}$

44. $\dfrac{3}{1 - i}$

45. $\dfrac{2 + i}{2 - i}$ **46.** $\dfrac{8 - 7i}{1 - 2i}$

47. $\dfrac{6 - 7i}{i}$ **48.** $\dfrac{8 + 20i}{2i}$

49. $\dfrac{1}{(2i)^3}$ **50.** $\dfrac{1}{(4 - 5i)^2}$

51. $\dfrac{5}{(1 + i)^3}$ **52.** $\dfrac{(2 - 3i)(5i)}{2 + 3i}$

53. $\dfrac{(21 - 7i)(4 + 3i)}{2 - 5i}$ **54.** $\dfrac{1}{i^3}$

55. $(2 + 3i)^2 + (2 - 3i)^2$ **56.** $(1 - 2i)^2 - (1 + 2i)^2$

In Exercises 57–64, use the quadratic formula to solve the quadratic equation.

57. $x^2 - 2x + 2 = 0$ **58.** $x^2 + 6x + 10 = 0$

59. $4x^2 + 16x + 17 = 0$ **60.** $9x^2 - 6x + 37 = 0$

61. $4x^2 + 16x + 15 = 0$ **62.** $9x^2 - 6x - 35 = 0$

63. $16t^2 - 4t + 3 = 0$ **64.** $5s^2 + 6s + 3 = 0$

65. Prove that the sum of a complex number and its conjugate is a real number.

66. Prove that the difference of a complex number and its conjugate is an imaginary number.

67. Prove that the product of a complex number and its conjugate is a real number.

68. Prove that the conjugate of the product of two complex numbers is the product of their conjugates.

69. Prove that the conjugate of the sum of two complex numbers is the sum of their conjugates.

5.2 Complex Solutions of Equations

The Fundamental Theorem of Algebra tells us that a polynomial equation of degree n has precisely n solutions in the complex number system. These solutions can be real or complex and may be repeated, as illustrated in Example 1.

EXAMPLE 1 Zeros of Polynomial Functions

(a) The first-degree function

$$f(x) = x - 2$$

has exactly *one* zero: $x = 2$.

(b) Counting multiplicity, the second-degree function

$$f(x) = x^2 - 6x + 9 = (x - 3)(x - 3)$$

has exactly *two* zeros: $x = 3$ and $x = 3$.

(c) The third-degree function

$$f(x) = x^3 + 4x = x(x - 2i)(x + 2i)$$

has exactly *three* zeros: $x = 0$, $x = -2i$, and $x = 2i$.

(d) The fourth-degree function

$$f(x) = x^4 - 1 = (x - 1)(x + 1)(x - i)(x + i)$$

has exactly *four* zeros: $x = -1$, $x = 1$, $x = -i$, and $x = i$.

In Section 1.2 we pointed out that the discriminant $(b^2 - 4ac)$ can be used to classify the two solutions given by the Quadratic Formula. In particular, if the discriminant is negative, then the quadratic equation $ax^2 + bx + c = 0$ has two complex solutions given by

$$x = \frac{-b \pm \sqrt{b^2 - 4ac}}{2a}.$$

EXAMPLE 2 Using the Discriminant

Use the discriminant to determine how many real solutions each of the following equations has.

(a) $4x^2 - 20x + 25 = 0$ (b) $13x^2 + 7x + 1 = 0$ (c) $5x^2 = 8x$

Complex Solutions of Equations

SOLUTION

(a) For the standard quadratic form $4x^2 - 20x + 25 = 0$, we have $a = 4$, $b = -20$, and $c = 25$. Therefore, the discriminant is

$$b^2 - 4ac = 400 - 4(4)(25) = 400 - 400 = 0.$$

Since the discriminant is zero, there is one repeated solution.

(b) In this case, $a = 13$, $b = 7$, and $c = 1$, with

$$b^2 - 4ac = 49 - 4(13)(1) = 49 - 52 = -3.$$

Since the discriminant is negative, there are *no* real solutions.

(c) In standard form the equation is

$$5x^2 - 8x = 0$$

with $a = 5$, $b = -8$, and $c = 0$. Thus, the discriminant is

$$b^2 - 4ac = 64 - 4(5)(0) = 64.$$

Since the discriminant is positive, there are *two* real solutions.

EXAMPLE 3   Using the Quadratic Formula

Use the Quadratic Formula to solve the following equation. If the discriminant is negative, list the complex solutions in $a + bi$ form.

$$6x^2 - 2x + 5 = 0$$

SOLUTION

Using $a = 6$, $b = -2$, and $c = 5$, we find the solutions to be

$$x = \frac{-b \pm \sqrt{b^2 - 4ac}}{2a} = \frac{-(-2) \pm \sqrt{4 - 4(6)(5)}}{2(6)}$$

$$= \frac{2 \pm \sqrt{-116}}{12}$$

$$= \frac{2 \pm 2\sqrt{-29}}{12}.$$

In $a + bi$ form, these complex solutions are

$$x = \frac{1}{6} + \frac{\sqrt{29}}{6}\, i \quad \text{and} \quad x = \frac{1}{6} - \frac{\sqrt{29}}{6}\, i.$$

In Example 3, note that the two complex zeros are **conjugates.** That is, they are of the form

$$a + bi \quad \text{and} \quad a - bi.$$

This is not a coincidence. In fact, the following result tells us that if a polynomial (with real coefficients) has one complex zero $(a + bi)$, then it *must* also have the conjugate $(a - bi)$ as a zero.

Complex Zeros Occur in Conjugate Pairs

If $a + bi$ $(b \neq 0)$ is a zero of a polynomial function, with real coefficients, then the conjugate $a - bi$ is also a zero of the function.

Remark: Be sure you see that this result is only true if the polynomial function has *real* coefficients. For instance, the result applies to the function $f(x) = x^2 + 1$, but not to the function $g(x) = x - i$.

EXAMPLE 4 *Finding a Polynomial with Given Zeros*

Find a *fourth-degree* polynomial function, with real coefficients, that has -1, -1, and $3i$ as zeros.

SOLUTION

Since $3i$ is a zero, we know that $-3i$ is also a zero. Thus, $f(x)$ can be written as

$$f(x) = a(x + 1)(x + 1)(x - 3i)(x + 3i).$$

For simplicity, we let $a = 1$, and we obtain

$$\begin{aligned} f(x) &= (x^2 + 2x + 1)(x^2 + 9) \\ &= x^4 + 2x^3 + 10x^2 + 18x + 9. \end{aligned}$$

EXAMPLE 5 *Finding a Polynomial with Given Zeros*

Find a *cubic* polynomial function f, with real coefficients, that has 2 and $1 - i$ as zeros, and such that $f(1) = 3$.

Complex Solutions of Equations

SOLUTION

Because $1 - i$ is a zero of f, so is $1 + i$. Therefore, we have

$$
\begin{aligned}
f(x) &= a(x - 2)[x - (1 - i)][x - (1 + i)] \\
&= a(x - 2)[x^2 - x(1 - i) - x(1 + i) + 1 - i^2] \\
&= a(x - 2)(x^2 - 2x + 2) \\
&= a(x^3 - 4x^2 + 6x - 4).
\end{aligned}
$$

To find the value of a, we use the fact that $f(1) = 3$ and obtain $f(1) = a(1 - 4 + 6 - 4) = 3$. Thus, $a = -3$ and we conclude that

$$
\begin{aligned}
f(x) &= -3(x^3 - 4x^2 + 6x - 4) \\
&= -3x^3 + 12x^2 - 18x + 12.
\end{aligned}
$$

EXAMPLE 6 *Solving a Polynomial Equation*

Find all solutions of $x^4 - x^2 - 20 = 0$.

SOLUTION

$$
\begin{aligned}
x^4 - x^2 - 20 &= 0 \\
(x^2 - 5)(x^2 + 4) &= 0 \\
(x + \sqrt{5})(x - \sqrt{5})(x^2 + 4) &= 0 \\
(x + \sqrt{5})(x - \sqrt{5})(x + 2i)(x - 2i) &= 0
\end{aligned}
$$

By setting each of these factors equal to 0, we find the solutions to be

$$x = -\sqrt{5}, \ x = \sqrt{5}, \ x = -2i, \text{ and } x = 2i.$$

EXAMPLE 7 *Finding the Zeros of a Polynomial Function*

Find all zeros of

$$f(x) = x^4 - 3x^3 + 6x^2 + 2x - 60$$

given that $1 + 3i$ is a zero of f.

SOLUTION

Since complex zeros occur in pairs, we know that $1 - 3i$ is also a zero of f. This means that both

$$[x - (1 + 3i)] \qquad \text{and} \qquad [x - (1 - 3i)]$$

are factors of $f(x)$. Multiplying these two factors produces

$$[x - (1 + 3i)][x - (1 - 3i)] = [(x - 1) - 3i][(x - 1) + 3i]$$
$$= (x - 1)^2 - 9i^2$$
$$= x^2 - 2x + 10.$$

Now, using long division, we can divide $x^2 - 2x + 10$ into $f(x)$ to obtain the following.

$$
\require{enclose}
\begin{array}{r}
x^2 - x - 6 \\
x^2 - 2x + 10 \enclose{longdiv}{x^4 - 3x^3 + 6x^2 + 2x - 60} \\
\underline{x^4 - 2x^3 + 10x^2} \\
-x^3 - 4x^2 + 2x \\
\underline{-x^3 + 2x^2 - 10x} \\
-6x^2 + 12x - 60 \\
\underline{-6x^2 + 12x - 60}
\end{array}
$$

Therefore, we have

$$f(x) = (x^2 - 2x + 10)(x^2 - x - 6)$$
$$= (x^2 - 2x + 10)(x - 3)(x + 2)$$

and we conclude that the zeros of f are

$$x = 1 + 3i, \ 1 - 3i, \ 3, \ \text{and} \ -2.$$

WARM UP

Write each complex number in standard form and give its complex conjugate.

1. $4 - \sqrt{-29}$
3. $-1 + \sqrt{-32}$

2. $-5 - \sqrt{-144}$
4. $6 + \sqrt{-1/4}$

Perform the indicated operations and write the answers in standard form.

5. $(-3 + 6i) - (10 - 3i)$

6. $(12 - 4i) + 20i$

7. $(4 - 2i)(3 + 7i)$

8. $(2 - 5i)(2 + 5i)$

9. $\dfrac{1 + i}{1 - i}$

10. $(3 + 2i)^3$

EXERCISES 5.2

In Exercises 1–8, use the discriminant to determine the number of real solutions.

1. $4x^2 - 4x + 1 = 0$ **2.** $2x^2 - x - 1 = 0$

3. $3x^2 + 4x + 1 = 0$ **4.** $x^2 + 2x + 4 = 0$

5. $2x^2 - 5x + 5 = 0$ **6.** $3x^2 - 6x + 3 = 0$

7. $\frac{1}{5}x^2 + \frac{6}{5}x - 8 = 0$ **8.** $\frac{1}{3}x^2 - 5x + 25 = 0$

In Exercises 9–20, solve the given quadratic equation. List any complex solutions in $a + bi$ form.

9. $x^2 - 5 = 0$ **10.** $3x^2 - 1 = 0$

11. $(x + 5)^2 - 6 = 0$ **12.** $16 - x^2 = 0$

13. $x^2 - 8x + 16 = 0$ **14.** $4x^2 + 4x + 1 = 0$

15. $x^2 + 2x + 5 = 0$ **16.** $54 + 16x - x^2 = 0$

17. $4x^2 - 4x + 5 = 0$ **18.** $4x^2 - 4x + 21 = 0$

19. $230 + 20x - 0.5x^2 = 0$

20. $6 - (x - 1)^2 = 0$

In Exercises 21–30, find a polynomial with integer coefficients that has the given zeros.

21. $1, 5i, -5i$ **22.** $4, 3i, -3i$

23. $2, 4 + i, 4 - i$ **24.** $6, -5 + 2i, -5 - 2i$

25. $i, -i, 6i, -6i$ **26.** $2, 2, 2, 4i, -4i$

27. $-5, -5, 1 + \sqrt{3}\,i$ **28.** $\frac{2}{3}, -1, 3 + \sqrt{2}\,i$

29. $\frac{3}{4}, -2, -\frac{1}{2} + i$ **30.** $0, 0, 4, 1 + i$

In Exercises 31–60, find all the zeros of the function and write the polynomial as a product of linear factors.

31. $f(x) = x^2 + 25$ **32.** $f(x) = x^2 - x + 56$

33. $h(x) = x^2 - 4x + 1$ **34.** $g(x) = x^2 + 10x + 23$

35. $f(x) = x^4 - 81$ **36.** $f(y) = y^4 - 625$

37. $f(z) = z^2 - 2z + 2$

38. $h(x) = x^3 - 3x^2 + 4x - 2$

39. $g(x) = x^3 - 6x^2 + 13x - 10$

40. $f(x) = x^3 - 2x^2 - 11x + 52$

41. $f(t) = t^3 - 3t^2 - 15t + 125$

42. $f(x) = x^3 + 11x^2 + 39x + 29$

43. $f(x) = x^3 + 24x^2 + 214x + 740$

44. $f(s) = 2s^3 - 5s^2 + 12s - 5$

45. $f(x) = 16x^3 - 20x^2 - 4x + 15$

46. $f(x) = 9x^3 - 15x^2 + 11x - 5$

47. $h(x) = x^3 - x + 6$

48. $h(x) = x^3 + 9x^2 + 27x + 35$

49. $f(x) = 5x^3 - 9x^2 + 28x + 6$

50. $g(x) = 3x^3 - 4x^2 + 8x + 8$

51. $g(x) = x^4 - 4x^3 + 8x^2 - 16x + 16$

52. $h(x) = x^4 + 6x^3 + 10x^2 + 6x + 9$

53. $f(x) = x^4 + 10x^2 + 9$

54. $f(x) = x^4 + 29x^2 + 100$

55. $f(x) = 2x^4 + 5x^3 + 4x^2 + 5x + 2$

56. $g(x) = x^5 - 8x^4 + 28x^3 - 56x^2 + 64x - 32$

57. $f(x) = x^4 + 6x^2 - 27$

58. $f(x) = x^4 - 2x^3 - 3x^2 + 12x - 18$
    [*Hint:* One factor is $x^2 - 6$.]

59. $f(x) = x^4 - 4x^3 + 5x^2 - 2x - 6$
    [*Hint:* One factor is $x^2 - 2x - 2$.]

60. $f(x) = x^4 - 3x^3 - x^2 - 12x - 20$
    [*Hint:* One factor is $x^2 + 4$.]

In Exercises 61–70, use the given zero of f to find all the zeros of f.

61. $f(x) = 2x^3 + 3x^2 + 50x + 75, \quad r = 5i$

62. $f(x) = x^3 + x^2 + 9x + 9, \quad r = 3i$

63. $f(x) = 2x^4 - x^3 + 7x^2 - 4x - 4, \quad r = 2i$

64. $g(x) = x^3 - 7x^2 - x + 87, \quad r = 5 + 2i$

65. $g(x) = 4x^3 + 23x^2 + 34x - 10, \quad r = -3 + i$

66. $h(x) = 3x^3 - 4x^2 + 8x + 8, \quad r = 1 - \sqrt{3}\,i$

67. $f(x) = x^4 + 3x^3 - 5x^2 - 21x + 22, \quad r = -3 + \sqrt{2}\,i$

68. $f(x) = x^3 + 4x^2 + 14x + 20, \quad r = -1 - 3i$

69. $h(x) = 8x^3 - 14x^2 + 18x - 9, \quad r = (1 - \sqrt{5}\,i)/2$

70. $f(x) = 25x^3 - 55x^2 - 54x - 18, \quad r = (-2 + \sqrt{2}\,i)/5$

71. Find a quadratic function f (with integer coefficients) that has $\pm\sqrt{b}\,i$ as zeros. Assume that b is a positive integer.

5.3   *Trigonometric Form of Complex Numbers*

In this section we develop the trigonometric form of a complex number. The usefulness of this form will not be fully apparent until we introduce DeMoivre's Theorem in Section 5.4.

Just as real numbers can be represented by points on the real number line, we can represent a complex number

$$z = a + bi$$

as the point (a, b) in a coordinate plane, called the **complex plane.** In this context, we call the horizontal axis the **real axis** and the vertical axis the **imaginary axis,** as shown in Figure 5.1.

The **absolute value** of the complex number $a + bi$ is defined to be the distance between the origin $(0, 0)$ and the point (a, b), as shown in Figure 5.2.

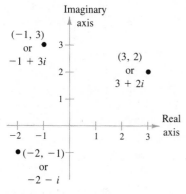

The Complex Plane

FIGURE 5.1

Definition of the Absolute Value of *a + bi*

The **absolute value** of the complex number $z = a + bi$ is given by

$$|a + bi| = \sqrt{a^2 + b^2}.$$

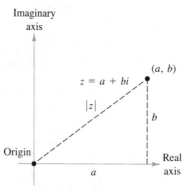

FIGURE 5.2 $|z| = \sqrt{a^2 + b^2}$

EXAMPLE 1 *Finding the Absolute Value of a Complex Number*

Plot the points corresponding to the following complex numbers and find the absolute value of each.

(a) $z = -3i$ (b) $z = -2 + 5i$

SOLUTION

The points are shown in Figure 5.3.

(a) The complex number $z = 0 + (-3)i$ has an absolute value of
$$|z| = \sqrt{0^2 + (-3)^2} = \sqrt{9} = 3.$$

(b) The complex number $z = -2 + 5i$ has an absolute value of
$$|z| = \sqrt{(-2)^2 + 5^2} = \sqrt{29}.$$

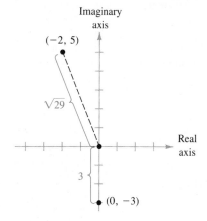

FIGURE 5.3

In Section 5.1 we discussed how to add, subtract, multiply, and divide complex numbers. To work effectively with *powers* and *roots* of complex numbers, it is helpful to write complex numbers in **trigonometric form.** In Figure 5.4, consider the nonzero complex number $a + bi$. By letting θ be the angle from the positive *x*-axis (measured counterclockwise) to the line segment connecting the origin and the point (a, b), we can write

$$a = r \cos \theta \qquad \text{and} \qquad b = r \sin \theta$$

where $r = \sqrt{a^2 + b^2}$. Consequently, we have

$$a + bi = (r \cos \theta) + (r \sin \theta)i$$

from which we obtain the following **trigonometric form of a complex number.**

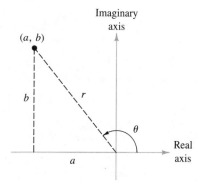

FIGURE 5.4   Complex Number: $a + bi$

Complex Numbers

Trigonometric Form of a Complex Number

Let $z = a + bi$ be a complex number. The **trigonometric form** of z is

$$z = r(\cos \theta + i \sin \theta)$$

where $a = r \cos \theta$, $b = r \sin \theta$, $r = \sqrt{a^2 + b^2}$, and $\tan \theta = b/a$. The number r is called the **modulus** of z, and θ is called an **argument** of z.

Remark: The trigonometric form of a complex number is also called the **polar form.** Because there are infinitely many choices for θ, the trigonometric form of a complex number is not unique. Normally, we use θ values in the interval $0 \leq \theta < 2\pi$, though on occasion we may use $\theta < 0$.

(a)

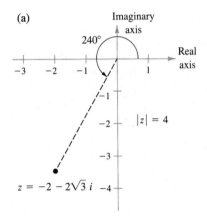

(b)

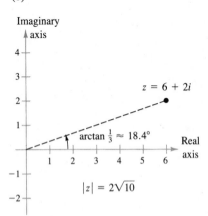

FIGURE 5.5

EXAMPLE 2 Writing Complex Numbers in Trigonometric Form

Write the following complex numbers in trigonometric form.

(a) $z = -2 - 2\sqrt{3}\, i$ \qquad\qquad (b) $z = 6 + 2i$

SOLUTION

(a) The absolute value of z is

$$r = |-2 - 2\sqrt{3}\, i| = \sqrt{(-2)^2 + (-2\sqrt{3})^2} = \sqrt{16} = 4$$

and the angle θ is given by

$$\tan \theta = \frac{b}{a} = \frac{-2\sqrt{3}}{-2} = \sqrt{3}.$$

Since $\tan 60° = \sqrt{3}$ and $z = -2 - 2\sqrt{3}\, i$ lies in Quadrant III, we choose θ to be $\theta = 180° + 60° = 240°$. Thus, the trigonometric form is

$$z = r(\cos \theta + i \sin \theta) = 4(\cos 240° + i \sin 240°).$$

See Figure 5.5(a).

(b) Here we have $r = |6 + 2i| = 2\sqrt{10}$ with θ given by

$$\tan \theta = \frac{2}{6} = \frac{1}{3} \qquad\qquad \theta \text{ in Quadrant I}$$

$$\theta = \arctan \frac{1}{3} \approx 18.4°.$$

Therefore the trigonometric form of z is

$$z = r(\cos \theta + i \sin \theta)$$

$$= 2\sqrt{10}\left[\cos\left(\arctan\frac{1}{3}\right) + i \sin\left(\arctan\frac{1}{3}\right)\right]$$

$$\approx 2\sqrt{10}(\cos 18.4° + i \sin 18.4°).$$

See Figure 5.5(b).

EXAMPLE 3 Writing a Complex Number in Standard Form

Write the following complex number in standard form $a + bi$.

$$z = \sqrt{8}\left[\cos\left(-\frac{\pi}{3}\right) + i \sin\left(-\frac{\pi}{3}\right)\right]$$

SOLUTION

Since $\cos(-\pi/3) = 1/2$ and $\sin(-\pi/3) = -\sqrt{3}/2$, we can write

$$z = \sqrt{8}\left[\cos\left(-\frac{\pi}{3}\right) + i \sin\left(-\frac{\pi}{3}\right)\right]$$

$$= \sqrt{8}\left[\frac{1}{2} - \frac{\sqrt{3}}{2}i\right]$$

$$= 2\sqrt{2}\left[\frac{1}{2} - \frac{\sqrt{3}}{2}i\right]$$

$$= \sqrt{2} - \sqrt{6}\,i.$$

The trigonometric form adapts nicely to multiplication and division of complex numbers. Suppose we are given two complex numbers

$$z_1 = r_1(\cos \theta_1 + i \sin \theta_1) \quad \text{and} \quad z_2 = r_2(\cos \theta_2 + i \sin \theta_2).$$

The product of z_1 and z_2 is

$$z_1 z_2$$

$$= r_1 r_2(\cos \theta_1 + i \sin \theta_1)(\cos \theta_2 + i \sin \theta_2)$$

$$= r_1 r_2[(\cos \theta_1 \cos \theta_2 - \sin \theta_1 \sin \theta_2) + i(\sin \theta_1 \cos \theta_2 + \cos \theta_1 \sin \theta_2)].$$

Using the sum and difference formulas for cosine and sine, we can rewrite this equation as

$$z_1 z_2 = r_1 r_2[\cos(\theta_1 + \theta_2) + i \sin(\theta_1 + \theta_2)].$$

This establishes the first part of the following rule. The second part is left to you (see Exercise 49).

Product and Quotient of Two Complex Numbers

Let $z_1 = r_1(\cos \theta_1 + i \sin \theta_1)$ and $z_2 = r_2(\cos \theta_2 + i \sin \theta_2)$ be complex numbers.

$$z_1 z_2 = r_1 r_2 [\cos(\theta_1 + \theta_2) + i \sin(\theta_1 + \theta_2)] \qquad \textit{Product}$$

$$\frac{z_1}{z_2} = \frac{r_1}{r_2} [\cos(\theta_1 - \theta_2) + i \sin(\theta_1 - \theta_2)], \quad z_2 \neq 0 \qquad \textit{Quotient}$$

Note that this rule says that to multiply two complex numbers we multiply moduli and add arguments, whereas to divide two complex numbers we divide moduli and subtract arguments.

EXAMPLE 4 Multiplying Complex Numbers in Trigonometric Form

Find the product of the following complex numbers.

$$z_1 = 2\left(\cos \frac{2\pi}{3} + i \sin \frac{2\pi}{3}\right)$$

$$z_2 = 8\left(\cos \frac{11\pi}{6} + i \sin \frac{11\pi}{6}\right)$$

SOLUTION

$$z_1 z_2 = 2\left(\cos \frac{2\pi}{3} + i \sin \frac{2\pi}{3}\right) \cdot 8\left(\cos \frac{11\pi}{6} + i \sin \frac{11\pi}{6}\right)$$

$$= 16\left[\cos\left(\frac{2\pi}{3} + \frac{11\pi}{6}\right) + i \sin\left(\frac{2\pi}{3} + \frac{11\pi}{6}\right)\right]$$

$$= 16\left[\cos \frac{5\pi}{2} + i \sin \frac{5\pi}{2}\right]$$

$$= 16\left[\cos \frac{\pi}{2} + i \sin \frac{\pi}{2}\right]$$

$$= 16[0 + i(1)] = 16i$$

Try checking this result by first converting to the standard forms $z_1 = -1 + \sqrt{3}\,i$ and $z_2 = 4\sqrt{3} - 4i$ and then multiplying.

EXAMPLE 5 Dividing Complex Numbers in Trigonometric Form

Find z_1/z_2 for the following two complex numbers.

$$z_1 = 24(\cos 300° + i \sin 300°)$$
$$z_2 = 8(\cos 75° + i \sin 75°)$$

SOLUTION

$$\frac{z_1}{z_2} = \frac{24(\cos 300° + i \sin 300°)}{8(\cos 75° + i \sin 75°)}$$

$$= \frac{24}{8}[\cos(300° - 75°) + i \sin(300° - 75°)]$$

$$= 3[\cos 225° + i \sin 225°]$$

$$= 3\left[\left(-\frac{\sqrt{2}}{2}\right) + i\left(-\frac{\sqrt{2}}{2}\right)\right]$$

$$= -\frac{3\sqrt{2}}{2} - \frac{3\sqrt{2}}{2} i$$

WARM UP

Write the complex numbers in standard form.

1. $-5 - \sqrt{-100}$　　　　　　　　　　　　　　**2.** $7 + \sqrt{-54}$

3. $-4i + i^2$　　　　　　　　　　　　　　　　　**4.** $3i^3$

Perform the indicated operations and write the answers in standard form.

5. $(3 - 10i) - (-3 + 4i)$　　　　　　　　　　　**6.** $(2 + \sqrt{-50}) + (4 - \sqrt{2}\,i)$

7. $(4 - 2i)(-6 + i)$　　　　　　　　　　　　　**8.** $(3 - 2i)(3 + 2i)$

9. $\dfrac{1 + 4i}{1 - i}$　　　　　　　　　　　　　　　**10.** $\dfrac{3 - 5i}{2i}$

EXERCISES 5.3

In Exercises 1–4, express the complex number in trigonometric form.

1.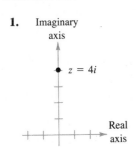

2.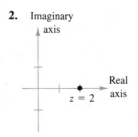

3.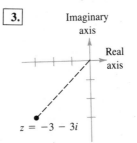

4.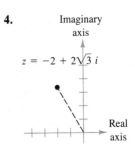

In Exercises 5–20, represent the complex numbers graphically, and find the trigonometric form of the number.

5. $3 - 3i$

6. $-2 - 2i$

7. $\sqrt{3} + i$

8. $-1 + \sqrt{3}\,i$

9. $-2(1 + \sqrt{3}\,i)$

10. $\frac{5}{2}(\sqrt{3} - i)$

11. $6i$

12. 4

13. $-7 + 4i$

14. $3 - i$

15. 7

16. $-2i$

17. $1 + 6i$

18. $2\sqrt{2} - i$

19. $-3 - i$

20. $1 + 3i$

In Exercises 21–30, represent the complex number graphically, and find the standard form of the number.

21. $2(\cos 150° + i \sin 150°)$

22. $5(\cos 135° + i \sin 135°)$

23. $\frac{3}{2}(\cos 300° + i \sin 300°)$

24. $\frac{3}{4}(\cos 315° + i \sin 315°)$

25. $3.75\left(\cos \dfrac{3\pi}{4} + i \sin \dfrac{3\pi}{4}\right)$

26. $8\left(\cos \dfrac{\pi}{12} + i \sin \dfrac{\pi}{12}\right)$

27. $4\left(\cos \dfrac{3\pi}{2} + i \sin \dfrac{3\pi}{2}\right)$

28. $6[\cos(230° \ 30') + i \sin(230° \ 30')]$

29. $3[\cos(18° \ 45') + i \sin(18° \ 45')]$

30. $7(\cos 0° + i \sin 0°)$

In Exercises 31–42, perform the indicated operation and leave the result in trigonometric form.

31. $[3(\cos 60° + i \sin 60°)][4(\cos 30° + i \sin 30°)]$

32. $[\frac{3}{2}(\cos 90° + i \sin 90°)][6(\cos 45° + i \sin 45°)]$

33. $[\frac{5}{3}(\cos 140° + i \sin 140°)][\frac{2}{3}(\cos 60° + i \sin 60°)]$

34. $[0.5(\cos 100° + i \sin 100°)][0.8(\cos 300° + i \sin 300°)]$

35. $[0.45(\cos 310° + i \sin 310°)][0.60(\cos 200° + i \sin 200°)]$

36. $(\cos 5° + i \sin 5°)(\cos 20° + i \sin 20°)$

37. $\dfrac{2(\cos 120° + i \sin 120°)}{4(\cos 40° + i \sin 40°)}$

38. $\dfrac{\cos 40° + i \sin 40°}{\cos 10° + i \sin 10°}$

39. $\dfrac{\cos(5\pi/3) + i \sin(5\pi/3)}{\cos \pi + i \sin \pi}$

40. $\dfrac{5[\cos(4.3) + i \sin(4.3)]}{4[\cos(2.1) + i \sin(2.1)]}$

41. $\dfrac{12(\cos 52° + i \sin 52°)}{3(\cos 110° + i \sin 110°)}$

42. $\dfrac{9(\cos 20° + i \sin 20°)}{5(\cos 75° + i \sin 75°)}$

In Exercises 43–48, (a) give the trigonometric form of the complex numbers, (b) perform the indicated operation using the trigonometric form, and (c) perform the indicated operation using the standard form and check your result with the answer in part (b).

43. $(2 + 2i)(1 - i)$

44. $(\sqrt{3} + i)(1 + i)$

45. $-2i(1 + i)$

46. $\dfrac{3 + 4i}{1 - \sqrt{3}\,i}$

47. $\dfrac{5}{2 + 3i}$

48. $\dfrac{4i}{-4 + 2i}$

49. Given two complex numbers $z_1 = r_1(\cos \theta_1 + i \sin \theta_1)$ and $z_2 = r_2(\cos \theta_2 + i \sin \theta_2)$, $z_2 \neq 0$, prove that

$$\frac{z_1}{z_2} = \frac{r_1}{r_2}[\cos(\theta_1 - \theta_2) + i \sin(\theta_1 - \theta_2)].$$

50. Show that the complex conjugate of

$$z = r(\cos \theta + i \sin \theta)$$

is

$$\bar{z} = r[\cos(-\theta) + i \sin(-\theta)].$$

51. Use the trigonometric form of z and $\bar{z}$ in Exercise 50 to find

(a) $z\bar{z}$ (b) $z/\bar{z}$, $z \neq 0$

52. Show that the negative of $z = r(\cos \theta + i \sin \theta)$ is $-z = r[\cos(\theta + \pi) + i \sin(\theta + \pi)]$.

53. Sketch the graph of all complex numbers z such that $|z| = 2$.

54. Sketch the graph of all complex numbers z such that the argument of each is $\theta = \pi/6$.

5.4 DeMoivre's Theorem and nth Roots

Our final look at complex numbers involves procedures for finding their powers and roots. Repeated use of the multiplication rule in the previous section yields

$$z = r(\cos\theta + i\sin\theta)$$

$$z^2 = r(\cos\theta + i\sin\theta)r(\cos\theta + i\sin\theta)$$
$$= r^2(\cos 2\theta + i\sin 2\theta)$$

$$z^3 = z^2(z) = r^2(\cos 2\theta + i\sin 2\theta)r(\cos\theta + i\sin\theta)$$
$$= r^3(\cos 3\theta + i\sin 3\theta).$$

Similarly,

$$z^4 = r^4(\cos 4\theta + i\sin 4\theta)$$
$$z^5 = r^5(\cos 5\theta + i\sin 5\theta)$$
$$\vdots$$

This pattern leads to the following important theorem.

DeMoivre's Theorem

If $z = r(\cos\theta + i\sin\theta)$ is a complex number and n is a positive integer, then

$$z^n = [r(\cos\theta + i\sin\theta)]^n = r^n(\cos n\theta + i\sin n\theta).$$

A complete proof of this theorem can be given using mathematical induction.

EXAMPLE 1 Finding Powers of a Complex Number

Use DeMoivre's Theorem to find $(-1 + \sqrt{3}\,i)^{12}$.

SOLUTION

We first convert to trigonometric form.

$$-1 + \sqrt{3}\,i = 2\left(\cos\frac{2\pi}{3} + i\sin\frac{2\pi}{3}\right)$$

Complex Numbers

Then, by DeMoivre's Theorem, we have

$$(-1 + \sqrt{3}\ i)^{12} = \left[2\left(\cos\frac{2\pi}{3} + i\ \sin\frac{2\pi}{3}\right)\right]^{12}$$

$$= 2^{12}\left[\cos(12)\frac{2\pi}{3} + i\ \sin(12)\frac{2\pi}{3}\right]$$

$$= 4096(\cos 8\pi + i\ \sin 8\pi)$$

$$= 4096(1 + 0) = 4096.$$

Are you surprised to see a real number as the answer?

Recall that a consequence of the Fundamental Theorem of Algebra (Section 5.2) is that a polynomial equation of degree n has n solutions in the complex number system. Hence, an equation like $x^6 = 1$ has six solutions, and in this particular case we can find the six solutions by factoring and using the quadratic formula.

$$x^6 - 1 = (x^3 - 1)(x^3 + 1)$$
$$= (x - 1)(x^2 + x + 1)(x + 1)(x^2 - x + 1) = 0$$

Consequently, the solutions are

$$x = \pm 1, \qquad x = \frac{-1 \pm \sqrt{3}\ i}{2}, \qquad \text{and} \qquad x = \frac{1 \pm \sqrt{3}\ i}{2}.$$

Each of these numbers is called a sixth root of 1. In general, we define the **nth root** of a complex number as follows.

Definition of nth Root of a Complex Number

The complex number $u = a + bi$ is an **nth root** of the complex number z if

$$z = u^n = (a + bi)^n.$$

To find a formula for an nth root of a complex number, we let u be an nth root of z, where

$$u = s(\cos \beta + i\ \sin \beta) \qquad \text{and} \qquad z = r(\cos \theta + i\ \sin \theta).$$

By DeMoivre's Theorem and the fact that $u^n = z$, we have

$$s^n(\cos n\beta + i\ \sin n\beta) = r(\cos \theta + i\ \sin \theta).$$

Now, taking the absolute value of both sides of this equation, it follows that $s^n = r$. Substituting back into the previous equation and dividing by r, we get

$$\cos n\beta + i\ \sin n\beta = \cos \theta + i\ \sin \theta.$$

Thus, it follows that

$$\cos n\beta = \cos \theta \quad \text{and} \quad \sin n\beta = \sin \theta.$$

Since both sine and cosine have a period of 2π, these last two equations have solutions if and only if the angles differ by a multiple of 2π. Consequently, there must exist an integer k such that

$$n\beta = \theta + 2\pi k$$

$$\beta = \frac{\theta + 2\pi k}{n}.$$

By substituting this value for β into the trigonometric form of u, we get the result stated in the following theorem.

---

nth Roots of a Complex Number

For a positive integer n, the complex number $z = r(\cos \theta + i \sin \theta)$ has exactly n distinct nth roots given by

$$\sqrt[n]{r}\left(\cos \frac{\theta + 2\pi k}{n} + i \sin \frac{\theta + 2\pi k}{n}\right)$$

where $k = 0, 1, 2, \ldots, n - 1$.

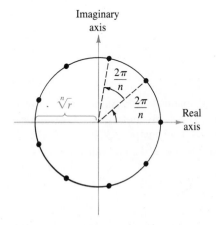

FIGURE 5.6

Remark: Note that when k exceeds $n - 1$ the roots begin to repeat. For instance, if $k = n$, the angle

$$\frac{\theta + 2\pi n}{n} = \frac{\theta}{n} + 2\pi$$

is coterminal with θ/n, which is also obtained when $k = 0$.

This formula for the nth roots of a complex number z has a nice geometrical intepretation, as shown in Figure 5.6. Note that because the nth roots of z all have the same magnitude $\sqrt[n]{r}$, they all lie on a circle of radius $\sqrt[n]{r}$ with center at the origin. Furthermore, the n roots are equally spaced along the circle, since successive nth roots have arguments that differ by $2\pi/n$.

We have already found the sixth roots of 1 by factoring and the quadratic formula. Now let's see how we can solve the same problem with the formula for nth roots.

EXAMPLE 2 Finding nth Roots of a Real Number

Find all the sixth roots of 1.

SOLUTION

First we write 1 in the trigonometric form

$$1 = 1(\cos 0 + i \sin 0).$$

Then, by the nth root formula, with $n = 6$ and $r = 1$, the roots have the form

$$\sqrt[6]{1}\left(\cos \frac{0 + 2\pi k}{6} + i \sin \frac{0 + 2\pi k}{6}\right)$$

or simply $\cos(\pi k/3) + i \sin(\pi k/3)$. Thus, for $k = 0, 1, 2, 3, 4, 5$, the sixth roots are

$$\cos 0 + i \sin 0 = 1$$

$$\cos \frac{\pi}{3} + i \sin \frac{\pi}{3} = \frac{1}{2} + \frac{\sqrt{3}}{2} i$$

$$\cos \frac{2\pi}{3} + i \sin \frac{2\pi}{3} = -\frac{1}{2} + \frac{\sqrt{3}}{2} i$$

$$\cos \pi + i \sin \pi = -1$$

$$\cos \frac{4\pi}{3} + i \sin \frac{4\pi}{3} = -\frac{1}{2} - \frac{\sqrt{3}}{2} i$$

$$\cos \frac{5\pi}{3} + i \sin \frac{5\pi}{3} = \frac{1}{2} - \frac{\sqrt{3}}{2} i$$

as shown in Figure 5.7.

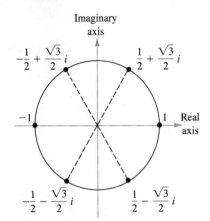

FIGURE 5.7

In Figure 5.7, notice that the roots obtained in Example 2 all have a magnitude of 1 and are equally spaced around this unit circle. Also, notice that the complex roots occur in conjugate pairs, as previously discussed in Section 5.2. We refer to the special case of the n distinct nth roots of 1 as the **nth roots of unity.**

EXAMPLE 3 Finding the nth Roots of a Complex Number

Find the three cube roots of $z = -2 + 2i$.

SOLUTION

Because z lies in the second quadrant, the trigonometric form for z is

$$z = -2 + 2i = \sqrt{8}(\cos 135° + i \sin 135°).$$

By our formula for nth roots, the cube roots have the form

$$\sqrt[6]{8}\left(\cos \frac{135° + 360°k}{3} + i \sin \frac{135° + 360°k}{3}\right).$$

Finally, for $k = 0, 1, 2$, we obtain the roots

$$\sqrt{2}(\cos 45° + i \sin 45°) = 1 + i$$
$$\sqrt{2}(\cos 165° + i \sin 165°) \approx -1.3660 + 0.3660\,i$$
$$\sqrt{2}(\cos 285° + i \sin 285°) \approx 0.3660 - 1.3660\,i.$$

The nth roots of a complex number can be useful for solving polynomial equations, as shown in Example 4.

EXAMPLE 4   Finding the Roots of a Polynomial Equation

Find all solutions to the equation $x^4 + 16 = 0$.

SOLUTION

The given equation can be written as

$$x^4 = -16 = 16(\cos \pi + i \sin \pi)$$

which means that we can solve the equation by finding the four fourth roots of -16. Each of these roots has the form

$$\sqrt[4]{16}\left(\cos \frac{\pi + 2\pi k}{4} + i \sin \frac{\pi + 2\pi k}{4}\right).$$

Finally, using $k = 0, 1, 2, 3$, we obtain the roots

$$2\left(\cos \frac{\pi}{4} + i \sin \frac{\pi}{4}\right) = 2\left(\frac{\sqrt{2}}{2} + \frac{\sqrt{2}}{2}\,i\right) = \sqrt{2} + \sqrt{2}\,i$$

$$2\left(\cos \frac{3\pi}{4} + i \sin \frac{3\pi}{4}\right) = 2\left(-\frac{\sqrt{2}}{2} + \frac{\sqrt{2}}{2}\,i\right) = -\sqrt{2} + \sqrt{2}\,i$$

$$2\left(\cos \frac{5\pi}{4} + i \sin \frac{5\pi}{4}\right) = 2\left(-\frac{\sqrt{2}}{2} - \frac{\sqrt{2}}{2}\,i\right) = -\sqrt{2} - \sqrt{2}\,i$$

$$2\left(\cos \frac{7\pi}{4} + i \sin \frac{7\pi}{4}\right) = 2\left(\frac{\sqrt{2}}{2} - \frac{\sqrt{2}}{2}\,i\right) = \sqrt{2} - \sqrt{2}\,i.$$

WARM UP

Simplify the expressions.

1. $\sqrt[3]{54}$

2. $\sqrt[4]{16 + 48}$

Write the complex numbers in trigonometric form.

3. $-5 + 5i$

4. $-3i$

5. -12

6. 12

Perform the indicated operations. Leave the results in trigonometric form.

7. $\left(\cos \dfrac{\pi}{4} + i \sin \dfrac{\pi}{4}\right)\left(\cos \dfrac{\pi}{2} + i \sin \dfrac{\pi}{2}\right)$

8. $\left(\cos \dfrac{\pi}{12} + i \sin \dfrac{\pi}{12}\right)\left(\cos \dfrac{5\pi}{6} + i \sin \dfrac{5\pi}{6}\right)$

9. $\dfrac{6[\cos(2\pi/3) + i \sin(2\pi/3)]}{3[\cos(\pi/6) + i \sin(\pi/6)]}$

10. $\dfrac{2(\cos 55° + i \sin 55°)}{3(\cos 10° + i \sin 10°)}$

EXERCISES 5.4

In Exercises 1–12, use DeMoivre's Theorem to find the indicated powers of the given complex number. Express the result in standard form.

1. $(1 + i)^5$

2. $(2 + 2i)^6$

3. $(-1 + i)^{10}$

4. $(1 - i)^{12}$

5. $2(\sqrt{3} + i)^7$

6. $4(1 - \sqrt{3}\,i)^3$

7. $[5(\cos 20° + i \sin 20°)]^3$

8. $[3(\cos 150° + i \sin 150°)]^4$

9. $\left(\cos \dfrac{5\pi}{4} + i \sin \dfrac{5\pi}{4}\right)^{10}$

10. $\left[2\left(\cos \dfrac{\pi}{2} + i \sin \dfrac{\pi}{2}\right)\right]^8$

11. $[5(\cos 3.2 + i \sin 3.2)]^4$

12. $(\cos 0 + i \sin 0)^{20}$

In Exercises 13–24, (a) use DeMoivre's Theorem to find the indicated roots, (b) represent each of the roots graphically, and (c) express each of the roots in standard form.

13. Square roots:
$9(\cos 120° + i \sin 120°)$

14. Square roots:
$16(\cos 60° + i \sin 60°)$

15. Fourth roots:
$16\left(\cos \dfrac{4\pi}{3} + i \sin \dfrac{4\pi}{3}\right)$

16. Fifth roots:
$32\left(\cos \dfrac{5\pi}{6} + i \sin \dfrac{5\pi}{6}\right)$

17. Square roots:
$-25i$

18. Fourth roots:
$625i$

19. Cube roots:
$-\dfrac{125}{2}(1 + \sqrt{3}\,i)$

20. Cube roots:
$-4\sqrt{2}(1 - i)$

21. Cube roots:
8

22. Fourth roots:
i

23. Fifth roots:
1

24. Cube roots:
1000

In Exercises 25–32, find all the solutions to the given equation and represent your solutions graphically.

25. $x^4 - i = 0$

26. $x^3 + 1 = 0$

27. $x^5 + 243 = 0$

28. $x^4 - 81 = 0$

29. $x^3 + 64i = 0$

30. $x^6 - 64i = 0$

31. $x^3 - (1 - i) = 0$

32. $x^4 + (1 + i) = 0$

CHAPTER 5 REVIEW EXERCISES

In Exercises 1–12, perform the indicated operations and write the result in standard form.

1. $(7 + 5i) + (-4 + 2i)$

2. $-(6 - 2i) + (-8 + 3i)$

3. $\left(\dfrac{\sqrt{2}}{2} - \dfrac{\sqrt{2}}{2}i\right) - \left(\dfrac{\sqrt{2}}{2} + \dfrac{\sqrt{2}}{2}i\right)$

4. $(13 - 8i) - 5i$

5. $5i(13 - 8i)$

6. $(1 + 6i)(5 - 2i)$

7. $(10 - 8i)(2 - 3i)$

8. $i(6 + i)(3 - 2i)$

9. $\dfrac{6 + i}{i}$

10. $\dfrac{3 + 2i}{5 + i}$

11. $\dfrac{4}{(-3i)}$

12. $\dfrac{1}{(2 + i)^4}$

In Exercises 13–20, use the discriminant to determine the number of real solutions of the equation.

13. $6x^2 + x - 2 = 0$

14. $20x^2 - 21x + 4 = 0$

15. $9x^2 - 12x + 4 = 0$

16. $3x^2 + 5x + 3 = 0$

17. $0.13x^2 - 0.45x + 0.65 = 0$

18. $4x^2 + \frac{4}{3}x + \frac{1}{9} = 0$

19. $15 + 2x - x^2 = 0$

20. $x(x + 12) = -46$

In Exercises 21 and 22, find a fourth-degree polynomial with the given zeros.

21. $-1, -1, \frac{1}{3}, -\frac{1}{2}$

22. $\frac{2}{3}, 4, \sqrt{3}i, -\sqrt{3}i$

In Exercises 23–32, find all zeros of the function.

23. $f(x) = x^2 + 6x - 16$

24. $f(x) = 2x^2 + 5x + 4$

25. $f(x) = x^3 - 18x^2 + 106x - 200$
[*Hint:* One zero is $x = 7 + i$.]

26. $f(x) = x^3 - 5x^2 - 9x + 45$
[*Hint:* One zero is $x = -3$.]

27. $f(x) = x^3 + 3x^2 - 5x + 25$
[*Hint:* One zero is $x = -5$.]

28. $f(x) = x^4 - 6x^3 + 23x^2 - 34x + 26$
[*Hint:* One zero is $x = 1 + i$.]

29. $f(x) = x^4 + 5x^3 + 2x^2 - 50x - 84$
[*Hint:* One zero is $x = -3 + \sqrt{5}i$.]

30. $f(x) = x^4 - 8x^3 + 24x^2 - 36x + 27$
[*Hint:* One zero is $x = 1 + \sqrt{2}i$.]

31. $f(x) = x^5 + x^4 - 3x^3 + 39x^2 + 92x - 130$
[*Hint:* Two zeros are $x = 1$ and $x = 2 + 3i$.]

32. $f(x) = x^6 - 6x^5 + 16x^4 - 24x^3 + 25x^2 - 18x + 10$
[*Hint:* Two zeros are $x = i$ and $x = 1 + i$.]

In Exercises 33–36, find the trigonometric form of the complex number.

33. $5 - 5i$

34. $-3\sqrt{3} + 3i$

35. $5 + 12i$

36. -7

In Exercises 37 and 38, write the complex number in standard form.

37. $100(\cos 240° + i \sin 240°)$

38. $8\left(\cos \dfrac{5\pi}{6} + i \sin \dfrac{5\pi}{6}\right)$

In Exercises 39–42, (a) express the two given complex numbers in trigonometric form, and then (b) find $z_1 z_2$ and z_1/z_2 using the trigonometric form.

39. $z_1 = -6, z_2 = 5i$

40. $z_1 = 2\sqrt{3} - 2i, z_2 = -10i$

41. $z_1 = -3(1 + i), z_2 = 2(\sqrt{3} + i)$

42. $z_1 = 5i, z_2 = 2 - 2i$

In Exercises 43–46, use DeMoivre's Theorem to find the indicated power of the given complex number. Express the result in standard form.

43. $\left[5\left(\cos \dfrac{\pi}{12} + i \sin \dfrac{\pi}{12} \right) \right]^4$

44. $\left[2\left(\cos \dfrac{4\pi}{15} + i \sin \dfrac{4\pi}{15} \right) \right]^5$

45. $(2 + 3i)^6$

46. $(1 - i)^8$

In Exercises 47–50, use DeMoivre's Theorem to find the roots of the given complex numbers.

47. The sixth roots of $-729i$.

48. The fourth roots of 256.

49. The cube roots of -1.

50. The fourth roots of $-1 + i$.

C H A P T E R   6

Exponential and Logarithmic Functions

6.1 Exponential Functions

Exponential functions are widely used in describing economic and physical phenomena such as compound interest, population growth, memory retention, and decay of radioactive material. Exponential functions involve a *constant base* and a *variable exponent* such as

$$f(x) = 2^x \quad \text{or} \quad g(x) = 3^{-x}.$$

The general definition of the **exponential function with base a** is as follows.

Definition of Exponential Function

The **exponential function f with base a** is denoted by

$$f(x) = a^x$$

where $a > 0$, $a \neq 1$, and x is any real number.

Remark: We exclude the base $a = 1$ because it yields $f(x) = 1^x = 1$. This is a constant function, not an exponential function.

Before discussing the behavior of exponential functions, we list some familiar properties of exponents.

303

Properties of Exponents (a, b > 0)

1. $a^0 = 1$ 2. $a^x a^y = a^{x+y}$ 3. $\dfrac{a^x}{a^y} = a^{x-y}$

4. $(a^x)^y = a^{xy}$ 5. $(ab)^x = a^x b^x$ 6. $\left(\dfrac{a}{b}\right)^x = \dfrac{a^x}{b^x}$

Although we have generally used these six properties with integer and rational values for x and y, it is important to realize that the properties hold for *any* real values for x and y. With a calculator we can readily obtain approximate values for a^x when x is not an integer or irrational, such as in $a^{\sqrt{2}}$ and a^π.

A technical definition of such forms is beyond the scope of this text. For our purposes it is sufficient to think of

$$a^{\sqrt{2}} \quad \text{(where } \sqrt{2} \approx 1.414214\text{)}$$

as that value which has the successively closer approximations

$$a^{1.4}, \; a^{1.41}, \; a^{1.414}, \; a^{1.4142}, \; a^{1.41421}, \; \ldots$$

Consequently, we assume in this text that a^x exists for all real x and that the properties of exponents can be extended to cover exponential functions. For instance,

$$a^{-x} = \frac{1}{a^x} = \left(\frac{1}{a}\right)^x.$$

EXAMPLE 1 *Graphs of y = aˣ*

On the same coordinate plane, sketch the graphs of the following functions.

(a) $f(x) = 2^x$ (b) $g(x) = 4^x$

SOLUTION

Table 6.1 lists some values for each function, and Figure 6.1 shows their graphs.

TABLE 6.1

x	-2	-1	0	1	2	3
(a) 2^x	$\frac{1}{4}$	$\frac{1}{2}$	1	2	4	8
(b) 4^x	$\frac{1}{16}$	$\frac{1}{4}$	1	4	16	64

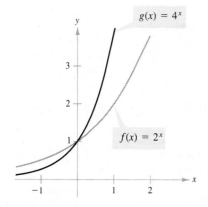

FIGURE 6.1

EXAMPLE 2 Graphs of y = a⁻ˣ

On the same coordinate plane, sketch the graphs of the following functions.

(a) $F(x) = 2^{-x}$ (b) $G(x) = \left(\dfrac{1}{4}\right)^{x}$

SOLUTION

To evaluate $F(x)$, we write

$$F(x) = 2^{-x} = \frac{1}{2^x} = \left(\frac{1}{2}\right)^{x}.$$

Table 6.2 lists some values for each function, and Figure 6.2 shows their graphs.

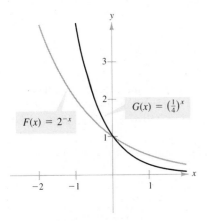

$G(x) = (\frac{1}{4})^x$

$F(x) = 2^{-x}$

FIGURE 6.2

TABLE 6.2

x	-3	-2	-1	0	1	2
(a) 2^{-x}	8	4	2	1	$\dfrac{1}{2}$	$\dfrac{1}{4}$
(b) $\left(\dfrac{1}{4}\right)^{x}$	64	16	4	1	$\dfrac{1}{4}$	$\dfrac{1}{16}$

Comparing the functions in Examples 1 and 2, observe that

$$F(x) = 2^{-x} = f(-x)$$

and

$$G(x) = \left(\frac{1}{4}\right)^{x} = 4^{-x} = g(-x).$$

Consequently, the graph of F is a reflection (in the y-axis) of the graph of f. The graphs of G and g have the same relationship. This is verified by a comparison of the graphs in Figures 6.1 and 6.2.

Remark: Examples 1 and 2 point out that the function $f(x) = a^x$ *increases* if $a > 1$ and *decreases* if $0 < a < 1$.

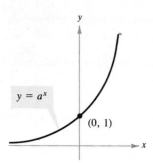

Graph of $y = a^x$
- Domain: $(-\infty, \infty)$
- Range: $(0, \infty)$
- Intercept: $(0, 1)$
- Increasing
- x-axis is a horizontal asymptote
  ($a^x \to 0$ as $x \to -\infty$)
- Continuous

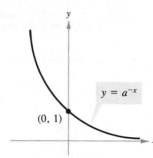

Graph of $y = a^{-x}$
- Domain: $(-\infty, \infty)$
- Range: $(0, \infty)$
- Intercept: $(0, 1)$
- Decreasing
- x-axis is a horizontal asymptote
  ($a^{-x} \to 0$ as $x \to \infty$)
- Continuous
- Reflection of graph of
  $y = a^x$ about y-axis

Characteristics of the Exponential Functions a^x and a^{-x} ($a > 1$)

FIGURE 6.3

(a)

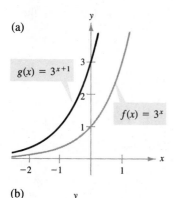

(b)

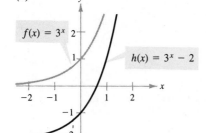

The graphs in Figures 6.1 and 6.2 are typical of the exponential functions a^x and a^{-x}. They have one y-intercept and one horizontal asymptote (the x-axis), and they are continuous. The basic characteristics of these exponential functions are summarized in Figure 6.3.

In the following example, we use the graph of a^x to sketch the graphs of functions of the form $f(x) = b \pm a^{x+c}$.

EXAMPLE 3 *Sketching Graphs of Exponential Functions*

Sketch the graph of each of the following.

(a) $g(x) = 3^{x+1}$

(b) $h(x) = 3^x - 2$

(c) $k(x) = -3^x$

SOLUTION

The graph of each of these three functions is similar to the graph of $f(x) = 3^x$, as shown in Figure 6.4.

(a) Because $g(x) = 3^{x+1} = f(x + 1)$, the graph of g can be obtained by shifting the graph of f one unit to the left.

(b) Because $h(x) = 3^x - 2 = f(x) - 2$, the graph of h can be obtained by shifting the graph of f down two units.

(c) Because $k(x) = -3^x = -f(x)$, the graph of k can be obtained by reflecting the graph of f in the x-axis.

(c)

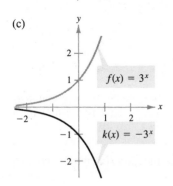

FIGURE 6.4

Exponential Functions

To evaluate exponential functions with a calculator, you need to use the exponential key $\boxed{y^x}$, where y is the base and x is the exponent. The base is entered first, then the exponent, as shown in the following keystroke sequences.

Number	Keystrokes	Display
$(1.085)^3$	1.085 $\boxed{y^x}$ 3 $\boxed{=}$	1.277289
$12^{5/7}$	12 $\boxed{y^x}$ $\boxed{(}$ 5 $\boxed{\div}$ 7 $\boxed{)}$ $\boxed{=}$	5.899888
$2^{-\pi}$	2 $\boxed{y^x}$ π $\boxed{+/-}$ $\boxed{=}$	0.1133147

EXAMPLE 4 *Sketching the Graph of an Exponential Function*

Sketch a graph of $f(x) = 2^{1-x^2}$.

SOLUTION

First, note that the graph of f is symmetric with respect to the y-axis because

$$f(-x) = 2^{1-(-x)^2} = 2^{1-x^2} = f(x).$$

Using this symmetry and the table of values

x	0	1	2
$f(x)$	2	1	$\dfrac{1}{8}$

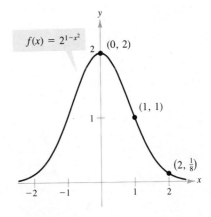

FIGURE 6.5

we obtain the graph shown in Figure 6.5. (Use your calculator to verify the shape of the graph between 0 and 1.)

Exponential and Logarithmic Functions

EXAMPLE 5 An Application Involving Radioactive Decay

Let y represent the mass of a particular radioactive element whose half-life is 25 years. After t years, the mass (in grams) is given by

$$y = 10\left(\frac{1}{2}\right)^{t/25}.$$

(a) What is the initial mass (when $t = 0$)?
(b) How much of the initial mass is present after 80 years?

SOLUTION

(a) When $t = 0$, the mass is

$$y = 10\left(\frac{1}{2}\right)^{0} = 10(1) = 10 \text{ grams.}$$

(b) When $t = 80$, the mass is

$$y = 10\left(\frac{1}{2}\right)^{80/25}$$
$$= 10(0.5)^{3.2}$$
$$\approx 1.088 \text{ grams.}$$

The graph of this function is shown in Figure 6.6.

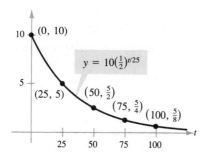

Radioactive Half-life of 25 Years

FIGURE 6.6

The Natural Base e

We used an unspecified base a to introduce exponential functions. It happens that in many applications the convenient choice for a base is the irrational number

$$e \approx 2.71828. \ldots$$

called the **natural base.** It is also the base most frequently used in calculus. The function $f(x) = e^x$ is called the **natural exponential function.** Its graph is shown in Figure 6.7.

Remark: Be sure you see that for the exponential function $f(x) = e^x$, e is the constant $2.71828. \ldots$, whereas x is the variable.

Some calculators have a natural exponential key $\boxed{e^x}$. On such a calculator you can evaluate e^2 by entering the sequence 2 $\boxed{e^x}$. Other calculators require the two-key sequence $\boxed{\text{INV}}$ $\boxed{\ln x}$ to evaluate exponential functions. On such a calculator you can evaluate e^2 using the following sequence.

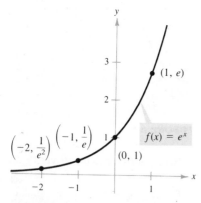

FIGURE 6.7

2 $\boxed{\text{INV}}$ $\boxed{\ln x}$ *Display* 7.3890561

Exponential Functions

Similarly, to evaluate e^{-1}, enter the sequence

1 $\boxed{+/-}$ $\boxed{\text{INV}}$ $\boxed{\text{ln } x}$.

Display 0.3678794

Example 6 gives an indication of how the irrational number e arises in applications. We will refer to this example when we develop the formula for continuous compounding of interest.

EXAMPLE 6 *Approximation of the Number e*

Evaluate the expression

$$\left(1 + \frac{1}{n}\right)^n$$

for several large values of n to see that the values approach $e \approx 2.71828$ as n increases without bound.

SOLUTION

Using the keystroke sequence

n $\boxed{1/x}$ $\boxed{+}$ 1 $\boxed{=}$ $\boxed{y^x}$ n $\boxed{=}$

we obtain the values shown in the following table.

n	10	100	1000	10,000	100,000	1,000,000
$\left(1 + \dfrac{1}{n}\right)^n$	2.59374	2.70481	2.71692	2.71815	2.71827	2.71828

From this table, it seems reasonable to conclude that

$$\left(1 + \frac{1}{n}\right)^n \to e \quad \text{as} \quad n \to \infty.$$

Compound Interest

One of the most familiar examples of exponential growth is that of an investment earning *compound interest*. Suppose a principal P is invested at an annual percentage rate r, compounded once a year. If the interest is added to the principal at the end of the year, then the balance is

$$P_1 = P + Pr = P(1 + r).$$

This pattern of multiplying the previous principal by $1 + r$ is then repeated each successive year, as shown in Table 6.3.

Exponential and Logarithmic Functions

TABLE 6.3

Time in years	Balance after each compounding
0	$P = P$
1	$P_1 = P(1 + r)$
2	$P_2 = P_1(1 + r) = P(1 + r)(1 + r) = P(1 + r)^2$
3	$P_3 = P_2(1 + r) = P(1 + r)^2(1 + r) = P(1 + r)^3$
$\vdots$	$\vdots$
n	$P_n = P(1 + r)^n$

To accommodate more frequent (quarterly, monthly, or daily) compounding of interest, we let n be the number of compoundings per year and t the number of years. Then the rate per compounding is r/n and the account balance after t years is

$$A = P\left(1 + \frac{r}{n}\right)^{nt}.$$

Amount with n compoundings per year

If we let the number of compoundings, n, increase without bound, we approach what is called **continuous compounding.** In the formula for n compoundings per year, let $m = n/r$. This produces

$$A = P\left(1 + \frac{r}{n}\right)^{nt} = P\left(1 + \frac{1}{m}\right)^{mrt} = P\left[\left(1 + \frac{1}{m}\right)^{m}\right]^{rt}.$$

As m increases without bound, we know from Example 6 that

$$\left(1 + \frac{1}{m}\right)^{m} \to e.$$

Hence, for continuous compounding, it follows that

$$P\left[\left(1 + \frac{1}{m}\right)^{m}\right]^{rt} \to P[e]^{rt}$$

and we write $A = Pe^{rt}$.

We summarize the two formulas for compound interest as follows.

Formulas for Compound Interest

After t years, the balance A in an account with principal P and annual percentage rate r (expressed as a decimal) is given by the following formulas.

1. For n compoundings per year: $A = P\left(1 + \frac{r}{n}\right)^{nt}$
2. For continuous compounding: $A = Pe^{rt}$

EXAMPLE 7 Finding the Balance for Compound Interest

A sum of $9000 is invested at an annual percentage rate of 8.5%, compounded annually. Find the balance in the account after three years.

SOLUTION

In this case, $P = 9000$, $r = 8.5\% = 0.085$, $n = 1$, and $t = 3$. Using the formula

$$A = P\left(1 + \frac{r}{n}\right)^{nt}$$

we have

$$A = 9000(1 + 0.085)^3 = 9000(1.085)^3 \approx \$11,495.60.$$

EXAMPLE 8 Compounding n Times and Continuously

A total of $12,000 is invested at an annual percentage rate of 9%. Find the balance after five years if it is compounded

(a) quarterly   (b) continuously.

SOLUTION

(a) For quarterly compoundings, we have $n = 4$. Thus, in five years at 9%, the balance is

$$A = P\left(1 + \frac{r}{n}\right)^{nt} = 12,000\left(1 + \frac{0.09}{4}\right)^{4(5)} = \$18,726.11.$$

(b) Compounded continuously, the balance is

$$A = Pe^{rt} = 12,000e^{0.09(5)} = \$18,819.75.$$

Note that continuous compounding yields

$$\$18,819.75 - \$18,726.11 = \$93.64$$

more than quarterly compounding.

Exponential and Logarithmic Functions

EXAMPLE 9 Population Growth

The number of fruit flies in an experimental population after t hours is given by

$$Q(t) = 20e^{0.03t}, \qquad t \geq 0.$$

(a) Find the initial number of fruit flies in the population.
(b) How large is the population of fruit flies after 72 hours?
(c) Sketch the graph of Q.

SOLUTION

(a) To find the initial population, we evaluate $Q(t)$ at $t = 0$.

$$Q(0) = 20e^{0.03(0)} = 20e^0 = 20(1) = 20 \text{ flies}$$

(b) After 72 hours, the population size is

$$Q(72) = 20e^{(0.03)(72)} = 20e^{2.16} \approx 173 \text{ flies}.$$

(c) To sketch the graph of Q, we evaluate $Q(t)$ for several values of t (rounded to the nearest integer), and plot the corresponding points, as shown in Figure 6.8.

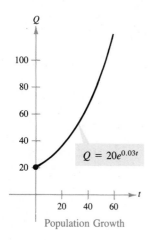

Population Growth

FIGURE 6.8

t	0	5	10	20	40	60
$20e^{0.03t}$	20	23	27	36	66	121

Remark: Many animal populations have a growth pattern described by the function $Q(t) = ce^{kt}$ where c is the original population, $Q(t)$ is the population at time t, and k is a constant determined by the rate of growth. Informally, we can write this as $(Then) = (Now)(e^{kt})$.

WARM UP

Use the properties of exponents to simplify the expressions.

1. $5^{2x}(5^{-x})$

2. $3^{-x}(3^{3x})$

3. $\dfrac{4^{5x}}{4^{2x}}$

4. $\dfrac{10^{2x}}{10^x}$

5. $(4^x)^2$

6. $(4^{2x})^5$

7. $\left(\dfrac{2^x}{3^x}\right)^{-1}$

8. $(4^{6x})^{1/2}$

9. $(2^{3x})^{-1/3}$

10. $(16^x)^{1/4}$

EXERCISES 6.1

In Exercises 1–14, use a calculator to evaluate the given quantity. Round your answers to three decimal places.

1. $(3.4)^{5.6}$

2. $(1.005)^{400}$

3. $1000(1.06)^{-5}$

4. $5000(2^{-1.5})$

5. $\sqrt[4]{763}$

6. $\sqrt[3]{4395}$

7. $8^{2\pi}$

8. $5^{-\pi}$

9. $100^{\sqrt{2}}$

10. $0.6^{\sqrt{3}}$

11. e^2

12. $e^{1/2}$

13. $e^{-3/4}$

14. $e^{3.2}$

In Exercises 15–22, match the exponential function with its graph. [The graphs are labeled (a)–(h).]

15. $f(x) = 3^x$

16. $f(x) = -3^x$

17. $f(x) = 3^{-x}$

18. $f(x) = -3^{-x}$

19. $f(x) = 3^x - 4$

20. $f(x) = 3^x + 1$

21. $f(x) = -3^{x-2}$

22. $f(x) = 3^{x-2}$

(a)

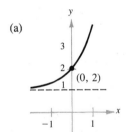

(b)

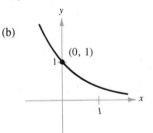

(c)

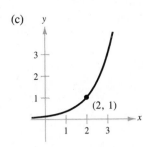

(d)

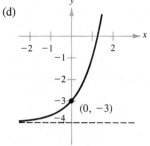

(e)

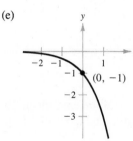

(f)

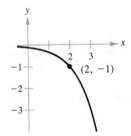

(g)

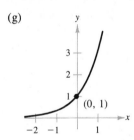

(h)

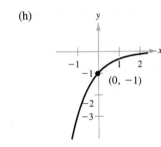

In Exercises 23–40, sketch the graph of the given exponential function.

23. $g(x) = 5^x$

24. $f(x) = \left(\frac{3}{2}\right)^x$

25. $f(x) = \left(\frac{1}{5}\right)^x = 5^{-x}$

26. $h(x) = \left(\frac{3}{2}\right)^{-x}$

27. $h(x) = 5^{x-2}$

28. $g(x) = \left(\frac{3}{2}\right)^{x+2}$

29. $g(x) = 5^{-x} - 3$

30. $f(x) = \left(\frac{3}{2}\right)^{-x} + 2$

31. $y = 2^{-x^2}$

32. $y = 3^{-x^2}$

33. $y = 3^{|x|}$

34. $y = 3^{-|x|}$

35. $s(t) = 2^{-t} + 3$

36. $s(t) = \frac{1}{4}(3^{-t})$

37. $f(x) = e^{2x}$

38. $h(x) = e^{x-2}$

39. $g(x) = 1 + e^{-x}$

40. $N(t) = 1000e^{-0.2t}$

In Exercises 41–44, complete the following table to determine the balance A for P dollars invested at rate r for t years and compounded n times per year.

n	1	2	4	12	365	Continuous compounding
A						

41. $P = \$2500$, $r = 12\%$, $t = 10$ years

42. $P = \$1000$, $r = 10\%$, $t = 10$ years

43. $P = \$2500$, $r = 12\%$, $t = 20$ years

44. $P = \$1000$, $r = 10\%$, $t = 40$ years

In Exercises 45–48, complete the following table to determine the amount of money P that should be invested at rate r to produce a final balance of $100,000 in t years.

t	1	10	20	30	40	50
P						

45. $r = 9\%$, compounded continuously

46. $r = 12\%$, compounded continuously

47. $r = 10\%$, compounded monthly

48. $r = 7\%$, compounded daily

49. The demand equation for a certain product is given by

$$p = 500 - 0.5e^{0.004x}.$$

Find the price p for a demand of (a) $x = 1000$ units and (b) $x = 1500$ units.

50. The demand equation for a certain product is given by

$$p = 5000\left(1 - \frac{4}{4 + e^{-0.002x}}\right).$$

Find the price p for a demand of (a) $x = 100$ units and (b) $x = 500$ units.

51. A certain type of bacteria increases according to the model

$$P(t) = 100e^{0.2197t}$$

where t is time in hours. Find (a) $P(0)$, (b) $P(5)$, and (c) $P(10)$.

52. The population of a town increases according to the model

$$P(t) = 2500e^{0.0293t}$$

where t is time in years, with $t = 0$ corresponding to 1980. Use the model to approximate the population in (a) 1985, (b) 1990, and (c) 2000.

53. Given the exponential function $f(x) = a^x$, show that
(a) $f(u + v) = f(u) \cdot f(v)$ (b) $f(2x) = [f(x)]^2$.

6.2 *Logarithmic Functions*

In Section 6.1, we saw that the exponential function a^x increases if $a > 1$ and decreases if $0 < a < 1$. In either case, a^x is one-to-one and must have an inverse function. We call this inverse function the **logarithmic function with base a.**

Definition of Logarithmic Function

For $x > 0$ and $0 < a \neq 1$,

$\quad y = \log_a x$ if and only if $x = a^y$.

The function given by

$\quad f(x) = \log_a x$

is called the **logarithmic function with base a.**

Remark: Note that the equations $y = \log_a x$ and $x = a^y$ are equivalent. The first equation is in logarithmic form and the second is in exponential form.

Since $\log_a x$ is the inverse function of a^x, it follows that the domain of $\log_a x$ is the range of a^x, $(0, \infty)$. In other words, $\log_a x$ is defined only if x is positive.

When evaluating logarithms, remember that

A LOGARITHM IS AN EXPONENT.

This means that $\log_a x$ is the exponent to which a must be raised to obtain x. For instance,

$$\log_2 8 = 3$$

because 2 must be raised to the third power to obtain 8.

EXAMPLE 1   *Evaluating Logarithms*

(a) $\log_2 32 = 5$ because $2^5 = 32$.

(b) $\log_3 27 = 3$ because $3^3 = 27$.

(c) $\log_4 2 = \dfrac{1}{2}$ because $4^{1/2} = \sqrt{4} = 2$.

(d) $\log_{10} \dfrac{1}{100} = -2$ because $10^{-2} = \dfrac{1}{10^2} = \dfrac{1}{100}$.

(e) $\log_3 1 = 0$ because $3^0 = 1$.

(f) $\log_2 2 = 1$ because $2^1 = 2$.

The following properties follow directly from the definition of the logarithmic function with base a.

Properties of Logarithms

1. $\log_a 1 = 0$ because $a^0 = 1$.
2. $\log_a a = 1$ because $a^1 = a$.
3. $\log_a a^x = x$ because $a^x = a^x$.

To sketch the graph of $y = \log_a x$, we can use the fact that the graphs of inverse functions are reflections of each other in the line $y = x$, as demonstrated in Example 2.

EXAMPLE 2 *Graphs of Exponential and Logarithmic Functions*

On the same coordinate plane, sketch the graphs of

(a) $f(x) = 2^x$ (b) $g(x) = \log_2 x$.

SOLUTION

(a) For $f(x) = 2^x$, we make up the following table of values.

x	-2	-1	0	1	2	3
$f(x) = 2^x$	$\dfrac{1}{4}$	$\dfrac{1}{2}$	1	2	4	8

$f(x) = 2^x$

$(0, 1)$

$(1, 0)$

$g(x) = \log_2 x$

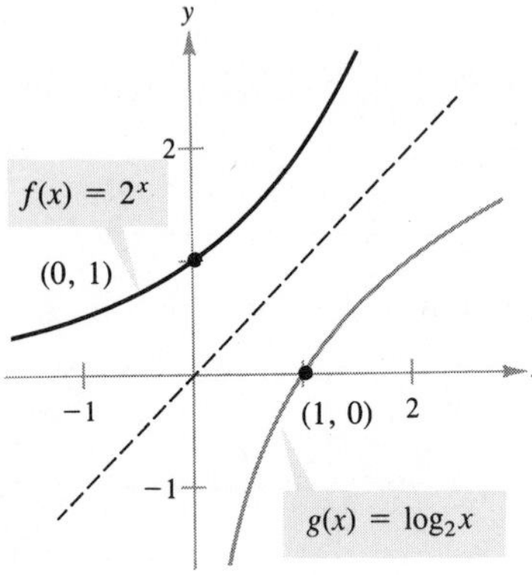

Inverse Functions

FIGURE 6.9

By plotting these points and connecting them with a smooth curve, we have the graph shown in Figure 6.9.

(b) Since $g(x) = \log_2 x$ is the inverse of $f(x) = 2^x$, the graph of g is obtained by reflecting the graph of f in the line $y = x$, as shown in Figure 6.9.

The logarithmic function with base 10 is called the **common logarithmic function.** On most calculators, this function is denoted by $\boxed{\textbf{log}}$. You can tell whether this key denotes base 10 by entering 10 $\boxed{\textbf{log}}$. The display should be 1.

EXAMPLE 3 *Sketching the Graph of a Logarithmic Function*

Sketch the graph of the common logarithmic function $y = \log_{10} x$.

SOLUTION

We begin by making the following table of values. Note that some of the values can be obtained without a calculator, while others require a calculator. We plot the corresponding points and sketch the graph shown in Figure 6.10.

	Without Calculator				With Calculator		
x	$\dfrac{1}{100}$	$\dfrac{1}{10}$	1	10	2	5	8
$\log_{10} x$	-2	-1	0	1	0.301	0.699	0.903

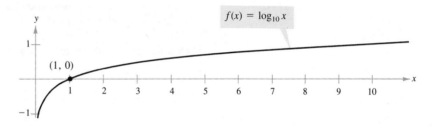

FIGURE 6.10

The nature of the graph in Figure 6.10 is typical of functions of the form $f(x) = \log_a x$, $a > 1$. They have one x-intercept and one vertical asymptote. Notice how slowly the graph rises for $x > 1$. In Figure 6.10 we would need to move out to $x = 1000$ before the graph rises to $y = 3$. We summarize the basic characteristics of logarithmic graphs in Figure 6.11.

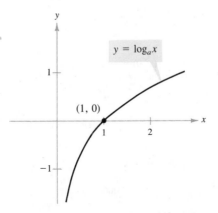

Graph of $y = \log_a x$, $a > 1$
- Domain: $(0, \infty)$
- Range: $(-\infty, \infty)$
- Intercept: $(1, 0)$
- Increasing
- y-axis is a vertical asymptote $(\log_a x \rightarrow -\infty$ as $x \rightarrow 0^+)$
- Continuous
- Reflection of graph of $y = a^x$ about the line $y = x$

FIGURE 6.11

Remark: In Figure 6.11, note that the vertical asymptote occurs at $x = 0$, where $\log_a x$ is undefined.

In the following example we use the graph of $\log_a x$ to sketch the graphs of functions of the form $y = b \pm \log_a(x + c)$.

Exponential and Logarithmic Functions

EXAMPLE 4 *Sketching the Graphs of Logarithmic Functions*

Sketch the graphs of the following functions.

(a) $g(x) = \log_4(x - 1)$

(b) $h(x) = 2 + \log_4 x$

SOLUTION

The graph of each of these functions is similar to the graph of $f(x) = \log_4 x$, as shown in Figure 6.12.

(a) Because $g(x) = \log_4(x - 1) = f(x - 1)$ the graph of g can be obtained by shifting the graph of f one unit to the right.

(b) Because $h(x) = 2 + \log_4 x = 2 + f(x)$ the graph of h can be obtained by shifting the graph of f two units up.

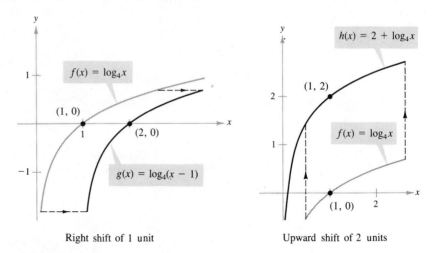

Right shift of 1 unit   Upward shift of 2 units

FIGURE 6.12

The function $f(x) = \log_a(bx + c)$ has a domain which consists of all x such that $bx + c > 0$. The vertical asymptote occurs when $bx + c = 0$, and the x-intercept occurs when $bx + c = 1$.

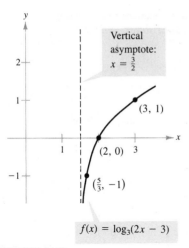

$$f(x) = \log_3(2x - 3)$$

FIGURE 6.13

EXAMPLE 5 *Sketching the Graph of a Logarithmic Function*

Sketch the graph of $f(x) = \log_3(2x - 3)$.

SOLUTION

The function f is defined for all x such that $2x - 3 > 0$, which implies that $x > 3/2$. Thus, the domain of f is the interval $(3/2, \infty)$. The vertical asymptote is $x = 3/2$ (the solution to $2x - 3 = 0$), and the x-intercept occurs when $x = 2$ (the solution to $2x - 3 = 1$). Using this information together with a couple of additional points, we obtain the graph shown in Figure 6.13.

The Natural Logarithmic Function

As with exponential functions, the most widely used base for logarithmic functions is the number e. We call the logarithmic function with base e the **natural logarithmic function** and denote it by the special symbol $\ln x$, read as "el en of x."

The Natural Logarithmic Function

The function defined by

$$f(x) = \log_e x = \ln x, \qquad x > 0$$

is called the **natural logarithmic function.**

The three properties of logarithms listed at the beginning of this section are also valid for natural logarithms.

Properties of Natural Logarithms

1. $\ln 1 = 0$ because $e^0 = 1$.
2. $\ln e = 1$ because $e^1 = e$.
3. $\ln e^x = x$ because $e^x = e^x$.

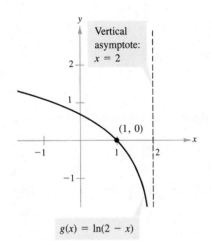

$f(x) = \ln x$

$(e, 1)$

$(1, 0)$

$\left(\dfrac{1}{e}, -1\right)$

FIGURE 6.14

EXAMPLE 6 Evaluating the Natural Logarithmic Function

(a) $\ln \dfrac{1}{e} = \ln e^{-1} = -1$ *Property 3*

(b) $\ln e^2 = 2$ *Property 3*

On most calculators, the natural logarithm is denoted by $\boxed{\text{ln } x}$.

Number	*Calculator Steps*	*Display*
(c) $\ln 2$	2 $\boxed{\text{ln } x}$	0.6931472
(d) $\ln 0.3$	$.3$ $\boxed{\text{ln } x}$	-1.2039728
(e) $\ln(-1)$	1 $\boxed{+/-}$ $\boxed{\text{ln } x}$	

Be sure you see that $\ln(-1)$ gives an error. This occurs because the domain of $\ln x$ is the set of positive real numbers. (See Figure 6.14.) Hence, $\ln(-1)$ is undefined.

The graph of the natural logarithmic function is shown in Figure 6.14.

EXAMPLE 7 Finding the Domain of Logarithmic Functions

Find the domain of the following functions.

(a) $f(x) = \ln(x - 2)$

(b) $g(x) = \ln(2 - x)$

(c) $h(x) = \ln x^2$

SOLUTION

(a) Because $\ln(x - 2)$ is defined only if $x - 2 > 0$, it follows that the domain of f is $(2, \infty)$.

(b) Because $\ln(2 - x)$ is defined only if $2 - x > 0$, it follows that the domain of g is $(-\infty, 2)$. The graph of g is shown in Figure 6.15.

(c) Because $\ln x^2$ is defined only if $x^2 > 0$, it follows that the domain of h is all real numbers except $x = 0$.

Vertical asymptote: $x = 2$

$(1, 0)$

$g(x) = \ln(2 - x)$

FIGURE 6.15

EXAMPLE 8   An Application

Students participating in a psychological experiment attended several lectures on a subject. Every month for a year after that, the students were tested to see how much of the material they remembered. The average scores for the group were given by the *human memory model*

$$f(t) = 75 - 6 \ln(t + 1), \qquad 0 \le t \le 12$$

where t is the time in months.

(a) What was the average score on the original ($t = 0$) exam?
(b) What was the average score at the end of $t = 2$ months?
(c) What was the average score at the end of $t = 6$ months?
(d) Sketch the graph of f.

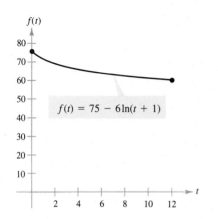

FIGURE 6.16

SOLUTION

(a) The original average score was

$$f(0) = 75 - 6 \ln(0 + 1) = 75 - 6(0) = 75.$$

(b) After two months, the average score was

$$f(2) = 75 - 6 \ln 3 \approx 75 - 6(1.0986) \approx 68.4.$$

(c) After six months, the average score was

$$f(6) = 75 - 6 \ln 7 \approx 75 - 6(1.9459) \approx 63.3.$$

(d) Several points are shown in the following table, and the graph of f is shown in Figure 6.16.

t	0	1	2	6	8	12
$f(t)$	75	70.8	68.4	63.3	61.8	59.6

Change of Base

Although 10 and e are the most frequently used bases, we occasionally need to evaluate logarithms to other bases. In such cases the following *change of base formula* is useful. (This formula is derived in Example 10, Section 6.4.)

Exponential and Logarithmic Functions

Change of Base Formula

Let a, b, and x be positive real numbers such that $a \neq 1$ and $b \neq 1$. Then $\log_a x$ is given by

$$\log_a x = \frac{\log_b x}{\log_b a}.$$

Remark: One way to look at the change of base formula is that logarithms to base a are simply *constant multiples* of logarithms to base b. The constant multiplier is $1/(\log_b a)$.

EXAMPLE 9 *Changing Bases*

Use *common* logarithms to evaluate the following.

(a) $\log_4 30$ (b) $\log_2 14$

SOLUTION

(a) Using the change of base formula with $a = 4$, $b = 10$, and $x = 30$, we convert to common logarithms and obtain

$$\log_4 30 = \frac{\log_{10} 30}{\log_{10} 4} \approx \frac{1.47712}{0.60206} \approx 2.4534.$$

(b) Using the change of base formula with $a = 2$, $b = 10$, and $x = 14$, we convert to common logarithms and obtain

$$\log_2 14 = \frac{\log_{10} 14}{\log_{10} 2} \approx \frac{1.14613}{0.30103} \approx 3.8074.$$

EXAMPLE 10 *Changing Bases*

Use *natural* logarithms to evaluate the following.

(a) $\log_4 30$ (b) $\log_2 14$

SOLUTION

(a) Using the change of base formula with $a = 4$, $b = e$, and $x = 30$, we convert to natural logarithms and obtain

$$\log_4 30 = \frac{\ln 30}{\ln 4} \approx \frac{3.40120}{1.38629} \approx 2.4534.$$

(b) Using the change of base formula with $a = 2$, $b = e$, and $x = 14$, we convert to natural logarithms and obtain

$$\log_2 14 = \frac{\ln 14}{\ln 2} \approx \frac{2.63906}{0.693147} \approx 3.8074.$$

Note that the results agree with those obtained in Example 9, using common logarithms.

WARM UP

Solve for x.

1. $2^x = 8$ **2.** $4^x = 1$
3. $10^x = 0.1$ **4.** $e^x = e$

Evaluate the given expressions. (Round your answers to three decimal places.)

5. e^2 **6.** e^{-1}

Describe how the graph of g is related to the graph of f.

7. $g(x) = f(x + 2)$ **8.** $g(x) = -f(x)$
9. $g(x) = -1 + f(x)$ **10.** $g(x) = f(-x)$

EXERCISES 6.2

In Exercises 1–16, evaluate the given expression without using a calculator.

1. $\log_2 16$

2. $\log_4 64$

3. $\log_5\left(\frac{1}{25}\right)$

4. $\log_2\left(\frac{1}{8}\right)$

5. $\log_{16} 4$

6. $\log_{27} 9$

7. $\log_7 1$

8. $\log_{10} 1000$

9. $\log_{10} 0.01$

10. $\log_{10} 10$

11. $\ln e^3$

12. $\ln \dfrac{1}{e}$

13. $\ln e^{-2}$

14. $\ln 1$

15. $\log_a a^2$

16. $\log_a \dfrac{1}{a}$

In Exercises 17–26, use the definition of a logarithm to write the given equation in logarithmic form. For instance, the logarithmic form of $2^3 = 8$ is $\log_2 8 = 3$.

17. $5^3 = 125$

18. $8^2 = 64$

19. $81^{1/4} = 3$

20. $9^{3/2} = 27$

21. $6^{-2} = \frac{1}{36}$

22. $10^{-3} = 0.001$

23. $e^3 = 20.0855.\ldots$

24. $e^0 = 1$

25. $e^x = 4$

26. $u^v = w$

Exponential and Logarithmic Functions

In Exercises 27–32, use a calculator to evaluate the logarithm. Round your answer to three decimal places.

27. $\log_{10} 345$ **28.** $\log_{10}\left(\frac{4}{5}\right)$

29. $\ln 18.42$ **30.** $\ln \sqrt{42}$

31. $\ln(1 + \sqrt{3})$ **32.** $\ln(\sqrt{5} - 2)$

In Exercises 33–36, demonstrate that f and g are inverses of each other by sketching their graphs on the same coordinate plane.

33. $f(x) = 3^x$, $g(x) = \log_3 x$

34. $f(x) = 5^x$, $g(x) = \log_5 x$

35. $f(x) = e^x$, $g(x) = \ln x$

36. $f(x) = 10^x$, $g(x) = \log_{10} x$

In Exercises 37–42, use the graph of $y = \ln x$ to match the given function to its graph. [The graphs are labeled (a)–(f).]

37. $f(x) = \ln x + 2$ **38.** $f(x) = -\ln x$

39. $f(x) = -\ln(x + 2)$ **40.** $f(x) = \ln(x - 1)$

41. $f(x) = \ln(1 - x)$ **42.** $f(x) = -\ln(-x)$

(a)
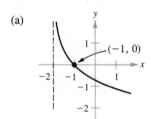

(b)
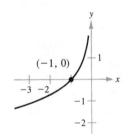

(c)
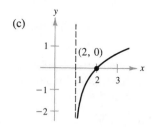

(d)
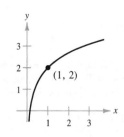

(e)
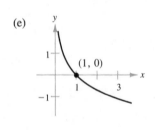

(f)
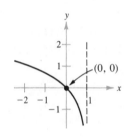

In Exercises 43–50, find the domain, vertical asymptote, and x-intercept of the logarithmic function, and sketch its graph.

43. $f(x) = \log_4 x$ **44.** $g(x) = \log_6 x$

45. $h(x) = \log_4(x - 3)$ **46.** $f(x) = -\log_6(x + 2)$

47. $f(x) = \ln(x - 2)$ **48.** $h(x) = \ln(x + 1)$

49. $g(x) = \ln(-x)$ **50.** $f(x) = \ln(3 - x)$

In Exercises 51–54, use the change of base formula to write the given logarithm as a multiple of a common logarithm. For instance, $\log_2 3 = (1/\log_{10} 2)\log_{10} 3$.

51. $\log_3 5$ **52.** $\log_4 10$

53. $\log_2 x$ **54.** $\ln 5$

In Exercises 55–58, use the change of base formula to write the given logarithm as a multiple of a natural logarithm. For instance, $\log_2 3 = (1/\ln 2)\ln 3$.

55. $\log_3 5$ **56.** $\log_4 10$

57. $\log_2 x$ **58.** $\log_{10} 5$

In Exercises 59–66, evaluate the logarithm using the change of base formula. Do the problem twice, once with common logarithms and once with natural logarithms. Round your answer to three decimal places.

59. $\log_3 7$ **60.** $\log_7 4$

61. $\log_{1/2} 4$ **62.** $\log_4(0.55)$

63. $\log_9(0.4)$ **64.** $\log_{20} 125$

65. $\log_{15} 1250$ **66.** $\log_{1/3}(0.015)$

67. Students in a mathematics class were given an exam and then retested monthly with an equivalent exam. The average score for the class was given by the human memory model

$$f(t) = 80 - 17 \log_{10}(t + 1), \qquad 0 \le t \le 12$$

where t is the time in months.
(a) What was the average score on the original exam ($t = 0$)?
(b) What was the average score after four months?
(c) What was the average score after ten months?

68. The time (in hours) necessary for a certain object to cool $10°$ is

$$t = \frac{10 \ln(1/2)}{\ln(3/4)}.$$

Find t.

69. The population of a town will double in

$$t = \frac{10 \ln 2}{\ln 67 - \ln 50}$$

years. Find t.

70. The work (in foot-pounds) done in compressing an initial volume of 9 cubic feet at a pressure of 15 pounds per square inch to a volume of 3 cubic feet is

$$W = 19,440(\ln 9 - \ln 3).$$

Find W.

71. The time in years for the world population to double if it is increasing at a continuous rate of r is given by

$$t = \frac{\ln 2}{r}.$$

Complete the following table.

r	0.005	0.010	0.015	0.020	0.025	0.030
t						

72. A principal P invested at $9\frac{1}{2}\%$ and compounded continuously, increases to an amount KP after t years, where t is given by

$$t = \frac{\ln K}{0.095}.$$

(a) Complete the following table.

K	1	2	3	4	6	8	10	12
t								

(b) Use the table in part (a) to graph this function.

73. (a) Use a calculator to complete the following table for the function

$$f(x) = \frac{\ln x}{x}.$$

x	1	5	10	10^2	10^4	10^6
$f(x)$						

(b) Use the table in part (a) to determine what $f(x)$ approaches as x increases without bound.

6.3 *Properties of Logarithms*

We know from the previous section that the logarithmic function with base a is the *inverse* of the exponential function with base a. Thus, it makes sense that the properties of exponents should have corresponding properties involving logarithms. For instance, the exponential property

$$a^0 = 1$$

has the corresponding logarithmic property

$$\log_a 1 = 0.$$

In this section we will study the logarithmic properties that correspond to the following three exponential properties.

1. $a^n a^m = a^{n+m}$

2. $\dfrac{a^n}{a^m} = a^{n-m}$

3. $(a^n)^m = a^{nm}$

Properties of Logarithms

Let a be a positive real number such that $a \neq 1$, and let n be a real number. If u and v are positive real numbers, then the following properties are true.

1. $\log_a(uv) = \log_a u + \log_a v$

2. $\log_a \dfrac{u}{v} = \log_a u - \log_a v$

3. $\log_a u^n = n \log_a u$

PROOF

We give a proof of Property 1 and leave the other two proofs for you. To prove Property 1, let

$$x = \log_a u \qquad \text{and} \qquad y = \log_a v.$$

The corresponding exponential forms of these two equations are

$$a^x = u \qquad \text{and} \qquad a^y = v.$$

Multiplying u and v produces $uv = a^x a^y = a^{x+y}$. The corresponding logarithmic form of $uv = a^{x+y}$ is

$$\log_a(uv) = x + y.$$

Hence, $\log_a(uv) = \log_a u + \log_a v$.

Remark: Note that there is no general property that can be used to simplify $\log_a(u \pm v)$. Specifically,

$$\log_a(x + y) \text{ DOES NOT EQUAL } \log_a x + \log_a y.$$

The natural logarithmic versions of these three properties are as follows.

1. $\ln(uv) = \ln u + \ln v$

2. $\ln \dfrac{u}{v} = \ln u - \ln v$

3. $\ln u^n = n \ln u$

EXAMPLE 1 *Using Properties of Logarithms*

Given $\ln 2 \approx 0.693$, $\ln 3 \approx 1.099$, and $\ln 7 \approx 1.946$, use the properties of logarithms to approximate the following.

(a) $\ln 6$ 　　　　　　　　　　　　　　　　(b) $\ln \dfrac{7}{27}$

SOLUTION

(a) $\ln 6 = \ln(2 \cdot 3)$

$\qquad = \ln 2 + \ln 3$ 　　　　　　　　　　　　*Property 1*

$\qquad \approx 0.693 + 1.099$

$\qquad = 1.792$

(b) $\ln \dfrac{7}{27} = \ln 7 - \ln 27$ 　　　　　　　　　　*Property 2*

$\qquad\quad = \ln 7 - \ln 3^3$

$\qquad\quad = \ln 7 - 3 \ln 3$ 　　　　　　　　　　*Property 3*

$\qquad\quad \approx 1.946 - 3(1.099)$

$\qquad\quad = -1.351$

EXAMPLE 2 *Using Properties of Logarithms*

Use the properties of logarithms to verify that

$$-\ln \frac{1}{2} = \ln 2.$$

SOLUTION

$$-\ln \frac{1}{2} = -\ln(2^{-1}) = -(-1) \ln 2 = \ln 2$$

Try checking this result on your calculator.

　　　The properties of logarithms are useful for rewriting logarithmic expressions in forms that simplify the operations of algebra and calculus. This is true because they convert complicated products, quotients, and exponential forms into simpler sums, differences, and products, respectively. Examples 3, 4, and 5 illustrate some cases.

EXAMPLE 3 *Rewriting the Logarithm of a Product*

Use the properties of logarithms to rewrite

$$\log_{10} 5x^2y$$

as the sum of logarithms.

SOLUTION

$$
\begin{aligned}
\log_{10} 5x^2y &= \log_{10} 5 + \log_{10} x^2y \\
&= \log_{10} 5 + \log_{10} x^2 + \log_{10} y \\
&= \log_{10} 5 + 2\log_{10} x + \log_{10} y
\end{aligned}
$$

EXAMPLE 4 *Rewriting the Logarithm of a Quotient*

Use the properties of logarithms to rewrite

$$\ln \frac{\sqrt{3x-5}}{7x^3}$$

as the sum and/or difference of logarithms.

SOLUTION

$$
\begin{aligned}
\ln \frac{\sqrt{3x-5}}{7x^3} &= \ln(3x-5)^{1/2} - \ln 7x^3 \\
&= \ln(3x-5)^{1/2} - (\ln 7 + \ln x^3) \\
&= \frac{1}{2}\ln(3x-5) - \ln 7 - 3\ln x
\end{aligned}
$$

EXAMPLE 5 *Expanding a Logarithmic Expression*

Use the properties of logarithms to rewrite

$$\ln\left(\frac{5x^2}{y}\right)^3$$

as the sum and/or difference of logarithms.

Properties of Logarithms

SOLUTION

$$\ln\left(\frac{5x^2}{y}\right)^3 = 3 \ln \frac{5x^2}{y}$$
$$= 3(\ln 5x^2 - \ln y)$$
$$= 3(\ln 5 + \ln x^2 - \ln y)$$
$$= 3(\ln 5 + 2 \ln x - \ln y)$$
$$= 3 \ln 5 + 6 \ln x - 3 \ln y$$

In Examples 3, 4, and 5, we used the properties of logarithms to *expand* logarithmic expressions. In Examples 6 and 7, we reverse the procedure and use the properties of logarithms to *condense* logarithmic expressions.

EXAMPLE 6 Condensing a Logarithmic Expression

Rewrite as the logarithm of a single quantity.

$$\frac{1}{2} \log_{10} x - 3 \log_{10}(x + 1)$$

SOLUTION

$$\frac{1}{2} \log_{10} x - 3 \log_{10}(x + 1) = \log_{10} x^{1/2} - \log_{10}(x + 1)^3$$
$$= \log_{10} \frac{\sqrt{x}}{(x + 1)^3}$$

EXAMPLE 7 Condensing a Logarithmic Expression

Rewrite as the logarithm of a single quantity.

$$2 \ln(x + 2) - \frac{1}{3}(\ln x + \ln y)$$

SOLUTION

$$2 \ln(x + 2) - \frac{1}{3}(\ln x + \ln y) = \ln(x + 2)^2 - (\ln x^{1/3} + \ln y^{1/3})$$
$$= \ln(x + 2)^2 - \ln(xy)^{1/3}$$
$$= \ln \frac{(x + 2)^2}{\sqrt[3]{xy}}$$

Exponential and Logarithmic Functions

When applying the properties of logarithms to a logarithmic function, you should be careful to check the domain of the function. For example, the domain of $f(x) = \ln x^2$ is all real $x \neq 0$, whereas the domain of $g(x) = 2 \ln x$ is all real $x > 0$.

WARM UP

Evaluate the expressions without using a calculator.

1. $\log_7 49$

2. $\log_2\left(\frac{1}{32}\right)$

3. $\ln \dfrac{1}{e^2}$

4. $\log_{10} 0.001$

Simplify the expressions.

5. $e^2 e^3$

6. $\dfrac{e^2}{e^3}$

7. $(e^2)^3$

8. $(e^2)^0$

Rewrite the expressions in exponential form.

9. $\dfrac{1}{x^2}$

10. $\sqrt{x}$

EXERCISES 6.3

In Exercises 1–20, use the properties of logarithms to write the expression as a sum, difference, or multiple of logarithms.

1. $\log_2 5x$

2. $\log_4 10z$

3. $\log_3 \dfrac{5}{x}$

4. $\log_5 \dfrac{y}{2}$

5. $\log_8 x^4$

6. $\log_6 z^{-3}$

7. $\ln \sqrt{z}$

8. $\ln \sqrt[3]{t}$

9. $\log_2 xyz$

10. $\ln \dfrac{xy}{z}$

11. $\ln \sqrt{a-1}$

12. $\log_3 \left(\dfrac{x^2-1}{x^3}\right)^3$

13. $\ln z(z-1)^2$

14. $\log_4 \sqrt{\dfrac{x^2}{y^3}}$

15. $\log_b \dfrac{x^2}{y^2 z^3}$

16. $\log_b \dfrac{\sqrt{x} y^4}{z^4}$

17. $\ln \sqrt[3]{x/y}$

18. $\ln \dfrac{x}{\sqrt{x^2+1}}$

19. $\log_9 \dfrac{x^4 \sqrt{y}}{z^5}$

20. $\ln \sqrt{x^2(x+2)}$

In Exercises 21–40, write the expression as the logarithm of a single quantity.

21. $\ln x + \ln 2$

22. $\ln y + \ln z$

23. $\log_4 z - \log_4 y$

24. $\log_5 8 - \log_5 t$

25. $2 \log_2(x+4)$

26. $-4 \log_6 2x$

27. $\ln x - 3 \ln(x+1)$

28. $2 \ln 8 + 5 \ln z$

29. $\frac{1}{3} \log_3 5x$

30. $\frac{3}{2} \log_7(z-2)$

31. $\log_3(x-2) - \log_3(x+2)$

32. $3 \ln x + 2 \ln y - 4 \ln z$

33. $\ln x - 2[\ln(x+2) + \ln(x-2)]$

34. $4[\ln z + \ln(z+5)] - 2 \ln(z-5)$

35. $\frac{1}{3}[2\ln(x+3) + \ln x - \ln(x^2 - 1)]$

36. $2[\ln x - \ln(x+1) - \ln(x-1)]$

37. $\frac{1}{3}[\ln y + 2\ln(y+4)] - \ln(y-1)$

38. $\frac{1}{2}[\ln(x+1) + 2\ln(x-1)] + 3\ln x$

39. $2\ln 3 - \frac{1}{2}\ln(x^2 + 1)$

40. $\frac{3}{2}\log_4 5t^6 - \frac{3}{4}\log_4 t^4$

In Exercises 41–56, approximate the logarithm using the properties of logarithms, given $\log_b 2 \approx 0.3562$, $\log_b 3 \approx 0.5646$, and $\log_b 5 \approx 0.8271$.

41. $\log_b 6$

42. $\log_b 15$

43. $\log_b\left(\frac{3}{2}\right)$

44. $\log_b\left(\frac{5}{3}\right)$

45. $\log_b 25$

46. $\log_b 18$ $(18 = 2 \cdot 3^2)$

47. $\log_b \sqrt{2}$

48. $\log_b\left(\frac{9}{2}\right)$

49. $\log_b 40$

50. $\log_b \sqrt[3]{75}$

51. $\log_b\left(\frac{1}{4}\right)$

52. $\log_b(3b^2)$

53. $\log_b \sqrt{5b}$

54. $\log_b 1$

55. $\log_b \dfrac{(4.5)^3}{\sqrt{3}}$

56. $\log_b \dfrac{4}{15^2}$

In Exercises 57–62, find the exact value of the logarithm.

57. $\log_3 9$

58. $\log_6 \sqrt[3]{6}$

59. $\log_4 16^{1.2}$

60. $\log_5\left(\frac{1}{125}\right)$

61. $\ln e^{4.5}$

62. $\ln \sqrt[4]{e^3}$

In Exercises 63–70, use the properties of logarithms to simplify the given logarithmic expression.

63. $\log_4 8$

64. $\log_5\left(\frac{1}{15}\right)$

65. $\log_7 \sqrt{70}$

66. $\log_2(4^2 \cdot 3^4)$

67. $\log_5\left(\frac{1}{250}\right)$

68. $\log_{10}\left(\frac{9}{300}\right)$

69. $\ln(5e^6)$

70. $\ln \dfrac{6}{e^2}$

71. Prove that $\log_b \dfrac{u}{v} = \log_b u - \log_b v$.

72. Prove that $\log_b u^n = n\log_b u$.

73. Use a calculator to demonstrate that

$$\frac{\ln x}{\ln y} \neq \ln \frac{x}{y} = \ln x - \ln y$$

by completing the following table.

x	y	$\dfrac{\ln x}{\ln y}$	$\ln \dfrac{x}{y}$	$\ln x - \ln y$
1	2			
3	4			
10	5			
4	0.5			

6.4 *Solving Exponential and Logarithmic Equations*

So far in this chapter, we have focused our study on the definitions, graphs, and properties of exponential and logarithmic functions. Here we will concentrate on procedures for *solving equations* involving these exponential and logarithmic functions. As a simple example, consider the exponential equation

$$2^x = 32.$$

We can obtain the solution by rewriting the equation in the form

$$2^x = 2^5.$$

Since exponential functions are one-to-one, we can equate exponents of like bases to obtain

$$x = 5.$$

Though this method works in some cases, it does not work for an equation as simple as

$$2^x = 7.$$

Because $2^2 = 4$ and $2^3 = 8$, we see that x lies between 2 and 3. To actually solve for x, we take the logarithm (with base 2) of both sides to obtain

$$\log_2 2^x = \log_2 7$$
$$x = \log_2 7$$
$$= \frac{\ln 7}{\ln 2} \qquad \text{*Change of base*}$$
$$\approx 2.81.$$

Guidelines for Solving Exponential and Logarithmic Equations

1. *To solve an exponential equation*, first isolate the exponential expression, then take the logarithm of both sides and solve for the variable.
2. *To solve a logarithmic equation*, rewrite the equation in exponential form and solve for the variable.

Note that these two guidelines are based on the following inverse properties of exponential and logarithmic functions.

Base a	*Base e*
1. $\log_a a^x = x$	$\ln e^x = x$
2. $a^{\log_a x} = x$	$e^{\ln x} = x$

Solving Exponential Equations

EXAMPLE 1 Solving an Exponential Equation

Solve for x in the equation $e^x = 72$.

SOLUTION

Taking the natural logarithm of both sides, we obtain

$$\ln e^x = \ln 72$$
$$x = \ln 72 \approx 4.277.$$

EXAMPLE 2 *Solving an Exponential Equation*

Solve for x in the equation $3^x = 0.026$.

SOLUTION

Taking the logarithm (with base 3) of both sides, we obtain

$$\log_3 3^x = \log_3 0.026$$

$$x = \log_3 0.026$$

$$= \frac{\ln 0.026}{\ln 3} \qquad\qquad \textit{Change of base}$$

$$\approx -3.322.$$

ALTERNATIVE SOLUTION

Rather than use logarithms with base 3, we could just as easily have taken the *natural logarithm* of both sides to obtain

$$\ln 3^x = \ln 0.026$$

$$x \ln 3 = \ln 0.026$$

$$x = \frac{\ln 0.026}{\ln 3} \approx -3.322.$$

Remember that the first step in solving an exponential equation is to isolate the exponential expression. This is demonstrated in Example 3.

EXAMPLE 3 *Solving an Exponential Equation*

Solve for x in the equation $4e^{2x} = 5$.

SOLUTION

We isolate the exponential by dividing both sides of the equation by 4 to obtain

$$e^{2x} = \frac{5}{4}.$$

Then, taking the natural logarithm of both sides, we have

$$\ln e^{2x} = \ln \frac{5}{4}$$

$$2x = \ln \frac{5}{4}$$

$$x = \frac{1}{2} \ln \frac{5}{4} \approx 0.112.$$

Exponential and Logarithmic Functions

When an equation involves two or more exponential expressions, we can still use a procedure similar to that demonstrated in the first three examples. However, the algebra is a bit more complicated. Study the next two examples carefully.

EXAMPLE 4 *Solving an Exponential Equation*

Solve the equation

$$4^{x+3} = 7^x.$$

SOLUTION

Taking the natural logarithm of both sides, we obtain the following.

$$\ln 4^{x+3} = \ln 7^x \qquad \textit{Take } \ln \textit{ of both sides}$$
$$(x + 3) \ln 4 = x \ln 7 \qquad \textit{Property 3}$$
$$x \ln 4 + 3 \ln 4 = x \ln 7 \qquad \textit{Distributive Property}$$
$$x \ln 4 - x \ln 7 = -3 \ln 4 \qquad \textit{Collect like terms}$$
$$x(\ln 4 - \ln 7) = -3 \ln 4 \qquad \textit{Factor out } x$$
$$x = \frac{-3 \ln 4}{\ln 4 - \ln 7} \qquad \textit{Divide}$$
$$x \approx 7.432$$

EXAMPLE 5 *Solving an Exponential Equation*

Solve for x in the equation

$$e^x + 2e^{-x} = 3.$$

SOLUTION

Some preliminary algebra is helpful. By multiplying both sides of the equation by e^x, we can eliminate the negative exponent and obtain an equation of quadratic type. Note how this works.

$$e^x + 2e^{-x} = 3 \qquad \textit{Given}$$
$$e^x(e^x + 2e^{-x}) = 3e^x \qquad \textit{Multiply both sides by } e^x$$
$$(e^x)^2 + 2 = 3e^x \qquad \textit{Distributive Property}$$
$$(e^x)^2 - 3e^x + 2 = 0 \qquad \textit{Quadratic form}$$
$$(e^x - 2)(e^x - 1) = 0 \qquad \textit{Factor}$$
$$e^x - 2 = 0 \qquad e^x - 1 = 0 \qquad \textit{Set factors to zero}$$
$$e^x = 2 \qquad \qquad e^x = 1$$
$$x = \ln 2 \qquad \qquad x = 0 \qquad \textit{Solutions}$$

Solving Logarithmic Equations

To solve a logarithmic equation such as

$$\ln x = 3 \qquad\qquad \textit{Logarithmic form}$$

we write the equation in exponential form as follows.

$$e^{\ln x} = e^3 \qquad\qquad \textit{Exponentiate both sides}$$
$$x = e^3 \qquad\qquad \textit{Exponential form}$$

This procedure is sometimes called *exponentiating* both sides of an equation.

EXAMPLE 6 Solving a Logarithmic Equation

Solve for x in the equation $2 \ln 3x = 4$.

SOLUTION

$$2 \ln 3x = 4 \qquad\qquad \textit{Given}$$
$$\ln 3x = 2 \qquad\qquad \textit{Divide both sides by 2}$$
$$3x = e^2 \qquad\qquad \textit{Exponential form}$$
$$x = \frac{1}{3}e^2 \qquad\qquad \textit{Divide both sides by 3}$$

The techniques used to solve equations involving logarithmic expressions can produce extraneous solutions, as demonstrated in Example 7.

EXAMPLE 7 Solving a Logarithmic Equation

Solve for x in the equation $\ln(x - 2) + \ln(2x - 3) = 2 \ln x$.

SOLUTION

$$\ln(x - 2) + \ln(2x - 3) = 2 \ln x \qquad \textit{Given}$$
$$\ln(x - 2)(2x - 3) = \ln x^2 \qquad \textit{Properties 1 and 3}$$
$$\ln(2x^2 - 7x + 6) = \ln x^2$$

Now, because the natural logarithmic function is one-to-one, we can write

$$2x^2 - 7x + 6 = x^2$$
$$x^2 - 7x + 6 = 0 \qquad\qquad \textit{Quadratic form}$$
$$(x - 6)(x - 1) = 0 \qquad\qquad \textit{Factor}$$
$$x - 6 = 0 \qquad x - 1 = 0 \qquad \textit{Set factors to zero}$$
$$x = 6 \qquad\qquad x = 1$$

Finally, by checking these two "solutions" in the original equation, we find that $x = 1$ is not valid. Can you see why? Thus, the only solution is $x = 6$.

EXAMPLE 8 *Solving a Logarithmic Equation*

Solve for x in the equation

$$\log_{10}(4x + 2) - \log_{10}(x - 1) = 1.$$

SOLUTION

$$\log_{10}(4x + 2) - \log_{10}(x - 1) = 1$$

$$\log_{10}\left(\frac{4x + 2}{x - 1}\right) = 1 \qquad \text{\textit{Logarithmic form}}$$

$$\frac{4x + 2}{x - 1} = 10 \qquad \text{\textit{Exponential form}}$$

$$4x + 2 = 10(x - 1)$$
$$4x - 10x = -10 - 2$$
$$-6x = -12$$
$$x = 2$$

EXAMPLE 9 *Finding the Zeros of a Logarithmic Function*

Find the zeros of the function

$$f(x) = 2 \ln(2x - 1) - \ln 9.$$

SOLUTION

We begin by setting $f(x)$ equal to zero and solving the resulting equation for x.

$$2 \ln(2x - 1) - \ln 9 = 0$$

$$2 \ln(2x - 1) = \ln 9$$

$$\ln(2x - 1) = \frac{1}{2} \ln 9$$

$$\ln(2x - 1) = \ln 9^{1/2}$$
$$\ln(2x - 1) = \ln 3$$
$$2x - 1 = 3$$
$$2x = 4$$
$$x = 2$$

Thus, f has one zero, $x = 2$. Try checking this by substituting $x = 2$ in the original function.

Solving Exponential and Logarithmic Equations

In the next example, we prove the *change of base formula* presented in Section 6.2.

EXAMPLE 10   The Change of Base Formula

Prove the change of base formula given in Section 6.2.

$$\log_a x = \frac{\log_b x}{\log_b a}$$

SOLUTION

We begin by letting

$$y = \log_a x$$

and writing the equivalent exponential form

$$a^y = x.$$

Now, taking the logarithm *with base b* of both sides, we have

$$\log_b a^y = \log_b x$$

$$y \log_b a = \log_b x$$

$$y = \frac{\log_b x}{\log_b a}$$

$$\log_a x = \frac{\log_b x}{\log_b a}.$$

When solving exponential or logarithmic equations, the following properties are useful.

1. $x = y$ if and only if $\log_a x = \log_a y$.
2. $x = y$ if and only if $a^x = a^y$, $a > 0$, $a \neq 1$.
3. $a = b$ if and only if $a^x = b^x$, $x \neq 0$.

Can you see where these properties were used in the examples in this section?

Exponential and Logarithmic Functions

WARM UP

Solve for x.

1. $x \ln 2 = \ln 3$
3. $2xe^2 = e^3$
5. $x^2 - 4x + 5 = 0$

2. $(x - 1) \ln 4 = 2$
4. $4xe^{-1} = 8$
6. $2x^2 - 3x + 1 = 0$

Simplify the expressions.

7. $\log_{10} 100^x$
9. $\ln e^{2x}$

8. $\log_4 64^x$
10. $\ln e^{-x^2}$

EXERCISES 6.4

In Exercises 1–10, solve for x.

1. $4^x = 16$
3. $7^x = \frac{1}{49}$
5. $\left(\frac{3}{4}\right)^x = \frac{27}{64}$
7. $\log_4 x = 3$
9. $\log_{10} x = -1$

2. $3^x = 243$
4. $8^x = 4$
6. $3^{x-1} = 27$
8. $\log_5 5x = 2$
10. $\ln(2x - 1) = 0$

In Exercises 11–16, apply the inverse properties of ln x and e^x to simplify the given expression.

11. $\ln e^{x^2}$
13. $e^{\ln(5x+2)}$
15. $e^{\ln x^2}$

12. $\ln e^{2x-1}$
14. $-1 + \ln e^{2x}$
16. $-8 + e^{\ln x^3}$

In Exercises 17–48, solve the given exponential equation.

17. $10^x = 42$
19. $\frac{1}{3}10^{2x} = 12$
21. $3(10^{x-1}) = 2$
23. $e^x = 10$
25. $500e^{-x} = 300$
27. $3e^{3x/2} = 40$
29. $25e^{2x+1} = 962$
31. $100\left(1 + \frac{0.09}{4}\right)^{4x} = 300$
32. $\frac{1250}{(1.04)^x} = 500$

18. $10^x = 570$
20. $8(10^{3x}) = 12$
22. $2^{3x} = 565$
24. $e^x = 6500$
26. $1000e^{-4x} = 75$
28. $6e^{1-x} = 25$
30. $(1.003)^{365t} = 2$

33. $e^{0.09t} = 3$
35. $\left(1 + \frac{0.10}{12}\right)^{12t} = 2$
37. $\frac{10,000}{1 + 19e^{-t/5}} = 2000$
39. $\left(\frac{1}{1.0775}\right)^N = 0.2247$
41. $10^{7-x} = 5^{x+1}$
43. $3(1 + e^{2x}) = 4$
45. $\frac{400}{1 + e^{-x}} = 200$
47. $\frac{e^x + e^{-x}}{e^x - e^{-x}} = 2$

34. $e^{0.125t} = 8$
36. $\left(1 + \frac{0.065}{365}\right)^{365t} = 4$
38. $80e^{-t/2} + 20 = 70$
40. $3^{2x+1} = 5^{x+2}$
42. $4^{x^2} = 100$
44. $20(100 - e^{x/2}) = 500$
46. $\frac{3000}{2 + e^{-2x}} = 1200$
48. $\frac{e^x + e^{-x}}{2} = 2$

In Exercises 49–64, solve the given logarithmic equation.

49. $\ln x = 5$
51. $2 \ln x = 7$
53. $2 \ln 4x = 0$
55. $\log_{10}(z - 3) = 2$
57. $\ln x + \ln(x - 2) = 1$
59. $\log_{10}(x + 4) - \log_{10} x = \log_{10}(x + 2)$
60. $\log_{10} x - \log_{10}(2x - 1) = 0$
61. $\ln x + \ln(x + 3) = 1$
62. $\log_2(x + 5) - \log_2(x - 2) = 3$
63. $\ln x^2 = (\ln x)^2$
64. $\log_4 x - \log_4(x - 1) = \frac{1}{2}$

50. $\ln 2x = -1$
52. $3 \ln 5x = 10$
54. $6 \ln(x + 1) = 2$
56. $\log_{10} x^2 = 20$
58. $\ln \sqrt{x + 2} = 1$

In Exercises 65 and 66, find the time required for a $1000 investment to double at interest rate r, compounded continuously. Solve for t in the exponential equation $2000 = 1000e^{rt}$.

65. $r = 0.085$ **66.** $r = 0.12$

67. The demand equation for a certain product is given by

$$p = 500 - 0.5(e^{0.004x}).$$

Find the demand x for a price of (a) $p = 350 and (b) $p = 300.

68. The demand equation for a certain product is given by

$$p = 5000\left(1 - \frac{4}{4 + e^{-0.002x}}\right).$$

Find the demand x for a price of (a) $p = 600 and (b) $p = 400.

69. The yield V (in millions of cubic feet per acre) for a forest at age t years is given by

$$V = 6.7e^{-48.1/t}.$$

Find the time necessary to have a yield of (a) 1.3 million cubic feet and (b) 2 million cubic feet.

70. In a group project in learning theory, a mathematical model for the proportion P of correct responses after n trials was found to be

$$P = \frac{0.83}{1 + e^{-0.2n}}.$$

After how many trials will 60% of the responses be correct?

6.5 Applications of Exponential and Logarithmic Functions

The behavior of many physical, economic, and social phenomena can be described by exponential and logarithmic functions. In this section, we look at four basic types of applications: (1) Compound Interest, (2) Growth and Decay, (3) Logistics Models, and (4) Intensity Models. The problems presented in this section require the full range of solution techniques studied in this chapter.

Compound Interest

From Section 6.1, recall the following two compound interest formulas, where A is the account balance, P is the principal, r is the annual percentage rate, and t is the time in years.

n Compoundings per Year	*Continuous Compounding*
$A = P\left(1 + \dfrac{r}{n}\right)^{nt}$	$A = Pe^{rt}$

Exponential and Logarithmic Functions

EXAMPLE 1 *Doubling Time for an Investment at Quarterly Compounding*

An investment is made in a trust fund at an annual percentage rate of 9.5%, compounded quarterly. How long will it take for the investment to double in value?

SOLUTION

For quarterly compounding, we use the formula

$$A = P\left(1 + \frac{r}{4}\right)^{4t}.$$

Using $r = 0.095$, the time required for the investment to double is given by solving for t in the equation $2P = A$.

$$2P = P\left(1 + \frac{0.095}{4}\right)^{4t} \qquad \text{2P = A}$$

$$2 = (1.02375)^{4t} \qquad \text{Divide both sides by P}$$

$$\ln 2 = \ln(1.02375)^{4t} \qquad \text{Take ln of both sides}$$

$$\ln 2 = 4t \ln(1.02375)$$

$$t = \frac{\ln 2}{4 \ln(1.02375)} \approx 7.4$$

Therefore, it will take approximately 7.4 years for the investment to double in value with quarterly compounding.

———

Try reworking Example 1 using continuous compounding. To do this you will need to solve the equation

$$2P = Pe^{0.095t}.$$

The solution is $t \approx 7.3$ years, which makes sense because the principal should double more quickly with continuous compounding than with quarterly compounding.

From Example 1, we see that the time required for an investment to double in value is independent of the amount invested. In general, the **doubling time** is as follows.

n Compoundings per Year	*Continuous Compounding*
$t = \dfrac{\ln 2}{n \ln[1 + (r/n)]}$	$t = \dfrac{\ln 2}{r}$

EXAMPLE 2 Finding an Annual Percentage Rate

An investment of $10,000 is compounded continuously. What annual percentage rate will produce a balance of $25,000 in ten years?

SOLUTION

We use the formula

$$A = Pe^{rt}$$

with $P = 10,000$, $A = 25,000$, and $t = 10$, and solve the following equation for r.

$$10,000e^{10r} = 25,000$$
$$e^{10r} = 2.5$$
$$10r = \ln 2.5$$
$$r = \frac{1}{10} \ln 2.5 \approx 0.0916$$

Thus, the annual percentage rate must be approximately 9.16%.

EXAMPLE 3 The Effective Yield for an Investment

A deposit is compounded continuously at an annual percentage rate of 7.5%. Find the **effective yield.** That is, find the simple interest rate that would yield the same balance at the end of one year.

SOLUTION

Using the formula $A = Pe^{rt}$ with $r = 0.075$ and $t = 1$, the balance at the end of one year is

$$A = Pe^{0.075(1)}$$
$$\approx P(1.0779)$$
$$= P(1 + 0.0779). \qquad\qquad A = P(1 + r)$$

Since the formula for simple interest after one year is

$$A = P(1 + r)$$

we conclude that the effective yield is approximately 7.79%.

Growth and Decay

The balance in an account earning *continuously compounded* interest is one example of a quantity that increases over time according to the **exponential growth model**

$$Q(t) = Ce^{kt}.$$

In this model, $Q(t)$ is the size of the population (balance, weight, and so forth) at any time t, C is the original population (when $t = 0$), and k is a constant determined by the rate of growth. If $k > 0$, the population *grows* (increases) over time, and if $k < 0$ it *decays* (decreases) over time. Example 9 of Section 6.1 is an example of population growth. Recall from Section 6.1 that we can remember this growth model as *(Then)* = *(Now)*(e^{kt}).

EXAMPLE 4 *Exponential Decay*

Radioactive iodine is a by-product of some types of nuclear reactors. Its **half-life** is 60 days. That is, after 60 days, a given amount of radioactive iodine will have decayed to half the original amount. Suppose a contained nuclear accident occurs and gives off an initial amount C of radioactive iodine.

(a) Write an equation for the amount of radioactive iodine present at any time t, following the accident.
(b) How long will it take for the radioactive iodine to decay to a level of 20% of the original amount?

SOLUTION

(a) We first need to find the rate k, in the exponential model $Q(t) = Ce^{kt}$. Knowing that half the original amount remains after $t = 60$ days, we obtain

$$Q(60) = Ce^{k(60)} = \frac{1}{2}C$$

$$e^{60k} = \frac{1}{2}$$

$$60k = -\ln 2$$

$$k = \frac{-\ln 2}{60} \approx -0.0116.$$

Thus, the exponential model is

$$Q(t) = Ce^{-0.0116t}.$$

(b) The time required to decay to 20% of the original amount is given by

$$Q(t) = Ce^{-0.0116t} = (0.2)C$$
$$e^{-0.0116t} = 0.2$$
$$-0.0116t = \ln 0.2$$
$$t = \frac{\ln 0.2}{-0.0116} \approx 139 \text{ days.}$$

In living organic material the ratio of radioactive carbon isotopes (Carbon 14) to the number of nonradioactive carbon isotopes (Carbon 12) is about 1 to 10^{12}. When organic material dies, its Carbon 12 content remains fixed, whereas its radioactive Carbon 14 begins to decay with a half-life of about 5700 years. To estimate the age of dead organic material, scientists use the following formula, which denotes the ratio of Carbon 14 to Carbon 12 present at any time t (in years).

$$R(t) = \frac{1}{10^{12}} 2^{-t/5700}$$

The graph of R is shown in Figure 6.17. Note that R decreases as the time t increases.

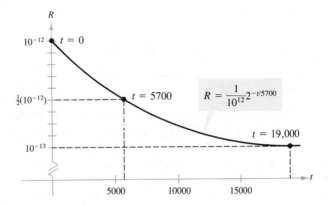

FIGURE 6.17

EXAMPLE 5 Carbon Dating

Suppose that the Carbon 14/Carbon 12 ratio of a newly discovered fossil is

$$R = \frac{1}{10^{13}}.$$

Estimate the age of the fossil.

SOLUTION

In the carbon dating model, we substitute the given value of R to obtain

$$\frac{1}{10^{13}} = \frac{1}{10^{12}} 2^{-t/5700}$$

$$\frac{1}{10} = 2^{-t/5700}$$

$$10 = 2^{t/5700}$$

$$\ln 10 = \ln 2^{t/5700} = \frac{t}{5700} \ln 2$$

$$t = 5700 \frac{\ln 10}{\ln 2} \approx 19{,}000 \text{ years.} \qquad \textit{Nearest thousand}$$

Thus, we estimate the age of the fossil to be 19,000 years.

Remark: The carbon dating model in Example 5 assumed that the Carbon 14/Carbon 12 ratio was one part in 10,000,000,000,000. Suppose that an error in measurement occurred and the actual ratio was only one part in 8,000,000,000,000. The fossil age corresponding to the actual ratio would then be approximately 17,000 years. Try checking this result.

Logistics Model

Some populations initially have rapid growth, followed by a declining rate of growth, as indicated by the graph in Figure 6.18. One model for describing this type of growth pattern is the **logistics curve** given by the function

$$Q(t) = \frac{M}{1 + (M/Q_0 - 1)e^{-kt}}$$

where M is the maximum value of Q and Q_0 is the initial population size. An example would be a bacteria culture allowed to grow initially under ideal conditions, followed by less favorable conditions that inhibit growth.

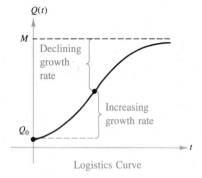

Logistics Curve

FIGURE 6.18

Applications of Exponential and Logarithmic Functions

EXAMPLE 6 *Logistics Curve*

On a college campus of 5000 students, one student returned from vacation with a contagious flu virus. The spread of the virus through the student body is given by

$$s(t) = \frac{5000}{1 + 4999e^{-0.8t}}$$

where $s(t)$ is the total number infected after t days. The college will cancel classes when 40% or more of the students are ill.

(a) How many are infected after five days?
(b) After how many days will the college cancel classes?

SOLUTION

(a) After five days, the number infected is

$$s(5) = \frac{5000}{1 + 4999e^{-0.8(5)}}$$

$$= \frac{5000}{1 + 4999e^{-4}}$$

$$\approx 54.$$

(b) In this case, the number infected is $(0.40)(5000) = 2000$. Therefore, we solve for t in the following equation.

$$2000 = \frac{5000}{1 + 4999e^{-0.8t}}$$

$$2000 + 9{,}998{,}000e^{-0.8t} = 5000$$

$$e^{-0.8t} = \frac{3000}{9{,}998{,}000} = \frac{3}{9998}$$

$$\ln(e^{-0.8t}) = \ln\frac{3}{9998}$$

$$-0.8t = \ln\frac{3}{9998}$$

$$t = \frac{\ln(3/9998)}{-0.8} \approx 10.1$$

Hence, after ten days, at least 40% of the students will be infected, and the college will cancel classes.

Exponential and Logarithmic Functions

Intensity Model

Sound and shock waves can be measured by the **intensity model**

$$S = K \log_{10} \frac{I}{I_0}$$

where I is the intensity of the stimulus wave, I_0 is the **threshold intensity** (the smallest value of I that can be detected by the listening device), and K determines the units in which S is measured. Sound heard by the human ear is measured in decibels. One **decibel** is considered the smallest detectable difference in the loudness of two sounds.

For earthquakes, the shock is measured by units on the *Richter Scale*, as demonstrated in Example 7.

EXAMPLE 7 *Magnitude of Earthquakes*

On the Richter Scale, the magnitude R of an earthquake of intensity I is given by

$$R = \log_{10} \frac{I}{I_0}$$

where $I_0 = 1$ is the minimum intensity used for comparison. Find the intensity per unit of area for the following earthquakes. (Intensity is a measure of the wave energy of an earthquake.)

(a) San Francisco in 1906, $R = 8.3$
(b) Mexico City in 1978, $R = 7.85$
(c) Predicted in 1990, $R = 6.3$

SOLUTION

(a) Since $I_0 = 1$ and $R = 8.3$, we have

$$8.3 = \log_{10} I$$
$$I = 10^{8.3} \approx 199{,}526{,}000.$$

(b) For Mexico City, we have $7.85 = \log_{10} I$, and

$$I = 10^{7.85} \approx 70{,}795{,}000.$$

(c) For $R = 6.3$, we have $6.3 = \log_{10} I$, and

$$I = 10^{6.3} \approx 1{,}995{,}000.$$

Note that an increase of two units on the Richter Scale (from 6.3 to 8.3) represents an intensity change by a factor of

$$\frac{199{,}526{,}000}{1{,}995{,}000} \approx 100.$$

EXERCISES 6.5

In Exercises 1–10, complete the table for a savings account in which interest is compounded continuously.

	Initial investment	Annual % rate	Effective yield	Time to double	Amount after 10 years
1.	$1000	12%			
2.	$20,000	$10\frac{1}{2}$%			
3.	$750			$7\frac{3}{4}$ yr	
4.	$10,000			5 yr	
5.	$500				$1,292.85
6.	$2000		4.5%		
7.		11%			$19,205.00
8.		8%			$20,000.00
9.	$5000		8.33%		
10.	$250		12.19%		

In Exercises 11 and 12, determine the principal P which must be invested at rate r, compounded monthly, so that $500,000 will be available for retirement in t years.

11. $r = 7\frac{1}{2}$%, $t = 20$

12. $r = 12$%, $t = 40$

In Exercises 13 and 14, determine the time necessary for $1000 to double if it is invested at interest rate r compounded (a) annually, (b) monthly, (c) daily, and (d) continuously.

13. $r = 11$%

14. $r = 10\frac{1}{2}$%

15. Complete the following table for the time t necessary for P dollars to triple if interest is compounded continuously at rate r.

r	2%	4%	6%	8%	10%	12%
t						

16. Repeat Exercise 15 for interest that is compounded annually.

In Exercises 17–20, consider making monthly deposits of P dollars into a savings account at an annual interest rate r, compounded monthly. Find the balance A after t years given that

$$A = \frac{P(e^{rt} - 1)}{e^{r/12} - 1}.$$

17. $P = $50, r = 7$%, $t = 20$ years

18. $P = $75, r = 9$%, $t = 25$ years

19. $P = $100, r = 10$%, $t = 40$ years

20. $P = $20, r = 6$%, $t = 50$ years

21. The population P of a city is given by

$$P = 105,300e^{0.015t}$$

where t is the time in years, with $t = 0$ corresponding to 1985. According to this model, in what year will the city have a population of 150,000?

22. The population P of a city is given by

$$P = 2500e^{kt}$$

where t is the time in years, with $t = 0$ corresponding to the year 1980. In 1935, the population was 3350. Find the value of k and use this result to predict the population in the year 2000.

23. Assume that the world population at time t is given by $P = P_0 e^{rt}$. How long will it take the world population to double? To triple?

24. The number of bacteria N in a culture is given by the model

$$N = 100e^{kt}$$

where t is the time in hours, with $t = 0$ corresponding to the time when $N = 100$. When $t = 5$, $N = 300$. How long does it take the population to double in size?

In Exercises 25–30, complete the table for the given radioactive isotope.

Isotope	Half-life (years)	Initial quantity	Amount after 1000 years	Amount after 10,000 years
25. Ra^{226}	1,620	10 g		
26. Ra^{226}	1,620		1.5 g	
27. C^{14}	5,730			2 g
28. C^{14}	5,730	3 g		
29. Pu^{230}	24,360		2.1 g	
30. Pu^{230}	24,360			0.4 g

31. What percentage of a present amount of radioactive radium (Ra^{226}) will remain after 100 years?

32. Find the half-life of a radioactive material if, after one year, 99.57% of the initial amount remains.

In Exercises 33–36, find the constant k such that the exponential function $y = Ce^{kt}$ passes through the given points on the graph.

33.
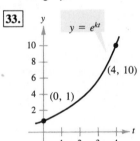

$y = e^{kt}$

(4, 10)

(0, 1)

34.
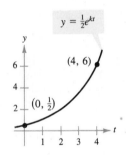

$y = \frac{1}{2}e^{kt}$

(4, 6)

$(0, \frac{1}{2})$

35.
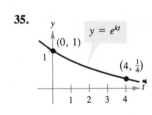

$y = e^{kt}$

(0, 1)

$(4, \frac{1}{4})$

36.
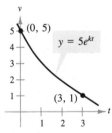

(0, 5)

$y = 5e^{kt}$

(3, 1)

37. The sales S (in thousands of units) of a new product after it has been on the market t years are given by

$$S(t) = 100(1 - e^{kt}).$$

(a) Find S as a function of t if 15,000 units have been sold after one year.

(b) How many units will be sold after five years?

38. The management at a factory has found that the maximum number of units a worker can produce in a day is 30. The learning curve for the number of units N produced per day after a new employee has worked t days is given by

$$N = 30(1 - e^{kt}).$$

After 20 days on the job, a particular worker produced 19 units.

(a) Find the learning curve for this worker (that is, find the value of k).

(b) How many days should pass before this worker is producing 25 units per day?

39. A certain lake is stocked with 500 fish, and the fish population increases according to the logistics curve

$$p(t) = \frac{10,000}{1 + 10e^{-t/5}}$$

where t is measured in months.

(a) Find $p(5)$.

(b) After how many months will the fish population be 2000?

40. A conservation organization releases 100 animals of an endangered species into a game preserve. The organization believes that the preserve has a carrying capacity of 1000 animals and that the growth of the herd will be modeled by the logistics curve

$$p(t) = \frac{1000}{1 + 9e^{-kt}}$$

where t is measured in years.

(a) Find k if the herd size is 134 after two years.

(b) Find $p(5)$.

41. The intensity level β, in decibels, of a sound wave is defined by

$$\beta(I) = 10 \log_{10} \frac{I}{I_0}$$

where I_0 is an intensity of 10^{-16} watts per square centimeter, corresponding roughly to the faintest sound that can be heard. Determine $\beta(I)$ for the following conditions.

(a) $I = 10^{-14}$ watts per centimeter (whisper)

(b) $I = 10^{-9}$ watts per centimeter (busy street corner)

(c) $I = 10^{-6.5}$ watts per centimeter (air hammer)

(d) $I = 10^{-4}$ watts per centimeter (threshold of pain)

42. Due to the installation of noise suppression materials, the noise level in an auditorium was reduced from 93 to 80 decibels. Find the percentage decrease in the intensity levels of the noise because of the installation of these materials.

In Exercises 43 and 44, use the Richter Scale (Example 7) for measuring the magnitude of earthquakes.

43. Find the magnitude R of an earthquake of intensity I (let $I_0 = 1$).
  (a) $I = 80,500,000$ (b) $I = 48,275,000$

44. Find the intensity I of an earthquake measuring R on the Richter Scale (let $I_0 = 1$).
  (a) Columbia in 1906, $R = 8.6$
  (b) Los Angeles in 1971, $R = 6.7$

In Exercises 45–48, use the acidity model given by $pH = -\log_{10}[H^+]$, where acidity (pH) is a measure of the hydrogen ion concentration $[H^+]$ (measured in moles of hydrogen per liter) of a solution.

45. Find the pH if $[H^+] = 2.3 \times 10^{-5}$.

46. Find the pH if $[H^+] = 11.3 \times 10^{-6}$.

47. Compute $[H^+]$ for a solution in which pH = 5.8.

48. Compute $[H^+]$ for a solution in which pH = 3.2.

In Exercises 49 and 50, use **Newton's Law of Cooling,** which states that the rate of change in the temperature of an object is proportional to the difference between its temperature and the temperature of its environment. If $T(t)$ is the temperature of the object at time t in minutes, T_0 is the initial temperature, and T_e is the constant temperature of the environment, then

$$T(t) = T_e + (T_0 - T_e)e^{-kt}.$$

49. An object in a room at 70° F cools from 350° F to 150° F in 45 minutes.
  (a) Find the temperature of the object as a function of time.
  (b) Find the temperature after it has cooled for one hour.
  (c) Find the time necessary for the object to cool to 80° F.

50. A thermometer is taken from a room at 72° F to the outdoors, where the temperature is 20° F. The reading on the thermometer drops to 48° F after one minute. Determine the reading after five minutes.

CHAPTER 6 REVIEW EXERCISES

In Exercises 1–16, sketch the graph of the function.

1. $f(x) = 6^x$
2. $f(x) = 0.3^x$
3. $g(x) = 6^{-x}$
4. $g(x) = 0.3^{-x}$
5. $h(x) = e^{-x/2}$
6. $h(x) = 2 - e^{-x/2}$
7. $f(x) = e^{x+2}$
8. $s(t) = 4e^{-2/t}, \quad t > 0$
9. $f(x) = \ln(x - 3)$
10. $f(x) = \ln|x|$
11. $f(x) = \ln x + 3$
12. $f(x) = \frac{1}{4}\ln x$
13. $g(x) = \log_2(-x)$
14. $g(x) = \log_5 x$
15. $h(x) = \ln(e^{x-1})$
16. $f(x) = e^{\ln x^2}$

In Exercises 17–24, use the properties of logarithms to write the expression as a sum, difference, or multiple of logarithms.

17. $\log_5 5x^2$
18. $\log_7 \dfrac{\sqrt{x}}{4}$
19. $\log_{10} \dfrac{5\sqrt{y}}{x^2}$
20. $\ln\left|\dfrac{x-1}{x+1}\right|$
21. $\ln\left|\dfrac{x^2+1}{x}\right|$
22. $\ln\sqrt{\dfrac{x^2+1}{x^4}}$
23. $\ln[(x^2+1)(x-1)]$
24. $\ln\sqrt[5]{\dfrac{4x^2-1}{4x^2+1}}$

In Exercises 25–32, write the expression as the logarithm of a single quantity.

25. $\log_{10} 5 - 2 \log_{10}(x + 4)$

26. $6 \log_8 z + \frac{1}{2} \log_8(z^2 + 4)$

27. $\frac{1}{2} \ln|2x - 1| - 2 \ln|x + 1|$

28. $5 \ln|x - 2| - \ln|x + 2| - 3 \ln|x|$

29. $2(\ln x + \frac{1}{3} \ln \sqrt{x})$

30. $\frac{1}{2} \ln(x^2 + 4x) - \ln 2 - \ln x$

31. $\ln 3 + \frac{1}{3} \ln(4 - x^2) - \ln x$

32. $3[\ln x - 2 \ln(x^2 + 1)] + 2 \ln 5$

In Exercises 33–40, determine whether the statement or equation is true or false.

33. The domain of the function $f(x) = \ln|x|$ is the set of all real numbers.

34. The range of the function $g(x) = e^{-x}$ is the set of all positive real numbers.

35. $\ln(x + y) = \ln x + \ln y$

36. $\dfrac{\ln x}{\ln y} = \ln x - \ln y$

37. $\ln \sqrt{x^4 + 2x^2} = \ln(|x| \sqrt{x^2 + 2})$

38. $\log_b b^{2x} = 2x$

39. $\dfrac{e^{2x} - 1}{e^x - 1} = e^x + 1$ **40.** $e^{x-1} = \dfrac{e^x}{e}$

41. A solution of a certain drug contained 500 units per milliliter when prepared. It was analyzed after 40 days and found to contain 300 units per milliliter. Assuming that the rate of decomposition is proportional to the amount present, the equation giving the amount A after t days is

$A = 500e^{-0.013t}$.

Use this model to find A when $t = 60$.

42. The number of miles s of roads cleared of snow is approximated by the model

$$s = 25 - \frac{13 \ln(h/12)}{\ln 3}, \qquad 2 \le h \le 15$$

where h is the depth of the snow in inches. Use this model to find s when $h = 10$ inches.

In Exercises 43 and 44, find the probability of waiting less than t units of time until the next occurrence of an event if that probability is approximated by the model

$F(t) = 1 - e^{-t/\lambda}$

where λ is the average time between successive occurrences of the event.

43. The average time between incoming calls at a switchboard is three minutes. If a call has just come in, find the probability that the next call will be within
(a) $\frac{1}{2}$ minute   (b) 2 minutes   (c) 5 minutes.

44. Trucks arrive at a terminal at an average of 3 per hour (therefore $\lambda = 20$ minutes). If a truck has just arrived, find the probability that the next arrival will be within
(a) 10 minutes   (b) 30 minutes   (c) 1 hour.

45. A certain automobile gets 28 miles per gallon of gasoline for speeds up to 50 miles per hour. Over 50 miles per hour, the number of miles per gallon drops at the rate of 12% for each 10 miles per hour. If s is the speed and y is the number of miles per gallon, then

$$y = 28e^{0.6 - 0.012s}, \qquad s \ge 50.$$

Use this function to complete the following table.

Speed	50	55	60	65	70
Miles per gallon					

46. Find the balance after 25 years if $25,000 is invested at 10% compounded
(a) monthly   (b) daily   (c) continuously.

In Exercises 47–50, find the exponential function $y = Ce^{kt}$ that passes through the two points.

47.
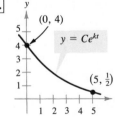

48.
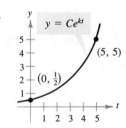

49.

50.
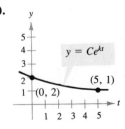

51. The demand equation for a certain product is given by

$$p = 500 - 0.5e^{0.004x}.$$

Find the demand x for a price of (a) $p = \$450$ and (b) $p = \$400$.

52. In a typing class, the average number of words per minute typed after t weeks of lessons was found to be

$$N = \frac{157}{1 + 5.4e^{-0.12t}}.$$

Find the time necessary to type (a) 50 words per minute and (b) 75 words per minute.

53. A deposit of $750 is made in a savings account for which the interest is compounded continuously. The balance will double in $7\frac{3}{4}$ years.
(a) What is the annual percentage rate for this account?
(b) Find the balance in the account after ten years.
(c) Find the effective yield.

54. A deposit of $10,000 is made in a savings account for which the interest is compounded continuously. The balance will double in five years.
(a) What is the annual percentage rate for this account?
(b) Find the balance after one year.
(c) Find the effective yield.

55. Complete the following table for the function $f(x) = e^{-x}$.

x	0	1	2	5	10	15
e^{-x}	1	0.36788				

This table demonstrates that for $k > 0$, e^{-kx} approaches 0 as x increases without bound.

56. Use the result of Exercise 55 to determine what each function $(k > 0)$ approaches as x increases.
(a) $f(x) = 30(1 - e^{-0.025x})$

(b) $f(x) = \dfrac{50}{2 + 3e^{-0.2x}}$

(c) $f(x) = (100 - a)e^{-kx} + a$

(d) $f(x) = 5000\left(1 - \dfrac{4}{4 + e^{-0.002x}}\right)$

57. In calculus it can be shown that

$$e^x \approx 1 + x + \frac{x^2}{2} + \frac{x^3}{6} + \frac{x^4}{24}.$$

Use this equation to approximate the following, and compare the results to those obtained with a calculator.
(a) e (b) $e^{1/2}$ (c) $e^{-1/2}$

CHAPTER 7

Lines in the Plane and Conics

7.1 Lines in the Plane: Slope

In this section, we study lines and their equations. Throughout this text, we follow the convention of using the term **line** to mean a *straight* line.

The Slope of a Line

The **slope** of a nonvertical line represents the number of units a line rises or falls vertically for each unit of horizontal change from left to right.

For instance, consider the two points (x_1, y_1) and (x_2, y_2) on the line shown in Figure 7.1. As we move from left to right along this line, a change of $(y_2 - y_1)$ units in the vertical direction corresponds to a change of $(x_2 - x_1)$ units in the horizontal direction. That is,

$$y_2 - y_1 = \text{the change in } y$$

and

$$x_2 - x_1 = \text{the change in } x.$$

The slope of the line is given by the ratio of these two changes.

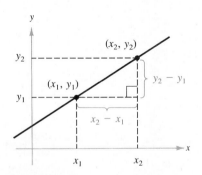

FIGURE 7.1

352

(a)

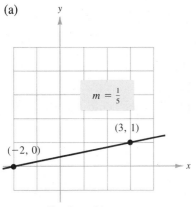

If *m* is positive,
the line rises.

(b)

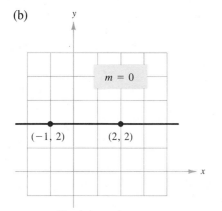

If *m* is zero, the
line is horizontal.

(c)

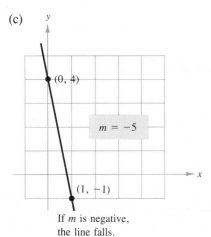

If *m* is negative,
the line falls.

FIGURE 7.2

Definition of the Slope of a Line

The **slope** m of the nonvertical line passing through the points (x_1, y_1) and (x_2, y_2) is

$$m = \frac{y_2 - y_1}{x_2 - x_1}$$

where $x_1 \neq x_2$.

When this formula is used, the *order of subtraction* is important. Given two points on a line, we are free to label either one of them as (x_1, y_1) and the other as (x_2, y_2). However, once this is done, we must form the numerator and denominator using the same order of subtraction.

$$\underbrace{m = \frac{y_2 - y_1}{x_2 - x_1}}_{\text{Correct}} \qquad \underbrace{m = \frac{y_1 - y_2}{x_1 - x_2}}_{\text{Correct}} \qquad \underbrace{m = \frac{y_2 - y_1}{x_1 - x_2}}_{\text{Incorrect}}$$

EXAMPLE 1   Finding the Slope of a Line Passing Through Two Points

Find the slopes of the lines passing through the following pairs of points.

(a) $(-2, 0)$ and $(3, 1)$ (b) $(-1, 2)$ and $(2, 2)$ (c) $(0, 4)$ and $(1, -1)$

SOLUTION

(a) The slope of the line through $(-2, 0)$ and $(3, 1)$ is

$$m = \frac{y_2 - y_1}{x_2 - x_1} \quad \longleftarrow \text{ Difference in } y\text{-values}$$
$$ \quad \longleftarrow \text{ Difference in } x\text{-values}$$

$$= \frac{1 - 0}{3 - (-2)}$$

$$= \frac{1}{3 + 2} = \frac{1}{5}.$$

(b) The slope of the line through $(-1, 2)$ and $(2, 2)$ is

$$m = \frac{2 - 2}{2 - (-1)} = \frac{0}{3} = 0.$$

(c) The slope of the line through $(0, 4)$ and $(1, -1)$ is

$$m = \frac{-1 - 4}{1 - 0} = \frac{-5}{1} = -5.$$

The graphs of the three lines are shown in Figure 7.2.

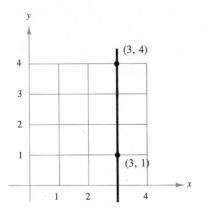

If the line is vertical,
the slope is undefined.

FIGURE 7.3

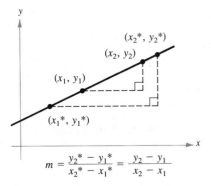

$$m = \frac{y_2{}^* - y_1{}^*}{x_2{}^* - x_1{}^*} = \frac{y_2 - y_1}{x_2 - x_1}$$

Any two points on a line can be
used to determine the slope of the line.

FIGURE 7.4

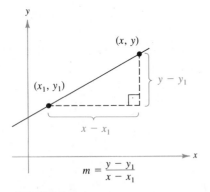

$$m = \frac{y - y_1}{x - x_1}$$

FIGURE 7.5

In Example 1, note that as we move *from left to right:*

1. A line with positive slope ($m > 0$) *rises*.
2. A line with negative slope ($m < 0$) *falls*.
3. A line with zero slope ($m = 0$) is *horizontal*.

Note that the definition of slope does not apply to vertical lines. For instance, consider the points $(3, 4)$ and $(3, 1)$ on the vertical line shown in Figure 7.3. In attempting to find the slope of this line, we obtain

$$m = \frac{4 - 1}{3 - 3}.$$ *Undefined division by zero*

Since division by zero is not defined, we do not define the slope of a vertical line.

Any two points on a line can be used to calculate its slope. This can be verified from the similar triangles shown in Figure 7.4. Recall that the ratios of corresponding sides of similar triangles are equal.

Equations of Lines

If we know the slope of a line and one point on the line, we can determine the equation of the line. For instance, in Figure 7.5, let (x_1, y_1) be a point on the line whose slope is m. If (x, y) is any *other* point on the line, then

$$\frac{y - y_1}{x - x_1} = m.$$

This equation, involving the two variables x and y, can be rewritten in the form

$$y - y_1 = m(x - x_1)$$

which is called the **point-slope form** of the equation of a line.

Point-Slope Form of the Equation of a Line

An equation of the line with slope m and passing through the point (x_1, y_1) is given by

$$y - y_1 = m(x - x_1).$$

EXAMPLE 2   **The Point-Slope Form of the Equation of a Line**

Find an equation of the line with slope 3 and passing through the point $(1, -2)$.

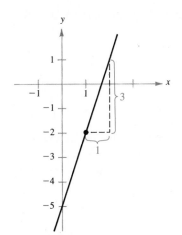

FIGURE 7.6

SOLUTION

Using the point-slope form with $(x_1, y_1) = (1, -2)$ and $m = 3$, we have

$$y - y_1 = m(x - x_1) \qquad \textit{Point-slope form}$$
$$y - (-2) = 3(x - 1)$$
$$y + 2 = 3x - 3$$
$$y = 3x - 5.$$

See Figure 7.6.

The point-slope form can be used to find the equation of a line passing through two points (x_1, y_1) and (x_2, y_2). First, we find the slope of the line

$$m = \frac{y_2 - y_1}{x_2 - x_1}.$$

Then, we use the point-slope form to obtain the equation

$$y - y_1 = \frac{y_2 - y_1}{x_2 - x_1}(x - x_1).$$

This is sometimes called the **two-point form** of the equation of a line.

EXAMPLE 3 A Linear Model for Sales Prediction

The total United States sales (including inventories) during the first two quarters of 1984 were 1596 and 1649 billion dollars, respectively. (a) Write a linear equation giving the total sales y in terms of the quarter x. (b) Use the equation to predict the total sales during the fourth quarter of 1984.

SOLUTION

(a) In Figure 7.7 we let $(1, 1596)$ and $(2, 1649)$ be two points on the line representing the total United States sales. The slope of the line passing through these two points is

$$m = \frac{1649 - 1596}{2 - 1} = 53.$$

By the point-slope form, the equation of this line is

$$y - y_1 = m(x - x_1)$$
$$y - 1596 = 53(x - 1)$$
$$y = 53x - 53 + 1596$$
$$y = 53x + 1543.$$

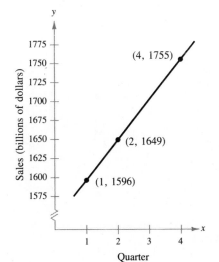

Total U.S. Sales in 1984

FIGURE 7.7

Lines in the Plane and Conics

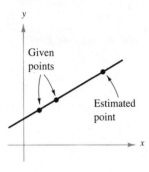

Linear Extrapolation

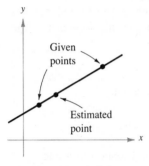

Linear Interpolation

FIGURE 7.8

(b) Using the equation from part (a), we estimate the fourth quarter ($x = 4$) sales to be

$$y = (53)(4) + 1543 = 1755 \text{ billion dollars.}$$

The prediction method illustrated in Example 3 is called **linear extrapolation.** Note in Figure 7.8 that the estimated point does not lie between the given points. When the estimated point lies *between* two given points, we call the procedure **linear interpolation.**

Sketching Graphs of Lines

Many problems in coordinate (or analytic) geometry can be classified in two basic categories.

1. Given a graph (or parts of it), find its equation.
2. Given an equation, find its graph.

For lines, the first problem is solved easily by using the point-slope form. However, this formula is not convenient for solving the second type of problem. The form better suited to graphing linear equations is the **slope-intercept form** of the equation of a line. To derive the slope-intercept form, we write

$$y - y_1 = m(x - x_1) \qquad \textit{Point-slope form}$$
$$y = mx - mx_1 + y_1$$
$$y = mx + b \qquad \textit{Slope-intercept form}$$

where $b = y_1 - mx_1$.

The Slope-Intercept Form of the Equation of a Line

The graph of the equation

$$y = mx + b$$

is a line with *slope m* and *y-intercept* $(0, b)$.

EXAMPLE 4 Graphing Linear Equations

Sketch the graphs of the following linear equations.

(a) $y = 2x + 1$ (b) $y = 2$ (c) $x + y = 2$

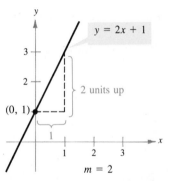

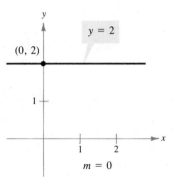

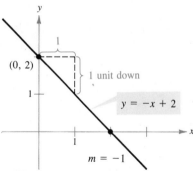

(a) When *m* is positive, the line rises.   (b) When *m* is zero, the line is horizontal.   (c) When *m* is negative, the line falls.

FIGURE 7.9

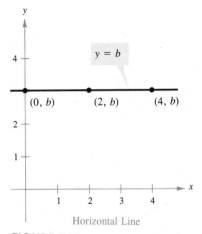

FIGURE 7.10

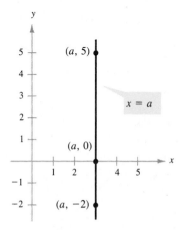

FIGURE 7.11

SOLUTION

(a) Since $b = 1$, the y-intercept is $(0, 1)$. Moreover, since the slope is $m = 2$, this line *rises* two units for each unit the line moves to the right, as shown in Figure 7.9(a).

(b) By writing the equation $y = 2$ in the form $y = (0)x + 2$, we see that the y-intercept is $(0, 2)$ and the slope is zero. A zero slope implies that the line is horizontal, as shown in Figure 7.9(b).

(c) By writing the equation $x + y = 2$ in slope-intercept form, $y = -x + 2$, we see that the y-intercept is $(0, 2)$. Moreover, since the slope is $m = -1$, this line *falls* one unit for each unit the line moves to the right, as shown in Figure 7.9(c).

From the slope-intercept form of the equation of a line, we see that a horizontal line ($m = 0$) has an equation of the form

$$y = (0)x + b \quad \text{or} \quad y = b. \qquad \textit{Horizontal line}$$

This is consistent with the fact that each point on a horizontal line through $(0, b)$ has a y-coordinate of b, as shown in Figure 7.10.

Similarly, each point on a vertical line through $(a, 0)$ has an x-coordinate of a, as shown in Figure 7.11. Hence, a vertical line has an equation of the form

$$x = a. \qquad \textit{Vertical line}$$

This equation cannot be written in the slope-intercept form, because the slope of a vertical line is undefined. However, *every* line has an equation that can be written in the **general form**

$$Ax + By + C = 0 \qquad \textit{General form}$$

where A and B are not *both* zero. If $A = 0$ (and $B \neq 0$), the equation can be reduced to the form $y = b$, which represents a horizontal line. If $B = 0$ (and

$A \neq 0$), the general equation can be reduced to the form $x = a$, which represents a vertical line.

For convenience, we list the five most common forms of equations of lines.

Summary of Equations of Lines

1. General form: $Ax + By + C = 0$
2. Vertical line: $x = a$
3. Horizontal line: $y = b$
4. Slope-intercept form: $y = mx + b$
5. Point-slope form: $y - y_1 = m(x - x_1)$

WARM UP

Simplify the expressions.

1. $\dfrac{4 - (-5)}{-3 - (-1)}$

2. $\dfrac{-5 - 8}{0 - (-3)}$

3. $\dfrac{5 - 5}{-2 + 3}$

4. $\dfrac{-1 + 1}{3 - (-4)}$

Solve for y in terms of x.

5. $2x - 3y = 5$

6. $4x + 2y = 0$

7. $y - (-4) = 3[x - (-1)]$

8. $y - 7 = \frac{2}{3}(x - 3)$

9. $y - (-1) = \dfrac{3 - (-1)}{2 - 4}(x - 4)$

10. $y - 5 = \dfrac{3 - 5}{0 - 2}(x - 2)$

EXERCISES 7.1

In Exercises 1–6, estimate the slope of the given line from its graph.

1.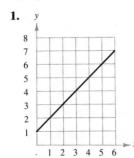

2.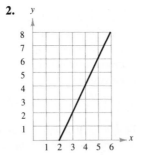

3.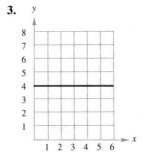

4.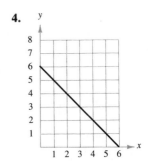

5.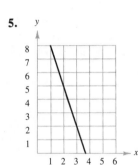

6.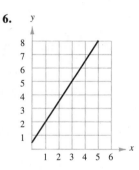

In Exercises 7 and 8, sketch the graph of the lines through the given point with the indicated slope. Make the sketches on the same set of coordinate axes.

Point	Slopes
7. (2, 3)	(a) 0 (b) 1 (c) 2 (d) -3
8. $(-4, 1)$	(a) 3 (b) -3 (c) $\frac{1}{2}$ (d) undefined

In Exercises 9–14, plot the points and find the slope of the line passing through each pair of points.

9. $(-3, -2)$, $(1, 6)$ **10.** $(2, 4)$, $(4, -4)$

11. $(-6, -1)$, $(-6, 4)$ **12.** $(0, -10)$, $(-4, 0)$

13. $(1, 2)$, $(-2, -2)$ **14.** $\left(\frac{7}{8}, \frac{3}{4}\right)$, $\left(\frac{5}{4}, -\frac{1}{4}\right)$

In Exercises 15–20, use the given point on the line and the slope of the line to find three additional points that the line passes through. (The solution is not unique.)

Point	Slope
15. (2, 1)	$m = 0$
16. $(-3, 4)$	m undefined
17. $(5, -6)$	$m = 1$
18. $(10, -6)$	$m = -1$
19. $(-8, 1)$	m undefined
20. $(-3, -1)$	$m = 0$

In Exercises 21–26, find the slope and y-intercept (if possible) of the line specified by the given equation.

21. $5x - y + 3 = 0$ **22.** $2x + 3y - 9 = 0$

23. $5x - 2 = 0$ **24.** $3y + 5 = 0$

25. $7x + 6y - 30 = 0$ **26.** $x - y - 10 = 0$

In Exercises 27–34, find an equation for the line passing through the given points and sketch a graph of the line.

27. $(5, -1)$, $(-5, 5)$ **28.** $(4, 3)$, $(-4, -4)$

29. $\left(2, \frac{1}{2}\right)$, $\left(\frac{1}{2}, \frac{5}{4}\right)$ **30.** $(-1, 4)$, $(6, 4)$

31. $(-8, 1)$, $(-8, 7)$ **32.** $(1, 1)$, $\left(6, -\frac{2}{3}\right)$

33. $(1, 0.6)$, $(-2, -0.6)$ **34.** $(-8, 0.6)$, $(2, -2.4)$

In Exercises 35–44, find an equation of the line that passes through the given point and has the indicated slope. Sketch the graph of the line.

Point	Slope
35. $(0, -2)$	$m = 3$
36. $(0, 10)$	$m = -1$
37. $(-3, 6)$	$m = -2$
38. $(0, 0)$	$m = 4$
39. $(4, 0)$	$m = -\frac{1}{3}$
40. $(-2, -5)$	$m = \frac{3}{4}$
41. $(6, -1)$	m undefined
42. $(-10, 4)$	$m = 0$
43. $\left(4, \frac{5}{2}\right)$	$m = \frac{4}{3}$
44. $\left(-\frac{1}{2}, \frac{3}{2}\right)$	$m = -3$

45. Prove that the line with intercepts $(a, 0)$ and $(0, b)$ has the following equation:

$$\frac{x}{a} + \frac{y}{b} = 1, \qquad a \neq 0, \quad b \neq 0.$$

46. Prove that the slope of the line given by $ax + by + c = 0$ is $-a/b$, provided $b \neq 0$.

In Exercises 47–52, use the result of Exercise 45 to write an equation of the indicated line.

47. x-intercept: (2, 0)   **48.** x-intercept: $(-3, 0)$
y-intercept: (0, 3)   y-intercept: (0, 4)

49. x-intercept: $\left(-\frac{1}{6}, 0\right)$ **50.** x-intercept: $\left(-\frac{2}{3}, 0\right)$
y-intercept: $\left(0, -\frac{2}{3}\right)$ y-intercept: (0, -2)

51. point on line: (1, 2)   **52.** point on line: $(-3, 4)$
x-intercept: (a, 0)   x-intercept: (a, 0)
y-intercept: (0, a)   y-intercept: (0, a)
$(a \neq 0)$ $(a \neq 0)$

53. Find the equation of the line giving the relationship between the temperature in degrees Celsius, C, and degrees Fahrenheit, F. Remember that water freezes at $0°$ Celsius ($32°$ Fahrenheit) and boils at $100°$ Celsius ($212°$ Fahrenheit).

54. Use the result of Exercise 53 to complete the following table.

C		$-10°$	$10°$			$177°$
F	$0°$			$68°$	$90°$	

55. An individual buys a $1000 corporate bond paying $9\frac{1}{2}\%$ simple interest. Write a linear equation giving the total interest I earned if the bond is held for t years.

56. A small business purchases a piece of equipment for $875. After five years the equipment will be outdated and have no value. Write a linear equation giving the value V of the equipment during the five years it will be used.

57. A store is offering a 15% discount on all items in its inventory. Write a linear equation giving the sale price S for an item with a list price L.

58. A manufacturer pays its assembly line workers $11.50 per hour. In addition, workers receive a piecework rate of $0.75 per unit produced. Write a linear equation for the hourly wages W in terms of the number of units produced per hour x.

59. A salesperson receives a monthly salary of $2500 plus a commission of 7% of his sales. Write a linear equation for the salesperson's monthly wage W in terms of his monthly sales S.

60. A sales representative uses her own car as she travels for her company. The cost to the company is $95 per day for lodging and meals plus $0.22 per mile driven. Write a linear equation giving the daily cost C to the company in terms of x, the number of miles driven.

61. A contractor purchases a piece of equipment for $36,500. The equipment requires an average expenditure of $5.25 per hour for fuel and maintenance, and the operator is paid $11.50 per hour.
   (a) Write a linear equation giving the total cost C of operating this equipment for t hours. (Include the purchase cost for the equipment.)
   (b) If customers are charged $27 per hour of machine use, write an equation for the revenue R derived from t hours of use.
   (c) Use the formula for profit $(P = R - C)$ to write an equation for the profit derived from t hours of use.
   (d) (*Break-Even Point*)   Use the result of part (c) to find the number of hours this equipment must be used to yield a profit of zero dollars.

62. A real estate office handles an apartment complex with 50 units. When the rent per unit is $380 per month, all 50 units are occupied. However, when the rent is $425 per month, the average number of occupied units drops to 47. Assume that the relationship between the monthly rent p and the demand x is linear.
   (a) Write the equation of the line giving the demand x in terms of the rent p.
   (b) (*Linear Extrapolation*)   Use this equation to predict the number of units occupied if the rent is raised to $455.
   (c) (*Linear Interpolation*)   Predict the number of units occupied if the rent is lowered to $395.

7.2 Additional Properties of Lines

Parallel and Perpendicular Lines

The slope of a line is a convenient tool for determining whether two lines are parallel or perpendicular.

Parallel Lines

Two distinct nonvertical lines are **parallel** if and only if their slopes are equal.

Additional Properties of Lines

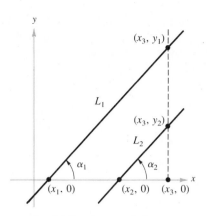

Parallel lines have equal slopes.

FIGURE 7.12

PROOF

Recall that the phrase "if and only if" is a way of stating two rules in one. One rule says, "If two nonvertical lines are parallel, then they must have equal slopes." The other rule is the converse, which says, "If two distinct lines have equal slopes, they must be parallel." We will prove the first of these two rules and leave the converse for you to prove.

Assume that we have two parallel lines L_1 and L_2 with slopes m_1 and m_2. If these lines are both horizontal, then $m_1 = m_2 = 0$, and the rule is established. If L_1 and L_2 are not horizontal, then they must intersect the x-axis at points $(x_1, 0)$ and $(x_2, 0)$, as shown in Figure 7.12. Since L_1 and L_2 are parallel, their intersection with the x-axis must produce equal angles α_1 and α_2. Therefore, the two right triangles with vertices

$$(x_1, 0), (x_3, 0), (x_3, y_1) \qquad \text{and} \qquad (x_2, 0), (x_3, 0), (x_3, y_2)$$

must be similar. From this we conclude that the ratios of their corresponding sides must be equal, and thus

$$m_1 = \frac{y_1 - 0}{x_3 - x_1} = \frac{y_2 - 0}{x_3 - x_2} = m_2.$$

Hence, the lines L_1 and L_2 must have equal slopes. ▬

EXAMPLE 1 **Equations of Parallel Lines**

Find an equation of the line that passes through the point $(2, -1)$ and is parallel to the line $2x - 3y = 5$, as shown in Figure 7.13.

SOLUTION

Writing the given equation in slope-intercept form, we have

$$2x - 3y = 5 \qquad \qquad \textit{Given equation}$$
$$3y = 2x - 5$$
$$y = \frac{2}{3}x - \frac{5}{3}. \qquad \qquad \textit{Slope-intercept form}$$

Therefore, the given line has a slope of $m = 2/3$. Since any line parallel to the given line must also have a slope of $2/3$, the required line through $(2, -1)$ has the equation

$$y - (-1) = \frac{2}{3}(x - 2)$$
$$3(y + 1) = 2(x - 2)$$
$$3y + 3 = 2x - 4$$
$$-2x + 3y = -7$$
$$2x - 3y = 7.$$

Note the similarity to the original equation $2x - 3y = 5$. ▬

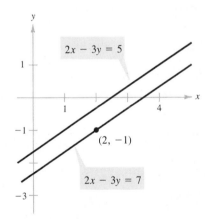

FIGURE 7.13

Perpendicular Lines

Two nonvertical lines are **perpendicular** if and only if their slopes are related by the equation

$$m_1 = -\frac{1}{m_2}.$$

PROOF

As in the previous rule, we will prove only one direction of the rule and leave the proof of the converse as an exercise.

Assume that we are given two nonvertical perpendicular lines L_1 and L_2 with slopes m_1 and m_2. For simplicity's sake let these two lines intersect at the origin, as shown in Figure 7.14. The vertical line $x = 1$ will intersect L_1 and L_2 at the respective points $(1, m_1)$ and $(1, m_2)$. Since L_1 and L_2 are perpendicular, the triangle formed by these two points and the origin is a right triangle. Thus, we can apply the Pythagorean Theorem and conclude that

$$\left(\begin{array}{c}\text{distance between}\\(0,0)\text{ and }(1, m_1)\end{array}\right)^2 + \left(\begin{array}{c}\text{distance between}\\(0,0)\text{ and }(1, m_2)\end{array}\right)^2 = \left(\begin{array}{c}\text{distance between}\\(1, m_1)\text{ and }(1, m_2)\end{array}\right)^2.$$

Using the Distance Formula, we have

$$(\sqrt{1 + m_1^2})^2 + (\sqrt{1 + m_2^2})^2 = (\sqrt{0^2 + (m_1 - m_2)^2})^2$$
$$1 + m_1^2 + 1 + m_2^2 = (m_1 - m_2)^2$$
$$2 + m_1^2 + m_2^2 = m_1^2 - 2m_1m_2 + m_2^2$$
$$2 = -2m_1m_2$$
$$-1 = m_1m_2$$
$$-\frac{1}{m_2} = m_1.$$

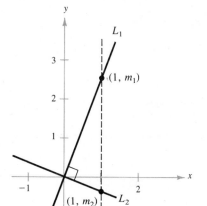

The slopes of perpendicular lines are negative reciprocals of each other.

FIGURE 7.14

EXAMPLE 2 *Equations of Perpendicular Lines*

Find an equation of the line that passes through the point $(2, -1)$ and is perpendicular to the line $2x - 3y = 5$.

Additional Properties of Lines

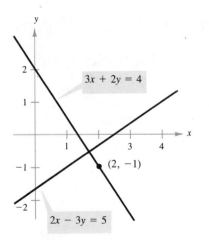

FIGURE 7.15

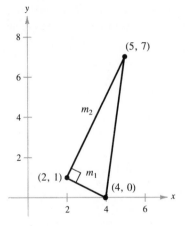

FIGURE 7.16

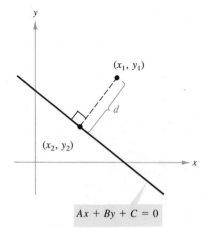

FIGURE 7.17

SOLUTION

From Example 1, the given line has slope $2/3$. Hence, any line perpendicular to this line must have a slope of $-3/2$. Therefore, the required line through $(2, -1)$ has the equation

$$y - (-1) = -\frac{3}{2}(x - 2)$$
$$2(y + 1) = -3(x - 2)$$
$$3x + 2y = 4.$$

The graphs of both lines are shown in Figure 7.15.

EXAMPLE 3   *Verifying a Right Triangle*

Show that the points $(2, 1)$, $(4, 0)$, and $(5, 7)$ are the vertices of a right triangle.

SOLUTION

From Figure 7.16 we find the slope of the line connecting $(4, 0)$ and $(2, 1)$ to be

$$m_1 = \frac{1 - 0}{2 - 4} = -\frac{1}{2}.$$

Furthermore, the slope of the line connecting $(2, 1)$ and $(5, 7)$ is

$$m_2 = \frac{7 - 1}{5 - 2} = \frac{6}{3} = 2.$$

Since m_1 and m_2 are negative reciprocals, we know that two sides of the triangle are perpendicular and hence the triangle is a right triangle.

Finding the distance between a line and a point not on the line is another application of perpendicular lines. We define this distance to be the length of the perpendicular line segment joining the point to the given line, as shown in Figure 7.17.

Distance Between a Point and a Line

The distance between the point (x_1, y_1) and the line given by $Ax + By + C = 0$ is

$$d = \frac{|Ax_1 + By_1 + C|}{\sqrt{A^2 + B^2}}.$$

PROOF

For simplicity's sake, we assume that the given line is neither horizontal nor vertical. By writing the equation $Ax + By + C = 0$ in the slope-intercept form

$$y = -\frac{A}{B}x - \frac{C}{B}$$

we see that the line has slope of

$$m = -\frac{A}{B}.$$

Thus, the slope of the line passing through (x_1, y_1) and perpendicular to the given line is B/A, and its equation is

$$y - y_1 = \frac{B}{A}(x - x_1).$$

These two lines intersect at the point (x_2, y_2), where

$$x_2 = \frac{B(Bx_1 - Ay_1) - AC}{A^2 + B^2} \quad \text{and} \quad y_2 = \frac{A(-Bx_1 + Ay_1) - BC}{A^2 + B^2}.$$

Finally, the distance between (x_1, y_1) and (x_2, y_2) is

$$d = \sqrt{(x_2 - x_1)^2 + (y_2 - y_1)^2}$$

$$= \sqrt{\left(\frac{B^2x_1 - ABy_1 - AC}{A^2 + B^2} - x_1\right)^2 + \left(\frac{-ABx_1 + A^2y_1 - BC}{A^2 + B^2} - y_1\right)^2}$$

$$= \sqrt{\frac{A^2(Ax_1 + By_1 + C)^2 + B^2(Ax_1 + By_1 + C)^2}{(A^2 + B^2)^2}}$$

$$= \frac{|Ax_1 + By_1 + C|}{\sqrt{A^2 + B^2}}.$$

EXAMPLE 4 *Finding the Distance Between a Point and a Line*

Find the distance between the point $(4, 1)$ and the line $y = 2x + 1$.

SOLUTION

The general form of the given equation is

$$-2x + y - 1 = 0.$$

Hence, the distance between the point and the line is

$$d = \frac{|-2(4) + 1(1) - 1|}{\sqrt{(-2)^2 + 1^2}} = \frac{8}{\sqrt{5}} \approx 3.58.$$

EXAMPLE 5 *An Application of Two Distance Formulas*

Figure 7.18 shows a triangle with vertices $A = (-3, 0)$, $B = (0, 4)$, and $C = (5, 2)$.

(a) Find the altitude from vertex B to side AC.
(b) Find the area of the triangle.

SOLUTION

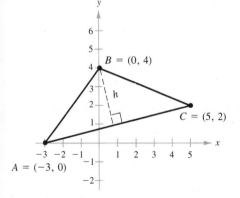

FIGURE 7.18

(a) To find the altitude, we can use the formula for the distance between line AC and the point $(0, 4)$. The equation of line AC is obtained as follows.

$$\text{Slope:} \quad m = \frac{2 - 0}{5 + 3} = \frac{1}{4}$$

$$\text{Equation:} \quad y - 0 = \frac{1}{4}(x + 3)$$

$$x - 4y + 3 = 0$$

Therefore, the distance between this line and the point $(0, 4)$ is

$$\text{altitude} = h = \frac{|1(0) - 4(4) + 3|}{\sqrt{1^2 + (-4)^2}} = \frac{13}{\sqrt{17}}.$$

(b) Using the formula for the *distance between two points*, we find the length of the base AC to be

$$b = \sqrt{(5 + 3)^2 + (2 - 0)^2} = \sqrt{68} = 2\sqrt{17}.$$

Finally, the area of the triangle in Figure 7.18 is

$$A = \frac{1}{2}bh = \frac{1}{2}(2\sqrt{17})\left(\frac{13}{\sqrt{17}}\right) = 13.$$

Inclination of a Line

Every nonhorizontal line must intersect the x-axis. The angle formed by such an intersection determines the **inclination** of the line, as specified in the following definition.

Definition of Inclination

The **inclination** of a nonhorizontal line is the positive angle θ (less than 180°) measured counterclockwise from the x-axis to the line. (See Figure 7.19.)

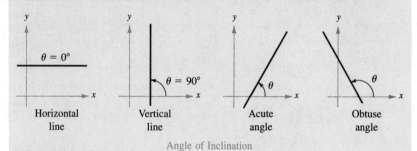

Horizontal line   Vertical line   Acute angle   Obtuse angle

Angle of Inclination

FIGURE 7.19

A horizontal line has an **inclination** of zero.

The inclination of a line is related to its slope in the following manner.

Inclination and Slope

If a line has inclination θ and slope m, then

$$m = \tan \theta.$$

PROOF

If $m = 0$, then the line is *horizontal* and $\theta = 0$. Thus, the result is true since $m = 0 = \tan 0$.

If the line has a *positive* slope, then it will intersect the x-axis. We label this point $(x_1, 0)$, as shown in Figure 7.20. If (x_2, y_2) is a second point on the line, then the slope is given by

$$m = \frac{y_2 - 0}{x_2 - x_1} = \frac{y_2}{x_2 - x_1} = \tan \theta.$$

We leave the case in which the line has a *negative* slope for you to prove.

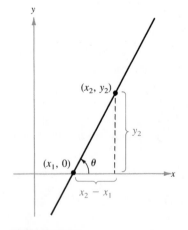

FIGURE 7.20

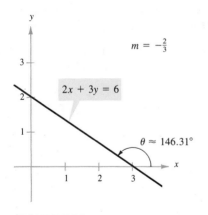

FIGURE 7.21

EXAMPLE 6 *Finding the Inclination of a Line*

Find the inclination of the line given by $2x + 3y = 6$.

SOLUTION

The slope of this line is $m = -2/3$. Thus, its inclination is determined from the equation

$$\tan \theta = -\frac{2}{3}.$$

From Figure 7.21, we see that $90° < \theta < 180°$. This means that

$$\theta = 180° + \arctan\left(-\frac{2}{3}\right)$$

$$\approx 180° - 33.69° \approx 146.31°.$$

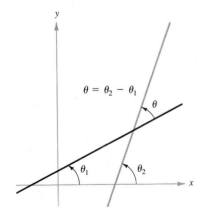

FIGURE 7.22

Two lines in a plane are either parallel or they intersect. If they intersect, then we can use their inclinations to find the angle between the lines. Specifically, if two lines have inclinations of θ_1 and θ_2, then the angle between the two lines is

$$\theta = \theta_2 - \theta_1$$

as shown in Figure 7.22. Note that this definition implies that the angle between two parallel lines is zero.

We can use the formula for the tangent of the difference of two angles

$$\tan \theta = \tan(\theta_2 - \theta_1) = \frac{\tan \theta_2 - \tan \theta_1}{1 + \tan \theta_1 \tan \theta_2}$$

to obtain the following convenient formula for the angle between two lines.

Angle Between Two Lines

If two nonperpendicular lines have slopes m_1 and m_2, then the angle between the two lines is given by

$$\tan \theta = \left| \frac{m_2 - m_1}{1 + m_1 m_2} \right|.$$

Lines in the Plane and Conics

EXAMPLE 7   Finding the Angle Between Two Lines

Find the angle between the following two lines.

Line 1: $2x - y - 4 = 0$
Line 2: $3x + 4y - 12 = 0$

SOLUTION

The two lines have slopes of

$$m_1 = 2 \quad \text{and} \quad m_2 = -\frac{3}{4}$$

respectively. Thus, the angle between the two lines is given by

$$\tan \theta = \left| \frac{m_2 - m_1}{1 + m_1 m_2} \right|$$

$$= \left| \frac{(-3/4) - 2}{1 + (-3/4)(2)} \right|$$

$$= \left| \frac{-11/4}{-2/4} \right| = \frac{11}{2}.$$

Finally, we conclude that the angle is

$$\theta = \arctan \frac{11}{2} \approx 79.70°$$

as shown in Figure 7.23.

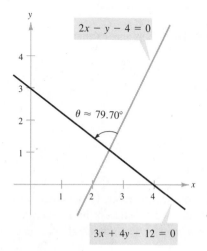

$2x - y - 4 = 0$

$\theta \approx 79.70°$

$3x + 4y - 12 = 0$

FIGURE 7.23

WARM UP

Find the slopes of the following lines.

1. $y = -2$ **2.** $y = -2x + 3$
3. $x + 2y = 6$ **4.** $2x - y = 4$

Find the slopes of the lines passing through the indicated points.

5. $(2, -3), (4, 5)$ **6.** $(-4, 5), (0, 7)$
7. Find $-1/m$ for $m = 4/5$.   **8.** Find $-1/m$ for $m = -2$.

Find the angle θ, $0° \leq \theta \leq 90°$. Round your answers to two decimal places.

9. $\tan \theta = \frac{4}{5}$ **10.** $\tan \theta = \frac{7}{3}$

EXERCISES 7.2

In Exercises 1–4, determine whether the lines L_1 and L_2 passing through the given pairs of points are parallel, perpendicular, or neither.

1. L_1: $(0, -1)$, $(5, 9)$
  L_2: $(0, 3)$, $(4, 1)$

2. L_1: $(-2, -1)$, $(1, 5)$
  L_2: $(1, 3)$, $(5, -5)$

3. L_1: $(3, 6)$, $(-6, 0)$
  L_2: $(0, -1)$, $(5, \frac{7}{3})$

4. L_1: $(4, 8)$, $(-4, 2)$
  L_2: $(3, -5)$, $(-1, \frac{1}{3})$

In Exercises 5–10, write an equation of the line through the given point (a) parallel to the given line and (b) perpendicular to the given line.

	Point	Line
5.	$(2, 1)$	$4x - 2y = 3$
6.	$(-3, 2)$	$x + y = 7$
7.	$(-6, 4)$	$3x + 4y = 7$
8.	$\left(\frac{7}{8}, \frac{3}{4}\right)$	$5x + 3y = 0$
9.	$(-1, 0)$	$y = -3$
10.	$(2, 5)$	$x = 4$

In Exercises 11–16, determine whether the three points are vertices of a right triangle.

11. $(4, 0)$, $(2, 1)$, $(-1, -5)$
12. $(-1, 1)$, $(5, -3)$, $(8, \frac{3}{2})$
13. $(1, -3)$, $(3, 2)$, $(-2, 4)$
14. $(1, 2)$, $(2, 1)$, $(3, 3)$
15. $(-2, -2)$, $(5, 1)$, $(3, \frac{17}{3})$
16. $(3, 7)$, $(4, 4)$, $\left(\frac{1}{3}, -\frac{2}{3}\right)$

In Exercises 17–20, determine whether the three points are collinear (lie on the same line).

17. $(0, -4)$, $(2, 0)$, $(3, 2)$
18. $(0, 4)$, $(7, -6)$, $(-5, 11)$
19. $(-2, 1)$, $(-1, 0)$, $(2, -2)$
20. $(1, 1)$, $(2, \frac{7}{2})$, $(-1, -4)$

In Exercises 21–28, find the distance between the point and the line.

21. $(0, 0)$, $4x + 3y = 10$
22. $(2, 3)$, $4x + 3y = 10$
23. $(-2, 1)$, $x - y - 2 = 0$
24. $(6, 2)$, $x + 1 = 0$
25. $\left(\frac{1}{2}, \frac{2}{3}\right)$, $y - 2 = 0$
26. $\left(\frac{1}{2}, 4\right)$, $2x - 5y = -4$
27. $\left(\frac{3}{2}, \frac{1}{3}\right)$, $8x + 9y = 15$
28. $(-6, 4)$, $3x + 4y = 1$

In Exercises 29 and 30, find the distance between the given parallel lines.

29. $x + y = 1$, $x + y = 5$
30. $3x - 4y = 1$, $3x - 4y = 10$

In Exercises 31–38, find the inclination of the line.

31. $x - \sqrt{3}y = 0$
32. $\sqrt{3}x + y = 3$
33. $x - y = 4$
34. $x - 3y = 11$
35. $5x + 3y = 10$
36. $-\frac{1}{3}x + \frac{5}{6}y = 1$
37. $0.02x + 0.15y = 0.25$
38. $4x - 2y = 3$

In Exercises 39–44, find the angle between the given lines.

39. $y = x + 2$, $y = 3$
40. $2x - 3y = 1$, $x + 5y = 2$
41. $4x + 3y + 2 = 0$, $3x + 4y - 7 = 0$
42. $2x - y + 7 = 0$, $x + y + 2 = 0$
43. $2x + 3y = 9$, $\frac{4}{3}x + 2y = 4$
44. $5x - 6y + 12 = 0$, $6x + 5y - 16 = 0$

In Exercises 45 and 46, find an equation of the line having the given inclination and passing through the indicated point.

45. $\theta = 60°$, (0, 6)

46. $\theta = 150°$, (−3, 4)

47. Write an equation of the line that bisects the angle between the lines given by $2x - y + 7 = 0$ and $x + y + 2 = 0$.

48. Find the distance between the origin and the line $4x + 3y = 10$ by first finding the point of intersection of the given line and the line through the origin perpendicular to the given line. Then find the distance between the origin and the point of intersection. Compare the result with that of Exercise 21.

49. Prove that if two distinct lines have equal slopes, they must be parallel.

50. Prove that if two nonvertical lines have slopes that are negative reciprocals, they must be perpendicular.

7.3 *Introduction to Conics: Parabolas*

Conic sections were discovered during the classical Greek period, which lasted from 600 to 300 B.C. By the beginning of the Alexandrian period, enough was known of conics for Apollonius (262–190 B.C.) to produce an eight-volume work on the subject.

The early Greeks were concerned largely with the geometrical properties of conics. It was not until the early seventeenth century that the broad applicability of conics became apparent, and they then played a prominent role in the early development of calculus.

Each **conic section** (or simply **conic**) can be described as the intersection of a plane and a double-napped cone. Notice from Figure 7.24 that in the formation of the four basic conics, the intersecting plane does not pass through

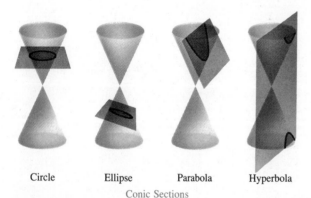

Circle   Ellipse   Parabola   Hyperbola

Conic Sections

FIGURE 7.24

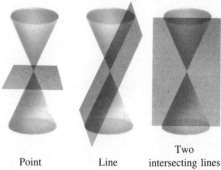

Point   Line   Two intersecting lines

Degenerate Conics

FIGURE 7.25

the vertex of the cone. When the plane does pass through the vertex, we call the resulting figure a **degenerate conic,** as shown in Figure 7.25.

There are several ways to approach a study of conics. We could begin by defining conics in terms of the intersections of planes and cones, as the Greeks did, or we could define them algebraically in terms of the general second-degree equation

$$Ax^2 + Bxy + Cy^2 + Dx + Ey + F = 0.$$

However, we will use a third approach, in which each of the conics is defined as a **locus** (collection) of points satisfying a certain geometric property. For example, in Section 1.4, we saw how the definition of a circle as the collection of all points (x, y) that are equidistant from a fixed point (h, k) led easily to the standard equation of a circle,

$$(x - h)^2 + (y - k)^2 = r^2.$$

In this and the following two sections, we give similar definitions to the other three types of conics. We will also identify practical geometric properties of conics used in the construction of objects such as bridges, searchlights, telescopes, and radar detectors.

Definition of a Parabola

A **parabola** is the set of all points (x, y) that are equidistant from a fixed line **(directrix)** and a fixed point **(focus)** not on the line.

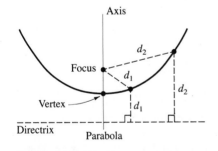

FIGURE 7.26

The midpoint between the focus and the directrix is called the **vertex,** and the line passing through the focus and the vertex is called the **axis** of the parabola. Note in Figure 7.26 that a parabola is symmetric with respect to its axis.

Using the definition of a parabola, we derive the following **standard form** of the equation of a parabola whose directrix is parallel to the *x*-axis or to the *y*-axis.

Standard Equation of a Parabola

The **standard form** of the equation of a parabola with vertex at (h, k) and directrix $y = k - p$ is

$$(x - h)^2 = 4p(y - k), \quad p \neq 0. \qquad \textit{Vertical axis}$$

For directrix $x = h - p$ the equation is

$$(y - k)^2 = 4p(x - h), \quad p \neq 0. \qquad \textit{Horizontal axis}$$

The focus lies on the axis p units (*directed distance*) from the vertex.

Lines in the Plane and Conics

PROOF

We prove only the case for which the directrix is parallel to the x-axis and the focus lies above the vertex, as shown in Figure 7.27(a). If (x, y) is any point on the parabola, then by definition it is equidistant from the focus $(h, k + p)$ and the directrix $y = k - p$, and we have

$$\sqrt{(x - h)^2 + [y - (k + p)]^2} = y - (k - p)$$
$$(x - h)^2 + [y - (k + p)]^2 = [y - (k - p)]^2$$
$$(x - h)^2 + y^2 - 2y(k + p) + (k + p)^2 = y^2 - 2y(k - p) + (k - p)^2$$
$$(x - h)^2 - 2py + 2pk = 2py - 2pk$$
$$(x - h)^2 = 4p(y - k).$$

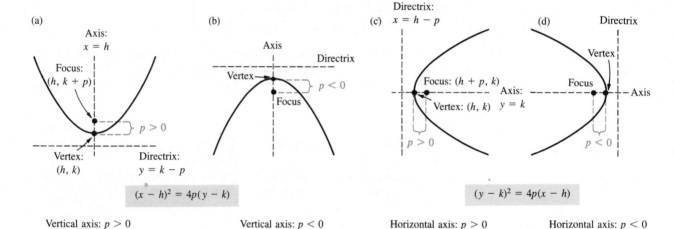

FIGURE 7.27

Parabolic Orientations

EXAMPLE 1 *Finding the Standard Equation of a Parabola*

Find the standard form of the equation of the parabola with vertex $(2, 1)$ and focus $(2, 4)$.

SOLUTION

Since the axis of the parabola is vertical, we consider the equation

$$(x - h)^2 = 4p(y - k)$$

where $h = 2$, $k = 1$, and $p = 4 - 1 = 3$. Thus, the standard form is

$$(x - 2)^2 = 12(y - 1).$$

The graph of this parabola is shown in Figure 7.28.

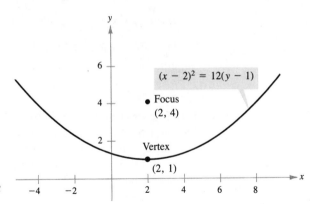

FIGURE 7.28

Remark: By expanding the standard equation in Example 1, we obtain the more common quadratic form $y = \frac{1}{12}(x^2 - 4x + 16)$.

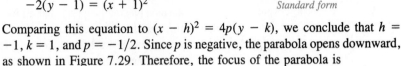

EXAMPLE 2 **Finding the Focus of a Parabola**

Find the focus of the parabola given by

$$y = -\frac{1}{2}x^2 - x + \frac{1}{2}.$$

SOLUTION

To find the focus we convert to standard form by completing the square.

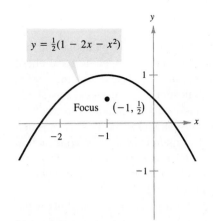

$$y = -\frac{1}{2}x^2 - x + \frac{1}{2} \qquad \text{\textit{Given equation}}$$

$$-2y = x^2 + 2x - 1 \qquad \text{\textit{Multiply by} } -2$$
$$1 - 2y = x^2 + 2x \qquad \text{\textit{Group terms}}$$
$$2 - 2y = x^2 + 2x + 1 \qquad \text{\textit{Add 1 to both sides}}$$
$$-2(y - 1) = (x + 1)^2 \qquad \text{\textit{Standard form}}$$

Comparing this equation to $(x - h)^2 = 4p(y - k)$, we conclude that $h = -1$, $k = 1$, and $p = -1/2$. Since p is negative, the parabola opens downward, as shown in Figure 7.29. Therefore, the focus of the parabola is

$$(h, k + p) = \left(-1, \frac{1}{2}\right). \qquad \text{\textit{Focus}}$$

FIGURE 7.29

Lines in the Plane and Conics

If the vertex of a parabola is at the origin, then the standard form

$$(x - h)^2 = 4p(y - k) \qquad \text{Vertex at (h, k)}$$

simplifies to

$$x^2 = 4py. \qquad \text{Vertex at origin}$$

This is consistent with the discussion of *horizontal* and *vertical shifts* in Section 1.6. That is, the factor $(x - h)$ indicates a horizontal shift, and $(y - k)$ represents a vertical shift from the origin. For instance, the equation $(x + 1)^2 = -2(y - 1)$ in Example 2 indicates a horizontal left shift of one unit and an upward shift of one unit (from the origin), as indicated in Figure 7.29.

EXAMPLE 3 Vertex at the Origin

Find the standard equation of the parabola with vertex at the origin and focus at (2, 0).

SOLUTION

The axis of the parabola is horizontal, passing through (0, 0) and (2, 0), as shown in Figure 7.30. Thus, we consider the standard form

$$y^2 = 4px$$

where $h = k = 0$ and $p = 2$. Therefore, the equation is

$$y^2 = 8x.$$

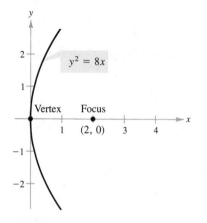

FIGURE 7.30

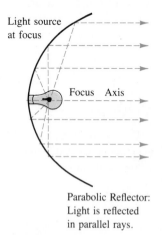

Parabolic Reflector:
Light is reflected
in parallel rays.

FIGURE 7.31

A line segment that passes through the focus of a parabola and has endpoints on the parabola is called a **focal chord.** The specific focal chord perpendicular to the axis of the parabola is called the **latus rectum.**

Parabolas occur in a wide variety of applications. For instance, a parabolic reflector can be formed by revolving a parabola about its axis. The resulting surface has the property that all incoming rays parallel to the axis are reflected through the focus of the parabola; this is the principle behind the construction of the parabolic mirrors used in reflecting telescopes. Conversely, the light rays emanating from the focus of a parabolic reflector used in a flashlight are all parallel to one another, as shown in Figure 7.31.

We say a line is **tangent** to a parabola at a point on the parabola if the line intersects, but does not cross, the parabola at the point. Tangent lines to parabolas have special properties related to the use of parabolas in constructing reflective surfaces.

Reflective Property of a Parabola

The tangent line to a parabola at a point P makes equal angles with the following two lines. (See Figure 7.32.)

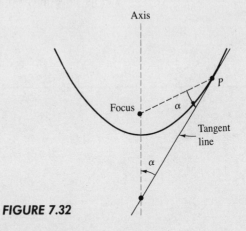

FIGURE 7.32

1. The line passing through P and the focus.
2. The axis of the parabola.

EXAMPLE 4 *Finding the Tangent Line at a Point on a Parabola*

Find the equation of the tangent line to the parabola given by $y = x^2$ at the point (1, 1).

SOLUTION

For this parabola, $p = 1/4$ and the focus is $(0, 1/4)$, as shown in Figure 7.33. We can find the y-intercept $(0, b)$ of the tangent line by equating the lengths of the two sides of the isosceles triangle

$$d_1 = \frac{1}{4} - b$$

and

$$d_2 = \sqrt{(1 - 0)^2 + [1 - (1/4)]^2} = \frac{5}{4}$$

shown in Figure 7.33. Setting $d_1 = d_2$ produces

$$\frac{1}{4} - b = \frac{5}{4}$$
$$b = -1.$$

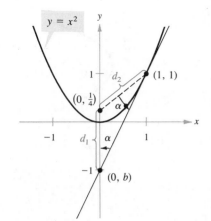

FIGURE 7.33

Lines in the Plane and Conics

Thus, the slope of the tangent line is

$$m = \frac{1 - (-1)}{1 - 0} = 2$$

and its slope-intercept equation is

$$y = 2x - 1.$$

WARM UP

Expand and simplify the expressions.

1. $(x - 5)^2 - 20$

2. $(x + 3)^2 - 1$

3. $10 - (x + 4)^2$

4. $4 - (x - 2)^2$

Complete the square on each quadratic expression.

5. $x^2 + 6x + 8$

6. $x^2 - 10x + 21$

7. $-x^2 + 2x + 1$

8. $-2x^2 + 4x - 2$

In each case find an equation of the line with given slope and passing through the given point.

9. $m = -\frac{2}{3}$, $(1, 6)$

10. $m = \frac{3}{4}$, $(3, -2)$

EXERCISES 7.3

In Exercises 1–6, match the equation with the correct graph. [The graphs are labeled (a)–(f).]

1. $y^2 = 4x$

2. $x^2 = -2y$

3. $x^2 = 8y$

4. $y^2 = -12x$

5. $(y - 1)^2 = 4(x - 2)$

6. $(x + 3)^2 = -2(y - 2)$

(a)

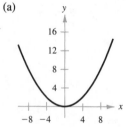

(b)

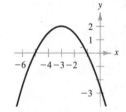

(c)

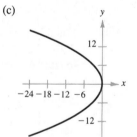

(d)

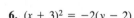

(e)

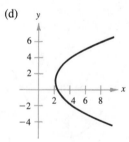

(f)

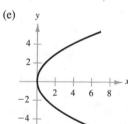

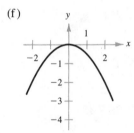

In Exercises 7–26, find the vertex, focus, and directrix of the parabola.

7. $y = 4x^2$

8. $y = 2x^2$

9. $y^2 = -6x$

10. $y^2 = 3x$

11. $x^2 + 8y = 0$

12. $x + y^2 = 0$

13. $(x - 1)^2 + 8(y + 2) = 0$

14. $(x + 3) + (y - 2)^2 = 0$

15. $\left(y + \frac{1}{2}\right)^2 = 2(x - 5)$

16. $\left(x + \frac{1}{2}\right)^2 - 4(y - 3) = 0$

17. $y = \frac{1}{4}(x^2 - 2x + 5)$

18. $y = -\frac{1}{6}(x^2 + 4x - 2)$

19. $4x - y^2 - 2y - 33 = 0$

20. $y^2 + x + y = 0$

21. $y^2 + 6y + 8x + 25 = 0$

22. $x^2 - 2x + 8y + 9 = 0$

23. $y^2 - 4y - 4x = 0$

24. $y^2 - 4x - 4 = 0$

25. $x^2 + 4x + 4y - 4 = 0$

26. $y^2 + 4y + 8x - 12 = 0$

In Exercises 27–42, find an equation of the specified parabola.

27. Vertex: $(0, 0)$
Focus: $\left(0, -\frac{3}{2}\right)$

28. Vertex: $(0, 0)$
Focus: $(2, 0)$

29. Vertex: $(0, 0)$
Focus: $(-2, 0)$

30. Vertex: $(0, 0)$
Focus: $(0, -2)$

31. Vertex: $(0, 0)$
Directrix: $y = -1$

32. Vertex: $(0, 0)$
Directrix: $x = 3$

33. Vertex: $(3, 2)$
Focus: $(1, 2)$

34. Vertex: $(-1, 2)$
Focus: $(-1, 0)$

35. Vertex: $(0, 4)$
Directrix: $y = 2$

36. Vertex: $(-2, 1)$
Directrix: $x = 1$

37. Focus: $(0, 0)$
Directrix: $y = 4$

38. Focus: $(2, 2)$
Directrix: $x = -2$

39. Axis: parallel to y-axis
Passes through points:
$(0, 3), (3, 4), (4, 11)$

40. Axis: parallel to x-axis
Passes through points:
$(4, -2), (0, 0), (3, -3)$

41.

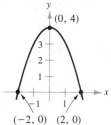

42.

43. Each cable of a particular suspension bridge is suspended (in the shape of a parabola) between two towers that are 400 feet apart and 50 feet above the roadway (see figure). The cables touch the roadway midway between the towers. Find an equation for the parabolic shape of each cable.

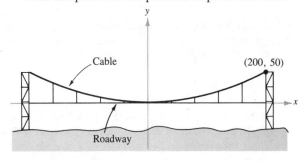

FIGURE FOR 43

44. An earth satellite in a 100-mile-high circular orbit around the earth has a velocity of approximately 17,500 miles per hour. If this velocity is multiplied by $\sqrt{2}$, then the satellite will have the minimum velocity necessary to escape the earth's gravity and it will follow a parabolic path with the center of the earth as the focus (see figure).
(a) Find the escape velocity of the satellite.
(b) Find an equation of its path (assume that the radius of the earth is 4000 miles).

45. The receiver in a parabolic television dish antenna is 3 feet from the vertex and is located at the focus (see figure). Find an equation of a cross section of the reflector. (Assume the dish is directed upward and the vertex is at the origin.)

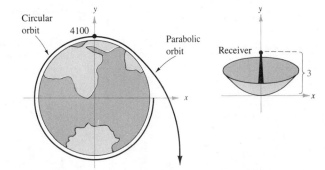

FIGURE FOR 44   ***FIGURE FOR 45***

46. Find the equation of the tangent line to the parabola $y = ax^2$ at $x = x_0$. Prove that the x-intercept of this tangent line is $(x_0/2, 0)$.

In Exercises 47–50, find the equation of the tangent line to the parabola at the given point.

47. $y = \frac{1}{2}x^2$, $(4, 8)$ **48.** $y = \frac{1}{2}x^2$, $\left(-3, \frac{9}{2}\right)$

49. $y = -2x^2$, $(-1, -2)$ **50.** $y = -2x^2$, $(3, -18)$

In Exercises 51–53, find an equation of the path of a projectile thrown horizontally with a velocity of v feet per second at a height of s feet, where the model is

$$y = -\frac{16}{v^2}x^2 + s.$$

51. Water is flowing from a horizontal pipe 48 feet above the ground at a rate of 10 feet per second (see figure). The falling stream of water has the shape of a parabola whose vertex, $(0, 48)$, is at the end of the pipe.
(a) Find the equation of the parabola.
(b) Where does the water hit the ground?

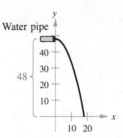

Water pipe

48

FIGURE FOR 51

52. A bomber flying due east at 550 miles per hour at an altitude of 42,000 feet releases a bomb. Neglecting air resistance, determine how far the bomb travels horizontally before striking the ground.

53. A ball is thrown horizontally from the top of a 75-foot tower with a velocity of 32 feet per second.
(a) Find the equation of the parabolic path.
(b) How far does the ball travel horizontally before striking the ground?

7.4 Ellipses

The second type of conic is called an **ellipse,** and is defined as follows.

Definition of an Ellipse

An **ellipse** is the set of all points (x, y) the sum of whose distances from two distinct fixed points (**foci**) is constant. (See Figure 7.34.)

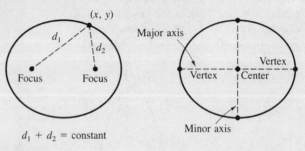

$d_1 + d_2 = $ constant

FIGURE 7.34

Ellipses

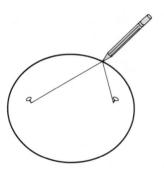

FIGURE 7.35

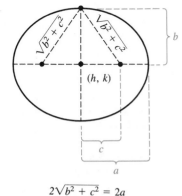

$$2\sqrt{b^2 + c^2} = 2a$$
$$b^2 + c^2 = a^2$$

FIGURE 7.36

The line through the foci intersects the ellipse at two points called the **vertices.** The chord joining the vertices is called the **major axis,** and its midpoint is called the **center** of the ellipse. The chord perpendicular to the major axis at the center is called the **minor axis** of the ellipse.

You can visualize the definition of an ellipse by imagining two thumbtacks placed at the foci, as shown in Figure 7.35. By fastening the ends of a fixed string are fastened to the thumbtacks and the string is drawn taut with a pencil, the path traced by the pencil will be an ellipse.

To derive the standard form of the equation of an ellipse, consider the ellipse in Figure 7.36 with the following points.

Center: (h, k)
Vertices: $(h \pm a, k)$
Foci: $(h \pm c, k)$

The sum of the distances from any point on the ellipse to the two foci is constant. At a vertex, this constant sum is

$$(a + c) + (a - c) = 2a \qquad \textit{Length of major axis}$$

or simply the length of the major axis. Now, if we let (x, y) be *any* point on the ellipse, the sum of the distances between (x, y) and the two foci must also be $2a$. That is,

$$\sqrt{[x - (h - c)]^2 + (y - k)^2} + \sqrt{[x - (h + c)]^2 + (y - k)^2} = 2a$$

which, after expanding and regrouping, reduces to

$$(a^2 - c^2)(x - h)^2 + a^2(y - k)^2 = a^2(a^2 - c^2).$$

Finally, in Figure 7.36, we can see that $b^2 = a^2 - c^2$, which implies that the equation of the ellipse is

$$b^2(x - h)^2 + a^2(y - k)^2 = a^2 b^2$$
$$\frac{(x - h)^2}{a^2} + \frac{(y - k)^2}{b^2} = 1.$$

Had we chosen a vertical major axis, we would have obtained a similar equation. Both results are summarized as follows.

Standard Equation of an Ellipse

The standard form of the equation of an ellipse, with center (h, k) and major and minor axes of lengths $2a$ and $2b$, where $0 < b < a$, is

$$\frac{(x - h)^2}{a^2} + \frac{(y - k)^2}{b^2} = 1 \qquad \textit{Major axis is horizontal}$$

$$\frac{(x - h)^2}{b^2} + \frac{(y - k)^2}{a^2} = 1. \qquad \textit{Major axis is vertical}$$

The foci lie on the major axis, c units from the center, with $c^2 = a^2 - b^2$.

Figure 7.37 shows both the vertical and horizontal orientations for an ellipse.

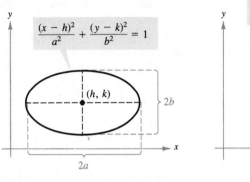

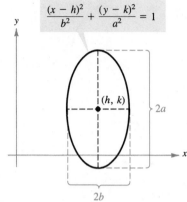

FIGURE 7.37

Standard Equations for Ellipses

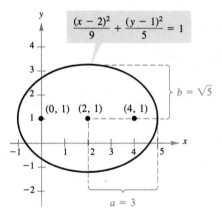

FIGURE 7.38

EXAMPLE 1 *Finding the Standard Equation of an Ellipse*

Find the standard form of the equation of the ellipse having foci at $(0, 1)$ and $(4, 1)$ and with a major axis of length 6, as shown in Figure 7.38.

SOLUTION

Since the foci occur at $(0, 1)$ and $(4, 1)$, the center of the ellipse is $(2, 1)$. This implies that the distance from the center to one of the foci is $c = 2$, and since $2a = 6$ we know that $a = 3$. Now, using $c^2 = a^2 - b^2$, we have

$$b = \sqrt{a^2 - c^2} = \sqrt{9 - 4} = \sqrt{5}.$$

Since the major axis is horizontal, the standard equation is

$$\frac{(x - 2)^2}{9} + \frac{(y - 1)^2}{5} = 1.$$

EXAMPLE 2 *Writing an Equation in Standard Form*

Sketch the graph of the ellipse whose equation is

$$x^2 + 4y^2 + 6x - 8y + 9 = 0.$$

SOLUTION

We begin by writing the given equation in standard form.

$x^2 + 4y^2 + 6x - 8y + 9 = 0$	*Given equation*
$(x^2 + 6x +\quad) + (4y^2 - 8y +\quad) = -9$	*Group terms*
$(x^2 + 6x +\quad) + 4(y^2 - 2y +\quad) = -9$	*Factor 4 out of y-terms*
$(x^2 + 6x + 9) + 4(y^2 - 2y + 1) = -9 + 9 + 4(1)$	*Add 9 and 4 to both sides*
$(x + 3)^2 + 4(y - 1)^2 = 4$	*Completed square form*
$\dfrac{(x + 3)^2}{4} + \dfrac{(y - 1)^2}{1} = 1$	*Standard form*

Now we see that the center occurs at $(h, k) = (-3, 1)$. Since the denominator of the x-term is $a^2 = 2^2$, we locate the endpoints of the major axis 2 units to the right and left of the center. Similarly, since the denominator of the y-term is $b^2 = 1^2$, we locate the endpoints of the minor axis 1 unit up and down from the center. The graph of this ellipse is shown in Figure 7.39.

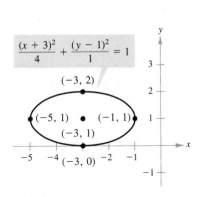

$$\frac{(x + 3)^2}{4} + \frac{(y - 1)^2}{1} = 1$$

FIGURE 7.39

EXAMPLE 3 *Analyzing an Ellipse*

Find the center, vertices, and foci of the ellipse given by

$$4x^2 + y^2 - 8x + 4y - 8 = 0.$$

SOLUTION

By completing the square, we can write the given equation in standard form.

$$4x^2 + y^2 - 8x + 4y - 8 = 0$$
$$4(x^2 - 2x + 1) + (y^2 + 4y + 4) = 8 + 4 + 4$$
$$4(x - 1)^2 + (y + 2)^2 = 16$$
$$\frac{(x - 1)^2}{4} + \frac{(y + 2)^2}{16} = 1$$

Thus, the major axis is vertical, where $h = 1$, $k = -2$, $a = 4$, $b = 2$, and $c = \sqrt{16 - 4} = 2\sqrt{3}$. Therefore, we have the following.

Center: $(1, -2)$ Vertices: $(1, -6)$ Foci: $(1, -2 - 2\sqrt{3})$
                                  $(1, 2)$ $(1, -2 + 2\sqrt{3})$

The graph of the ellipse is shown in Figure 7.40.

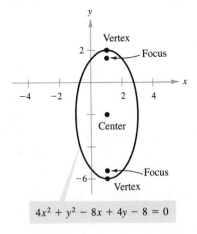

$$4x^2 + y^2 - 8x + 4y - 8 = 0$$

FIGURE 7.40

Remark: If the constant term in the equation in Example 3 had been $F \geq$ 8, then we would have obtained one of the following degenerate cases.

1. Single Point: $\dfrac{(x-1)^2}{4} + \dfrac{(y+2)^2}{16} = 0$ $F = 8$

2. No Solution Points: $\dfrac{(x-1)^2}{4} + \dfrac{(y+2)^2}{16} < 0$ $F > 8$

Ellipses have many practical and aesthetic uses. For instance, machine gears, supporting arches, and acoustical designs often involve elliptical shapes. The orbits of satellites and planets are also ellipses. In Example 4 we investigate the elliptical orbit of the moon about the earth.

EXAMPLE 4 **An Application Involving an Elliptical Orbit**

The moon travels about the earth in an elliptical orbit with the earth at one focus, as shown in Figure 7.41. The major and minor axes of the orbit have lengths of 768,806 kilometers and 767,746 kilometers, respectively. Find the greatest and least distances (the apogee and perigee) from the earth's center to the moon's center.

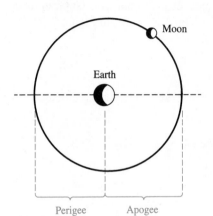

Earth

Perigee   Apogee

FIGURE 7.41

SOLUTION

Since $2a = 768{,}806$ and $2b = 767{,}746$, we have $a = 384{,}403$, $b = 383{,}873$, and

$$c = \sqrt{a^2 - b^2} \approx 20{,}179.$$

Therefore, the greatest distance between the center of the earth and the center of the moon is

$$a + c \approx 404{,}582 \text{ km}$$

and the least distance is

$$a - c \approx 364{,}224 \text{ km}.$$

Eccentricity

One of the reasons it was difficult for early astronomers to detect that the orbits of the planets are ellipses is that the foci of the planetary orbits are relatively close to their centers, thus making the orbits nearly circular. To measure the ovalness of an ellipse, we use the concept of **eccentricity.**

Ellipses

Definition of Eccentricity

The **eccentricity** e of an ellipse is given by the ratio

$$e = \frac{c}{a}.$$

To see how this ratio is used to describe the shape of an ellipse, note that since the foci of an ellipse are located along the major axis between the vertices and the center, it follows that

$$0 < c < a.$$

For an ellipse that is nearly circular, the foci are close to the center and the ratio c/a is small, as shown in Figure 7.42. On the other hand, for an elongated ellipse, the foci are close to the vertices, and the ratio c/a is close to 1.

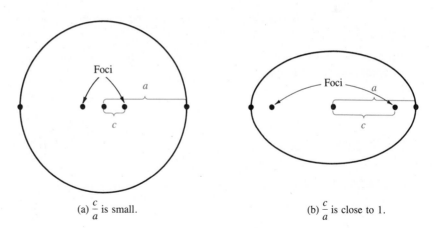

FIGURE 7.42

(a) $\dfrac{c}{a}$ is small.

(b) $\dfrac{c}{a}$ is close to 1.

Remark: Note that $0 < e < 1$ for every ellipse.

The orbit of the moon has an eccentricity of $e = 0.0549$, and the eccentricities of the nine planetary orbits are as follows.

Mercury: $e = 0.2056$	Saturn: $e = 0.0543$
Venus: $e = 0.0068$	Uranus: $e = 0.0460$
Earth: $e = 0.0167$	Neptune: $e = 0.0082$
Mars: $e = 0.0934$	Pluto: $e = 0.2481$
Jupiter: $e = 0.0484$	

WARM UP

Sketch the graph of each equation.

1. $x^2 = 9y$ **2.** $y^2 = 9x$

3. $y^2 = -9x$ **4.** $x^2 = -9y$

In each case find the unknown in the equation $c^2 = a^2 - b^2$. (Assume a, b, and c are positive.)

5. $a = 13$, $b = 5$ **6.** $a = \sqrt{10}$, $c = 3$

7. $b = 6$, $c = 8$ **8.** $a = 7$, $b = 5$

Simplify the compound fractions.

9. $\dfrac{x^2}{1/4} + \dfrac{y^2}{1/3} = 1$ **10.** $\dfrac{(x-1)^2}{4/9} + \dfrac{(y+2)^2}{1/9} = 1$

EXERCISES 7.4

In Exercises 1–6, match the equation with the correct graph. [The graphs are labeled (a)–(f).]

1. $\dfrac{x^2}{1} + \dfrac{y^2}{9} = 1$

2. $\dfrac{x^2}{9} + \dfrac{y^2}{1} = 1$

3. $\dfrac{x^2}{9} + \dfrac{y^2}{4} = 1$

4. $\dfrac{x^2}{9} + \dfrac{y^2}{9} = 1$

5. $\dfrac{(x-2)^2}{16} + \dfrac{(y+1)^2}{4} = 1$

6. $\dfrac{(x+2)^2}{4} + \dfrac{(y+2)^2}{25} = 1$

(c)

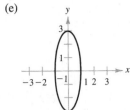

(d)

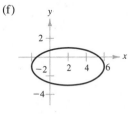

(e)   (f)

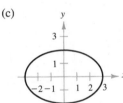

In Exercises 7–26, find the center, foci, vertices, and eccentricity of the ellipse and sketch its graph.

7. $\dfrac{x^2}{25} + \dfrac{y^2}{16} = 1$ **8.** $\dfrac{x^2}{144} + \dfrac{y^2}{169} = 1$

9. $\dfrac{x^2}{16} + \dfrac{y^2}{25} = 1$ **10.** $\dfrac{x^2}{169} + \dfrac{y^2}{144} = 1$

11. $\dfrac{x^2}{9} + \dfrac{y^2}{5} = 1$ **12.** $\dfrac{x^2}{28} + \dfrac{y^2}{64} = 1$

(a)   (b)

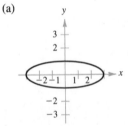

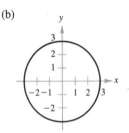

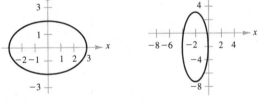

13. $x^2 + 4y^2 = 4$

14. $5x^2 + 3y^2 = 15$

15. $3x^2 + 2y^2 = 6$

16. $5x^2 + 7y^2 = 70$

17. $4x^2 + y^2 = 1$

18. $16x^2 + 25y^2 = 1$

19. $\dfrac{(x-1)^2}{9} + \dfrac{(y-5)^2}{25} = 1$ **20.** $(x+2)^2 + 4(y+4)^2 = 1$

21. $9x^2 + 4y^2 + 36x - 24y + 36 = 0$

22. $9x^2 + 4y^2 - 36x + 8y + 31 = 0$

23. $16x^2 + 25y^2 - 32x + 50y + 16 = 0$

24. $9x^2 + 25y^2 - 36x - 50y + 61 = 0$

25. $12x^2 + 20y^2 - 12x + 40y - 37 = 0$

26. $36x^2 + 9y^2 + 48x - 36y + 43 = 0$

In Exercises 27–42, find an equation of the specified ellipse.

27. Vertices: $(\pm 6, 0)$
Foci: $(\pm 5, 0)$

28. Vertices: $(\pm 2, 0)$
Minor axis of length 3

29. Vertices: $(\pm 5, 0)$
Foci: $(\pm 2, 0)$

30. Vertices: $(0, \pm 8)$
Foci: $(0, \pm 4)$

31. Vertices: $(0, \pm 2)$
Minor axis of length 2

32. Foci: $(\pm 2, 0)$
Major axis of length 8

33. Vertices: $(0, 2)$, $(4, 2)$
Minor axis of length 2

34. Foci: $(0, 0)$, $(4, 0)$
Major axis of length 8

35. Foci: $(0, 0)$, $(0, 8)$
Major axis of length 16

36. Vertices: $(2, \frac{1}{2})$, $(2, -\frac{5}{2})$
Minor axis of length 2

37. Vertices: $(0, \pm 5)$
Solution point: $(4, 2)$

38. Vertices: $(5, 0)$, $(5, 12)$
Endpoints of minor axis: $(0, 6)$, $(10, 6)$

39. Center: $(3, 2)$, $a = 3c$
Foci: $(1, 2)$, $(5, 2)$

40. Center: $(0, 4)$, $a = 2c$
Vertices: $(\pm 4, 4)$

41. Vertices: $(\pm 5, 0)$
Eccentricity: $\frac{3}{5}$

42. Vertices: $(0, \pm 8)$
Eccentricity: $\frac{1}{2}$

43. A fireplace arch is to be constructed in the shape of a semiellipse. The opening is to have a height of 2 feet at the center and a width of 5 feet along the base (see figure). The contractor will first draw the form of the ellipse by the method shown in Figure 7.35. Where should the tacks be placed and how long should the piece of string be?

44. A line segment through a focus with endpoints on the ellipse and perpendicular to the major axis is called a **latus rectum** of the ellipse (see figure). Knowing the lengths of the latus recta is helpful in sketching an ellipse because it yields four additional points on the curve. Show that the length of each latus rectum is $2b^2/a$.

45. Sketch the graph of each ellipse, making use of the endpoints of the latus recta (see Exercise 44).

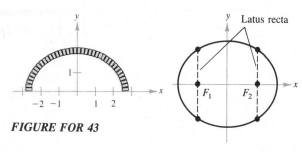

FIGURE FOR 43

FIGURE FOR 44

(a) $\dfrac{x^2}{4} + \dfrac{y^2}{1} = 1$

(b) $6x^2 + 4y^2 = 1$

(c) $5x^2 + 3y^2 = 15$

46. Show that the eccentricity of the ellipse

$$\frac{x^2}{a^2} + \frac{y^2}{b^2} = 1$$

is identical to the eccentricity of

$$\frac{(tx)^2}{a^2} + \frac{(ty)^2}{b^2} = 1$$

for any nonzero real number t.

47. The earth moves in an elliptical orbit with the sun at one of the foci. The length of half the major axis is 93 million miles, and the eccentricity is 0.017. Find the least and greatest distances between the earth and the sun.

48. If the distances to the apogee and the perigee of an elliptical orbit of an earth satellite are measured from the center of the earth, show that the eccentricity of the orbit is given by $e = (A - P)/(A + P)$, where A and P are the apogee and perigee distances, respectively.

49. *Sputnik I*, launched by the Russians in October 1957, had 583 miles and 132 miles above the earth's surface as the highest and lowest points of its elliptical orbit. What was the eccentricity of this orbit?

50. On November 26, 1963, the United States launched *Explorer 18*. Its low point over the surface of the earth was 119 miles, and its high point was 122,000 miles from the surface of the earth.
(a) Find the eccentricity of its elliptical orbit.
(b) Find an equation that describes its orbit.

51. Show that the equation of an ellipse can be written

$$\frac{(x-h)^2}{a^2} + \frac{(y-k)^2}{a^2(1-e^2)} = 1.$$

Note that as e approaches zero, with a remaining fixed, the ellipse approaches a circle of radius a.

Lines in the Plane and Conics

7.5 Hyperbolas

The definition of a hyperbola parallels that of an ellipse. The distinction is that for an ellipse the *sum* of the distances between the foci and a point on the ellipse is fixed, while for a hyperbola the *difference* of these distances is fixed.

Definition of a Hyperbola

A **hyperbola** is the set of all points (x, y) the difference of whose distances from two distinct fixed points (foci) is constant. (See Figure 7.43.)

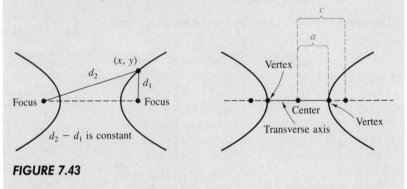

FIGURE 7.43

Every hyperbola has two disconnected parts called **branches.** The line through the two foci intersects a hyperbola at two points, called the **vertices.** The line segment connecting the vertices is called the **transverse axis,** and the midpoint of the transverse axis is called the **center** of the hyperbola.

The development of the standard form of the equation of a hyperbola is similar to that of an ellipse, and we list the following result without proof.

Standard Equation of a Hyperbola

The standard form of the equation of a hyperbola with center at (h, k) is

$$\frac{(x - h)^2}{a^2} - \frac{(y - k)^2}{b^2} = 1 \qquad \text{\textit{Transverse axis is horizontal}}$$

$$\frac{(y - k)^2}{a^2} - \frac{(x - h)^2}{b^2} = 1. \qquad \text{\textit{Transverse axis is vertical}}$$

The vertices are a units from the center, and the foci are c units from the center. Moreover, $b^2 = c^2 - a^2$.

Figure 7.44 shows both the vertical and horizontal orientations for a hyperbola.

Hyperbolas

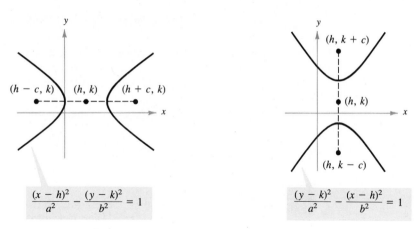

FIGURE 7.44   Standard Equations for Hyperbolas

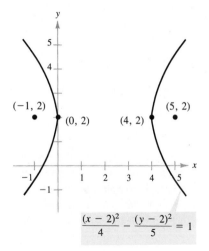

$$\frac{(x-2)^2}{4} - \frac{(y-2)^2}{5} = 1$$

FIGURE 7.45

EXAMPLE 1   *Finding the Standard Equation of a Hyperbola*

Find the standard form of the equation of the hyperbola with foci at $(-1, 2)$ and $(5, 2)$ and vertices at $(0, 2)$ and $(4, 2)$.

SOLUTION

By the Midpoint Formula, the center of the hyperbola occurs at the point $(2, 2)$. Furthermore, $c = 3$ and $a = 2$, and it follows that

$$b^2 = 3^2 - 2^2 = 9 - 4 = 5.$$

Thus, the equation of the hyperbola is

$$\frac{(x-2)^2}{4} - \frac{(y-2)^2}{5} = 1.$$

Figure 7.45 shows the graph of the hyperbola.

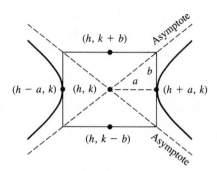

FIGURE 7.46

An important aid in sketching the graph of a hyperbola is the determination of its **asymptotes,** as shown in Figure 7.46. Each hyperbola has two asymptotes that intersect at the center of the hyperbola. The asymptotes pass through the vertices of a rectangle of dimension $2a$ by $2b$, with its center at (h, k). The line segment of length $2b$, joining $(h, k + b)$ and $(h, k - b)$, is referred to as the **conjugate axis** of the hyperbola. The following result identifies the equations for the asymptotes.

Lines in the Plane and Conics

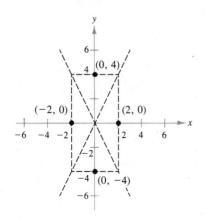

FIGURE 7.47

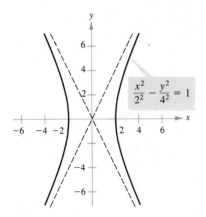

FIGURE 7.48

Asymptotes of a Hyperbola

For a *horizontal* transverse axis the equations of the asymptotes are

$$y = k + \frac{b}{a}(x - h) \quad \text{and} \quad y = k - \frac{b}{a}(x - h).$$

For a *vertical* transverse axis the equations of the asymptotes are

$$y = k + \frac{a}{b}(x - h) \quad \text{and} \quad y = k - \frac{a}{b}(x - h).$$

EXAMPLE 2 *Using Asymptotes to Sketch a Hyperbola*

Sketch the hyperbola whose equation is

$$4x^2 - y^2 = 16.$$

SOLUTION

Rewriting this equation in standard form, we have

$$\frac{4x^2}{16} - \frac{y^2}{16} = \frac{16}{16}$$

$$\frac{x^2}{2^2} - \frac{y^2}{4^2} = 1. \qquad \textit{Standard form}$$

From this, we conclude that the transverse axis is horizontal and the vertices occur at $(-2, 0)$ and $(2, 0)$. Moreover, the ends of the conjugate axis occur at $(0, -4)$ and $(0, 4)$, and we are able to sketch the rectangle shown in Figure 7.47. Finally, by drawing the asymptotes through the corners of this rectangle, we complete the sketch, as shown in Figure 7.48.

EXAMPLE 3 *Finding the Asymptotes of a Hyperbola*

Sketch the hyperbola given by $4x^2 - 3y^2 + 8x + 16 = 0$ and find the equations of its asymptotes.

SOLUTION

$$4x^2 - 3y^2 + 8x + 16 = 0 \qquad \textit{Given equation}$$
$$4(x^2 + 2x) - 3y^2 = -16$$
$$-4(x^2 + 2x + 1) + 3y^2 = 16 - 4$$
$$-4(x + 1)^2 + 3y^2 = 12$$
$$\frac{y^2}{4} - \frac{(x + 1)^2}{3} = 1 \qquad \textit{Standard form}$$

Hyperbolas

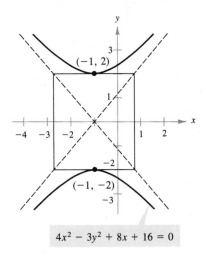

$$4x^2 - 3y^2 + 8x + 16 = 0$$

FIGURE 7.49

From this equation we conclude that the hyperbola is centered at $(-1, 0)$, has vertices at $(-1, 2)$ and $(-1, -2)$, and the ends of the conjugate axis occur at $(-1 - \sqrt{3}, 0)$ and $(-1 + \sqrt{3}, 0)$. To sketch the graph of the hyperbola, we draw a rectangle through these four points. The asymptotes are the lines passing through the corners of the rectangle, as shown in Figure 7.49. Finally using $a = 2$ and $b = \sqrt{3}$, we conclude that the equations of the asymptotes are

$$y = \frac{2}{\sqrt{3}}(x + 1)$$

and

$$y = -\frac{2}{\sqrt{3}}(x + 1).$$

Remark: If the constant term F in the equation in Example 3 had been $F = -4$ instead of 16, then we would have obtained the following degenerate case.

Two Intersecting Lines: $\quad \dfrac{y^2}{4} - \dfrac{(x + 1)^2}{3} = 0$

EXAMPLE 4 *Using Asymptotes to Find the Standard Equation*

Find the standard form of the equation of the hyperbola having vertices at $(3, -5)$ and $(3, 1)$ and with asymptotes $y = 2x - 8$ and $y = -2x + 4$, as shown in Figure 7.50.

SOLUTION

By the Midpoint Formula, the center of the hyperbola is at $(3, -2)$. Furthermore, the hyperbola has a vertical transverse axis with $a = 3$. From the given equations, we determine the slopes of the asymptotes to be

$$m_1 = 2 = \frac{a}{b} \quad \text{and} \quad m_2 = -2 = -\frac{a}{b}$$

and since $a = 3$, we conclude that $b = 3/2$. Thus, the standard equation is

$$\frac{(y + 2)^2}{9} - \frac{(x - 3)^2}{9/4} = 1.$$

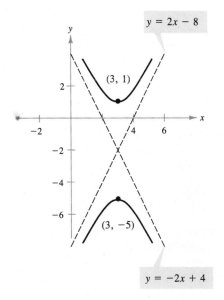

FIGURE 7.50

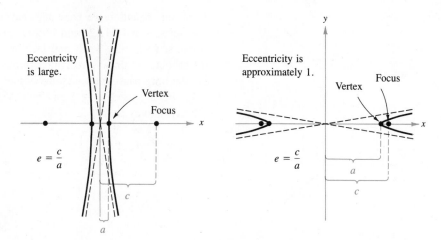

FIGURE 7.51

As with ellipses, the **eccentricity** of a hyperbola is $e = c/a$, and because $c > a$ it follows that $e > 1$. If the eccentricity is large, then the branches of the hyperbola are nearly flat. If the eccentricity is close to 1, then the branches of the hyperbola are more pointed, as shown in Figure 7.51.

The following application was developed during World War II. It shows how the properties of hyperbolas can be used in radar and other detection systems.

EXAMPLE 5 *An Application Involving Hyperbolas*

Two microphones, 1 mile apart, record an explosion. Microphone A received the sound 2 seconds before microphone B. Where did the explosion come from?

SOLUTION

Assuming sound travels at 1100 feet per second, we know that the explosion took place 2200 feet further from B than from A, as shown in Figure 7.52. The locus of all points that are 2200 feet closer to A than to B is one branch of the hyperbola $(x^2/a^2) - (y^2/b^2) = 1$, where

$$c = \frac{5280}{2} = 2640 \quad \text{and} \quad a = \frac{2200}{2} = 1100.$$

Thus, $b^2 = c^2 - a^2 = 5,759,600$ and we conclude that the explosion occurred somewhere on the right branch of the hyperbola given by

$$\frac{x^2}{1,210,000} - \frac{y^2}{5,759,600} = 1.$$

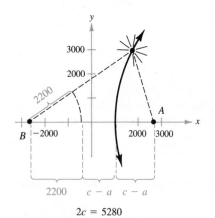

$$2c = 5280$$
$$2200 + 2(c - a) = 5280$$

FIGURE 7.52

(b) For the general equation

$$4x^2 - y^2 + 8x - 6y + 4 = 0$$

we have $AC = 4(-1) < 0$. Thus, the graph is a hyperbola.

(c) For the general equation

$$2x^2 + 4y^2 - 4x + 12y = 0$$

we have $AC = (2)(4) > 0$. Thus, the graph is an ellipse. ▬

WARM UP

Find the distance between the two points.

1. (4, 1), (10, 6)

2. (−1, 5), (3, −2)

Sketch the graph of the lines on the same set of coordinate axes.

3. $y = \pm\frac{1}{2}x$

4. $y = 3 \pm \frac{1}{2}x$

5. $y = 3 \pm \frac{1}{2}(x - 4)$

6. $y = \pm\frac{1}{2}(x - 4)$

Identify the graph of each equation.

7. $x^2 + 4y = 4$

8. $x^2 + 4y^2 = 4$

9. $4x^2 + 4y^2 = 4$

10. $x + 4y^2 = 4$

EXERCISES 7.5

In Exercises 1–6, match the equation with the correct graph.
[The graphs are labeled (a)–(f).]

1. $\dfrac{x^2}{9} - \dfrac{y^2}{4} = 1$

2. $\dfrac{y^2}{9} - \dfrac{x^2}{4} = 1$

3. $\dfrac{y^2}{1} - \dfrac{x^2}{16} = 1$

4. $\dfrac{y^2}{16} - \dfrac{x^2}{1} = 1$

5. $\dfrac{(x - 2)^2}{9} - \dfrac{y^2}{4} = 1$

6. $\dfrac{(x + 1)^2}{16} - \dfrac{(y - 3)^2}{9} = 1$

(c)

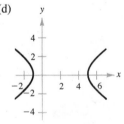

(d)

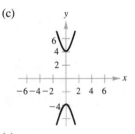

(a)

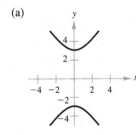

(b)

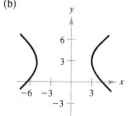

(e)

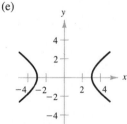

(f)

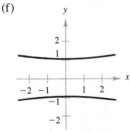

In Exercises 7–26, find the center, vertices, and foci of the hyperbola and sketch its graph, using asymptotes as an aid.

7. $x^2 - y^2 = 1$

8. $\dfrac{x^2}{9} - \dfrac{y^2}{16} = 1$

9. $y^2 - \dfrac{x^2}{4} = 1$

10. $\dfrac{y^2}{9} - x^2 = 1$

11. $\dfrac{y^2}{25} - \dfrac{x^2}{144} = 1$

12. $\dfrac{x^2}{36} - \dfrac{y^2}{4} = 1$

13. $2x^2 - 3y^2 = 6$

14. $3y^2 = 5x^2 + 15$

15. $5y^2 = 4x^2 + 20$

16. $7x^2 - 3y^2 = 21$

17. $\dfrac{(x-1)^2}{4} - \dfrac{(y+2)^2}{1} = 1$

18. $\dfrac{(x+1)^2}{144} - \dfrac{(y-4)^2}{25} = 1$

19. $(y+6)^2 - (x-2)^2 = 1$

20. $\dfrac{(y-1)^2}{1/4} - \dfrac{(x+3)^2}{1/9} = 1$

21. $9x^2 - y^2 - 36x - 6y + 18 = 0$

22. $x^2 - 9y^2 + 36y - 72 = 0$

23. $9y^2 - x^2 + 2x + 54y + 62 = 0$

24. $16y^2 - x^2 + 2x + 64y + 63 = 0$

25. $x^2 - 9y^2 + 2x - 54y - 80 = 0$

26. $9x^2 - y^2 + 54x + 10y + 55 = 0$

In Exercises 27–40, find an equation for the specified hyperbola.

27. Vertices: $(0, \pm 2)$
Foci: $(0, \pm 4)$

28. Vertices: $(\pm 3, 0)$
Foci: $(\pm 5, 0)$

29. Vertices: $(\pm 1, 0)$
Asymptotes: $y = \pm 3x$

30. Vertices: $(0, \pm 3)$
Asymptotes: $y = \pm 3x$

31. Vertices: $(2, 0), (6, 0)$
Foci: $(0, 0), (8, 0)$

32. Vertices: $(2, 3), (2, -3)$
Foci: $(2, 5), (2, -5)$

33. Vertices: $(4, 1), (4, 9)$
Foci: $(4, 0), (4, 10)$

34. Vertices: $(-2, 1), (2, 1)$
Foci: $(-3, 1), (3, 1)$

35. Vertices: $(0, \pm 3)$
Solution point: $(-2, 5)$

36. Vertices: $(\pm 2, 0)$
Solution point: $(3, \sqrt{3})$

37. Vertices: $(-2, 1), (2, 1)$
Solution point: $(4, 3)$

38. Vertices: $(2, 3), (2, -3)$
Solution point: $(0, 5)$

39. Vertices: $(0, 2), (6, 2)$
Asymptotes: $y = \frac{2}{3}x$
$y = 4 - \frac{2}{3}x$

40. Vertices: $(3, 0), (3, 4)$
Asymptotes: $y = \frac{2}{3}x$
$y = 4 - \frac{2}{3}x$

41. Three listening stations located at $(4400, 0)$, $(4400, 1100)$, and $(-4400, 0)$ hear an explosion. If the latter two stations heard the sound 1 second and 5 seconds after the first, respectively, where did the explosion occur? Assume that the coordinate system is measured in feet and that sound travels at the rate of 1100 feet per second.

42. Use the definition of a hyperbola to derive its standard form. (The development of the equation is similar to the development of the standard form for the equation of the ellipse given in Section 11.2.)

In Exercises 43–50, classify the graph of each equation as a circle, a parabola, an ellipse, or a hyperbola. (There are no degenerate cases.)

43. $x^2 + y^2 - 6x + 4y + 9 = 0$

44. $x^2 + 4y^2 - 6x + 16y + 21 = 0$

45. $4x^2 - y^2 - 4x - 3 = 0$

46. $y^2 - 4y - 4x = 0$

47. $4x^2 + 3y^2 + 8x - 24y + 51 = 0$

48. $4y^2 - 2x^2 - 4y - 8x - 15 = 0$

49. $25x^2 - 10x - 200y - 119 = 0$

50. $4x^2 + 4y^2 - 16y + 15 = 0$

Lines in the Plane and Conics

7.6 Rotation and the General Second-Degree Equation

In the previous three sections we have shown that the equation of a conic with axes parallel to one of the coordinate axes has a standard form that can be written in the general form

$$Ax^2 + Cy^2 + Dx + Ey + F = 0. \qquad \textit{Horizontal or vertical axes}$$

In this section we investigate the equations of conics whose axes are rotated so that they are not parallel to either the x-axis or the y-axis. The general equation for such conics contains an *xy-term*.

$$Ax^2 + Bxy + Cy^2 + Dx + Ey + F = 0 \qquad \textit{Equation in xy-plane}$$

To eliminate this xy-term, we use a procedure called **rotation of axes.** Our objective is to rotate the x- and y-axes until they are parallel to the axes of the conic. We denote the rotated axes as the x'-axis and the y'-axis, as shown in Figure 7.54. Having accomplished this, the equation of the conic in the in the new $x'y'$-plane will have the form

$$A'(x')^2 + C'(y')^2 + D'x' + E'y' + F' = 0. \qquad \textit{Equation in x'y'-plane}$$

Since this equation has no xy-term, we can obtain a standard form by completing the square.

The following theorem identifies how much to rotate the axes to eliminate the xy-term and also the equations for determining the new coefficients A', C', D', E', and F'.

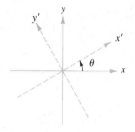

Rotated: x'-axis
y'-axis

FIGURE 7.54

Rotation of Axes to Eliminate an xy-Term

The general second-degree equation

$$Ax^2 + Bxy + Cy^2 + Dx + Ey + F = 0$$

can be rewritten as

$$A'(x')^2 + C'(y')^2 + D'x' + E'y' + F' = 0$$

by rotating the coordinate axes through an angle θ, where

$$\cot 2\theta = \frac{A - C}{B}.$$

The coefficients of the new equation are obtained by making the substitutions

$$x = x' \cos \theta - y' \sin \theta$$
$$y = x' \sin \theta + y' \cos \theta.$$

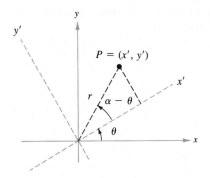

Rotated: $x' = r \cos(\alpha - \theta)$
$y' = r \sin(\alpha - \theta)$

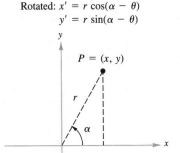

Original: $x = r \cos \alpha$
$y = r \sin \alpha$

FIGURE 7.55

PROOF

We need to discover how the coordinates in the xy-system are related to the coordinates in the $x'y'$-system. To do this, we choose a point $P = (x, y)$ in the original system and attempt to find its coordinates (x', y') in the rotated system. In either system the distance r between the point P and the origin is the same; thus, the equations for x, y, x', and y' are those given in Figure 7.55. Using the formulas for the sine and cosine of the difference of two angles, we have the following.

$$\begin{aligned}
x' &= r \cos(\alpha - \theta) \\
&= r(\cos \alpha \cos \theta + \sin \alpha \sin \theta) \\
&= r \cos \alpha \cos \theta + r \sin \alpha \sin \theta \\
&= x \cos \theta + y \sin \theta \\
y' &= r \sin(\alpha - \theta) \\
&= r(\sin \alpha \cos \theta - \cos \alpha \sin \theta) \\
&= r \sin \alpha \cos \theta - r \cos \alpha \sin \theta \\
&= y \cos \theta - x \sin \theta
\end{aligned}$$

Solving this system for x and y yields

$$\begin{aligned}
x &= x' \cos \theta - y' \sin \theta \\
y &= x' \sin \theta + y' \cos \theta.
\end{aligned}$$

Finally, by substituting these values for x and y into the original equation and collecting terms, we obtain

$$\begin{aligned}
A' &= A \cos^2 \theta + B \cos \theta \sin \theta + C \sin^2 \theta \\
C' &= A \sin^2 \theta - B \cos \theta \sin \theta + C \cos^2 \theta \\
D' &= D \cos \theta + E \sin \theta \\
E' &= -D \sin \theta + E \cos \theta \\
F' &= F.
\end{aligned}$$

Now, in order to eliminate the $x'y'$-term, we must select θ so that $B' = 0$, as follows.

$$\begin{aligned}
B' &= 2(C - A) \sin \theta \cos \theta + B(\cos^2 \theta - \sin^2 \theta) \\
&= (C - A) \sin 2\theta + B \cos 2\theta \\
&= B(\sin 2\theta)\left(\frac{C - A}{B} + \cot 2\theta\right) = 0, \quad \sin 2\theta \neq 0
\end{aligned}$$

If $B = 0$, no rotation is necessary since the xy-term is not present in the original equation. If $B \neq 0$, then the only way to make $B' = 0$ is to let

$$\cot 2\theta = \frac{A - C}{B}, \quad B \neq 0.$$

Thus, we have established the desired results.

EXAMPLE 1 *Rotation of Axes for a Hyperbola*

Write the equation $xy - 1 = 0$ in standard form.

SOLUTION

Since $A = 0$, $B = 1$, and $C = 0$, we have

$$\cot 2\theta = \frac{A - C}{B} = 0 \implies 2\theta = \frac{\pi}{2} \implies \theta = \frac{\pi}{4}$$

which implies that

$$x = x' \cos \frac{\pi}{4} - y' \sin \frac{\pi}{4}$$

$$= x'\left(\frac{\sqrt{2}}{2}\right) - y'\left(\frac{\sqrt{2}}{2}\right) = \frac{x' - y'}{\sqrt{2}}$$

and

$$y = x' \sin \frac{\pi}{4} + y' \cos \frac{\pi}{4}$$

$$= x'\left(\frac{\sqrt{2}}{2}\right) + y'\left(\frac{\sqrt{2}}{2}\right) = \frac{x' + y'}{\sqrt{2}}.$$

The equation in the $x'y'$-system is obtained by substituting these expressions into the equation $xy - 1 = 0$.

$$\left(\frac{x' - y'}{\sqrt{2}}\right)\left(\frac{x' + y'}{\sqrt{2}}\right) - 1 = 0$$

$$\frac{(x')^2 - (y')^2}{2} - 1 = 0$$

$$\frac{(x')^2}{(\sqrt{2})^2} - \frac{(y')^2}{(\sqrt{2})^2} = 1 \qquad \text{\textit{Standard form}}$$

This is the equation of a hyperbola centered at the origin with vertices at $(\pm\sqrt{2}, 0)$ in the $x'y'$-system, as shown in Figure 7.56. To find the coordinates of the vertices in the xy-system, we substitute the coordinates $(\pm\sqrt{2}, 0)$ into the equations

$$x = \frac{x' - y'}{\sqrt{2}} \qquad \text{and} \qquad y = \frac{x' + y'}{\sqrt{2}}.$$

This substitution yields the vertices $(1, 1)$ and $(-1, -1)$ in the xy-system. Note also that the asymptotes of the hyperbola have equations $y' = \pm x'$, which correspond to the original x- and y-axes.

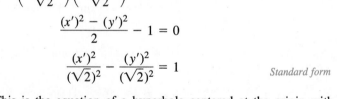

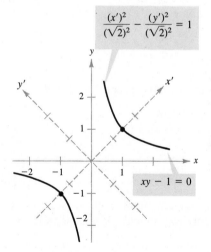

Vertices:
In $x'y'$-system: $(\sqrt{2}, 0)$, $(-\sqrt{2}, 0)$
In xy-system: $(1, 1)$, $(-1, -1)$

FIGURE 7.56

EXAMPLE 2 Rotation of Axes for an Ellipse

Sketch the graph of the equation

$$7x^2 - 6\sqrt{3}xy + 13y^2 - 16 = 0.$$

SOLUTION

Since $A = 7$, $B = -6\sqrt{3}$, and $C = 13$, we have

$$\cot 2\theta = \frac{A - C}{B} = \frac{7 - 13}{-6\sqrt{3}} = \frac{1}{\sqrt{3}} \implies \theta = \frac{\pi}{6}.$$

Therefore, the equation in the $x'y'$-system is obtained by making the substitutions

$$x = x' \cos \frac{\pi}{6} - y' \sin \frac{\pi}{6}$$

$$= x'\left(\frac{\sqrt{3}}{2}\right) - y'\left(\frac{1}{2}\right)$$

$$= \frac{\sqrt{3}x' - y'}{2}$$

and

$$y = x' \sin \frac{\pi}{6} + y' \cos \frac{\pi}{6}$$

$$= x'\left(\frac{1}{2}\right) + y'\left(\frac{\sqrt{3}}{2}\right)$$

$$= \frac{x' + \sqrt{3}y'}{2}$$

into the original equation. Thus, we have

$$7x^2 - 6\sqrt{3}xy + 13y^2 - 16 = 0$$

$$7\left(\frac{\sqrt{3}x' - y'}{2}\right)^2 - 6\sqrt{3}\left(\frac{\sqrt{3}x' - y'}{2}\right)\left(\frac{x' + \sqrt{3}y'}{2}\right) + 13\left(\frac{x' + \sqrt{3}y'}{2}\right)^2 - 16 = 0$$

which simplifies to

$$4(x')^2 + 16(y')^2 - 16 = 0$$

$$\frac{(x')^2}{4} + \frac{(y')^2}{1} = 1. \qquad \text{_Standard form}$$

This is the equation of an ellipse centered at the origin with vertices at $(\pm 2, 0)$ in the $x'y'$-system, as shown in Figure 7.57.

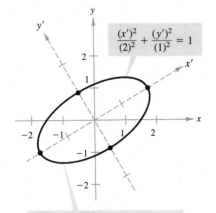

$$\frac{(x')^2}{(2)^2} + \frac{(y')^2}{(1)^2} = 1$$

$$7x^2 - 6\sqrt{3}xy + 13y^2 - 16 = 0$$

Vertices:
In $x'y'$-system: $(\pm 2, 0)$, $(0, \pm 1)$
In xy-system: $(\pm\sqrt{3}, \pm 1)$, $\left(\pm\frac{1}{2}, \mp\frac{\sqrt{3}}{2}\right)$

FIGURE 7.57

Lines in the Plane and Conics

Remark: Remember that the substitutions

$$x = x' \cos \theta - y' \sin \theta \qquad \text{and} \qquad y = x' \cos \theta + y' \sin \theta$$

were developed to eliminate the $x'y'$-term in the rotated system. You can use this as a check on your work. In other words, if your final equation contains an $x'y'$-term, you know that you made a mistake.

In constructing Examples 1 and 2 we carefully chose the equations so that θ would turn out to be one of the common angles 30°, 45°, and so forth. Of course, many second-degree equations do not yield such common solutions to the equation

$$\cot 2\theta = \frac{A - C}{B}.$$

Example 3 illustrates such a case.

EXAMPLE 3   *Rotation of Axes for a Parabola*

Sketch the graph of $x^2 - 4xy + 4y^2 + 5\sqrt{5}y + 1 = 0$.

SOLUTION

Since $A = 1$, $B = -4$, and $C = 4$, we have

$$\cot 2\theta = \frac{A - C}{B} = \frac{1 - 4}{-4} = \frac{3}{4}.$$

Using the identity $\cot 2\theta = (\cot^2 \theta - 1)/(2 \cot \theta)$ produces

$$\cot 2\theta = \frac{3}{4} = \frac{\cot^2 \theta - 1}{2 \cot \theta}$$

from which we obtain the equation

$$4 \cot^2 \theta - 4 = 6 \cot \theta$$
$$4 \cot^2 \theta - 6 \cot \theta - 4 = 0$$
$$(2 \cot \theta - 4)(2 \cot \theta + 1) = 0.$$

Considering $0 < \theta < \pi/2$, we have $2 \cot \theta = 4$. Thus,

$$\cot \theta = 2 \;\; \Longrightarrow \;\; \theta \approx 26.6°.$$

From the triangle in Figure 7.58 we obtain $\sin \theta = 1/\sqrt{5}$ and $\cos \theta = 2/\sqrt{5}$. Consequently, we use the substitutions

$$x = x' \cos \theta - y' \sin \theta = x'\left(\frac{2}{\sqrt{5}}\right) - y'\left(\frac{1}{\sqrt{5}}\right) = \frac{2x' - y'}{\sqrt{5}}$$

$$y = x' \sin \theta + y' \cos \theta = x'\left(\frac{1}{\sqrt{5}}\right) + y'\left(\frac{2}{\sqrt{5}}\right) = \frac{x' + 2y'}{\sqrt{5}}.$$

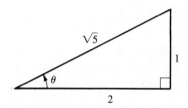

FIGURE 7.58

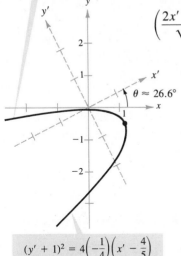

$x^2 - 4xy + 4y^2 + 5\sqrt{5}y + 1 = 0$

$(y' + 1)^2 = 4\left(-\dfrac{1}{4}\right)\left(x' - \dfrac{4}{5}\right)$

Vertex:

In $x'y'$-system: $\left(\dfrac{4}{5}, -1\right)$

In xy-system: $\left(\dfrac{13}{5\sqrt{5}}, -\dfrac{6}{5\sqrt{5}}\right)$

FIGURE 7.59

Substituting these expressions into the original equation, we have

$$x^2 - 4xy + 4y^2 + 5\sqrt{5}y + 1 = 0$$

$$\left(\frac{2x' - y'}{\sqrt{5}}\right)^2 - 4\left(\frac{2x' - y'}{\sqrt{5}}\right)\left(\frac{x' + 2y'}{\sqrt{5}}\right) + 4\left(\frac{x' + 2y'}{\sqrt{5}}\right)^2 + 5\sqrt{5}\left(\frac{x' + 2y'}{\sqrt{5}}\right) + 1 = 0$$

which simplifies to

$$5(y')^2 + 5x' + 10y' + 1 = 0$$

$$5(y' + 1)^2 = -5x' + 4 \qquad \textit{Complete the square}$$

$$(y' + 1)^2 = (-1)\left(x' - \frac{4}{5}\right). \qquad \textit{Standard form}$$

The graph of this equation is a parabola with its vertex at $(4/5, -1)$. Its axis is parallel to the x'-axis in the $x'y'$-system, as shown in Figure 7.59.

Invariants under Rotation

In the rotation of axes theorem listed at the beginning of this section, note that the constant term $F' = F$ is the same in both equations, and we say that it is **invariant under rotation.** The next theorem lists some other rotation invariants.

Rotation Invariants

The rotation of coordinate axes through an angle θ that transforms the equation $Ax^2 + Bxy + Cy^2 + Dx + Ey + F = 0$ into the form

$$A'(x')^2 + C'(y')^2 + D'x' + E'y' + F' = 0$$

has the following rotation invariants.

1. $F = F'$
2. $A + C = A' + C'$
3. $B^2 - 4AC = (B')^2 - 4A'C'$

We can use the results of this theorem to classify the graph of a second-degree equation *with* an xy-term in much the same way we did for second-degree equations *without* an xy-term. Note that since $B' = 0$, the invariant $B^2 - 4AC$ reduces to

$$B^2 - 4AC = -4A'C'. \qquad \textit{Discriminant}$$

Lines in the Plane and Conics

We call this quantity the **discriminant** of the equation

$$Ax^2 + Bxy + Cy^2 + Dx + Ey + F = 0.$$

Now, from the classification procedure given in Section 7.5, we know that the sign of $A'C'$ determines the type of graph for the equation

$$A'(x')^2 + C'(y')^2 + D'x' + E'y' + F' = 0.$$

Consequently, the sign of $B^2 - 4AC$ will determine the type of graph for the original equation, as given in the following result.

Classification of Conics by the Discriminant

The graph of the equation $Ax^2 + Bxy + Cy^2 + Dx + Ey + F = 0$ is, except in degenerate cases, determined by its discriminant as follows.

1. Ellipse or circle: $B^2 - 4AC < 0$
2. Parabola: $B^2 - 4AC = 0$
3. Hyperbola: $B^2 - 4AC > 0$

EXAMPLE 4 *Using the Discriminant*

Classify the graph of each of the following equations.

(a) $4xy - 9 = 0$

(b) $2x^2 - 3xy + 2y^2 - 2x = 0$

(c) $x^2 - 6xy + 9y^2 - 2y + 1 = 0$

(d) $3x^2 + 8xy + 4y^2 - 7 = 0$

SOLUTION

(a) Since $B^2 - 4AC = 16 - 0 > 0$, the graph is a hyperbola.

(b) Since $B^2 - 4AC = 9 - 16 < 0$, the graph is a circle or an ellipse.

(c) Since $B^2 - 4AC = 36 - 36 = 0$, the graph is a parabola.

(d) Since $B^2 - 4AC = 64 - 48 > 0$, the graph is a hyperbola.

WARM UP

Match each equation with its graph.

1. $\dfrac{x^2}{1} + \dfrac{y^2}{4} = 1$

2. $x^2 = 4y$

3. $\dfrac{x^2}{1} - \dfrac{y^2}{4} = 1$

4. $(x - 1)^2 + y^2 = 4$

5. $y^2 = -4x$

6. $x^2 - 5y^2 = -5$

(a)
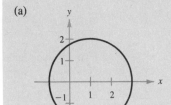

(b)
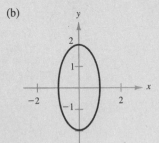

(c)
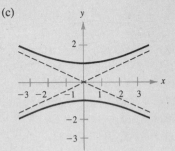

(d)
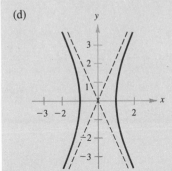

(e)
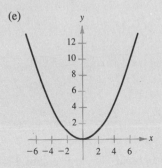

(f)
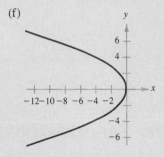

Evaluate the trigonometric functions and rewrite each expression.

7. $x \cos \dfrac{\pi}{3} - y \sin \dfrac{\pi}{3}$

8. $x \sin\left(-\dfrac{\pi}{6}\right) + y \cos\left(-\dfrac{\pi}{6}\right)$

Expand the expressions.

9. $\left(\dfrac{2x - 3y}{\sqrt{13}}\right)^2$

10. $\left(\dfrac{x - \sqrt{2}y}{\sqrt{3}}\right)^2$

EXERCISES 7.6

In Exercises 1–16, rotate the axes to eliminate the xy-term. Sketch the graph of the resulting equation, showing both sets of axes.

1. $xy + 1 = 0$ **2.** $xy - 4 = 0$

3. $9x^2 + 24xy + 16y^2 + 90x - 130y = 0$

4. $9x^2 + 24xy + 16y^2 + 80x - 60y = 0$

5. $x^2 - 10xy + y^2 + 1 = 0$

6. $xy + x - 2y + 3 = 0$

7. $xy - 2y - 4x = 0$

8. $2x^2 - 3xy - 2y^2 + 10 = 0$

9. $5x^2 - 2xy + 5y^2 - 12 = 0$

10. $13x^2 + 6\sqrt{3}xy + 7y^2 - 16 = 0$

11. $3x^2 - 2\sqrt{3}xy + y^2 + 2x + 2\sqrt{3}y = 0$

12. $16x^2 - 24xy + 9y^2 - 60x - 80y + 100 = 0$

13. $17x^2 + 32xy - 7y^2 = 75$

14. $40x^2 + 36xy + 25y^2 = 52$

15. $32x^2 + 50xy + 7y^2 = 52$

16. $4x^2 - 12xy + 9y^2 + (4\sqrt{13} - 12)x - (6\sqrt{13} + 8)y = 91$

In Exercises 17–24, use the discriminant to determine whether the graph of the equation is a parabola, an ellipse, or a hyperbola.

17. $16x^2 - 24xy + 9y^2 - 30x - 40y = 0$

18. $x^2 - 4xy - 2y^2 - 6 = 0$

19. $13x^2 - 8xy + 7y^2 - 45 = 0$

20. $2x^2 + 4xy + 5y^2 + 3x - 4y - 20 = 0$

21. $x^2 - 6xy - 5y^2 + 4x - 22 = 0$

22. $36x^2 - 60xy + 25y^2 + 9y = 0$

23. $x^2 + 4xy + 4y^2 - 5x - y - 3 = 0$

24. $x^2 + xy + 4y^2 + x + y = 0$

25. Show that the equation $x^2 + y^2 = r^2$ is invariant under rotation of axes.

CHAPTER 7 REVIEW EXERCISES

In Exercises 1–6, find an equation of the line through the two points.

1. $(0, 0)$, $(0, 10)$ **2.** $(-1, 4)$, $(2, 0)$

3. $(2, 1)$, $(14, 6)$ **4.** $(-2, 2)$, $(3, -10)$

5. $(-1, 0)$, $(6, 2)$ **6.** $(1, 6)$, $(4, 2)$

In Exercises 7–10, find t so that the three points are collinear.

7. $(-2, 5)$, $(0, t)$, $(1, 1)$ **8.** $(-6, 1)$, $(1, t)$, $(10, 5)$

9. $(1, -4)$, $(t, 3)$, $(5, 10)$ **10.** $(-3, 3)$, $(t, -1)$, $(8, 6)$

In Exercises 11–14, show that the given points form the vertices of the indicated polygon.

11. Parallelogram: $(1, 1)$, $(8, 2)$, $(9, 5)$, $(2, 4)$

12. Isosceles triangle: $(4, 5)$, $(1, 0)$, $(-1, 2)$

13. Right triangle: $(-1, -1)$, $(10, 7)$, $(2, 18)$

14. Square: $(-4, 0)$, $(1, -3)$, $(4, 2)$, $(-1, 5)$

In Exercises 15–22, match the equation with the correct graph. [The graphs are labeled (a)–(h).]

15. $x^2 = 4y$ **16.** $x^2 = -4y$

17. $y^2 = 4x$ **18.** $y^2 = -4x$

19. $\dfrac{x^2}{1} + \dfrac{y^2}{4} = 1$ **20.** $\dfrac{x^2}{4} + \dfrac{y^2}{1} = 1$

21. $\dfrac{x^2}{1} - \dfrac{y^2}{4} = 1$ **22.** $\dfrac{y^2}{4} - \dfrac{x^2}{1} = 1$

(a)

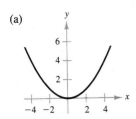

(b)

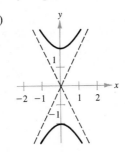

(c)

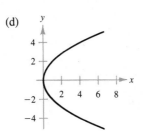

(d)

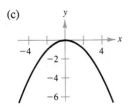

(e)

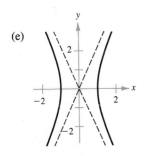

(f)

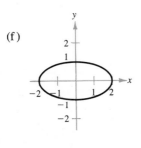

(g)

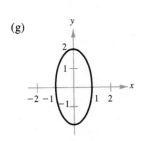

(h)

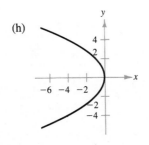

In Exercises 23–32, identify and sketch the graph of the equation.

23. $16x^2 + 16y^2 - 16x + 24y - 3 = 0$
24. $y^2 - 12y - 8x + 20 = 0$
25. $3x^2 - 2y^2 + 24x + 12y + 24 = 0$
26. $4x^2 + y^2 - 16x + 15 = 0$
27. $3x^2 + 2y^2 - 12x + 12y + 29 = 0$
28. $4x^2 - 4y^2 - 4x + 8y - 11 = 0$
29. $x^2 - 6x + 2y + 9 = 0$
30. $x^2 + y^2 - 2x - 4y + 5 = 0$
31. $x^2 + y^2 + 2xy + 2\sqrt{2}x - 2\sqrt{2}y + 2 = 0$
32. $9x^2 + 6y^2 + 4xy - 20 = 0$

In Exercises 33–36, find an equation of the specified parabola.

33. Vertex: (4, 2)   **34.** Vertex: (2, 0)
  Focus: (4, 0)   Focus: (0, 0)
35. Vertex: (0, 2)   **36.** Vertex: (2, 2)
  Directrix: $x = -3$ Directrix: $y = 0$

In Exercises 37–40, find an equation of the specified ellipse.

37. Vertices: $(-3, 0), (7, 0)$
  Foci: $(0, 0), (4, 0)$
38. Vertices: $(2, 0), (2, 4)$
  Foci: $(2, 1), (2, 3)$
39. Vertices: $(0, \pm 6)$
  Passes through $(2, 2)$
40. Vertices: $(0, 1), (4, 1)$
  Endpoints of minor axis: $(2, 0), (2, 2)$

In Exercises 41–44, find an equation of the specified hyperbola.

41. Vertices: $(0, \pm 1)$ **42.** Vertices: $(2, 2), (-2, 2)$
  Foci: $(0, \pm 3)$ Foci: $(4, 2), (-4, 2)$
43. Foci: $(0, 0), (8, 0)$
  Asymptotes: $y = \pm 2(x - 4)$
44. Foci: $(3, \pm 2)$
  Asymptotes: $y = \pm 2(x - 3)$

45. A large parabolic antenna is described as the surface formed by revolving the parabola $y = (1/200)x^2$ on the interval $[0, 100]$ about the y-axis. The receiving and transmitting equipment is positioned at the focus. Find the coordinates of the focus.

46. A semi-elliptical archway is to be formed over the entrance to an estate. The arch is to be set on pillars that are 10 feet apart and is to have a height (above the height of the pillars) of 4 feet. Where should the foci be placed in order to sketch the elliptical arch?

In Exercises 47–50, find an equation of the specified tangent line. The tangent line to the conic $(x^2/a^2) \pm (y^2/b^2) = 1$ at the point (x_0, y_0) is given by

$$\frac{x_0 x}{a^2} \pm \frac{y_0 y}{b^2} = 1.$$

Conic	Point
47. $\dfrac{x^2}{100} + \dfrac{y^2}{25} = 1$	$(-8, 3)$
48. $x^2 + 7y^2 = 16$	$(3, 1)$
49. $\dfrac{x^2}{9} - y^2 = 1$	$(6, \sqrt{3})$
50. $\dfrac{x^2}{4} - \dfrac{y^2}{2} = 1$	$(6, 4)$

In Exercises 51–54, find the points of intersection (if any) of the graphs of the equations.

51. $x + 3y = 15$
$\quad\ x^2 + y^2 = 25$

52. $y = 2x$
$\quad\ y = x^2 + 1$

53. $x^2 = 3y$
$\quad\ 3x^2 + y^2 - 9y - 25 = 0$

54. $\dfrac{x^2}{4} + \dfrac{y^2}{2} = 1$
$\quad\ x^2 - y^2 = 2$

CHAPTER 8

Polar Coordinates and Parametric Equations

8.1 Polar Coordinates

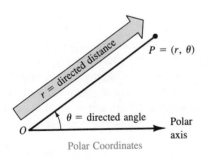

r = directed distance

$P = (r, \theta)$

θ = directed angle

O

Polar axis

Polar Coordinates

FIGURE 8.1

So far, we have been representing graphs of equations as collections of points (x, y) in the rectangular coordinate system, where x and y represent the directed distances from the coordinate axes to the point (x, y). In this section we introduce a second system called the **polar coordinate system.**

To form the polar coordinate system in the plane, we fix a point O, called the **pole** (or **origin**), and construct from O an initial ray called the **polar axis,** as shown in Figure 8.1. Then each point P in the plane can be assigned **polar coordinates** (r, θ) as follows.

1. $r = $ *directed distance* from O to P
2. $\theta = $ *directed angle*, counterclockwise from polar axis to segment $\overline{OP}$

In the polar coordinate system, it is convenient to locate points with respect to a grid of concentric circles intersected by **radial lines** through the pole. This procedure is shown in the following example.

Polar Coordinates and Parametric Equations

EXAMPLE 1 *Plotting Points in the Polar Coordinate System*

(a) The point $(r, \theta) = (2, \pi/3)$ lies at the intersection of a circle of radius $r = 2$ and a ray that is the terminal side of the angle $\theta = \pi/3$, as shown in Figure 8.2(a).

(b) The point $(r, \theta) = (3, -\pi/6)$ lies in the fourth quadrant, 3 units from the pole. Note that we measure negative angles *clockwise*, as shown in Figure 8.2(b).

(c) The point $(r, \theta) = (3, 11\pi/6)$ coincides with the point $(3, -\pi/6)$, as shown in Figure 8.2(c).

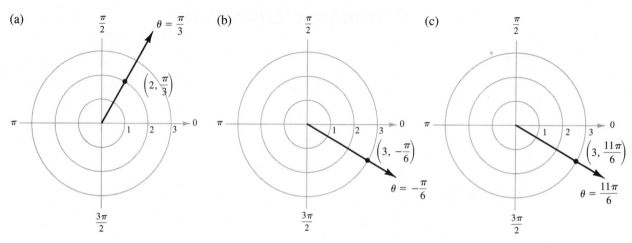

FIGURE 8.2

In rectangular coordinates, each point (x, y) has a unique representation. This is not true for polar coordinates. For instance, the coordinates (r, θ) and $(r, 2\pi + \theta)$ represent the same point as illustrated in Example 1. Another way to obtain multiple representations of a point is to use negative values for r. Since r is a *directed distance*, the coordinates (r, θ) and $(-r, \theta + \pi)$ represent the same point. In general, the point (r, θ) can be represented as

$$(r, \theta) = (r, \theta \pm 2n\pi)$$

or

$$(r, \theta) = (-r, \theta \pm (2n + 1)\pi)$$

where n is any integer. Moreover, the pole is represented by $(0, \theta)$ where θ is any angle.

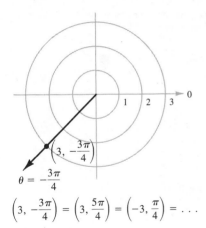

$\theta = -\dfrac{3\pi}{4}$

$\left(3, -\dfrac{3\pi}{4}\right) = \left(3, \dfrac{5\pi}{4}\right) = \left(-3, \dfrac{\pi}{4}\right) = \cdots$

FIGURE 8.3

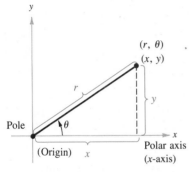

Relating Polar and Rectangular Coordinates

FIGURE 8.4

EXAMPLE 2 Multiple Representation of Points

Plot the point $(3, -3\pi/4)$ and find three additional polar representations of this point, using $-2\pi < \theta < 2\pi$.

SOLUTION

The point is shown in Figure 8.3. Three other representations are as follows.

$$\left(3, \dfrac{-3\pi}{4} + 2\pi\right) = \left(3, \dfrac{5\pi}{4}\right) \qquad \textit{Add } 2\pi \textit{ to } \theta$$

$$\left(-3, \dfrac{-3\pi}{4} - \pi\right) = \left(-3, \dfrac{-7\pi}{4}\right) \qquad \textit{Replace r by } -r; \textit{ subtract } \pi \textit{ from } \theta$$

$$\left(-3, \dfrac{-3\pi}{4} + \pi\right) = \left(-3, \dfrac{\pi}{4}\right) \qquad \textit{Replace r by } -r; \textit{ add } \pi \textit{ to } \theta$$

To establish the relationship between polar and rectangular coordinates, we let the polar axis coincide with the positive x-axis and the pole with the origin, as shown in Figure 8.4. Since (x, y) lies on a circle of radius r, it follows that $r^2 = x^2 + y^2$. Moreover, for $r > 0$, the definitions of the trigonometric functions imply that

$$\tan \theta = \dfrac{y}{x}, \qquad \cos \theta = \dfrac{x}{r}, \qquad \text{and} \qquad \sin \theta = \dfrac{y}{r}.$$

If $r < 0$, we can show that the same relationships hold. For example, consider the point (r, θ), where $r < 0$. Then, since $(-r, \theta + \pi)$ represents the same point and $-r > 0$, we have $-\sin \theta = \sin(\theta + \pi) = -y/r$, which implies that $\sin \theta = y/r$.

These relationships allow us to convert *coordinates* or *equations* from one system to the other as indicated in the following rule.

Coordinate Conversion

The polar coordinates (r, θ) are related to the rectangular coordinates (x, y) as follows.

$$x = r \cos \theta \qquad \text{and} \qquad y = r \sin \theta$$

$$\tan \theta = \dfrac{y}{x} \qquad \text{and} \qquad r^2 = x^2 + y^2$$

Hyperbolas

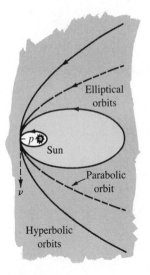

Elliptical
orbits

Sun

p

v

Parabolic
orbit

Hyperbolic
orbits

FIGURE 7.53

In Example 5, we were able to determine only the hyperbola on which the explosion occurred, but not the exact location of the explosion. If, however, we had received the sound from a third position C, then two other hyperbolas would have been determined. The exact location of the explosion would have been the point where these three hyperbolas intersected.

Another interesting application of conic sections involves the orbits of comets in our solar system. Of the 610 comets identified prior to 1970, 245 have elliptical orbits, 295 have parabolic orbits, and 70 have hyperbolic orbits. The center of the sun is a focus of each of these orbits, and each orbit has a vertex at the point where the comet is closest to the sun, as shown in Figure 7.53. Undoubtedly there have been many comets with parabolic or hyperbolic orbits that were not identified. We only get to see such comets *once*. Comets with elliptical orbits, such as Halley's comet, are the only ones that remain in our solar system.

If p is the distance between the vertex and the focus in meters, and v is the velocity of the comet at the vertex in meters per second, then the type of orbit is determined as follows.

1. Ellipse: $v < \sqrt{2GM/p}$
2. Parabola: $v = \sqrt{2GM/p}$
3. Hyperbola: $v > \sqrt{2GM/p}$

In each of these equations, $M \approx 1.991 \times 10^{30}$ kilograms (the mass of the sun) and $G \approx 6.67 \times 10^{-11}$ cubic meters per gram-second squared.

We conclude this section with a procedure for classifying a conic using the coefficients in the general form of its equation.

Classifying a Conic from Its General Equation

The graph of $Ax^2 + Cy^2 + Dx + Ey + F = 0$ is one of the following (except in degenerate cases).

1. Circle: $A = C$
2. Parabola: $AC = 0$ $A = 0$ or $C = 0$, but not both.
3. Ellipse: $AC > 0$ A and C have like signs.
4. Hyperbola: $AC < 0$ A and C have unlike signs.

EXAMPLE 6   Classifying Conics from General Equations

(a) For the general equation

$$4x^2 - 9x + y - 5 = 0$$

we have $AC = 4(0) = 0$. Thus, the graph is a parabola.

EXAMPLE 3 *Polar-to-Rectangular Conversion*

(a) For the point $(r, \theta) = (2, \pi)$, we have

$$x = r \cos \theta = 2 \cos \pi = -2$$

and

$$y = r \sin \theta = 2 \sin \pi = 0.$$

Thus, the rectangular coordinates are $(x, y) = (-2, 0)$, as shown in Figure 8.5.

(b) For the point $(r, \theta) = (\sqrt{3}, \pi/6)$, we have

$$x = \sqrt{3} \cos \frac{\pi}{6}$$

$$= \sqrt{3} \left(\frac{\sqrt{3}}{2} \right) = \frac{3}{2}$$

and

$$y = \sqrt{3} \sin \frac{\pi}{6}$$

$$= \sqrt{3} \left(\frac{1}{2} \right) = \frac{\sqrt{3}}{2}$$

and the rectangular coordinates are $(x, y) = (3/2, \sqrt{3}/2)$, as shown in Figure 8.5.

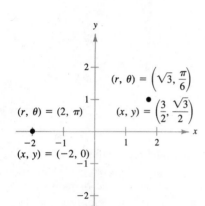

FIGURE 8.5

EXAMPLE 4 *Rectangular-to-Polar Conversion*

(a) For the second quadrant point $(x, y) = (-1, 1)$, we have

$$\tan \theta = \frac{y}{x} = -1 \implies \theta = \frac{3\pi}{4}.$$

Since θ lies in the same quadrant as (x, y), we use positive r.

$$r = \sqrt{x^2 + y^2} = \sqrt{(-1)^2 + (1)^2} = \sqrt{2}$$

Thus, *one* set of polar coordinates is $(r, \theta) = (\sqrt{2}, 3\pi/4)$.

(b) Since the point $(x, y) = (0, 2)$ lies on the positive y-axis, we choose $\theta = \pi/2$ and $r = 2$, and one set of polar coordinates is $(r, \theta) = (2, \pi/2)$.

(a)

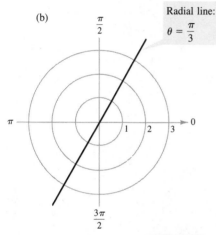

Circle:
$r = 2$

By comparing Examples 3 and 4, we see that point conversion from the polar to the rectangular system is straightforward, whereas point conversion from the rectangular to the polar system is more involved. For equations, the opposite is true. To convert a rectangular equation to polar form, we simply replace x by $r \cos \theta$ and y by $r \sin \theta$. For instance, the rectangular equation $y = x^2$ has the polar form

$$r \sin \theta = (r \cos \theta)^2.$$

On the other hand, converting a polar equation to rectangular form requires considerable ingenuity.

In the next example we demonstrate several polar-to-rectangular conversions that enable us to sketch the graphs of some polar equations.

(b)

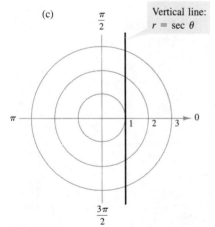

Radial line:
$\theta = \dfrac{\pi}{3}$

EXAMPLE 5 *Converting Polar Equations to Rectangular Form*

Describe the graphs of the following polar equations and find the corresponding rectangular equation.

(a) $r = 2$

(b) $\theta = \dfrac{\pi}{3}$

(c) $r = \sec \theta$

(c)

Vertical line:
$r = \sec \theta$

SOLUTION

(a) The graph of the polar equation $r = 2$ consists of all points that are 2 units from the pole. In other words, this graph is a circle centered at the origin and having a radius of 2, as shown in Figure 8.6(a). We can confirm this by converting to rectangular coordinates, using the relationship $r^2 = x^2 + y^2$.

$$r = 2 \quad \Longrightarrow \quad r^2 = 2^2 \quad \Longrightarrow \quad x^2 + y^2 = 2^2$$

Polar equation   Rectangular equation

(b) The graph of the polar equation $\theta = \pi/3$ consists of all points on the line that makes an angle of $\pi/3$ with the positive x-axis, as shown in Figure 8.6(b). To convert to rectangular form, we make use of the relationship $\tan \theta = y/x$.

$$\theta = \frac{\pi}{3} \quad \Longrightarrow \quad \tan \theta = \sqrt{3} \quad \Longrightarrow \quad y = \sqrt{3}x$$

Polar equation   Rectangular equation

FIGURE 8.6

(c) The graph of the polar equation $r = \sec \theta$ is not evident by simple inspection, so we convert to rectangular form by using the relationship $r \cos \theta = x$.

$$r = \sec \theta \implies r \cos \theta = 1 \implies x = 1$$

Polar equation   Rectangular equation

Now, we see that the graph is a vertical line, as shown in Figure 8.6(c).

Curve-sketching by converting to rectangular form is not always convenient. In the next section we demonstrate a straightforward point-plotting technique.

WARM UP

Find a positive angle coterminal with each of the given angles.

1. $\dfrac{11\pi}{4}$ **2.** $-\dfrac{5\pi}{6}$

In each case find $\sin \theta$ and $\cos \theta$ if θ is an angle in standard position with its terminal side passing through the given point.

3. $(2, 1)$ **4.** $(4, -3)$

Find the measure (in radians) of an angle in standard position if its terminal side passes through the given point.

5. $(-4, 4)$ **6.** $(3, 2)$

Evaluate the functions without using a calculator.

7. $\sin \dfrac{4\pi}{3}$ **8.** $\cos \dfrac{3\pi}{4}$

Use a calculator to evaluate the functions.

9. $\cos \dfrac{3\pi}{5}$ **10.** $\sin 1.34$

EXERCISES 8.1

In Exercises 1–8, the polar coordinates of a point are given. Plot the point and find the corresponding rectangular coordinates.

1. $(4, 3\pi/6)$ **2.** $(4, 3\pi/2)$

3. $(-1, 5\pi/4)$ **4.** $(0, -\pi)$

5. $(4, -\pi/3)$ **6.** $(-1, -3\pi/4)$

7. $(\sqrt{2}, 2.36)$ **8.** $(-3, -1.57)$

In Exercises 9–16, the rectangular coordinates of a point are given. Find two sets of polar coordinates for the point, using $0 \le \theta < 2\pi$.

9. $(1, 1)$ **10.** $(0, -5)$

11. $(-3, 4)$ **12.** $(3, -1)$

13. $(-\sqrt{3}, -\sqrt{3})$ **14.** $(-2, 0)$

15. $(4, 6)$ **16.** $(5, 12)$

In Exercises 17–30, convert the given rectangular equation to polar form.

17. $x^2 + y^2 = 9$ **18.** $x^2 + y^2 = a^2$

19. $x^2 + y^2 - 2ax = 0$ **20.** $x^2 + y^2 - 2ay = 0$

21. $y = 4$ **22.** $y = b$

23. $x = 10$ **24.** $x = a$

25. $3x - y + 2 = 0$ **26.** $4x + 7y - 2 = 0$

27. $xy = 4$ **28.** $y = x$

29. $(x^2 + y^2)^2 - 9(x^2 - y^2) = 0$

30. $y^2 - 8x - 16 = 0$

In Exercises 31–40, convert the given polar equation to rectangular form.

31. $r = 4 \sin \theta$ **32.** $r = 4 \cos \theta$

33. $\theta = \dfrac{\pi}{6}$ **34.** $r = 4$

35. $r = 2 \csc \theta$ **36.** $r^2 = \sin 2\theta$

37. $r = 2 \sin 3\theta$ **38.** $r = \dfrac{1}{1 - \cos \theta}$

39. $r = \dfrac{6}{2 - 3 \sin \theta}$ **40.** $r = \dfrac{6}{2 \cos \theta - 3 \sin \theta}$

41. Show that the distance between (r_1, θ_1) and (r_2, θ_2) is
$$\sqrt{r_1{}^2 + r_2{}^2 - 2r_1r_2 \cos(\theta_1 - \theta_2)}.$$

42. Show that the points (r_1, θ_1), (r_2, θ_2), and (r_3, θ_3) are collinear if and only if
$$r_2r_3 \sin(\theta_3 - \theta_2) + r_3r_1 \sin(\theta_1 - \theta_3) + r_1r_2 \sin(\theta_2 - \theta_1) = 0.$$

8.2   Graphs of Polar Equations

In previous chapters we spent a substantial amount of time learning how to sketch graphs in rectangular coordinates. We began with the basic point-plotting method which was then enhanced by sketching aids such as symmetry, intercepts, asymptotes, periods, and shifts. We approach curve-sketching in the polar coordinate system in a similar way, beginning with a demonstration of point-plotting.

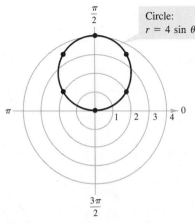

FIGURE 8.7

EXAMPLE 1   Graphing a Polar Equation by Point-Plotting

Sketch the graph of the polar equation $r = 4 \sin \theta$.

SOLUTION

The sine function is periodic, so we can get a full range of r-values by considering values of θ in the interval $0 \le \theta \le 2\pi$, shown in the following table.

Polar Coordinates and Parametric Equations

θ	0	$\dfrac{\pi}{6}$	$\dfrac{\pi}{3}$	$\dfrac{\pi}{2}$	$\dfrac{2\pi}{3}$	$\dfrac{5\pi}{6}$	π	$\dfrac{7\pi}{6}$	$\dfrac{3\pi}{2}$	$\dfrac{11\pi}{6}$	2π
r	0	2	$2\sqrt{3}$	4	$2\sqrt{3}$	2	0	-2	-4	-2	0

By plotting these points as shown in Figure 8.7, it appears that the graph a circle of radius 2 whose center is at the point $(x, y) = (0, 2)$.

In Figure 8.7 note that as θ increases from 0 to 2π the graph is traced out twice. Moreover, note that the graph is *symmetric with respect to the line* $\theta = \pi/2$. Had we known about this symmetry and retracing ahead of time we could have reduced our table of points to half its size.

Symmetry with respect to the line $\theta = \pi/2$ is one of three important types of symmetry to consider in polar curve-sketching. (See Figure 8.8.)

Tests for Symmetry in Polar Coordinates

The graph of a polar equation is symmetric with respect to the following if the given substitution yields an equivalent equation.

1. The line $\theta = \pi/2$: Replace (r, θ) by $(r, \pi - \theta)$ or $(-r, -\theta)$.
2. The polar axis: Replace (r, θ) by $(r, -\theta)$ or $(-r, \pi - \theta)$.
3. The pole: Replace (r, θ) by $(r, \pi + \theta)$ or $(-r, \theta)$.

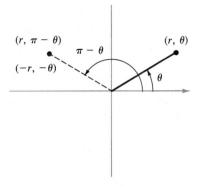

Symmetry with Respect
to the Line $\theta = \pi/2$

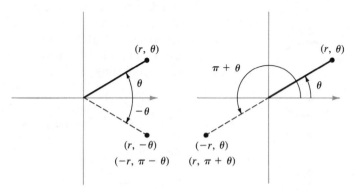

Symmetry with Respect
to the Polar Axis

Symmetry with Respect
to the Pole

FIGURE 8.8

Graphs of Polar Equations

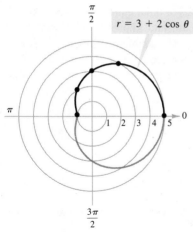

$r = 3 + 2 \cos \theta$

FIGURE 8.9

EXAMPLE 2 Using Symmetry to Sketch the Graph of a Polar Equation

Use symmetry to sketch the graph of $r = 3 + 2 \cos \theta$.

SOLUTION

Replacing (r, θ) by $(r, -\theta)$ produces

$$r = 3 + 2 \cos(-\theta)$$
$$= 3 + 2 \cos \theta.$$

Thus, we conclude that the curve is symmetric with respect to the polar axis. Plotting the points in the following table and using polar axis symmetry, we obtain the graph shown in Figure 8.9.

θ	0	$\pi/3$	$\pi/2$	$2\pi/3$	π
r	5	4	3	2	1

These three tests for symmetry in polar coordinates are sufficient to guarantee symmetry, but they are not necessary. For instance, Figure 8.10 shows the graph of $r = \theta + 2\pi$ to be symmetric with respect to the line $\theta = \pi/2$. Yet the test fails to indicate symmetry because neither of the following replacements yields an equivalent equation.

Original equation	Replacement	New equation
$r = \theta + 2\pi$	(r, θ) by $(-r, -\theta)$	$-r = -\theta + 2\pi$
$r = \theta + 2\pi$	(r, θ) by $(r, \pi - \theta)$	$r = -\theta + 3\pi$

The equations discussed in Examples 1 and 2 are of the form

$$r = 4 \sin \theta = f(\sin \theta) \qquad \text{\textit{r is a function of} sin } \theta$$

and

$$r = 3 + \cos \theta = g(\cos \theta). \qquad \text{\textit{r is a function of} cos } \theta$$

The graph of the first equation is symmetric with respect to the line $\theta = \pi/2$, and the graph of the second equation is symmetric with respect to the polar axis. This observation can be generalized to yield the following *quick test for symmetry*.

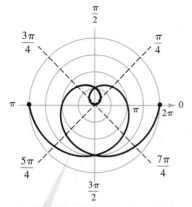

Spiral of Archimedes:
$r = \theta + 2\pi, -4\pi \le \theta \le 0$

FIGURE 8.10

1. The graph of $r = f(\sin \theta)$ is symmetric with respect to the line $\theta = \pi/2$.
2. The graph of $r = g(\cos \theta)$ is symmetric with respect to the polar axis.

Polar Coordinates and Parametric Equations

Two additional aids to sketching graphs of polar equations involve knowing the θ-values for which $|r|$ is maximum and knowing the θ-values for which $r = 0$. For instance, in Example 1, the maximum value of $|r|$ for $r = 4 \sin \theta$ is $|r| = 4$, and this occurs when $\theta = \pi/2$, as shown in Figure 8.7. Moreover, $r = 0$ when $\theta = 0$. We demonstrate the use of these two sketching aids in Examples 3 and 4.

EXAMPLE 3 *Sketching a Polar Graph*

Sketch the graph of $r = 1 - 2 \cos \theta$.

SOLUTION

From the equation $r = 1 - 2 \cos \theta$, we obtain the following information.

Symmetry: With respect to the polar axis

Maximum value of $|r|$: $r = 3$ when $\theta = \pi$

Zero of r: $r = 0$ when $\theta = \pi/3$

Making a table for several θ-values in the interval $[0, \pi]$ and plotting the corresponding points produces the graph shown in Figure 8.11.

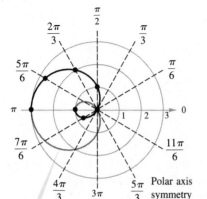

Limaçon:
$r = 1 - 2 \cos \theta$

FIGURE 8.11

θ	0	$\pi/6$	$\pi/3$	$\pi/2$	$2\pi/3$	$5\pi/6$	π
r	-1	-0.73	0	1	2	2.73	3

Remark: Note how the negative r-values determine the *inner loop* of the graph in Figure 8.11. This type of graph is called a limaçon.

Some curves reach their zeros and maximum r-values at more than one point. We show how to handle this situation in Example 4.

EXAMPLE 4 *Sketching a Polar Graph*

Sketch the graph of $r = 2 \cos 3\theta$.

SOLUTION

Symmetry: With respect to the polar axis

Maximum value of $|r|$: $|r| = 2$ when $3\theta = 0, \pi, 2\pi, 3\pi$
or $\quad \theta = 0, \pi/3, 2\pi/3, \pi$

Zeros of r: $r = 0$ when $3\theta = \pi/2, 3\pi/2, 5\pi/2$
or $\quad \theta = \pi/6, \pi/2, 5\pi/6$

θ	0	$\pi/12$	$\pi/6$	$\pi/4$	$\pi/3$	$5\pi/12$	$\pi/2$
r	2	$\sqrt{2}$	0	$-\sqrt{2}$	-2	$-\sqrt{2}$	0

By plotting these points and using the specified symmetry, zeros, and maximum values, we obtain the graph shown in Figure 8.12. Note how the entire curve is generated as θ increases from 0 to π.

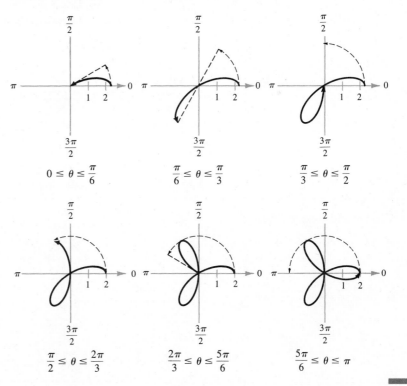

FIGURE 8.12

Remark: The graph shown in Figure 8.12 is called a **rose curve,** and each of the loops on the graph is called a *petal* of the rose curve.

Special Polar Graphs

Several important types of graphs have equations that are simpler in polar form than in rectangular form. For example, the circle $r = 4 \sin \theta$ in Example 1 has the more complicated rectangular equation $x^2 + (y - 2)^2 = 4$. The following list gives several other types of graphs that have simple polar equations.

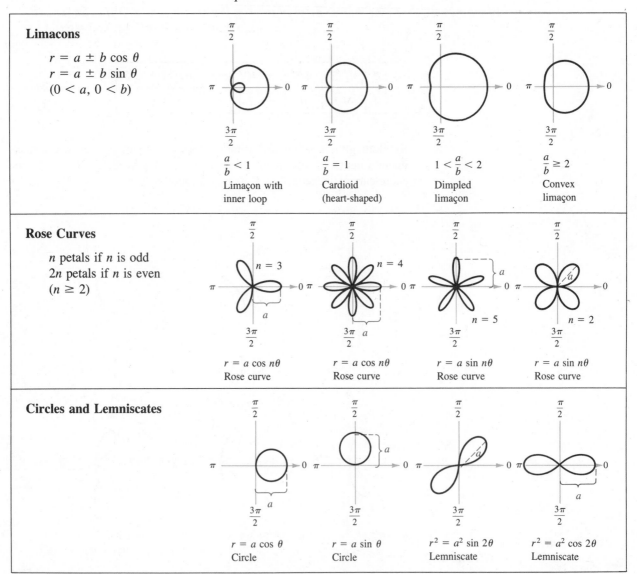

Limacons

$r = a \pm b \cos \theta$
$r = a \pm b \sin \theta$
$(0 < a, 0 < b)$

$\dfrac{a}{b} < 1$
Limaçon with inner loop

$\dfrac{a}{b} = 1$
Cardioid (heart-shaped)

$1 < \dfrac{a}{b} < 2$
Dimpled limaçon

$\dfrac{a}{b} \geq 2$
Convex limaçon

Rose Curves

n petals if *n* is odd
2*n* petals if *n* is even
$(n \geq 2)$

$n = 3$
$r = a \cos n\theta$
Rose curve

$n = 4$
$r = a \cos n\theta$
Rose curve

$n = 5$
$r = a \sin n\theta$
Rose curve

$n = 2$
$r = a \sin n\theta$
Rose curve

Circles and Lemniscates

$r = a \cos \theta$
Circle

$r = a \sin \theta$
Circle

$r^2 = a^2 \sin 2\theta$
Lemniscate

$r^2 = a^2 \cos 2\theta$
Lemniscate

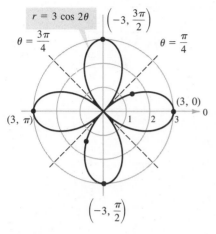

$r = 3 \cos 2\theta$ $\left(-3, \dfrac{3\pi}{2}\right)$

$\theta = \dfrac{3\pi}{4}$ $\theta = \dfrac{\pi}{4}$

$(3, 0)$

$(3, \pi)$ 1   2   3   0

$\left(-3, \dfrac{\pi}{2}\right)$

FIGURE 8.13

EXAMPLE 5 *Sketching a Rose Curve*

Sketch the graph of $r = 3 \cos 2\theta$.

SOLUTION

Type of curve:	Rose curve with $2n = 4$ petals
Symmetry:	With respect to polar axis and the line $\theta = \pi/2$
Maximum value of $|r|$:	$|r| = 3$ when $\theta = 0,\ \pi/2,\ \pi,\ 3\pi/2$
Zeros of r:	$r = 0$ when $\theta = \pi/4,\ 3\pi/4$

Using this information together with the additional points shown in the following table, we obtain the graph shown in Figure 8.13.

θ	0	$\pi/6$	$\pi/4$	$\pi/3$
r	3	3/2	0	−3/2

EXAMPLE 6 *Sketching a Lemniscate*

Sketch the graph of $r^2 = 9 \sin 2\theta$.

SOLUTION

Type of curve:	Lemniscate
Symmetry:	With respect to the pole
Maximum value of $|r|$:	$|r| = 3$ when $\theta = \pi/4$
Zeros of r:	$r = 0$ when $\theta = 0,\ \pi/2$

If $\sin 2\theta < 0$, then this equation has no solution points. Thus, we restrict the values of θ to those for which $\sin 2\theta \geq 0$.

$$0 \leq \theta \leq \frac{\pi}{2} \quad \text{or} \quad \pi \leq \theta \leq \frac{3\pi}{2}$$

Moreover, using symmetry, we need only consider the first of these two intervals. By finding a few additional points, we can obtain the graph shown in Figure 8.14.

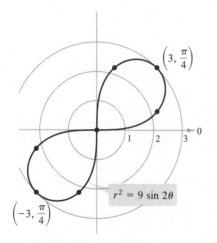

$\left(3, \dfrac{\pi}{4}\right)$

1   2   3   0

$r^2 = 9 \sin 2\theta$

$\left(-3, \dfrac{\pi}{4}\right)$

FIGURE 8.14

Polar Coordinates and Parametric Equations

θ	0	$\pi/12$	$\pi/4$	$5\pi/12$	$\pi/2$
$r = \pm3\sqrt{\sin 2\theta}$	0	$\pm3/\sqrt{2}$	±3	$\pm3/\sqrt{2}$	0

WARM UP

Determine the amplitude and period of each function.

1. $y = 5 \sin 4x$

2. $y = 3 \cos 2\pi x$

3. $y = -5 \cos \dfrac{\pi x}{2}$

4. $y = -\dfrac{1}{2} \sin \dfrac{x}{2}$

Identify and sketch the graph of each function through two periods.

5. $y = 2 \sin x$

6. $y = 3 \cos x$

7. $y = 4 \cos 2x$

8. $y = 2 \sin \pi x$

Use the identities for the sine and cosine of the sum and difference of angles to simplify the expressions.

9. $\sin\left(x - \dfrac{\pi}{6}\right)$

10. $\cos\left(x + \dfrac{\pi}{4}\right)$

EXERCISES 8.2

In Exercises 1–30, sketch the graph of the equation.

1. $r = 5$

2. $r = 2$

3. $\theta = \dfrac{\pi}{6}$

4. $\theta = -\dfrac{\pi}{4}$

5. $r = 3 \sin \theta$

6. $r = 3 \cos \theta$

7. $r = 3(1 - \cos \theta)$

8. $r = 2(1 - \sin \theta)$

9. $r = 4(1 + \sin \theta)$

10. $r = 1 + \cos \theta$

11. $r = 2 + 3 \sin \theta$

12. $r = 4 + 5 \cos \theta$

13. $r = 3 - 4 \cos \theta$

14. $r = 2(1 - 2 \sin \theta)$

15. $r = 3 - 2 \cos \theta$

16. $r = 5 - 4 \sin \theta$

17. $r = 2 + \sin \theta$

18. $r = 4 + 3 \cos \theta$

19. $r = 2 \cos 3\theta$

20. $r = -\sin 5\theta$

21. $r = 3 \sin 2\theta$

22. $r = 3 \cos 2\theta$

23. $r = 2 \sec \theta$

24. $r = 3 \csc \theta$

25. $r = \dfrac{3}{\sin \theta - 2 \cos \theta}$

26. $r = \dfrac{6}{2 \sin \theta - 3 \cos \theta}$

27. $r^2 = 4 \cos 2\theta$

28. $r^2 = 4 \sin \theta$

29. $r^2 = 4 \sin 2\theta$

30. $r = \theta$

31. The graph of $r = f(\theta)$ is rotated about the pole through an angle ϕ. Show that the equation for the rotated graph is $r = f(\theta - \phi)$.

32. Consider the graph of $r = f(\sin \theta)$.
(a) Show that if the graph is rotated counterclockwise $\pi/2$ radians about the pole the equation for the rotated graph will be $r = f(-\cos \theta)$.
(b) Show that if this graph is rotated counterclockwise π radians about the pole the equation for the rotated graph will be $r = f(-\sin \theta)$.
(c) Show that if this graph is rotated counterclockwise $3\pi/2$ radians about the pole the equation for the rotated graph will be $r = f(\cos \theta)$.

In Exercises 33–36, use the results of Exercises 31 and 32.

33. Write an equation for the limaçon

$$r = 2 - \sin \theta$$

after it has been rotated by the given amount.
(a) $\pi/4$ (b) $\pi/2$ (c) π (d) $3\pi/2$

34. Write an equation for the rose curve

$$r = 2 \sin 2\theta$$

after it has been rotated by the given amount.
(a) $\pi/6$ (b) $\pi/2$ (c) $2\pi/3$ (d) π

35. Sketch the graphs of the equations.
(a) $r = 1 - \sin \theta$ (b) $r = 1 - \sin\left(\theta - \dfrac{\pi}{4}\right)$

36. Sketch the graphs of the equations.
(a) $r = 3 \sec \theta$ (b) $r = 3 \sec\left(\theta - \dfrac{\pi}{4}\right)$

(c) $r = 3 \sec\left(\theta + \dfrac{\pi}{3}\right)$ (d) $r = 3 \sec\left(\theta - \dfrac{\pi}{2}\right)$

8.3 Polar Equations of Conics

In Sections 7.4 and 7.5, we saw that the rectangular equations of ellipses and hyperbolas take simple forms when the origin lies at their *center*. As it happens, there are many important applications of conics in which it is more convenient to use one of the *foci* as the origin for the coordinate system. For example, the sun lies at the focus of the earth's orbit. Similarly, the light source of a parabolic reflector lies at its focus. In this section we will see that polar equations of conics take simple forms if one of the foci lies at the pole.

We begin with an alternative definition of a conic using the concept of eccentricity.

Alternative Definition of Conic

The locus of a point in the plane that moves so that its distance from a fixed point (focus) is in constant ratio to its distance from a fixed line (directrix) is called a **conic**. The constant ratio is called the **eccentricity** of the conic and is denoted by e. Moreover, the conic is an **ellipse** if $e < 1$, a **parabola** if $e = 1$, and a **hyperbola** if $e > 1$.

In Figure 8.15, note that for each type of conic, the pole corresponds to the fixed point (focus) given in the definition. The benefit of this location is seen in the proof of the following theorem.

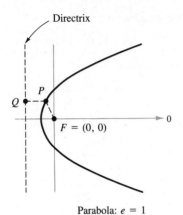

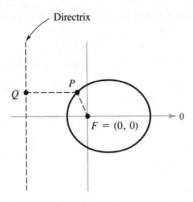

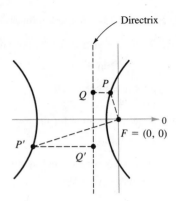

Parabola: $e = 1$
$PF = PQ$

Ellipse: $0 < e < 1$
$\dfrac{PF}{PQ} < 1$

Hyperbola: $e > 1$
$\dfrac{PF}{PQ} = \dfrac{P'F}{P'Q'} > 1$

FIGURE 8.15

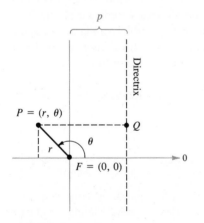

FIGURE 8.16

Polar Equations of Conics

The graph of a polar equation of the form

1. $r = \dfrac{ep}{1 \pm e \cos \theta}$

2. $r = \dfrac{ep}{1 \pm e \sin \theta}$

is a conic, where $e > 0$ is the eccentricity and $|p|$ is the distance between the focus (pole) and the directrix.

PROOF

We give a proof for $r = ep/(1 + e \cos \theta)$ with $p > 0$. The proofs of the other cases are similar. In Figure 8.16, consider a vertical directrix, p units to the right of the focus $F = (0, 0)$. If $P = (r, \theta)$ is a point on the graph of $r = ep/(1 + e \cos \theta)$, then the distance between P and the directrix is

$$PQ = |p - x| = |p - r \cos \theta|$$
$$= \left| p - \left(\frac{ep}{1 + e \cos \theta} \right) \cos \theta \right|$$
$$= \left| p \left(1 - \frac{e \cos \theta}{1 + e \cos \theta} \right) \right|$$
$$= \left| \frac{p}{1 + e \cos \theta} \right|$$
$$= \left| \frac{r}{e} \right|.$$

Moreover, since the distance between P and the pole is simply $PF = |r|$, the ratio of PF to PQ is

$$\frac{PF}{PQ} = \frac{|r|}{|r/e|} = |e| = e$$

and by definition, the graph of the equation must be a conic.

By completing the proofs of the other three cases, you can see that the equations

$$r = \frac{ep}{1 \pm e \cos \theta}$$

correspond to conics with vertical directrices and the equations

$$r = \frac{ep}{1 \pm e \sin \theta}$$

correspond to conics with horizontal directrices. Moreover, the converse is also true—that is, any conic with a focus at the pole and having a horizontal or vertical directrix can be represented by one of the given equations.

EXAMPLE 1 *Determining a Conic from its Equation*

Sketch the graph of the conic given by

$$r = \frac{15}{3 - 2 \cos \theta}.$$

SOLUTION

To determine the type of conic, we rewrite the equation as

$$r = \frac{15}{3 - 2 \cos \theta} = \frac{5}{1 - (2/3) \cos \theta}.$$

From this form we conclude that the graph is an ellipse with $e = 2/3$. We sketch the upper half of the ellipse by plotting points from $\theta = 0$ to $\theta = \pi$ as shown in Figure 8.17. Then, using symmetry with respect to the polar axis, we sketch the lower half.

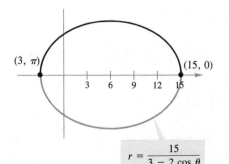

$$r = \frac{15}{3 - 2 \cos \theta}$$

FIGURE 8.17

For the ellipse in Figure 8.17, the major axis is horizontal and the vertices lie at $(15, 0)$ and $(3, \pi)$. Thus, the length of the *major* axis is $2a = 18$. To find the length of the *minor* axis, we use the equations $e = c/a$ and $b^2 = a^2 - c^2$ to conclude that

$$b^2 = a^2 - c^2 = a^2 - (ea)^2 = a^2(1 - e^2). \qquad \textit{Ellipse}$$

Since $e = 2/3$, we have $b^2 = 9^2[1 - (2/3)^2] = 45$, which implies that $b = \sqrt{45} = 3\sqrt{5}$. Thus, the length of the minor axis is $2b = 6\sqrt{5}$. A similar analysis for hyperbolas yields

$$b^2 = c^2 - a^2 = (ea)^2 - a^2 = a^2(e^2 - 1). \qquad \textit{Hyperbola}$$

EXAMPLE 2 Sketching a Conic from its Polar Equation

Sketch the graph of the polar equation

$$r = \frac{32}{3 + 5 \sin \theta}.$$

SOLUTION

Dividing each term by 3, we have

$$r = \frac{32/3}{1 + (5/3)\sin \theta}.$$

Since $e = 5/3 > 1$, the graph is a hyperbola. The transverse axis of the hyperbola lies on the line $\theta = \pi/2$ and the vertices occur at

$$\left(4, \frac{\pi}{2}\right)$$

and

$$\left(-16, \frac{3\pi}{2}\right).$$

Since the length of the transverse axis is 12, we see that $a = 6$. To find b, we write

$$b^2 = a^2(e^2 - 1) = 6^2\left[\left(\frac{5}{3}\right)^2 - 1\right] = 64.$$

Therefore, $b = 8$. Finally, we use a and b to determine the asymptotes of the hyperbola and obtain the sketch shown in Figure 8.18.

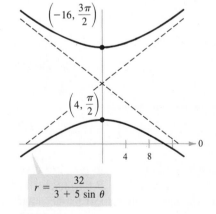

$$r = \frac{32}{3 + 5 \sin \theta}$$

FIGURE 8.18

In the next example we are asked to find a polar equation for a specified conic. To do this we let p be the distance between the pole and the directrix. With this interpretation of p, we suggest the following guidelines for finding a polar equation for a conic.

1. Horizontal directrix above the pole: $r = \dfrac{ep}{1 + e \sin \theta}$

2. Horizontal directrix below the pole: $r = \dfrac{ep}{1 - e \sin \theta}$

3. Vertical directrix to the right of the pole: $r = \dfrac{ep}{1 + e \cos \theta}$

4. Vertical directrix to the left of the pole: $r = \dfrac{ep}{1 - e \cos \theta}$

EXAMPLE 3 Finding the Polar Equation for a Conic

Find a polar equation for the parabola whose focus is the pole and whose directrix is the line $y = 3$.

SOLUTION

From Figure 8.19, we see that the directrix is horizontal and we choose an equation of the form

$$r = \frac{ep}{1 + e \sin \theta}.$$

Moreover, since the eccentricity of a parabola is $e = 1$ and the distance between the pole and the directrix is $p = 3$, we have the equation

$$r = \frac{3}{1 + \sin \theta}.$$

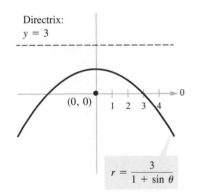

Directrix:
$y = 3$

$(0, 0)$

$r = \dfrac{3}{1 + \sin \theta}$

FIGURE 8.19

Kepler's Laws (listed below), named after the German astronomer Johannes Kepler (1571–1630), can be used to describe the orbits of the planets about the sun.

1. Each planet moves in an elliptical orbit with the sun as a focus.
2. A ray from the sun to the planet sweeps out equal areas of the ellipse in equal times.
3. The square of the period is proportional* to the cube of the mean distance between the planet and the sun.

*If we use the earth as a reference with a period of 1 year and a distance of 1 astronomical unit, the proportionality constant is 1. For example, since Mars has a mean distance to the sun of $d = 1.523$ AU, its period P is given by $d^3 = P^2$. Thus, the period for Mars is $P = 1.88$ years.

Although Kepler simply stated these laws on the basis of observation, they were later validated by Isaac Newton (1642–1727). In fact, Newton was able to show that each law can be deduced from a set of universal laws of motion and gravitation which govern the movement of all heavenly bodies, including comets and satellites. This is illustrated in the next example involving the comet named after the English mathematician and physicist Edmund Halley (1656–1742).

EXAMPLE 4 *An Application*

Halley's comet has an elliptical orbit with an eccentricity of $e \approx 0.97$. The length of the major axis of the orbit is approximately 36.18 astronomical units. (An astronomical unit is defined to be the mean distance between the earth and the sun, 93 million miles.) Find a polar equation for the orbit. How close does Halley's comet come to the sun?

SOLUTION

Using a vertical axis, as shown in Figure 8.20, we choose an equation of the form $r = ep/(1 + e \sin \theta)$. Since the vertices of the ellipse occur when $\theta = \pi/2$ and $\theta = 3\pi/2$, we can determine the length of the major axis to be the sum of the r-values of the vertices. That is,

$$2a = \frac{0.97p}{1 + 0.97} + \frac{0.97p}{1 - 0.97}$$
$$\approx 32.83p$$
$$\approx 36.18.$$

Thus, $p \approx 1.102$ and $ep \approx (0.97)(1.102) \approx 1.069$. Using this value in the equation, we have

$$r = \frac{1.069}{1 + 0.97 \sin \theta}$$

where r is measured in astronomical units. To find the closest point to the sun (the focus), we write

$$c = ea \approx (0.97)(18.09)$$
$$\approx 17.55.$$

Since c is the distance between the focus and the center, the closest point is

$$a - c \approx 18.09 - 17.55$$
$$\approx 0.54 \text{ AU}$$
$$\approx 50{,}000{,}000 \text{ miles.}$$

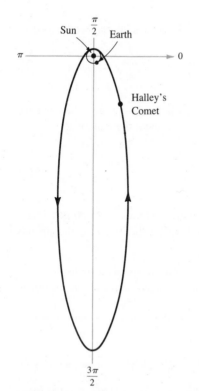

FIGURE 8.20

WARM UP

In each case the polar coordinates of a point are given. Plot the point and find the corresponding rectangular coordinates.

1. $(-3, 3\pi/4)$

2. $(4, -2\pi/3)$

In each case the rectangular coordinates of a point are given. Find two sets of polar coordinates for the point, using $0 \le \theta < 2\pi$.

3. $(0, -3)$

4. $(-5, 12)$

Convert the rectangular equations to polar form.

5. $x^2 + y^2 = 25$

6. $x^2 y = 4$

Convert the polar equations to rectangular form.

7. $r \sin \theta = -4$

8. $r = 4 \cos \theta$

Identify and sketch the graph of each polar equation.

9. $r = 1 - \sin \theta$

10. $r = 1 + 2 \cos \theta$

EXERCISES 8.3

In Exercises 1–6, match the polar equation with the correct graph. [The graphs are labeled (a)–(f).]

1. $r = \dfrac{6}{1 - \cos \theta}$

2. $r = \dfrac{2}{2 - \cos \theta}$

3. $r = \dfrac{3}{1 - 2 \sin \theta}$

4. $r = \dfrac{2}{1 + \sin \theta}$

5. $r = \dfrac{6}{2 - \sin \theta}$

6. $r = \dfrac{2}{2 + 3 \cos \theta}$

(c)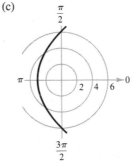

(d)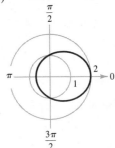

(a)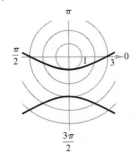

(b)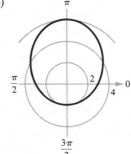

(e)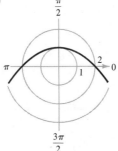

(f)

In Exercises 7–20, identify and sketch the graph of the polar equation.

7. $r = \dfrac{2}{1 - \cos \theta}$

8. $r = \dfrac{4}{1 + \sin \theta}$

9. $r = \dfrac{5}{1 + \sin \theta}$

10. $r = \dfrac{6}{1 + \cos \theta}$

11. $r = \dfrac{2}{2 + \cos \theta}$

12. $r = \dfrac{3}{3 - \sin \theta}$

13. $r = \dfrac{4}{2 + \sin \theta}$

14. $r = \dfrac{6}{3 - 2 \cos \theta}$

15. $r = \dfrac{5}{1 + 2 \cos \theta}$

16. $r = \dfrac{3}{2 + 4 \sin \theta}$

17. $r = \dfrac{1}{1 - \sin \theta}$

18. $r = \dfrac{3}{4 + 2 \cos \theta}$

19. $r = \dfrac{3}{2 - 6 \cos \theta}$

20. $r = \dfrac{3}{2 + 6 \sin \theta}$

In Exercises 21–36, find a polar equation of the specified conic with focus at (0, 0).

Conic	Eccentricity	Directrix
21. Parabola	$e = 1$	$x = -1$
22. Parabola	$e = 1$	$y = -2$
23. Ellipse	$e = \frac{1}{2}$	$y = 1$
24. Ellipse	$e = \frac{3}{4}$	$y = -2$
25. Hyperbola	$e = 2$	$x = 1$
26. Hyperbola	$e = \frac{3}{2}$	$x = -1$

27. Parabola, vertex $\left(1, -\dfrac{\pi}{2}\right)$

28. Ellipse, vertices $(2, 0)$, $(8, \pi)$

29. Ellipse, vertices $\left(2, \dfrac{\pi}{2}\right)$, $\left(4, \dfrac{3\pi}{2}\right)$

30. Parabola, vertex $(4, 0)$

31. Hyperbola, vertices $\left(1, \dfrac{3\pi}{2}\right)$, $\left(9, \dfrac{3\pi}{2}\right)$

32. Hyperbola, vertices $\left(4, \dfrac{\pi}{2}\right)$, $\left(-1, \dfrac{3\pi}{2}\right)$

33. Hyperbola, vertices $(2, 0)$, $(10, 0)$

34. Ellipse, vertices $(20, 0)$, $(4, \pi)$

35. Parabola, vertex $(5, \pi)$

36. Parabola, vertex $\left(10, \dfrac{\pi}{2}\right)$

37. Show that the polar equation of the ellipse

$$\frac{x^2}{a^2} + \frac{y^2}{b^2} = 1 \quad \text{is} \quad r^2 = \frac{b^2}{1 - e^2 \cos^2 \theta}.$$

38. Show that the polar equation of the hyperbola

$$\frac{x^2}{a^2} - \frac{y^2}{b^2} = 1 \quad \text{is} \quad r^{2\cdot} = \frac{-b^2}{1 - e^2 \cos^2 \theta}.$$

In Exercises 39–44, use the results of Exercises 37 and 38 to write the polar form of the equation of the conic.

39. $\dfrac{x^2}{169} + \dfrac{y^2}{144} = 1$

40. $\dfrac{x^2}{25} + \dfrac{y^2}{16} = 1$

41. $\dfrac{x^2}{9} - \dfrac{y^2}{16} = 1$

42. $\dfrac{x^2}{36} - \dfrac{y^2}{4} = 1$

43. Hyperbola One Focus: $\left(5, \dfrac{\pi}{2}\right)$

Vertices: $\left(4, \dfrac{\pi}{2}\right)$, $\left(4, -\dfrac{\pi}{2}\right)$

44. Ellipse One Focus: $(4, 0)$
Vertices: $(5, 0)$, $(5, \pi)$

45. On November 26, 1963, the United States launched *Explorer 18*. Its low point in its elliptical orbit over the surface of the earth was 119 miles, and its high point was 122,000 miles from the surface of the earth. (The center of the earth is the focus of the orbit.) Find the polar equation that describes its orbit.

46. Use the equation of Exercise 45 to find the distance from the earth's surface to the satellite when the angle between the line from the center of the earth to the satellite and the major axis of the ellipse is 60°.

47. An earth satellite in a 100-mile-high circular orbit around the earth has a velocity of approximately 17,500 miles per hour. If this velocity is multiplied by $\sqrt{2}$, then the satellite will have the minimum velocity necessary to escape the earth's gravity and it will follow a parabolic path with the center of the earth as the focus (see figure). Find a polar equation of the parabolic path of the satellite (assume the radius of the earth is 4000 miles).

FIGURE FOR 47

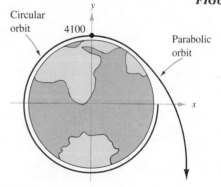

Circular orbit

4100

Parabolic orbit

8.4 *Plane Curves and Parametric Equations*

Up to this point we have been representing a graph by a single equation involving the *two* variables x and y. In this section, we look at situations in which it is useful to introduce a *third* variable to represent a curve in the plane.

To see the usefulness of this procedure, consider the path followed by an object that is propelled into the air at an angle of 45°. If the initial velocity of the object is 48 feet per second, it can be shown that it follows the parabolic path given by

$$y = -\frac{x^2}{72} + x \qquad \textit{Rectangular equation}$$

as shown in Figure 8.21. However, this equation does not tell the whole story. Although it does tell us *where* the object has been, it doesn't tell us *when* the object was at a given point (x, y) on the path. To determine this time, we introduce a third variable t, which we call a **parameter.** It is possible to write both x and y as functions of t to obtain the **parametric equations**

$$x = 24\sqrt{2}\,t \qquad \text{and} \qquad y = -16t^2 + 24\sqrt{2}\,t. \qquad \textit{Parametric equations}$$

From this set of equations we can determine that at time $t = 0$, the object is at the point $(0, 0)$. Similarly, at time $t = 1$, the object is at the point $(24\sqrt{2}, 24\sqrt{2} - 16)$, and so on.

For this particular motion problem, x and y are continuous functions of t, and we call the resulting path a **plane curve.** (For our purposes, it is sufficient to think of a *continuous function* as one whose graph can be traced without lifting the pencil from the paper.)

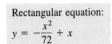

Rectangular equation:
$$y = -\frac{x^2}{72} + x$$

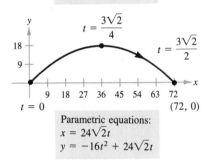

Parametric equations:
$$x = 24\sqrt{2}t$$
$$y = -16t^2 + 24\sqrt{2}t$$

Curvilinear Motion:
two variables for position
one variable for time

FIGURE 8.21

Definition of a Plane Curve

If f and g are continuous functions of t on an interval I, then the set of ordered pairs $(f(t), g(t))$ is called a **plane curve** C. The equations $x = f(t)$ and $y = g(t)$ are called **parametric equations** for C and t is called the **parameter.**

Remark: The set of points $(x, y) = (f(t), g(t))$ in the plane is called the **graph of the curve** C. For simplicity, we will not distinguish between a curve and its graph.

When sketching a curve represented by a pair of parametric equations, we still plot points in the xy-plane. Each set of coordinates (x, y) is determined from a value chosen for the parameter t. By plotting the resulting points in the order of *increasing* values of t, we trace the curve in a specific direction. This is called the **orientation** of the curve.

EXAMPLE 1 Sketching a Curve

Sketch the curve described by the parametric equations

$$x = t^2 - 4$$

and

$$y = \frac{t}{2}, \qquad -2 \le t \le 3.$$

SOLUTION

Using values of t on the given interval, the parametric equations yield the points (x, y) shown in the following table.

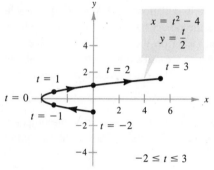

FIGURE 8.22

t	-2	-1	0	1	2	3
x	0	-3	-4	-3	0	5
y	-1	$-1/2$	0	$1/2$	1	$3/2$

By plotting these points in the order of increasing t we obtain the curve C shown in Figure 8.22. Note that the arrows on the curve indicate its orientation as t increases from -2 to 3.

Note that the graph shown in Figure 8.23 does not define y as a function of x. This points out one benefit of parametric equations—they can be used to represent graphs that are more general than graphs of functions.

It often happens that two different sets of parametric equations have the same graph. For example, the set of parametric equations

$$x = 4t^2 - 4$$

and

$$y = t, \qquad -1 \le t \le \frac{3}{2}$$

has the same graph as the set given in Example 1. However, by comparing the values of t in Figures 8.22 and 8.23, we see that this second graph is traced out more *rapidly* (considering t as time) than the first graph. Thus, in applications, different parametric representations can be used to represent various *speeds* at which objects travel along a given path.

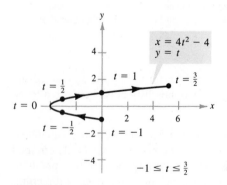

FIGURE 8.23

Eliminating the Parameter

In Example 1, we used simple point-plotting to sketch the given curve. This tedious process can sometimes be simplified by finding a rectangular equation (in x and y) that has the same graph. We call this process **eliminating the parameter.**

Parametric equations		Solve for t in one equation		Substitute into second equation		Rectangular equation
$x = t^2 - 4$	$\rightarrow$	$t = 2y$	$\rightarrow$	$x = (2y)^2 - 4$	$\rightarrow$	$x = 4y^2 - 4$
$y = t/2$						

Now, we recognize that the equation $x = 4y^2 - 4$ represents a parabola with a horizontal axis and vertex at $(-4, 0)$.

One word of caution is in order when converting equations from parametric to rectangular form: the range of x and y implied by the parametric equations may be altered by the change to rectangular form. In such instances it is necessary to adjust the domain of the rectangular equation so that its graph matches the graph of the parametric equations. Such a situation is demonstrated in the next example.

EXAMPLE 2   *Adjusting the Domain after Eliminating the Parameter*

Sketch the curve represented by the equations

$$x = \frac{1}{\sqrt{t+1}} \quad \text{and} \quad y = \frac{t}{t+1}$$

by eliminating the parameter and adjusting the domain of the resulting rectangular equation.

SOLUTION

Solving for t in the equation for x, we have

$$x = \frac{1}{\sqrt{t+1}} \quad\Longrightarrow\quad x^2 = \frac{1}{t+1}$$

which implies that

$$t = \frac{1 - x^2}{x^2}.$$

Now, substituting into the equation for y, we obtain

$$y = \frac{t}{t+1} = \frac{(1 - x^2)/x^2}{[(1 - x^2)/x^2] + 1} = 1 - x^2.$$

The rectangular equation, $y = 1 - x^2$, is defined for all values of x, but from the parametric equation for x we see that the curve is defined only when

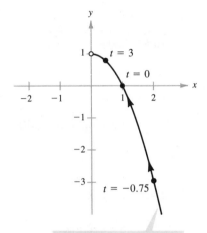

Parametric equations:
$$x = \frac{1}{\sqrt{t+1}}, \, y = \frac{t}{\sqrt{t+1}}$$

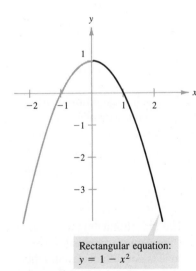

Rectangular equation:
$$y = 1 - x^2$$

FIGURE 8.24

Polar Coordinates and Parametric Equations

$-1 < t$. This implies that we should restrict the domain of x to positive values, as shown in Figure 8.24.

It is not necessary for the parameter in a set of parametric equations to represent time. Our next example uses an *angle* as the parameter. In this example, we use *trigonometric identities* to eliminate the parameter.

EXAMPLE 3 Using a Trigonometric Identity to Eliminate a Parameter

Sketch the curve represented by

$$x = 3 \cos \theta \quad \text{and} \quad y = 4 \sin \theta, \quad 0 \le \theta \le 2\pi$$

by eliminating the parameter and finding the corresponding rectangular equation.

SOLUTION

We begin by solving for $\cos \theta$ and $\sin \theta$ in the given equations.

$$\cos \theta = \frac{x}{3} \quad \text{and} \quad \sin \theta = \frac{y}{4} \qquad \textit{Solve for sin } \theta \textit{ and cos } \theta$$

Now, we make use of the identity $\sin^2 \theta + \cos^2 \theta = 1$ to form an equation involving only x and y.

$$\cos^2 \theta + \sin^2 \theta = 1 \qquad \textit{Trigonometric identity}$$

$$\cos^2 \theta + \sin^2 \theta = \left(\frac{x}{3}\right)^2 + \left(\frac{y}{4}\right)^2 = 1 \qquad \textit{Substitute}$$

$$\frac{x^2}{9} + \frac{y^2}{16} = 1 \qquad \textit{Rectangular equation}$$

From this rectangular equation, we see that the graph is an ellipse centered at $(0, 0)$, with vertices at $(0, 4)$ and $(0, -4)$, and minor axis of length $2b = 6$, as shown in Figure 8.25. Note that the elliptic curve is traced out *counterclockwise* as θ varies from 0 to 2π.

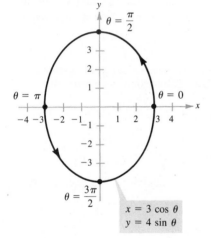

FIGURE 8.25

In Examples 2 and 3 it is important to realize that eliminating the parameter is primarily an *aid to curve-sketching*. If the parametric equations represent the path of a moving object, the graph alone is not sufficient to describe the object's motion. We still need the parametric equations to tell us the *position*, *direction*, and *speed* at a given time.

Finding Parametric Equations for a Graph

We have been looking at techniques for sketching the graph represented by a set of parametric equations. We now look at the reverse problem. How can we determine a set of parametric equations for a given graph or a given physical description? From the discussion following Example 1, we know that such a representation is not unique. This is further demonstrated in the following example in which we find two different parametric representations for a given graph.

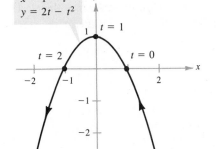

$x = 1 - t$
$y = 2t - t^2$

Graph of $y = 1 - x^2$

FIGURE 8.26

EXAMPLE 4 *Finding Parametric Equations for a Given Graph*

Find a set of parametric equations to represent the graph of $y = 1 - x^2$, using the following parameters.

(a) $t = x$ \qquad\qquad (b) $t = 1 - x$

SOLUTION

(a) Letting $x = t$, we obtain the parametric equations

$$x = t \quad \text{and} \quad y = 1 - x^2 = 1 - t^2.$$

(b) Letting $t = 1 - x$, we obtain

$$x = 1 - t \quad \text{and} \quad y = 1 - (1 - t)^2 = 2t - t^2.$$

In Figure 8.26, note how the resulting curve is oriented by the increasing values of t. For part (a) the curve would have the opposite orientation.

EXAMPLE 5 *Parametric Equations for a Cycloid*

Determine the curve traced out by a point P on the circumference of a circle of radius a as the circle rolls along a straight line in a plane. Such a curve is called a **cycloid.**

SOLUTION

As our parameter, we let θ be the measure of the circle's rotation, and we let the point $P = (x, y)$ begin at the origin. When $\theta = 0$, P is at the origin; when $\theta = \pi$, P is at a maximum point $(\pi a, 2a)$; and when $\theta = 2\pi$, P is back on the x-axis at $(2\pi a, 0)$. From Figure 8.27, we see that $\angle APC = 180° - \theta$. Hence, we have

$$\sin \theta = \sin(180° - \theta) = \sin(\angle APC) = \frac{AC}{a} = \frac{BD}{a}$$

$$\cos \theta = -\cos(180° - \theta) = -\cos(\angle APC) = \frac{AP}{-a}$$

Polar Coordinates and Parametric Equations

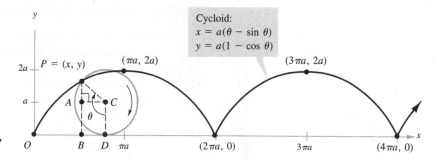

FIGURE 8.27

which implies that $AP = -a \cos \theta$ and $BD = a \sin \theta$. Now since the circle rolls along the x-axis, we know that $OD = \overparen{PD} = a\theta$. Furthermore, since $BA = DC = a$, we have

$$x = OD - BD = a\theta - a \sin \theta$$
$$y = BA + AP = a - a \cos \theta.$$

Therefore, the parametric equations are

$$x = a(\theta - \sin \theta) \qquad \text{and} \qquad y = a(1 - \cos \theta).$$

WARM UP

Sketch the graph of each equation.

1. $y = -\frac{1}{4}x^2$

2. $y = 4 - \frac{1}{4}(x - 2)^2$

3. $16x^2 + y^2 = 16$

4. $-16x^2 + y^2 = 16$

5. $x + y = 4$

6. $x^2 + y^2 = 16$

Simplify the expressions.

7. $10 \sin^2 \theta + 10 \cos^2 \theta$

8. $5 \sec^2 \theta - 5$

9. $\sec^4 x - \tan^4 x$

10. $\dfrac{\sin 2\theta}{4 \cos \theta}$

EXERCISES 8.4

In Exercises 1–26, sketch the curve represented by the parametric equations and write the corresponding rectangular equation by eliminating the parameter.

1. $x = 3t - 1, \quad y = 2t + 1$

2. $x = 3 - 2t, \quad y = 2 + 3t$

3. $x = t + 1, \quad y = t^2$

4. $x = t + 1, \quad y = t^3$

5. $x = \sqrt{t}, \quad y = 1 - t$

6. $x = \sqrt[3]{t}, \quad y = 1 - t$

7. $x = t^3, \quad y = \dfrac{t^2}{2}$

8. $x = 1 + \dfrac{1}{t}, \quad y = t - 1$

9. $x = t - 1, \quad y = \dfrac{t}{t - 1}$

10. $x = t^2 + t, \quad y = t^2 - t$

11. $x = 3 \cos \theta, \quad y = 3 \sin \theta$

12. $x = \cos \theta, \quad y = 3 \sin \theta$

13. $x = 4 \sin 2\theta, \quad y = 2 \cos 2\theta$

14. $x = 2 \cos 2\theta, \quad y = 2 \sin 2\theta$

15. $x = \cos \theta, \quad y = 2 \sin^2 \theta$

16. $x = 4 \cos^2 \theta, \quad y = 2 \sin \theta$

17. $x = \sec \theta, \quad y = \cos \theta$

18. $x = \tan^2 \theta, \quad y = \sec^2 \theta$

19. $x = 4 + 2 \cos \theta, \quad y = -1 + \sin \theta$

20. $x = 4 + 2 \cos \theta, \quad y = -1 + 2 \sin \theta$

21. $x = 4 + 2 \cos \theta, \quad y = -1 + 4 \sin \theta$

22. $x = \sec \theta, \quad y = \tan \theta$

23. $x = 4 \sec \theta, \quad y = 3 \tan \theta$

24. $x = e^{2t}, \quad y = e^t$

25. $x = e^{-t}, \quad y = e^{3t}$

26. $x = \ln 2t, \quad y = t^2$

In Exercises 27–30, eliminate the parameter and obtain the standard form of the rectangular equation of the curve.

27. Line through (x_1, y_1) and (x_2, y_2):

$$x = x_1 + t(x_2 - x_1), \quad y = y_1 + t(y_2 - y_1)$$

28. Circle:

$$x = h + r \cos \theta, \quad y = k + r \sin \theta$$

29. Ellipse:

$$x = h + a \cos \theta, \quad y = k + b \sin \theta$$

30. Hyperbola:

$$x = h + a \sec \theta, \quad y = k + b \tan \theta$$

In Exercises 31–38, find a set of parametric equations for the given line or conic.

31. Line through $(0, 0)$ and $(5, -2)$

32. Line through $(1, 4)$ and $(5, -2)$

33. Circle   Center: $(2, 1)$
    Radius: 4

34. Circle   Center: $(-3, 1)$
    Radius: 3

35. Ellipse   Vertices: $(\pm 5, 0)$
    Foci: $(\pm 4, 0)$

36. Ellipse   Vertices: $(4, 7), (4, -3)$
    Foci: $(4, 5), (4, -1)$

37. Hyperbola   Vertices: $(\pm 4, 0)$
    Foci: $(\pm 5, 0)$

38. Hyperbola   Vertices: $(0, \pm 1)$
    Foci: $(0, \pm 2)$

In Exercises 39 and 40, find two different sets of parametric equations for the given rectangular equation.

39. $y = x^3$ **40.** $y = x^2$

In Exercises 41–46, sketch the curve represented by the parametric equations.

41. Cycloid:

$$x = 2(\theta - \sin \theta), \quad y = 2(1 - \cos \theta)$$

42. Cycloid:

$$x = \theta + \sin \theta, \quad y = 1 - \cos \theta$$

43. Witch of Agnesi:

$$x = 2 \cot \theta, \quad y = 2 \sin^2 \theta$$

44. Curtate cycloid:

$$x = 2\theta - \sin \theta, \quad y = 2 - \cos \theta$$

45. Prolate cycloid:

$$x = \theta - \frac{3}{2} \sin \theta, \quad y = 1 - \frac{3}{2} \cos \theta$$

46. Folium of Descartes:

$$x = \frac{3t}{1 + t^3}, \quad y = \frac{3t^2}{1 + t^3}$$

47. A wheel of radius a rolls along a straight line without slipping (see figure). Find the parametric equation for the curve described by a point P that is b units from the center of the wheel. This curve is called a *curtate cycloid* when $b < a$.

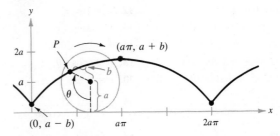

FIGURE FOR 47

48. A wheel of radius 1 rolls around the outside of a circle of radius 2 without slipping (see figure). Show that the parametric equations for the curve described by a point on the rolling wheel are $x = 3 \cos \theta - \cos 3\theta$, and $y = 3 \sin \theta - \sin 3\theta$. This curve is called an *epicycloid*.

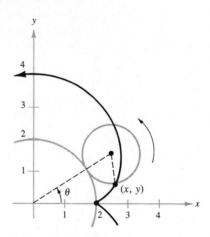

FIGURE FOR 48

CHAPTER 8 REVIEW EXERCISES

In Exercises 1–4, the polar coordinates of a point are given. Plot the point and find the rectangular coordinates of the point.

1. $\left(-2, \dfrac{4\pi}{3}\right)$ **2.** $\left(3, -\dfrac{5\pi}{6}\right)$

3. $\left(5, \dfrac{7\pi}{4}\right)$ **4.** $\left(-4, \dfrac{2\pi}{3}\right)$

In Exercises 5–8, the rectangular coordinates of a point are given. Find two sets of polar coordinates for the point, using $0 \le \theta < 2\pi$.

5. $(0, 8)$ **6.** $(3, 3)$

7. $(-1, 2)$ **8.** $(-3, 0)$

In Exercises 9–24, identify and sketch the graph of the polar equation.

9. $r = 4$ **10.** $\theta = \dfrac{\pi}{12}$

11. $r = 4 \sin 2\theta$ **12.** $r = 2\theta$

13. $r = -2(1 + \cos \theta)$ **14.** $r = 3 - 4 \cos \theta$

15. $r = 4 - 3 \cos \theta$ **16.** $r = \cos 5\theta$

17. $r = -3 \cos 2\theta$ **18.** $r^2 = \cos 2\theta$

19. $r^2 = 4 \sin^2 2\theta$ **20.** $r = 3 \csc \theta$

21. $r = \dfrac{3}{\cos[\theta - (\pi/4)]}$ **22.** $r = \dfrac{4}{5 - 3 \cos \theta}$

23. $r = \dfrac{2}{1 - \sin \theta}$ **24.** $r = \dfrac{1}{1 + 2 \sin \theta}$

In Exercises 25–30, convert the given polar equation to rectangular form.

25. $r = 3 \cos \theta$ **26.** $r = 4 \sec\left(\theta - \dfrac{\pi}{3}\right)$

27. $r = \dfrac{2}{1 + \sin \theta}$ **28.** $r = \dfrac{1}{2 - \cos \theta}$

29. $r^2 = \cos 2\theta$ **30.** $r = 10$

In Exercises 31 and 32, convert the given rectangular equation to polar form.

31. $(x^2 + y^2)^2 = ax^2y$ **32.** $x^2 + y^2 - 4x = 0$

In Exercises 33–38, find a polar equation for the given line or conic.

33. Parabola   Vertex: $(2, \pi)$
               Focus: $(0, 0)$

34. Ellipse   Vertices: $(5, 0), (1, \pi)$
               One focus: $(0, 0)$

35. Hyperbola   Vertices: $(1, 0), (7, 0)$
                   One focus: $(0, 0)$

36. Parabola   Vertex: $(2, \pi/2)$
               Focus: $(0, 0)$

37. Circle   Center: $(5, \pi/2)$
            Passes through $(0, 0)$

38. Line through $(0, 0)$ and slope $\sqrt{3}$

In Exercises 39–50, sketch the curve represented by the parametric equations and, if possible, write the corresponding rectangular equation by eliminating the parameter.

39. $x = 2t, \quad y = 4t$
40. $x = t^2, \quad y = \sqrt{t}$
41. $x = 1 + 4t, \quad y = 2 - 3t$
42. $x = t + 4, \quad y = t^2$
43. $x = e^t, \quad y = e^{-t}$
44. $x = 3 + 2\cos\theta, \quad y = 2 + 5\sin\theta$
45. $x = 6\cos\theta, \quad y = 6\sin\theta$
46. $x = \sec\theta, \quad y = \tan\theta$
47. $x = \dfrac{1}{t}, \quad y = t^2$
48. $x = \dfrac{1}{t}, \quad y = 2t + 3$
49. $x = \cos^3\theta, \quad y = 4\sin^3\theta$
50. $x = 2\theta - \sin\theta, \quad y = 2 - \cos\theta$

51. Find a parametric representation of the ellipse with center at $(-3, 4)$, major axis horizontal and 8 units in length, and minor axis 6 units in length.

52. Find a parametric representation of the hyperbola with vertices at $(0, \pm 4)$ and foci at $(0, \pm 5)$.

53. Show that the Cartesian equation of a cycloid is
$$x = a \arccos\left[(a - y)/a\right] \pm \sqrt{2ay - y^2}.$$

54. The involute of a circle is described by the endpoint P of a string that is held taut as it is unwound from a spool, as shown in the accompanying figure. (The spool does not rotate.) Show that a parametric representation of the involute of a circle is given by
$$x = r(\cos\theta + \theta\sin\theta)$$
and
$$y = r(\sin\theta - \theta\cos\theta).$$

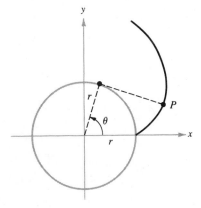

FIGURE FOR 54

Appendix A

Using Logarithmic and Trigonometric Tables

Although it is more efficient to use calculators than tables in computations with logarithms or trigonometric functions, it is instructive to see how to work with tables. We begin with a discussion of the use of tables of logarithms.

Using Logarithmic Tables

Using base 10, we note first that every positive real number can be written as a product $c \times 10^k$, where $1 \leq c < 10$ and k is an integer. For example, $1989 = 1.989 \times 10^3$, $5.37 = 5.37 \times 10^0$, and $0.0439 = 4.39 \times 10^{-2}$. Suppose we apply the properties of logarithms to the number 1989.

$$1989 = 1.989 \times 10^3$$
$$\begin{aligned} \log_{10} 1989 &= \log_{10} (1.989 \times 10^3) \\ &= \log_{10} (1.989) + \log_{10} (10^3) \\ &= \log_{10} (1.989) + 3 \log_{10} (10) \\ &= \log_{10} (1.989) + 3 \end{aligned}$$

In general, for any positive real number x (expressible as $x = c \times 10^k$), its common logarithm has the **standard form**

$$\log_{10} x = \log_{10} c + \log_{10} (10^k) = \log_{10} c + k$$

where $1 \leq c < 10$. We call $\log_{10} c$ the **mantissa** and k the **characteristic** of

$\log_{10} x$. Since the function $f(x) = \log_{10} x$ increases as x increases and since $1 \le c < 10$, it follows that

$$\log_{10} 1 \le \log_{10} c < \log_{10} 10$$
$$0 \le \log_{10} c < 1$$

which means that the *mantissa* of $\log_{10} x$ lies between 0 and 1.

The common logarithm table in Appendix D gives four-decimal-place approximations of the *mantissa* for the logarithm of every three-digit number between 1.00 and 9.99. The next example shows how to use the table in Appendix D to approximate common logarithms.

EXAMPLE 1 *Approximating Common Logarithms with Tables*

Use the tables in Appendix D to approximate the following.

(a) $\log_{10} 85.6$ (b) $\log_{10} 0.000329$

SOLUTION

(a) Since $85.6 = 8.56 \times 10^1$, the characteristic is 1. Using the common logarithm table, we see that the mantissa is $\log_{10} 8.56 \approx 0.9325$. Therefore,

$$\log_{10} 85.6 = (\text{mantissa}) + (\text{characteristic})$$
$$= \log_{10} 8.56 + 1$$
$$\approx 0.9325 + 1$$
$$= 1.9325.$$

(b) Since $0.000329 = 3.29 \times 10^{-4}$, the characteristic is -4. From the common logarithm table for the mantissa 3.29, we obtain

$$\log_{10} 0.000329 = \log_{10} 3.29 + (-4)$$
$$\approx 0.3598 - 4$$
$$= -3.6402.$$

The next example shows how to combine the use of properties of logarithms with tables to evaluate logarithms.

EXAMPLE 2 *Combining Properties of Logarithms with Tables*

Use the tables in Appendix D to approximate $\log_{10} \sqrt[3]{38.6}$.

SOLUTION

$$\log_{10} \sqrt[3]{38.6} = \frac{1}{3} \log_{10} 38.6$$

$$= \frac{1}{3}(\log_{10} 3.86 + 1)$$

$$\approx \frac{1}{3}(0.5866 + 1)$$

$$\approx 0.5289$$

           ━━

The table for common logarithms can be used in the *reverse* manner to find the number (called an **antilogarithm**) that has a given logarithm. We demonstrate this procedure in Example 3.

EXAMPLE 3 *Finding the Antilogarithm of a Number*

Use the tables in Appendix D to approximate the value of x in each of the following.

(a) $\log_{10} x = 2.6571$ (b) $x = 10^{-3.6364}$

SOLUTION

(a) We know that $\log_{10} x = 2.6571 = 0.6571 + 2$. Thus, the mantissa is 0.6571 and the characteristic is 2. From the table, we find that the mantissa 0.6571 corresponds approximately to $\log_{10} 4.54$. Since the characteristic is 2, it follows that x is given by

$$x \approx 4.54 \times 10^2 = 454.$$

(b) In logarithmic form, this exponential equation can be written as $\log_{10} x = -3.6364$. To obtain the standard form, we add and subtract 4 to obtain

$$\log_{10} x = (4 - 3.6364) - 4 = 0.3636 - 4.$$

Thus, the mantissa is 0.3636 and the characteristic is -4. From the table, we find that

$$x \approx 2.31 \times 10^{-4} = 0.000231.$$

           ━━

For numbers with more than three nonzero digits, we can still use the common logarithm tables by applying a procedure called **linear interpolation.** This procedure is based on the fact that changes in $\log_{10} x$ are approximately proportional to the corresponding changes in x. We demonstrate the procedure in Example 4.

EXAMPLE 4 *Linear Interpolation*

Use linear interpolation and the tables in Appendix D to approximate the value of $\log_{10} 5.382$.

SOLUTION

The three-digit x-values in the table that are closest to 5.382 are $x = 5.38$ and $x = 5.39$. We use the logarithms of these two values in the following arrangement.

$$0.01 \left\{ 0.002 \left\{ \begin{array}{l} \log_{10} 5.38 \approx 0.7308 \\ \log_{10} 5.382 \approx ? \\ \log_{10} 5.39 \approx 0.7316 \end{array} \right\} d \right\} 0.0008$$

Note that the differences between the x-values are beside the left braces and the differences between the corresponding logarithms are beside the right braces. From this arrangement, we can write the following proportion.

$$\frac{d}{0.0008} = \frac{0.002}{0.01}$$

$$d = (0.0008)\left(\frac{0.002}{0.01}\right) = 0.00016 \approx 0.0002$$

Therefore,

$$\log_{10} 5.382 \approx \log_{10} 5.38 + d \approx 0.7308 + 0.002 = 0.7310.$$

The next example shows how to perform numerical computations with logarithms.

EXAMPLE 5 *Using Logarithms to Perform Numerical Computations*

Use the tables in Appendix D to approximate the value of

$$x = \frac{(1.9)^3}{\sqrt{82.7}}.$$

SOLUTION

Using the properties of logarithms, we can write

$$\log_{10} x = 3 \log_{10} 1.9 - \frac{1}{2} \log_{10} 82.7$$

$$\approx 3(0.2788) - \frac{1}{2}(1.9175) \approx -0.12235.$$

By adding and subtracting 1, we obtain the standard form

$$\log_{10} x \approx (1 - 0.12235) - 1 = 0.87765 - 1.$$

Finally, since the antilogarithm of 0.87765 is approximately 7.54, we find that

$$x \approx 7.54 \times 10^{-1} = 0.754.$$

Using Trigonometric Tables

Linear interpolation can also be used with tables of trigonometric functions. We demonstrate this procedure in Example 6.

EXAMPLE 6 Approximating Values of Trigonometric Functions

Use the trigonometric tables in Appendix E to approximate the following.

(a) $\cos 132°\ 14'$ (b) $\tan(-20°\ 23')$

SOLUTION

(a) The given angle lies in Quadrant II with a reference angle of $\theta = 179°\ 60' - 132°\ 14' = 47°\ 46'$, so that

$$\cos 132°\ 14' = -\cos 47°\ 46'.$$

Using the table in Appendix E, we obtain the following arrangement.

$$10 \left\{ 6 \left\{ \begin{array}{l} \cos 47°\ 40' \approx 0.6734 \\ \cos 47°\ 46' \approx\ ? \\ \cos 47°\ 50' \approx 0.6713 \end{array} \right. {\Large\}} d \right\} 0.0021$$

Thus, $d = 0.0021(6/10) \approx 0.0013$, and we obtain

$$\begin{aligned} \cos 132°\ 14' &= -\cos 47°\ 46' \\ &\approx -(\cos 47°\ 40' - 0.0013) \\ &\approx -(0.6734 - 0.0013) \\ &= -0.6721. \end{aligned}$$

(b) We know that $\tan(-20° \ 23') = -\tan 20° \ 23'$, and we interpolate as follows.

$$10\left\{ 3\left\{ \begin{array}{l} \tan 20° \ 20' \approx 0.3706 \\ \tan 20° \ 23' \approx ? \\ \tan 20° \ 30' \approx 0.3739 \end{array} \right\}d \right\}0.0033$$

Thus, $d = 0.0033(3/10) \approx 0.0010$, and we obtain

$$
\begin{aligned}
\tan(-20° \ 23') &= -\tan 20° \ 23' \\
&\approx -(\tan 20° \ 20' + 0.0010) \\
&\approx -(0.3706 + 0.0010) \\
&= -0.3716.
\end{aligned}
$$

As with antilogarithms, the trigonometric tables can be used in the reverse manner, as demonstrated in Example 7.

EXAMPLE 7 *Approximating Angles Using Trigonometric Tables*

Approximate the value of t in the interval $0 \le t \le \pi/2$ such that $\sin t = 0.8619$.

SOLUTION

From the column corresponding to the sine in the tables in Appendix E, we choose the two values closest to 0.8619 and interpolate as follows. (Remember to use radians.)

$$0.0029\left\{ d\left\{ \begin{array}{l} \sin 1.0385 \approx 0.8616 \\ \sin (?) \approx 0.8619 \\ \sin 1.0414 \approx 0.8631 \end{array} \right\}3 \right\}15$$

Thus, $d = 0.0029(3/15) \approx 0.0006$, and we obtain

$$t \approx 1.0385 + 0.0006 = 1.0391 \text{ radians.}$$

EXERCISES FOR APPENDIX A

In Exercises 1–4, approximate the common logarithm of the given number by using the table in Appendix D.

1. (a) 417   (b) 0.0417
2. (a) 985   (b) 9.85
3. (a) 6300   (b) 1000
4. (a) 0.0001   (b) 41.3

In Exercises 5–8, approximate the common logarithm of the given quantity by using the table in Appendix D.

5. (a) $\dfrac{5.30}{21.5}$ (b) $(30500)(0.258)$

6. (a) $\sqrt[3]{5.33}$ (b) $(1.02)^{36}$
7. (a) $\sqrt[5]{7200}$ (b) $(3.4)^{8}$
8. (a) $\dfrac{(3.6)^{6}}{500}$ (b) $(0.245)^{4}(8.7)^{3}$

In Exercises 9–12, find N (antilogarithm) by using the table in Appendix D.

9. (a) $N = 10^{4.3979}$ (b) $N = 10^{-1.6021}$
10. (a) $\log_{10} N = 3.6702$ (b) $\log_{10} N = -2.3298$
11. (a) $\log_{10} N = 6.1335$ (b) $\log_{10} N = 8.1335 - 10$
12. (a) $\log_{10} N = 4.8420$ (b) $\log_{10} N = 7.8420 - 10$

In Exercises 13 and 14, use linear interpolation to approximate the common logarithm of the given number by using the table in Appendix D.

13. (a) 4385   (b) 0.6058
14. (a) 125.2   (b) 0.08675

In Exercises 15 and 16, use linear interpolation to approximate the antilogarithm N by using the table in Appendix D.

15. (a) $\log_{10} N = 5.6175$ (b) $\log_{10} N = -2.1503$
16. (a) $N = 10^{0.5743}$ (b) $N = 10^{9.9317}$

In Exercises 17–22, approximate the given quantity by using common logarithms.

17. $\dfrac{(86.4)(8.09)}{38.6}$ **18.** $\dfrac{1243}{(42.8)(67.9)}$
19. $\sqrt[3]{86.5}$ **20.** $\sqrt[4]{(4.705)(18.86)}$
21. $500(1.03)^{20}$ **22.** $(0.2313)^{6}$

In Exercises 23 and 24, approximate the natural logarithm of the given number by using the table in Appendix C.

23. (a) 6.24   (b) 9.55
24. (a) 2.605   (b) 3.005

In Exercises 25 and 26, approximate the antilogarithm N by using the table in Appendix C.

25. (a) $\ln N = 2.0096$ (b) $\ln N = 1.4422$
26. (a) $\ln N = 1.1233$ (b) $\ln N = 0.2271$

In Exercises 27 and 28, approximate the exponential by using the table in Appendix B.

27. (a) $e^{3.5}$ (b) $e^{-3.5}$
28. (a) $e^{6.2}$ (b) $e^{-6.2}$

In Exercises 29–32, approximate the trigonometric function by using the table in Appendix E.

29. (a) $\cos 34°$ (b) $\tan 56°$
30. (a) $\sin 12° \ 30'$ (b) $\cot 77° \ 30'$
31. (a) $\sin 1.0472$ (b) $\cot 0.5236$
32. (a) $\tan 233°$ (b) $\cos(-56°)$

In Exercises 33–37, approximate two values of θ in degrees $(0 \le \theta < 360°)$ corresponding to the given functions by using the table in Appendix E.

33. (a) $\sin \theta = 0.4226$ (b) $\cot \theta = 2.145$
34. (a) $\cos \theta = 0.7585$ (b) $\tan \theta = 1.921$
35. (a) $\cos \theta = 0.4230$ (b) $\cot \theta = 4$
36. (a) $\sin \theta = 0.5140$ (b) $\tan \theta = 0.5$

Appendix B

EXPONENTIAL TABLES

x	e^x	e^{-x}	x	e^x	e^{-x}	x	e^x	e^{-x}
0.0	1.0000	1.0000	3.5	33.115	0.0302	7.0	1096.63	0.0009
0.1	1.1052	0.9048	3.6	36.598	0.0273	7.1	1211.97	0.0008
0.2	1.2214	0.8187	3.7	40.447	0.0247	7.2	1339.43	0.0007
0.3	1.3499	0.7408	3.8	44.701	0.0224	7.3	1480.30	0.0007
0.4	1.4918	0.6703	3.9	49.402	0.0202	7.4	1635.98	0.0006
0.5	1.6487	0.6065	4.0	54.598	0.0183	7.5	1808.04	0.0006
0.6	1.8221	0.5488	4.1	60.340	0.0166	7.6	1998.20	0.0005
0.7	2.0138	0.4966	4.2	66.686	0.0150	7.7	2208.35	0.0005
0.8	2.2255	0.4493	4.3	73.700	0.0136	7.8	2440.60	0.0004
0.9	2.4596	0.4066	4.4	81.451	0.0123	7.9	2697.28	0.0004
1.0	2.7183	0.3679	4.5	90.017	0.0111	8.0	2980.96	0.0003
1.1	3.0042	0.3329	4.6	99.484	0.0101	8.1	3294.47	0.0003
1.2	3.3201	0.3012	4.7	109.95	0.0091	8.2	3640.95	0.0003
1.3	3.6693	0.2725	4.8	121.51	0.0082	8.3	4023.87	0.0002
1.4	4.0552	0.2466	4.9	134.29	0.0074	8.4	4447.07	0.0002
1.5	4.4817	0.2231	5.0	148.41	0.0067	8.5	4914.77	0.0002
1.6	4.9530	0.2019	5.1	164.02	0.0061	8.6	5431.66	0.0002
1.7	5.4739	0.1827	5.2	181.27	0.0055	8.7	6002.91	0.0002
1.8	6.0496	0.1653	5.3	200.34	0.0050	8.8	6634.24	0.0002
1.9	6.6859	0.1496	5.4	221.41	0.0045	8.9	7331.97	0.0001
2.0	7.3891	0.1353	5.5	244.69	0.0041	9.0	8103.08	0.0001
2.1	8.1662	0.1225	5.6	270.43	0.0037	9.1	8955.29	0.0001
2.2	9.0250	0.1108	5.7	298.87	0.0033	9.2	9897.13	0.0001
2.3	9.9742	0.1003	5.8	330.30	0.0030	9.3	10938.02	0.0001
2.4	11.023	0.0907	5.9	365.04	0.0027	9.4	12088.38	0.0001
2.5	12.182	0.0821	6.0	403.43	0.0025	9.5	13359.73	0.0001
2.6	13.464	0.0743	6.1	445.86	0.0022	9.6	14764.78	0.0001
2.7	14.880	0.0672	6.2	492.75	0.0020	9.7	16317.61	0.0001
2.8	16.445	0.0608	6.3	544.57	0.0018	9.8	18033.74	0.0001
2.9	18.174	0.0550	6.4	601.85	0.0017	9.9	19930.37	0.0001
3.0	20.086	0.0498	6.5	665.14	0.0015	10.0	22026.47	0.0000
3.1	22.198	0.0450	6.6	735.10	0.0014			
3.2	24.533	0.0408	6.7	812.41	0.0012			
3.3	27.113	0.0369	6.8	897.85	0.0011			
3.4	29.964	0.0334	6.9	992.27	0.0010			

Appendix C

NATURAL LOGARITHMIC TABLES

	0.00	0.01	0.02	0.03	0.04	0.05	0.06	0.07	0.08	0.09
1.0	0.0000	0.0100	0.0198	0.0296	0.0392	0.0488	0.0583	0.0677	0.0770	0.0862
1.1	0.0953	0.1044	0.1133	0.1222	0.1310	0.1398	0.1484	0.1570	0.1655	0.1740
1.2	0.1823	0.1906	0.1989	0.2070	0.2151	0.2231	0.2311	0.2390	0.2469	0.2546
1.3	0.2624	0.2700	0.2776	0.2852	0.2927	0.3001	0.3075	0.3148	0.3221	0.3293
1.4	0.3365	0.3436	0.3507	0.3577	0.3646	0.3716	0.3784	0.3853	0.3920	0.3988
1.5	0.4055	0.4121	0.4187	0.4253	0.4318	0.4383	0.4447	0.4511	0.4574	0.4637
1.6	0.4700	0.4762	0.4824	0.4886	0.4947	0.5008	0.5068	0.5128	0.5188	0.5247
1.7	0.5306	0.5365	0.5423	0.5481	0.5539	0.5596	0.5653	0.5710	0.5766	0.5822
1.8	0.5878	0.5933	0.5988	0.6043	0.6098	0.6152	0.6206	0.6259	0.6313	0.6366
1.9	0.6419	0.6471	0.6523	0.6575	0.6627	0.6678	0.6729	0.6780	0.6831	0.6881
2.0	0.6931	0.6981	0.7031	0.7080	0.7129	0.7178	0.7227	0.7275	0.7324	0.7372
2.1	0.7419	0.7467	0.7514	0.7561	0.7608	0.7655	0.7701	0.7747	0.7793	0.7839
2.2	0.7885	0.7930	0.7975	0.8020	0.8065	0.8109	0.8154	0.8198	0.8242	0.8286
2.3	0.8329	0.8372	0.8416	0.8459	0.8502	0.8544	0.8587	0.8629	0.8671	0.8713
2.4	0.8755	0.8796	0.8838	0.8879	0.8920	0.8961	0.9002	0.9042	0.9083	0.9123
2.5	0.9163	0.9203	0.9243	0.9282	0.9322	0.9361	0.9400	0.9439	0.9478	0.9517
2.6	0.9555	0.9594	0.9632	0.9670	0.9708	0.9746	0.9783	0.9821	0.9858	0.9895
2.7	0.9933	0.9969	1.0006	1.0043	1.0080	1.0116	1.0152	1.0188	1.0225	1.0260
2.8	1.0296	1.0332	1.0367	1.0403	1.0438	1.0473	1.0508	1.0543	1.0578	1.0613
2.9	1.0647	1.0682	1.0716	1.0750	1.0784	1.0818	1.0852	1.0886	1.0919	1.0953
3.0	1.0986	1.1019	1.1053	1.1086	1.1119	1.1151	1.1184	1.1217	1.1249	1.1282
3.1	1.1314	1.1346	1.1378	1.1410	1.1442	1.1474	1.1506	1.1537	1.1569	1.1600
3.2	1.1632	1.1663	1.1694	1.1725	1.1756	1.1787	1.1817	1.1848	1.1878	1.1909
3.3	1.1939	1.1969	1.2000	1.2030	1.2060	1.2090	1.2119	1.2149	1.2179	1.2208
3.4	1.2238	1.2267	1.2296	1.2326	1.2355	1.2384	1.2413	1.2442	1.2470	1.2499
3.5	1.2528	1.2556	1.2585	1.2613	1.2641	1.2669	1.2698	1.2726	1.2754	1.2782
3.6	1.2809	1.2837	1.2865	1.2892	1.2920	1.2947	1.2975	1.3002	1.3029	1.3056
3.7	1.3083	1.3110	1.3137	1.3164	1.3191	1.3218	1.3244	1.3271	1.3297	1.3324
3.8	1.3350	1.3376	1.3403	1.3429	1.3455	1.3481	1.3507	1.3533	1.3558	1.3584
3.9	1.3610	1.3635	1.3661	1.3686	1.3712	1.3737	1.3762	1.3788	1.3813	1.3838
4.0	1.3863	1.3888	1.3913	1.3938	1.3962	1.3987	1.4012	1.4036	1.4061	1.4085
4.1	1.4110	1.4134	1.4159	1.4183	1.4207	1.4231	1.4255	1.4279	1.4303	1.4327
4.2	1.4351	1.4375	1.4398	1.4422	1.4446	1.4469	1.4493	1.4516	1.4540	1.4563
4.3	1.4586	1.4609	1.4633	1.4656	1.4679	1.4702	1.4725	1.4748	1.4770	1.4793
4.4	1.4816	1.4839	1.4861	1.4884	1.4907	1.4929	1.4951	1.4974	1.4996	1.5019
4.5	1.5041	1.5063	1.5085	1.5107	1.5129	1.5151	1.5173	1.5195	1.5217	1.5239
4.6	1.5261	1.5282	1.5304	1.5326	1.5347	1.5369	1.5390	1.5412	1.5433	1.5454
4.7	1.5476	1.5497	1.5518	1.5539	1.5560	1.5581	1.5602	1.5623	1.5644	1.5665
4.8	1.5686	1.5707	1.5728	1.5748	1.5769	1.5790	1.5810	1.5831	1.5851	1.5872
4.9	1.5892	1.5913	1.5933	1.5953	1.5974	1.5994	1.6014	1.6034	1.6054	1.6074
5.0	1.6094	1.6114	1.6134	1.6154	1.6174	1.6194	1.6214	1.6233	1.6253	1.6273
5.1	1.6292	1.6312	1.6332	1.6351	1.6371	1.6390	1.6409	1.6429	1.6448	1.6467
5.2	1.6487	1.6506	1.6525	1.6544	1.6563	1.6582	1.6601	1.6620	1.6639	1.6658
5.3	1.6677	1.6696	1.6715	1.6734	1.6752	1.6771	1.6790	1.6808	1.6827	1.6845
5.4	1.6864	1.6882	1.6901	1.6919	1.6938	1.6956	1.6974	1.6993	1.7011	1.7029

NATURAL LOGARITHMIC TABLES (Continued)

	0.00	0.01	0.02	0.03	0.04	0.05	0.06	0.07	0.08	0.09
5.5	1.7047	1.7066	1.7084	1.7102	1.7120	1.7138	1.7156	1.7174	1.7192	1.7210
5.6	1.7228	1.7246	1.7263	1.7281	1.7299	1.7317	1.7334	1.7352	1.7370	1.7387
5.7	1.7405	1.7422	1.7440	1.7457	1.7475	1.7492	1.7509	1.7527	1.7544	1.7561
5.8	1.7579	1.7596	1.7613	1.7630	1.7647	1.7664	1.7681	1.7699	1.7716	1.7733
5.9	1.7750	1.7766	1.7783	1.7800	1.7817	1.7834	1.7851	1.7867	1.7884	1.7901
6.0	1.7918	1.7934	1.7951	1.7967	1.7984	1.8001	1.8017	1.8034	1.8050	1.8066
6.1	1.8083	1.8099	1.8116	1.8132	1.8148	1.8165	1.8181	1.8197	1.8213	1.8229
6.2	1.8245	1.8262	1.8278	1.8294	1.8310	1.8326	1.8342	1.8358	1.8374	1.8390
6.3	1.8405	1.8421	1.8437	1.8453	1.8469	1.8485	1.8500	1.8516	1.8532	1.8547
6.4	1.8563	1.8579	1.8594	1.8610	1.8625	1.8641	1.8656	1.8672	1.8687	1.8703
6.5	1.8718	1.8733	1.8749	1.8764	1.8779	1.8795	1.8810	1.8825	1.8840	1.8856
6.6	1.8871	1.8886	1.8901	1.8916	1.8931	1.8946	1.8961	1.8976	1.8991	1.9006
6.7	1.9021	1.9036	1.9051	1.9066	1.9081	1.9095	1.9110	1.9125	1.9140	1.9155
6.8	1.9169	1.9184	1.9199	1.9213	1.9228	1.9242	1.9257	1.9272	1.9286	1.9301
6.9	1.9315	1.9330	1.9344	1.9359	1.9373	1.9387	1.9402	1.9416	1.9430	1.9445
7.0	1.9459	1.9473	1.9488	1.9502	1.9516	1.9530	1.9544	1.9559	1.9573	1.9587
7.1	1.9601	1.9615	1.9629	1.9643	1.9657	1.9671	1.9685	1.9699	1.9713	1.9727
7.2	1.9741	1.9755	1.9769	1.9782	1.9796	1.9810	1.9824	1.9838	1.9851	1.9865
7.3	1.9879	1.9892	1.9906	1.9920	1.9933	1.9947	1.9961	1.9974	1.9988	2.0001
7.4	2.0015	2.0028	2.0042	2.0055	2.0069	2.0082	2.0096	2.0109	2.0122	2.0136
7.5	2.0149	2.0162	2.0176	2.0189	2.0202	2.0215	2.0229	2.0242	2.0255	2.0268
7.6	2.0281	2.0295	2.0308	2.0321	2.0334	2.0347	2.0360	2.0373	2.0386	2.0399
7.7	2.0412	2.0425	2.0438	2.0451	2.0464	2.0477	2.0490	2.0503	2.0516	2.0528
7.8	2.0541	2.0554	2.0567	2.0580	2.0592	2.0605	2.0618	2.0631	2.0643	2.0656
7.9	2.0669	2.0681	2.0694	2.0707	2.0719	2.0732	2.0744	2.0757	2.0769	2.0782
8.0	2.0794	2.0807	2.0819	2.0832	2.0844	2.0857	2.0869	2.0882	2.0894	2.0906
8.1	2.0919	2.0931	2.0943	2.0956	2.0968	2.0980	2.0992	2.1005	2.1017	2.1029
8.2	2.1041	2.1054	2.1066	2.1078	2.1090	2.1102	2.1114	2.1126	2.1138	2.1150
8.3	2.1163	2.1175	2.1187	2.1199	2.1211	2.1223	2.1235	2.1247	2.1258	2.1270
8.4	2.1282	2.1294	2.1306	2.1318	2.1330	2.1342	2.1353	2.1365	2.1377	2.1389
8.5	2.1401	2.1412	2.1424	2.1436	2.1448	2.1459	2.1471	2.1483	2.1494	2.1506
8.6	2.1518	2.1529	2.1541	2.1552	2.1564	2.1576	2.1587	2.1599	2.1610	2.1622
8.7	2.1633	2.1645	2.1656	2.1668	2.1679	2.1691	2.1702	2.1713	2.1725	2.1736
8.8	2.1748	2.1759	2.1770	2.1782	2.1793	2.1804	2.1815	2.1827	2.1838	2.1849
8.9	2.1861	2.1872	2.1883	2.1894	2.1905	2.1917	2.1928	2.1939	2.1950	2.1961
9.0	2.1972	2.1983	2.1994	2.2006	2.2017	2.2028	2.2039	2.2050	2.2061	2.2072
9.1	2.2083	2.2094	2.2105	2.2116	2.2127	2.2138	2.2148	2.2159	2.2170	2.2181
9.2	2.2192	2.2203	2.2214	2.2225	2.2235	2.2246	2.2257	2.2268	2.2279	2.2289
9.3	2.2300	2.2311	2.2322	2.2332	2.2343	2.2354	2.2364	2.2375	2.2386	2.2396
9.4	2.2407	2.2418	2.2428	2.2439	2.2450	2.2460	2.2471	2.2481	2.2492	2.2502
9.5	2.2513	2.2523	2.2534	2.2544	2.2555	2.2565	2.2576	2.2586	2.2597	2.2607
9.6	2.2618	2.2628	2.2638	2.2649	2.2659	2.2670	2.2680	2.2690	2.2701	2.2711
9.7	2.2721	2.2732	2.2742	2.2752	2.2762	2.2773	2.2783	2.2793	2.2803	2.2814
9.8	2.2824	2.2834	2.2844	2.2854	2.2865	2.2875	2.2885	2.2895	2.2905	2.2915
9.9	2.2925	2.2935	2.2946	2.2956	2.2966	2.2976	2.2986	2.2996	2.3006	2.3016

Appendix D

COMMON LOGARITHMIC TABLES

	0.00	0.01	0.02	0.03	0.04	0.05	0.06	0.07	0.08	0.09
1.0	0.0000	0.0043	0.0086	0.0128	0.0170	0.0212	0.0253	0.0294	0.0334	0.0374
1.1	0.0414	0.0453	0.0492	0.0531	0.0569	0.0607	0.0645	0.0682	0.0719	0.0755
1.2	0.0792	0.0828	0.0864	0.0899	0.0934	0.0969	0.1004	0.1038	0.1072	0.1106
1.3	0.1139	0.1173	0.1206	0.1239	0.1271	0.1303	0.1335	0.1367	0.1399	0.1430
1.4	0.1461	0.1492	0.1523	0.1553	0.1584	0.1614	0.1644	0.1673	0.1703	0.1732
1.5	0.1761	0.1790	0.1818	0.1847	0.1875	0.1903	0.1931	0.1959	0.1987	0.2014
1.6	0.2041	0.2068	0.2095	0.2122	0.2148	0.2175	0.2201	0.2227	0.2253	0.2279
1.7	0.2304	0.2330	0.2355	0.2380	0.2405	0.2430	0.2455	0.2480	0.2504	0.2529
1.8	0.2553	0.2577	0.2601	0.2625	0.2648	0.2672	0.2695	0.2718	0.2742	0.2765
1.9	0.2788	0.2810	0.2833	0.2856	0.2878	0.2900	0.2923	0.2945	0.2967	0.2989
2.0	0.3010	0.3032	0.3054	0.3075	0.3096	0.3118	0.3139	0.3160	0.3181	0.3201
2.1	0.3222	0.3243	0.3263	0.3284	0.3304	0.3324	0.3345	0.3365	0.3385	0.3404
2.2	0.3424	0.3444	0.3464	0.3483	0.3502	0.3522	0.3541	0.3560	0.3579	0.3598
2.3	0.3617	0.3636	0.3655	0.3674	0.3692	0.3711	0.3729	0.3747	0.3766	0.3784
2.4	0.3802	0.3820	0.3838	0.3856	0.3874	0.3892	0.3909	0.3927	0.3945	0.3962
2.5	0.3979	0.3997	0.4014	0.4031	0.4048	0.4065	0.4082	0.4099	0.4116	0.4133
2.6	0.4150	0.4166	0.4183	0.4200	0.4216	0.4232	0.4249	0.4265	0.4281	0.4298
2.7	0.4314	0.4330	0.4346	0.4362	0.4378	0.4393	0.4409	0.4425	0.4440	0.4456
2.8	0.4472	0.4487	0.4502	0.4518	0.4533	0.4548	0.4564	0.4579	0.4594	0.4609
2.9	0.4624	0.4639	0.4654	0.4669	0.4683	0.4698	0.4713	0.4728	0.4742	0.4757
3.0	0.4771	0.4786	0.4800	0.4814	0.4829	0.4843	0.4857	0.4871	0.4886	0.4900
3.1	0.4914	0.4928	0.4942	0.4955	0.4969	0.4983	0.4997	0.5011	0.5024	0.5038
3.2	0.5052	0.5065	0.5079	0.5092	0.5105	0.5119	0.5132	0.5145	0.5159	0.5172
3.3	0.5185	0.5198	0.5211	0.5224	0.5237	0.5250	0.5263	0.5276	0.5289	0.5302
3.4	0.5315	0.5328	0.5340	0.5353	0.5366	0.5378	0.5391	0.5403	0.5416	0.5428
3.5	0.5441	0.5453	0.5465	0.5478	0.5490	0.5502	0.5514	0.5527	0.5539	0.5551
3.6	0.5563	0.5575	0.5587	0.5599	0.5611	0.5623	0.5635	0.5647	0.5658	0.5670
3.7	0.5682	0.5694	0.5705	0.5717	0.5729	0.5740	0.5752	0.5763	0.5775	0.5786
3.8	0.5798	0.5809	0.5821	0.5832	0.5843	0.5855	0.5866	0.5877	0.5888	0.5899
3.9	0.5911	0.5922	0.5933	0.5944	0.5955	0.5966	0.5977	0.5988	0.5999	0.6010
4.0	0.6021	0.6031	0.6042	0.6053	0.6064	0.6075	0.6085	0.6096	0.6107	0.6117
4.1	0.6128	0.6138	0.6149	0.6160	0.6170	0.6180	0.6191	0.6201	0.6212	0.6222
4.2	0.6232	0.6243	0.6253	0.6263	0.6274	0.6284	0.6294	0.6304	0.6314	0.6325
4.3	0.6335	0.6345	0.6355	0.6365	0.6375	0.6385	0.6395	0.6405	0.6415	0.6425
4.4	0.6435	0.6444	0.6454	0.6464	0.6474	0.6484	0.6493	0.6503	0.6513	0.6522
4.5	0.6532	0.6542	0.6551	0.6561	0.6571	0.6580	0.6590	0.6599	0.6609	0.6618
4.6	0.6628	0.6637	0.6646	0.6656	0.6665	0.6675	0.6684	0.6693	0.6702	0.6712
4.7	0.6721	0.6730	0.6739	0.6749	0.6758	0.6767	0.6776	0.6785	0.6794	0.6803
4.8	0.6812	0.6821	0.6830	0.6839	0.6848	0.6857	0.6866	0.6875	0.6884	0.6893
4.9	0.6902	0.6911	0.6920	0.6928	0.6937	0.6946	0.6955	0.6964	0.6972	0.6981
5.0	0.6990	0.6998	0.7007	0.7016	0.7024	0.7033	0.7042	0.7050	0.7059	0.7067
5.1	0.7076	0.7084	0.7093	0.7101	0.7110	0.7118	0.7126	0.7135	0.7143	0.7152
5.2	0.7160	0.7168	0.7177	0.7185	0.7193	0.7202	0.7210	0.7218	0.7226	0.7235
5.3	0.7243	0.7251	0.7259	0.7267	0.7275	0.7284	0.7292	0.7300	0.7308	0.7316
5.4	0.7324	0.7332	0.7340	0.7348	0.7356	0.7364	0.7372	0.7380	0.7388	0.7396

COMMON LOGARITHMIC TABLES (Continued)

	0.00	0.01	0.02	0.03	0.04	0.05	0.06	0.07	0.08	0.09
5.5	0.7404	0.7412	0.7419	0.7427	0.7435	0.7443	0.7451	0.7459	0.7466	0.7474
5.6	0.7482	0.7490	0.7497	0.7505	0.7513	0.7520	0.7528	0.7536	0.7543	0.7551
5.7	0.7559	0.7566	0.7574	0.7582	0.7589	0.7597	0.7604	0.7612	0.7619	0.7627
5.8	0.7634	0.7642	0.7649	0.7657	0.7664	0.7672	0.7679	0.7686	0.7694	0.7701
5.9	0.7709	0.7716	0.7723	0.7731	0.7738	0.7745	0.7752	0.7760	0.7767	0.7774
6.0	0.7782	0.7789	0.7796	0.7803	0.7810	0.7818	0.7825	0.7832	0.7839	0.7846
6.1	0.7853	0.7860	0.7868	0.7875	0.7882	0.7889	0.7896	0.7903	0.7910	0.7917
6.2	0.7924	0.7931	0.7938	0.7945	0.7952	0.7959	0.7966	0.7973	0.7980	0.7987
6.3	0.7993	0.8000	0.8007	0.8014	0.8021	0.8028	0.8035	0.8041	0.8048	0.8055
6.4	0.8062	0.8069	0.8075	0.8082	0.8089	0.8096	0.8102	0.8109	0.8116	0.8122
6.5	0.8129	0.8136	0.8142	0.8149	0.8156	0.8162	0.8169	0.8176	0.8182	0.8189
6.6	0.8195	0.8202	0.8209	0.8215	0.8222	0.8228	0.8235	0.8241	0.8248	0.8254
6.7	0.8261	0.8267	0.8274	0.8280	0.8287	0.8293	0.8299	0.8306	0.8312	0.8319
6.8	0.8325	0.8331	0.8338	0.8344	0.8351	0.8357	0.8363	0.8370	0.8376	0.8382
6.9	0.8388	0.8395	0.8401	0.8407	0.8414	0.8420	0.8426	0.8432	0.8439	0.8445
7.0	0.8451	0.8457	0.8463	0.8470	0.8476	0.8482	0.8488	0.8494	0.8500	0.8506
7.1	0.8513	0.8519	0.8525	0.8531	0.8537	0.8543	0.8549	0.8555	0.8561	0.8567
7.2	0.8573	0.8579	0.8585	0.8591	0.8597	0.8603	0.8609	0.8615	0.8621	0.8627
7.3	0.8633	0.8639	0.8645	0.8651	0.8657	0.8663	0.8669	0.8675	0.8681	0.8686
7.4	0.8692	0.8698	0.8704	0.8710	0.8716	0.8722	0.8727	0.8733	0.8739	0.8745
7.5	0.8751	0.8756	0.8762	0.8768	0.8774	0.8779	0.8785	0.8791	0.8797	0.8802
7.6	0.8808	0.8814	0.8820	0.8825	0.8831	0.8837	0.8842	0.8848	0.8854	0.8859
7.7	0.8865	0.8871	0.8876	0.8882	0.8887	0.8893	0.8899	0.8904	0.8910	0.8915
7.8	0.8921	0.8927	0.8932	0.8938	0.8943	0.8949	0.8954	0.8960	0.8965	0.8971
7.9	0.8976	0.8982	0.8987	0.8993	0.8998	0.9004	0.9009	0.9015	0.9020	0.9025
8.0	0.9031	0.9036	0.9042	0.9047	0.9053	0.9058	0.9063	0.9069	0.9074	0.9079
8.1	0.9085	0.9090	0.9096	0.9101	0.9106	0.9112	0.9117	0.9122	0.9128	0.9133
8.2	0.9138	0.9143	0.9149	0.9154	0.9159	0.9165	0.9170	0.9175	0.9180	0.9186
8.3	0.9191	0.9196	0.9201	0.9206	0.9212	0.9217	0.9222	0.9227	0.9232	0.9238
8.4	0.9243	0.9248	0.9253	0.9258	0.9263	0.9269	0.9274	0.9279	0.9284	0.9289
8.5	0.9294	0.9299	0.9304	0.9309	0.9315	0.9320	0.9325	0.9330	0.9335	0.9340
8.6	0.9345	0.9350	0.9355	0.9360	0.9365	0.9370	0.9375	0.9380	0.9385	0.9390
8.7	0.9395	0.9400	0.9405	0.9410	0.9415	0.9420	0.9425	0.9430	0.9435	0.9440
8.8	0.9445	0.9450	0.9455	0.9460	0.9465	0.9469	0.9474	0.9479	0.9484	0.9489
8.9	0.9494	0.9499	0.9504	0.9509	0.9513	0.9518	0.9523	0.9528	0.9533	0.9538
9.0	0.9542	0.9547	0.9552	0.9557	0.9562	0.9566	0.9571	0.9576	0.9581	0.9586
9.1	0.9590	0.9595	0.9600	0.9605	0.9609	0.9614	0.9619	0.9624	0.9628	0.9633
9.2	0.9638	0.9643	0.9647	0.9652	0.9657	0.9661	0.9666	0.9671	0.9675	0.9680
9.3	0.9685	0.9689	0.9694	0.9699	0.9703	0.9708	0.9713	0.9717	0.9722	0.9727
9.4	0.9731	0.9736	0.9741	0.9745	0.9750	0.9754	0.9759	0.9764	0.9768	0.9773
9.5	0.9777	0.9782	0.9786	0.9791	0.9795	0.9800	0.9805	0.9809	0.9814	0.9818
9.6	0.9823	0.9827	0.9832	0.9836	0.9841	0.9845	0.9850	0.9854	0.9859	0.9863
9.7	0.9868	0.9872	0.9877	0.9881	0.9886	0.9890	0.9894	0.9899	0.9903	0.9908
9.8	0.9912	0.9917	0.9921	0.9926	0.9930	0.9934	0.9939	0.9943	0.9948	0.9952
9.9	0.9956	0.9961	0.9965	0.9969	0.9974	0.9978	0.9983	0.9987	0.9991	0.9996

TRIGONOMETRIC TABLES

1 degree ≈ 0.01745 radians
1 radian ≈ 57.29578 degrees

For $0 \le \theta \le 45$, read from upper left.
For $45 \le \theta \le 90$, read from lower right.
For $90 \le \theta \le 360$, use the identities:

θ	Quadrant II	Quadrant III	Quadrant IV
$\sin \theta$	$\sin(180-\theta)$	$-\sin(\theta-180)$	$-\sin(360-\theta)$
$\cos \theta$	$-\cos(180-\theta)$	$-\cos(\theta-180)$	$\cos(360-\theta)$
$\tan \theta$	$-\tan(180-\theta)$	$\tan(\theta-180)$	$-\tan(360-\theta)$
$\cot \theta$	$-\cot(180-\theta)$	$\cot(\theta-180)$	$-\cot(360-\theta)$

Degrees	Radians	sin	cos	tan	cot		
0°00′	.0000	.0000	1.0000	.0000	—	1.5708	90°00′
10	.0029	.0029	1.0000	.0029	343.774	1.5679	50
20	.0058	.0058	1.0000	.0058	171.885	1.5650	40
30	.0087	.0087	1.0000	.0087	114.589	1.5621	30
40	.0116	.0116	.9999	.0116	85.940	1.5592	20
50	.0145	.0145	.9999	.0145	68.750	1.5563	10
1°00′	.0175	.0175	.9998	.0175	57.290	1.5533	89°00′
10	.0204	.0204	.9998	.0204	49.104	1.5504	50
20	.0233	.0233	.9997	.0233	42.964	1.5475	40
30	.0262	.0262	.9997	.0262	38.188	1.5446	30
40	.0291	.0291	.9996	.0291	34.368	1.5417	20
50	.0320	.0320	.9995	.0320	31.242	1.5388	10
2°00′	.0349	.0349	.9994	.0349	28.636	1.5359	88°00′
10	.0378	.0378	.9993	.0378	26.432	1.5330	50
20	.0407	.0407	.9992	.0407	24.542	1.5301	40
30	.0436	.0436	.9990	.0437	22.904	1.5272	30
40	.0465	.0465	.9989	.0466	21.470	1.5243	20
50	.0495	.0494	.9988	.0495	20.206	1.5213	10
3°00′	.0524	.0523	.9986	.0524	19.081	1.5184	87°00′
10	.0553	.0552	.9985	.0553	18.075	1.5155	50
20	.0582	.0581	.9983	.0582	17.169	1.5126	40
30	.0611	.0610	.9981	.0612	16.350	1.5097	30
40	.0640	.0640	.9980	.0641	15.605	1.5068	20
50	.0669	.0669	.9978	.0670	14.924	1.5039	10
	cos	sin	cot	tan	Radians	Degrees	

Degrees	Radians	sin	cos	tan	cot		
4°00′	.0698	.0698	.9976	.0699	14.301	1.5010	86°00′
10	.0727	.0727	.9974	.0729	13.727	1.4981	50
20	.0756	.0756	.9971	.0758	13.197	1.4952	40
30	.0785	.0785	.9969	.0787	12.706	1.4923	30
40	.0814	.0814	.9967	.0816	12.251	1.4893	20
50	.0844	.0843	.9964	.0846	11.826	1.4864	10
5°00′	.0873	.0872	.9962	.0875	11.430	1.4835	85°00′
10	.0902	.0901	.9959	.0904	11.059	1.4806	50
20	.0931	.0929	.9957	.0934	10.712	1.4777	40
30	.0960	.0958	.9954	.0963	10.385	1.4748	30
40	.0989	.0987	.9951	.0992	10.078	1.4719	20
50	.1018	.1016	.9948	.1022	9.788	1.4690	10
6°00′	.1047	.1045	.9945	.1051	9.514	1.4661	84°00′
10	.1076	.1074	.9942	.1080	9.255	1.4632	50
20	.1105	.1103	.9939	.1110	9.010	1.4603	40
30	.1134	.1132	.9936	.1139	8.777	1.4573	30
40	.1164	.1161	.9932	.1169	8.556	1.4544	20
50	.1193	.1190	.9929	.1198	8.345	1.4515	10
7°00′	.1222	.1219	.9925	.1228	8.144	1.4486	83°00′
10	.1251	.1248	.9922	.1257	7.953	1.4457	50
20	.1280	.1276	.9918	.1287	7.770	1.4428	40
30	.1309	.1305	.9914	.1317	7.596	1.4399	30
40	.1338	.1334	.9911	.1346	7.429	1.4370	20
50	.1367	.1363	.9907	.1376	7.269	1.4341	10
	cos	sin	cot	tan	Radians	Degrees	

TRIGONOMETRIC TABLES (Continued)

Degrees	Radians	sin	cos	tan	cot		
8°00′	.1396	.1392	.9903	.1405	7.115	1.4312	82°00′
10	.1425	.1421	.9899	.1435	6.968	1.4283	50
20	.1454	.1449	.9894	.1465	6.827	1.4254	40
30	.1484	.1478	.9890	.1495	6.691	1.4224	30
40	.1513	.1507	.9886	.1524	6.561	1.4195	20
50	.1542	.1536	.9881	.1554	6.435	1.4166	10
9°00′	.1571	.1564	.9877	.1584	6.314	1.4137	81°00′
10	.1600	.1593	.9872	.1614	6.197	1.4108	50
20	.1629	.1622	.9868	.1644	6.084	1.4079	40
30	.1658	.1650	.9863	.1673	5.976	1.4050	30
40	.1687	.1679	.9858	.1703	5.871	1.4021	20
50	.1716	.1708	.9853	.1733	5.769	1.3992	10
10°00′	.1745	.1736	.9848	.1763	5.671	1.3963	80°00′
10	.1774	.1765	.9843	.1793	5.576	1.3934	50
20	.1804	.1794	.9838	.1823	5.485	1.3904	40
30	.1833	.1822	.9833	.1853	5.396	1.3875	30
40	.1862	.1851	.9827	.1883	5.309	1.3846	20
50	.1891	.1880	.9822	.1914	5.226	1.3817	10
11°00′	.1920	.1908	.9816	.1944	5.145	1.3788	79°00′
10	.1949	.1937	.9811	.1974	5.066	1.3759	50
20	.1978	.1965	.9805	.2004	4.989	1.3730	40
30	.2007	.1994	.9799	.2035	4.915	1.3701	30
40	.2036	.2022	.9793	.2065	4.843	1.3672	20
50	.2065	.2051	.9787	.2095	4.773	1.3643	10
12°00′	.2094	.2079	.9781	.2126	4.705	1.3614	78°00′
10	.2123	.2108	.9775	.2156	4.638	1.3584	50
20	.2153	.2136	.9769	.2186	4.574	1.3555	40
30	.2182	.2164	.9763	.2217	4.511	1.3526	30
40	.2211	.2193	.9757	.2247	4.449	1.3497	20
50	.2240	.2221	.9750	.2278	4.390	1.3468	10
13°00′	.2269	.2250	.9744	.2309	4.331	1.3439	77°00′
10	.2298	.2278	.9737	.2339	4.275	1.3410	50
20	.2327	.2306	.9730	.2370	4.219	1.3381	40
30	.2356	.2334	.9724	.2401	4.165	1.3352	30
40	.2385	.2363	.9717	.2432	4.113	1.3323	20
50	.2414	.2391	.9710	.2462	4.061	1.3294	10
14°00′	.2443	.2419	.9703	.2493	4.011	1.3265	76°00′
10	.2473	.2447	.9696	.2524	3.962	1.3235	50
20	.2502	.2476	.9689	.2555	3.914	1.3206	40
30	.2531	.2504	.9681	.2586	3.867	1.3177	30
40	.2560	.2532	.9674	.2617	3.821	1.3148	20
50	.2589	.2560	.9667	.2648	3.776	1.3119	10
15°00′	.2618	.2588	.9659	.2679	3.732	1.3090	75°00′
10	.2647	.2616	.9652	.2711	3.689	1.3061	50
20	.2676	.2644	.9644	.2742	3.647	1.3032	40
30	.2705	.2672	.9636	.2773	3.606	1.3003	30
40	.2734	.2700	.9628	.2805	3.566	1.2974	20
50	.2763	.2728	.9621	.2836	3.526	1.2945	10
16°00′	.2793	.2756	.9613	.2867	3.487	1.2915	74°00′
10	.2822	.2784	.9605	.2899	3.450	1.2886	50
20	.2851	.2812	.9596	.2931	3.412	1.2857	40
30	.2880	.2840	.9588	.2962	3.376	1.2828	30
40	.2909	.2868	.9580	.2994	3.340	1.2799	20
50	.2938	.2896	.9572	.3026	3.305	1.2770	10
17°00′	.2967	.2924	.9563	.3057	3.271	1.2741	73°00′
10	.2996	.2952	.9555	.3089	3.237	1.2712	50
20	.3025	.2979	.9546	.3121	3.204	1.2683	40
30	.3054	.3007	.9537	.3153	3.172	1.2654	30
40	.3083	.3035	.9528	.3185	3.140	1.2625	20
50	.3113	.3062	.9520	.3217	3.108	1.2595	10
		cos	sin	cot	tan	Radians	Degrees

Degrees	Radians	sin	cos	tan	cot		
18°00′	.3142	.3090	.9511	.3249	3.078	1.2566	72°00′
10	.3171	.3118	.9502	.3281	3.047	1.2537	50
20	.3200	.3145	.9492	.3314	3.018	1.2508	40
30	.3229	.3173	.9483	.3346	2.989	1.2479	30
40	.3258	.3201	.9474	.3378	2.960	1.2450	20
50	.3287	.3228	.9465	.3411	2.932	1.2421	10
19°00′	.3316	.3256	.9455	.3443	2.904	1.2392	71°00′
10	.3345	.3283	.9446	.3476	2.877	1.2363	50
20	.3374	.3311	.9436	.3508	2.850	1.2334	40
30	.3403	.3338	.9426	.3541	2.824	1.2305	30
40	.3432	.3365	.9417	.3574	2.798	1.2275	20
50	.3462	.3393	.9407	.3607	2.773	1.2246	10
20°00′	.3491	.3420	.9397	.3640	2.747	1.2217	70°00′
10	.3520	.3448	.9387	.3673	2.723	1.2188	50
20	.3549	.3475	.9377	.3706	2.699	1.2159	40
30	.3578	.3502	.9367	.3739	2.675	1.2130	30
40	.3607	.3529	.9356	.3772	2.651	1.2101	20
50	.3636	.3557	.9346	.3805	2.628	1.2072	10
21°00′	.3665	.3584	.9336	.3839	2.605	1.2043	69°00′
10	.3694	.3611	.9325	.3872	2.583	1.2014	50
20	.3723	.3638	.9315	.3906	2.560	1.1985	40
30	.3752	.3665	.9304	.3939	2.539	1.1956	30
40	.3782	.3692	.9293	.3973	2.517	1.1926	20
50	.3811	.3719	.9283	.4006	2.496	1.1897	10
22°00′	.3840	.3746	.9272	.4040	2.475	1.1868	68°00′
10	.3869	.3773	.9261	.4074	2.455	1.1839	50
20	.3898	.3800	.9250	.4108	2.434	1.1810	40
30	.3927	.3827	.9239	.4142	2.414	1.1781	30
40	.3956	.3854	.9228	.4176	2.394	1.1752	20
50	.3985	.3881	.9216	.4210	2.375	1.1723	10
23°00′	.4014	.3907	.9205	.4245	2.356	1.1694	67°00′
10	.4043	.3934	.9194	.4279	2.337	1.1665	50
20	.4072	.3961	.9182	.4314	2.318	1.1636	40
30	.4102	.3987	.9171	.4348	2.300	1.1606	30
40	.4131	.4014	.9159	.4383	2.282	1.1577	20
50	.4160	.4041	.9147	.4417	2.264	1.1548	10
24°00′	.4189	.4067	.9135	.4452	2.246	1.1519	66°00′
10	.4218	.4094	.9124	.4487	2.229	1.1490	50
20	.4247	.4120	.9112	.4522	2.211	1.1461	40
30	.4276	.4147	.9100	.4557	2.194	1.1432	30
40	.4305	.4173	.9088	.4592	2.177	1.1403	20
50	.4334	.4200	.9075	.4628	2.161	1.1374	10
25°00′	.4363	.4226	.9063	.4663	2.145	1.1345	65°00′
10	.4392	.4253	.9051	.4699	2.128	1.1316	50
20	.4422	.4279	.9038	.4734	2.112	1.1286	40
30	.4451	.4305	.9026	.4770	2.097	1.1257	30
40	.4480	.4331	.9013	.4806	2.081	1.1228	20
50	.4509	.4358	.9001	.4841	2.066	1.1199	10
26°00′	.4538	.4384	.8988	.4877	2.050	1.1170	64°00′
10	.4567	.4410	.8975	.4913	2.035	1.1141	50
20	.4596	.4436	.8962	.4950	2.020	1.1112	40
30	.4625	.4462	.8949	.4986	2.006	1.1083	30
40	.4654	.4488	.8936	.5022	1.991	1.1054	20
50	.4683	.4514	.8923	.5059	1.977	1.1025	10
27°00′	.4712	.4540	.8910	.5095	1.963	1.0996	63°00′
10	.4741	.4566	.8897	.5132	1.949	1.0966	50
20	.4771	.4592	.8884	.5169	1.935	1.0937	40
30	.4800	.4617	.8870	.5206	1.921	1.0908	30
40	.4829	.4643	.8857	.5243	1.907	1.0879	20
50	.4858	.4669	.8843	.5280	1.894	1.0850	10
		cos	sin	cot	tan	Radians	Degrees

Appendix E

TRIGONOMETRIC TABLES (Continued)

Degrees	Radians	sin	cos	tan	cot		Degrees	Radians	sin	cos	tan	cot	
28°00′	.4887	.4695	.8829	.5317	1.881	1.0821	62°00′	37°00′	.6458	.6018	.7986	.7536	1.327	.9250	53°00′
10	.4916	.4720	.8816	.5354	1.868	1.0792	50	10	.6487	.6041	.7969	.7581	1.319	.9221	50
20	.4945	.4746	.8802	.5392	1.855	1.0763	40	20	.6516	.6065	.7951	.7627	1.311	.9192	40
30	.4974	.4772	.8788	.5430	1.842	1.0734	30	30	.6545	.6088	.7934	.7673	1.303	.9163	30
40	.5003	.4797	.8774	.5467	1.829	1.0705	20	40	.6574	.6111	.7916	.7720	1.295	.9134	20
50	.5032	.4823	.8760	.5505	1.816	1.0676	10	50	.6603	.6134	.7898	.7766	1.288	.9105	10
29°00′	.5061	.4848	.8746	.5543	1.804	1.0647	61°00′	38°00′	.6632	.6157	.7880	.7813	1.280	.9076	52°00′
10	.5091	.4874	.8732	.5581	1.792	1.0617	50	10	.6661	.6180	.7862	.7860	1.272	.9047	50
20	.5120	.4899	.8718	.5619	1.780	1.0588	40	20	.6690	.6202	.7844	.7907	1.265	.9018	40
30	.5149	.4924	.8704	.5658	1.767	1.0559	30	30	.6720	.6225	.7826	.7954	1.257	.8988	30
40	.5178	.4950	.8689	.5696	1.756	1.0530	20	40	.6749	.6248	.7808	.8002	1.250	.8959	20
50	.5207	.4975	.8675	.5735	1.744	1.0501	10	50	.6778	.6271	.7790	.8050	1.242	.8930	10
30°00′	.5236	.5000	.8660	.5774	1.732	1.0472	60°00′	39°00′	.6807	.6293	.7771	.8098	1.235	.8901	51°00′
10	.5265	.5025	.8646	.5812	1.720	1.0443	50	10	.6836	.6316	.7753	.8146	1.228	.8872	50
20	.5294	.5050	.8631	.5851	1.709	1.0414	40	20	.6865	.6338	.7735	.8195	1.220	.8843	40
30	.5323	.5075	.8616	.5890	1.698	1.0385	30	30	.6894	.6361	.7716	.8243	1.213	.8814	30
40	.5325	.5100	.8601	.5930	1.686	1.0356	20	40	.6923	.6383	.7698	.8292	1.206	.8785	20
50	.5381	.5125	.8587	.5969	1.675	1.0327	10	50	.6952	.6406	.7679	.8342	1.199	.8756	10
31°00′	.5411	.5150	.8572	.6009	1.664	1.0297	59°00′	40°00′	.6981	.6428	.7660	.8391	1.192	.8727	50°00′
10	.5440	.5175	.8557	.6048	1.653	1.0268	50	10	.7010	.6450	.7642	.8441	1.185	.8698	50
20	.5469	.5200	.8542	.6088	1.643	1.0239	40	20	.7039	.6472	.7623	.8491	1.178	.8668	40
30	.5498	.5225	.8526	.6128	1.632	1.0210	30	30	.7069	.6494	.7604	.8541	1.171	.8639	30
40	.5527	.5250	.8511	.6168	1.621	1.0181	20	40	.7098	.6517	.7585	.8591	1.164	.8610	20
50	.5556	.5275	.8496	.6208	1.611	1.0152	10	50	.7127	.6539	.7566	.8642	1.157	.8581	10
32°00′	.5585	.5299	.8480	.6249	1.600	1.0123	58°00′	41°00′	.7156	.6561	.7547	.8693	1.150	.8552	49°00′
10	.5614	.5324	.8465	.6289	1.590	1.0094	50	10	.7185	.6583	.7528	.8744	1.144	.8523	50
20	.5643	.5348	.8450	.6330	1.580	1.0065	40	20	.7214	.6604	.7509	.8796	1.137	.8494	40
30	.5672	.5373	.8434	.6371	1.570	1.0036	30	30	.7243	.6626	.7490	.8847	1.130	.8465	30
40	.5701	.5398	.8418	.6412	1.560	1.0007	20	40	.7272	.6648	.7470	.8899	1.124	.8436	20
50	.5730	.5422	.8403	.6453	1.550	.9977	10	50	.7301	.6670	.7451	.8952	1.117	.8407	10
33°00′	.5760	.5446	.8387	.6494	1.540	.9948	57°00′	42°00′	.7330	.6691	.7431	.9004	1.111	.8378	48°00′
10	.5789	.5471	.8371	.6536	1.530	.9919	50	10	.7359	.6713	.7412	.9057	1.104	.8348	50
20	.5818	.5495	.8355	.6577	1.520	.9890	40	20	.7389	.6734	.7392	.9110	1.098	.8319	40
30	.5847	.5519	.8339	.6619	1.511	.9861	30	30	.7418	.6756	.7373	.9163	1.091	.8290	30
40	.5876	.5544	.8323	.6661	1.501	.9832	20	40	.7447	.6777	.7353	.9217	1.085	.8261	20
50	.5905	.5568	.8307	.6703	1.492	.9803	10	50	.7476	.6799	.7333	.9271	1.079	.8232	10
34°00′	.5934	.5592	.8290	.6745	1.483	.9774	56°00′	43°00′	.7505	.6820	.7314	.9325	1.072	.8203	47°00′
10	.5963	.5616	.8274	.6787	1.473	.9745	50	10	.7534	.6841	.7294	.9380	1.066	.8174	50
20	.5992	.5640	.8258	.6830	1.464	.9716	40	20	.7563	.6862	.7274	.9435	1.060	.8145	40
30	.6021	.5664	.8241	.6873	1.455	.9687	30	30	.7592	.6884	.7254	.9490	1.054	.8116	30
40	.6050	.5688	.8225	.6916	1.446	.9657	20	40	.7621	.6905	.7234	.9545	1.048	.8087	20
50	.6080	.5712	.8208	.6959	1.437	.9628	10	50	.7650	.6926	.7214	.9601	1.042	.8058	10
35°00′	.6109	.5736	.8192	.7002	1.428	.9599	55°00′	44°00′	.7679	.6947	.7193	.9657	1.036	.8029	46°00′
10	.6138	.5760	.8175	.7046	1.419	.9570	50	10	.7709	.6967	.7173	.9713	1.030	.7999	50
20	.6167	.5783	.8158	.7089	1.411	.9541	40	20	.7738	.6988	.7153	.9770	1.024	.7970	40
30	.6196	.5807	.8141	.7133	1.402	.9512	30	30	.7767	.7009	.7133	.9827	1.018	.7941	30
40	.6225	.5831	.8124	.7177	1.393	.9483	20	40	.7796	.7030	.7112	.9884	1.012	.7912	20
50	.6254	.5854	.8107	.7221	1.385	.9454	10	50	.7825	.7050	.7092	.9942	1.006	.7883	10
36°00′	.6283	.5878	.8090	.7265	1.376	.9425	54°00′	45°00′	.7854	.7071	.7071	1.0000	1.000	.7854	45°00′
10	.6312	.5901	.8073	.7310	1.368	.9396	50								
20	.6341	.5925	.8056	.7355	1.360	.9367	40								
30	.6370	.5948	.8039	.7400	1.351	.9338	30								
40	.6400	.5972	.8021	.7445	1.343	.9308	20								
50	.6429	.5995	.8004	.7490	1.335	.9279	10								
		cos	sin	cot	tan	Radians	Degrees			cos	sin	cot	tan	Radians	Degrees

Appendix F

Graphs of Inverse Trigonometric Functions

Domain: $[-1, 1]$
Range: $[-\pi/2, \pi/2]$

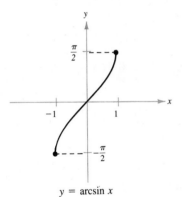

$y = \arcsin x$

Domain: $(-\infty, -1]$ and $[1, \infty)$
Range: $[-\pi/2, 0)$ and $(0, \pi/2]$

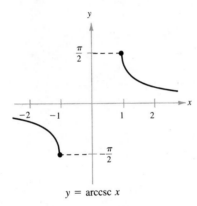

$y = \operatorname{arccsc} x$

Domain: $(-\infty, \infty)$
Range: $(-\pi/2, \pi/2)$

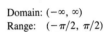
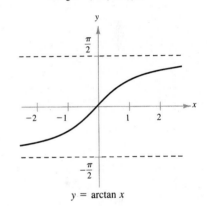

$y = \arctan x$

Domain: $[-1, 1]$
Range: $[0, \pi]$

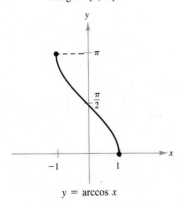

$y = \arccos x$

Domain: $(-\infty, -1]$ and $[1, \infty)$
Range: $[0, \pi/2)$ and $(\pi/2, \pi]$

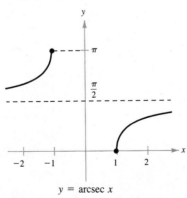

$y = \operatorname{arcsec} x$

Domain: $(-\infty, \infty)$
Range: $(0, \pi)$

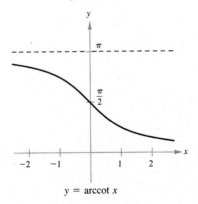

$y = \operatorname{arccot} x$

Graphs of Inverse Trigonometric Functions

Function	Domain	Range
$y = \arcsin x$ iff $\sin y = x$	$-1 \leq x \leq 1$	$-\pi/2 \leq y \leq \pi/2$
$y = \arccos x$ iff $\cos y = x$	$-1 \leq x \leq 1$	$0 \leq y \leq \pi$
$y = \arctan x$ iff $\tan y = x$	$-\infty < x < \infty$	$-\pi/2 < y < \pi/2$
$y = \text{arccot } x$ iff $\cot y = x$	$-\infty < x < \infty$	$0 < y < \pi$
$y = \text{arcsec } x$ iff $\sec y = x$	$|x| \geq 1$	$0 \leq y \leq \pi, \quad y \neq \pi/2$
$y = \text{arccsc } x$ iff $\csc y = x$	$|x| \geq 1$	$-\pi/2 \leq y \leq \pi/2, \quad y \neq 0$

Answers to Warm Ups and Odd-Numbered Exercises

CHAPTER 1
Section 1.1 (*page 9*)

1. $\frac{3}{2} < 7$

3. $-4 > -8$

5. $\frac{5}{6} > \frac{2}{3}$

7. $x < 0$ 9. $A \geq 30$ 11. $3.5\% \leq R \leq 6\%$
13. 10   15. $\pi - 3$ 17. -1 19. -6
21. 4   23. $\frac{5}{2}$ 25. 51   27. 14.99
29. $|x - 5| \leq 3$ 31. $|z - \frac{3}{2}| > 1$ 33. $|y| \geq 6$
35. (a) 0.625   (b) $0.3\overline{3}$ (c) $0.123\overline{123}$
37. $\frac{127}{90}, \frac{584}{413}, \frac{7071}{5000}, \sqrt{2}, \frac{47}{33}$ (b) $\frac{7071}{5000}$
39. c   40. h   41. f   42. e   43. g   44. a
45. b   46. d   47. $|x| \leq 2$ 49. $|x - 9| \geq 3$
51. $|x - 12| \leq 10$ 53. $|x + 3| > 5$

Section 1.2 (*page 21*)

1. (a) No   (b) No   (c) Yes   (d) No
3. (a) Yes   (b) Yes   (c) No   (d) No   5. 9
7. -4 9. 9   11. No solution   13. 10   15. 4
17. 3   19. 5   21. $\frac{11}{6}$ 23. 0

25. $\frac{1 + 4b}{2 + a}, a \neq -2$ 27. $0, -\frac{1}{2}$ 29. $4, -2$

31. $-\frac{7}{4}$ 33. $-\frac{3}{2}, 11$ 35. $\pm 2\sqrt{3}$ 37. $12 \pm 3\sqrt{2}$

39. $\frac{1}{2}, -1$ 41. $\frac{1}{4}, -\frac{3}{4}$ 43. $1 \pm \sqrt{3}$ 45. $\frac{2}{3} \pm \frac{\sqrt{7}}{3}$

47. $-\frac{1}{2} \pm \sqrt{2}$ 49. $6 \pm \sqrt{11}$ 51. $-\frac{1}{2} \pm \frac{\sqrt{21}}{6}$

53. $x \approx 0.976, -0.643$ 55. $x \approx 0.561, 0.126$

57. $0, \pm\frac{3\sqrt{2}}{2}$ 59. $3, -1, 0$ 61. ± 3 63. $-3, 0$
65. $3, 1, -1$ 67. ± 1 69. $\pm 3, \pm 1$ 71. ± 2
73. $\pm\frac{1}{2}, \pm 4$ 75. $1, -2$

Section 1.3 (*page 29*)

> **WARM UP**
>
> 1. 5   2. $3\sqrt{2}$ 3. 1   4. -2
> 5. $3(\sqrt{2} + \sqrt{5})$ 6. $2(\sqrt{3} + \sqrt{11})$
> 7. $-3, 11$ 8. $9, 1$ 9. 11   10. 4

1.

3.

5. 8   7. 5   9. $a = 4, b = 3, c = 5$
11. $a = 10, b = 3, c = \sqrt{109}$
13. (a)

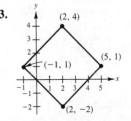

15. (a)

(b) 10   (c) (5, 4)

(b) 17   (c) $(0, \frac{5}{2})$

A19

17. (a)

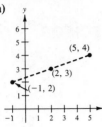

19. (a)

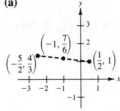

1. (a) Yes   **(b)** Yes   **3. (a)** No   **(b)** Yes
5. (a) Yes   **(b)** Yes   **7.** 2   **9.** 4
11. $(5, 0)$, $(0, -5)$ **13.** $(-2, 0)$, $(1, 0)$, $(0, -2)$
15. $(0, 0)$, $(-2, 0)$ **17.** $(1, 0)$, $\left(0, \frac{1}{2}\right)$
19. y-axis symmetry   **21.** x-axis symmetry
23. Origin symmetry   **25.** Origin symmetry
27. c   **28.** f   **29.** d   **30.** a   **31.** e   **32.** b

(b) $2\sqrt{10}$ **(c)** $(2, 3)$

(b) $\dfrac{\sqrt{82}}{3}$ **(c)** $\left(-1, \frac{7}{6}\right)$

21. (a)

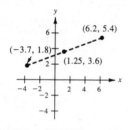

23. (a)

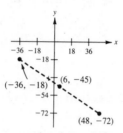

33. Intercepts:
$\left(\frac{2}{3}, 0\right)$, $(0, 2)$

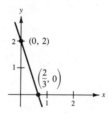

35. Intercepts:
$(-1, 0)$, $(1, 0)$, $(0, 1)$
Symmetry: y-axis

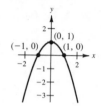

(b) $\sqrt{110.97}$
(c) $(1.25, 3.6)$

(b) $6\sqrt{277}$
(c) $(6, -45)$

25. $(\sqrt{5})^2 + (\sqrt{45})^2 = (\sqrt{50})^2$
27. All sides have a length of $\sqrt{5}$.   **29.** $x = 6, -4$
31. $y = \pm 15$ **33.** $3x - 2y - 1 = 0$
35. Quadrant IV   **37.** Quadrant I
39. Quadrant II   **41.** Quadrant III or IV
43. $(2x_m - x_1, 2y_m - y_1)$
45. $\left(\dfrac{3x_1 + x_2}{4}, \dfrac{3y_1 + y_2}{4}\right)$, $\left(\dfrac{x_1 + x_2}{2}, \dfrac{y_1 + y_2}{2}\right)$,
$\left(\dfrac{x_1 + 3x_2}{4}, \dfrac{y_1 + 3y_2}{4}\right)$

47. \$630,000
49. (a) \$200   **(b)** \$180   **(c)** \$173   **(d)** \$162
51. (a) \$12.6 billion   **(b)** \$14.5 billion

37. Intercepts:
$(3, 0)$, $(1, 0)$, $(0, 3)$

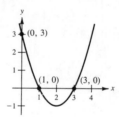

39. Intercepts:
$(-\sqrt[3]{2}, 0)$, $(0, 2)$

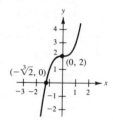

Section 1.4   *(page 42)*

WARM UP

1. $y = \dfrac{3x - 2}{5}$ **2.** $y = -\dfrac{(x - 5)(x + 1)}{2}$
3. 2   **4.** 1, -5 **5.** 0, ± 3 **6.** ± 2
7. $y = x^3 + 4x$ **8.** $x^2 + y^2 = 4$
9. $y = 4x^2 + 8$ **10.** $y^2 = 3x^2 + 4$

41. Intercepts: $(0, 0)$,
$(2, 0)$

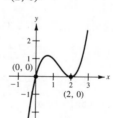

43. Intercept: $(3, 0)$

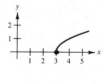

45. Intercept: (0, 0)
Symmetry: Origin

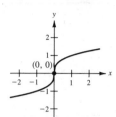

47. Intercepts: (2, 0), (0, 2)

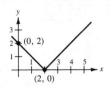

65. $\left(x - \frac{1}{2}\right)^2 + \left(y - \frac{1}{2}\right)^2 = 2$

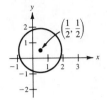

67. $\left(x + \frac{1}{2}\right)^2 + \left(y + \frac{5}{4}\right)^2 = \frac{9}{4}$

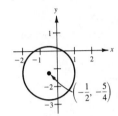

49. Intercepts: (−1, 0),
(0, 1), (0, −1)
Symmetry: *x*-axis

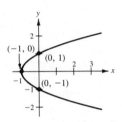

51. Intercepts: (−2, 0),
(2, 0), (0, 2), (0, −2)
Symmetry: *x*-axis,
y-axis, origin

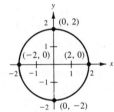

53. $x^2 + y^2 = 9$
55. $(x - 2)^2 + (y + 1)^2 = 16$
57. $(x + 1)^2 + (y - 2)^2 = 5$
59. $(x - 3)^2 + (y - 4)^2 = 25$
61. $(x - 1)^2 + (y + 3)^2 = 4$

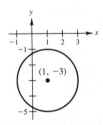

63. $(x - 1)^2 + (y + 3)^2 = 0$

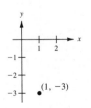

69. (a)

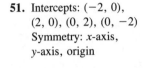

(b) $y = 399.55$

71. (a)

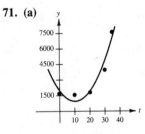

(b) $y = \$9,706.40$

Section 1.5   (*page* 54)

WARM UP

1. −73   **2.** 13   **3.** $2(x + 2)$ **4.** $-8(x - 2)$
5. $y = \frac{7}{5} - \frac{2}{5}x$ **6.** $y = \pm x$
7. $x \le -2, x \ge 2$ **8.** $-3 \le x \le 3$
9. All real numbers   **10.** $x \le 1, x \ge 2$

1. (a) $6 - 4(3)$ **(b)** $6 - 4(-7)$
    (c) $6 - 4t$ **(d)** $6 - 4(c + 1)$
3. (a) $\dfrac{1}{4 + 1}$ **(b)** $\dfrac{1}{0 + 1}$
    (c) $\dfrac{1}{4x + 1}$ **(d)** $\dfrac{1}{(x + h) + 1}$
5. (a) -1 **(b)** -9 **(c)** $2x - 5$ **(d)** $-\frac{5}{2}$
7. (a) 0 **(b)** 3 **(c)** $x^2 + 2x$ **(d)** -0.75

9. (a) 1   **(b)** -7 **(c)** $3 - 2|x|$ **(d)** 2.5

11. (a) $\frac{1}{7}$ **(b)** $-\frac{1}{9}$ **(c)** Undefined   **(d)** $\dfrac{1}{y^2 + 6y}$

13. (a) 1   **(b)** -1 **(c)** 1   **(d)** $\dfrac{|x - 1|}{x - 1}$

15. (a) -1 **(b)** 2   **(c)** 4   **(d)** 6

17. $3 + h$ **19.** $3x\Delta x + 3x^2 + (\Delta x)^2$

21. 3   **23.** 5   **25.** ± 3 **27.** $\frac{10}{7}$

29. All real numbers

31. All real numbers except $t = 0$

33. $y \geq 10$ **35.** $-1 \leq x \leq 1$

37. All real numbers except $x = 0, -2$

39. Not a function   **41.** Function   **43.** Function

45. Not a function   **47.** Function

49. (a) Function   **(b)** Not a function

(c) Function   **(d)** Not a function

51. $(-2, 4), (-1, 1), (0, 0), (1, 1), (2, 4)$

53. $(-2, 0), (-1, 1), (0, \sqrt{2}), (1, \sqrt{3}), (2, 2)$

55. $2, -1$ **57.** 3, 0   **59.** $V = e^3$

61. $A = \dfrac{C^2}{4\pi}$ **63.** $V = 4x(6 - x)^2$

65. $A = \dfrac{x^2}{x - 1}, x > 1$ **67.** $h = \sqrt{d^2 - 2000^2}$

69. (a) $C = 12.30x + 98,000$ **(b)** $R = 17.98x$

(c) $P = 5.68x - 98,000$

71. (a) $R = \dfrac{240n - n^2}{20}$

(b)

n	90	100	110	120
$R(n)$	\$675	\$700	\$715	\$720

n	130	140	150
$R(n)$	\$715	\$700	\$675

73.

y	5	10	20
$F(y)$	26,474	149,760	847,170

y	30	40
$F(y)$	2,334,527	4,792,320

Section 1.6   (*page* 67)

1. Domain: $[1, \infty)$

Range: $[0, \infty)$

3. Domain: $(-\infty, -2], [2, \infty)$

Range: $[0, \infty)$

5. Domain: $[-5, 5]$

Range: $[0, 5]$

7. Function   **9.** Not a function   **11.** Function

13. (a) Increasing on $(-\infty, \infty)$ **(b)** Odd function

15. (a) Increasing on $(-\infty, 0), (2, \infty)$, decreasing on $(0, 2)$

(b) Neither even nor odd

17. (a) Increasing on $(-1, 0), (1, \infty)$, decreasing on

$(-\infty, -1), (0, 1)$ **(b)** Even function

19. (a) Increasing on $(-2, \infty)$, decreasing on $(-3, -2)$

(b) Neither even nor odd

21. Even   **23.** Odd   **25.** Neither even nor odd

27. Even   **29.** Neither even nor odd

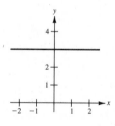

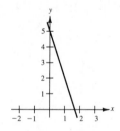

31. Odd   **33.** Neither even nor odd

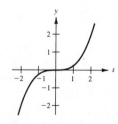

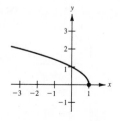

35. Neither even nor odd   **37.** Neither even nor odd

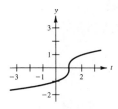

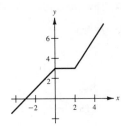

(c)

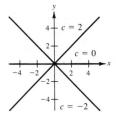

39. $(-\infty, 4]$ **41.** $(-\infty, -3], [3, \infty)$

51. (a)   **(b)**

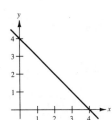

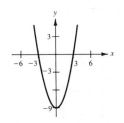

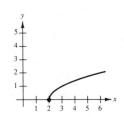

43. $[-1, 1]$ **45.** $(-\infty, \infty)$

(c)   **(d)**

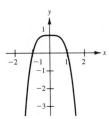

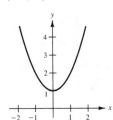

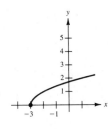

47. $f(x) < 0$ for all x

(e)   **(f)**

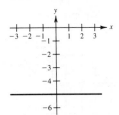

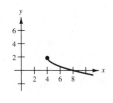

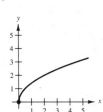

49. (a)   **(b)**

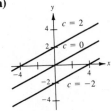

53. (a) $g(x) = (x - 1)^2 + 1$ **(b)** $g(x) = -(x + 1)^2$
55. $h = (4x - x^2) - 3$ **57.** $h = 4x - 2x^2$
59. $L = (4 - y^2) - (y + 2)$

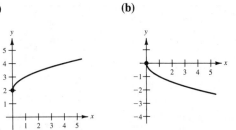

Section 1.7   *(page* 80)

1. (a) $2x$ (b) 2 (c) $x^2 - 1$ (d) $\dfrac{x + 1}{x - 1}, x \neq 1$

3. (a) $x^2 - x + 1$ (b) $x^2 + x - 1$
   (c) $x^2 - x^3$ (d) $\dfrac{x^2}{1 - x}, x \neq 1$

5. (a) $\dfrac{x + 1}{x^2}$ (b) $\dfrac{x - 1}{x^2}$ (c) $\dfrac{1}{x^3}$ (d) $x, x \neq 0$

7. 9   9. $4t^2 - 2t + 5$ 11. 0   13. 26   15. 5

17. (a) $(x - 1)^2$ (b) $x^2 - 1$ (c) x^4

19. (a) $20 - 3x$ (b) $-3x$ (c) $9x + 20$

21. (a) 3   (b) 0   23. (a) 0   (b) 4

25.
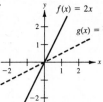

27.

29.

31.
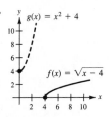

33.
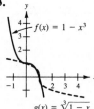

35. One-to-one

37. Not one-to-one   39. One-to-one

41. Not one-to-one   43. Not one-to-one

45. $f^{-1}(x) = \dfrac{x + 3}{2}$ 47. $f^{-1}(x) = \sqrt[5]{x}$

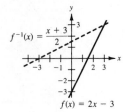

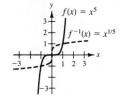

49. $f^{-1}(x) = x^2, x \geq 0$

51. $f^{-1}(x) = \sqrt{4 - x^2}, 0 \leq x \leq 2$

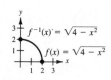

53. $f^{-1}(x) = x^3 + 1$

55. Not one-to-one

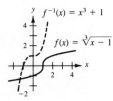

57. $g^{-1}(x) = 8x$ 59. Not one-to-one

61. $f^{-1}(x) = \sqrt{x} - 3, x \geq 0$ 63. $h^{-1}(x) = \dfrac{1}{x}$

65. $f^{-1}(x) = \dfrac{x^2 - 3}{2}, x \geq 0$ 67. Not one-to-one

69. $f^{-1}(x) = -\sqrt{25 - x}$

71.

x	0	1	2	3	4
$f^{-1}(x)$	-2	0	1	2	4

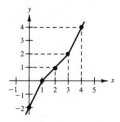

73. $(g^{-1} \circ f^{-1})(x) = \dfrac{x + 1}{2}$

75. $(f \circ g)^{-1}(x) = \dfrac{x + 1}{2}$

Chapter 1 Review Exercises   (*page* 82)

1. $-4 < \frac{4}{3}$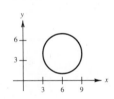

3. $-6 < -4$

5. 2   **7.** -10

9.

11.

13. $|x - 7| \geq 4$ **15.** $|x - 5| \leq 2$ **17.** 20

19. $0, \frac{12}{5}$ **21.** $0, 2$ **23.** $\frac{1}{5}$ **25.** $\pm\dfrac{\sqrt{2}}{2}$

27. $0, 1, 2$ **29.** $1, \frac{5}{3}$

31. (a) 10   **(b)** $(0, 5)$ **(c)** $x = 0$

33. (a) 13   **(b)** $\left(8, \frac{7}{2}\right)$ **(c)** $5x - 12y + 2 = 0$

35. Intercept: $(0, 0)$
Symmetry: x-axis

37. Intercepts: $(0, 0)$,
$(2\sqrt{2}, 0)$, $(-2\sqrt{2}, 0)$
Symmetry: y-axis

39. Intercepts: $(0, 0)$,
$(-2, 0)$, $(2, 0)$
Symmetry: Origin

41. $(6, 4)$, $r = 3$

43. $\left(\frac{1}{2}, 5\right)$, $r = \frac{3}{2}$

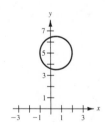

45.

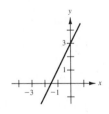

47.

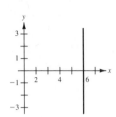

49.

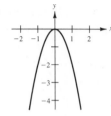

51.

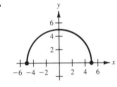

53.

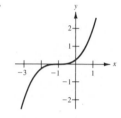

55. (a) 5   **(b)** 17   **(c)** $t^4 + 1$ **(d)** $x^2 + 4x + 5$

57. $[-5, 5]$ **59.** All real numbers except $s = 3$

61. (a) $f^{-1}(x) = 2x + 6$

(b)

63. (a) $f^{-1}(x) = x^2 - 1$, $x \geq 0$

(b)

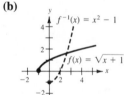

65. (a) $f^{-1}(x) = \sqrt{x + 5}$, $x \geq -5$

(b)

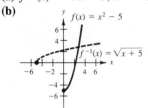

67. $x \geq 4$, $f^{-1}(x) = \sqrt{\dfrac{x}{2} + 4}$

69. $x \geq 2$, $f^{-1}(x) = \sqrt{x^2 + 4}$

71. -7 **73.** 23   **75.** 9

77. $A = x(12 - x)$, $(0, 6]$ **79.** 4 farmers

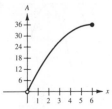

CHAPTER 2

Section 2.1   *(page 95)*

WARM UP

1. 45   **2.** 70   **3.** $\dfrac{\pi}{6}$ **4.** $\dfrac{\pi}{3}$ **5.** $\dfrac{\pi}{4}$

6. $\dfrac{4\pi}{3}$ **7.** $\dfrac{\pi}{9}$ **8.** $\dfrac{11\pi}{6}$ **9.** 45   **10.** 45

1. (a) Quadrant I   **(b)** Quadrant III
3. (a) Quadrant IV   **(b)** Quadrant II
5. (a) Quadrant III   **(b)** Quadrant II
7. (a) Quadrant II   **(b)** Quadrant IV
9. (a) Quadrant III   **(b)** Quadrant I
11. (a)   **(b)**

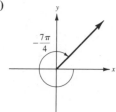

13. (a)   **(b)**

15. (a)   **(b)**

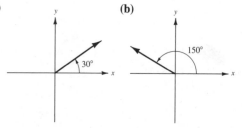

17. (a) $\dfrac{19\pi}{9}$, $-\dfrac{17\pi}{9}$ **(b)** $\dfrac{10\pi}{3}$, $-\dfrac{2\pi}{3}$

19. (a) $\dfrac{7\pi}{4}$, $-\dfrac{\pi}{4}$ **(b)** $\dfrac{28\pi}{15}$, $-\dfrac{32\pi}{15}$

21. (a) 396°, $-324°$ **(b)** 315°, $-405°$
23. (a) 660°, $-60°$ **(b)** 20°, $-340°$

25. (a) $\dfrac{\pi}{6}$ **(b)** $\dfrac{5\pi}{6}$ **27. (a)** $-\dfrac{\pi}{9}$ **(b)** $-\dfrac{4\pi}{3}$

29. (a) 270°   **(b)** 210°   **31. (a)** 420°   **(b)** $-66°$
33. 2.007   **35.** -3.776 **37.** 9.285   **39.** -0.014
41. 25.714°   **43.** 337.5°   **45.** $-756°$
47. $-114.592°$ **49. (a)** 54.75°   **(b)** $-128.5°$
51. (a) 85.308°   **(b)** 330.007°
53. (a) 240° 36′   **(b)** $-145°$ 48′
55. (a) 143° 14′ 22″   **(b)** $-205°$ 7′ 8″
57. $\frac{2}{5}$ rad   **59.** 1.724 rad
61. 15π in ≈ 47.12 in   **63.** 12 m
65. 591.72 mi   **67.** 1141.02 mi   **69.** 4.655°
71. 171.89°
73. (a) 560.2 rev/min   **(b)** 3520 rad/min
75. (a) 2 in: 3400π rad/min; 4 in: 1700π rad/min
     (b) 850 rev/min

Section 2.2   *(page 105)*

WARM UP

1. $\sqrt{3}$ **2.** -1 **3.** $-\dfrac{\sqrt{3}}{3}$ **4.** 1   **5.** $\dfrac{2\pi}{3}$

6. $\dfrac{5\pi}{4}$ **7.** $\dfrac{7\pi}{4}$ **8.** $\dfrac{4\pi}{3}$ **9.** 2π **10.** π

1. $\left(\dfrac{\sqrt{2}}{2}, \dfrac{\sqrt{2}}{2}\right)$ **3.** $\left(-\dfrac{\sqrt{3}}{2}, \dfrac{1}{2}\right)$ **5.** $\left(-\dfrac{1}{2}, -\dfrac{\sqrt{3}}{2}\right)$

7. $(0, -1)$ **9. (a)** $\dfrac{\sqrt{2}}{2}$ **(b)** $\dfrac{\sqrt{2}}{2}$ **(c)** 1

11. (a) $-\dfrac{1}{2}$ **(b)** $\dfrac{\sqrt{3}}{2}$ **(c)** $-\dfrac{\sqrt{3}}{3}$

13. (a) 0   **(b)** -1 **(c)** 0

15. (a) $\dfrac{\sqrt{2}}{2}$ **(b)** $-\dfrac{\sqrt{2}}{2}$ **(c)** -1

17. (a) $-\dfrac{1}{2}$ **(b)** $\dfrac{\sqrt{3}}{2}$ **(c)** $-\dfrac{\sqrt{3}}{3}$

19. (a) $-\dfrac{\sqrt{3}}{2}$ **(b)** $-\dfrac{1}{2}$ **(c)** $\sqrt{3}$

21. $\sin \dfrac{\pi}{4} = \dfrac{\sqrt{2}}{2}$ $\csc \dfrac{\pi}{4} = \sqrt{2}$

$\cos \dfrac{\pi}{4} = \dfrac{\sqrt{2}}{2}$ $\sec \dfrac{\pi}{4} = \sqrt{2}$

$\tan \dfrac{\pi}{4} = 1$ $\cot \dfrac{\pi}{4} = 1$

23. $\sin \dfrac{\pi}{2} = 1$ $\csc \dfrac{\pi}{2} = 1$

$\cos \dfrac{\pi}{2} = 0$ $\sec \dfrac{\pi}{2}$ is undefined.

$\tan \dfrac{\pi}{2}$ is undefined. $\cos \dfrac{\pi}{2} = 0$

25. $\sin\left(-\dfrac{4\pi}{3}\right) = \dfrac{\sqrt{3}}{2}$ $\csc\left(-\dfrac{4\pi}{3}\right) = \dfrac{2\sqrt{3}}{3}$

$\cos\left(-\dfrac{4\pi}{3}\right) = -\dfrac{1}{2}$ $\sec\left(-\dfrac{4\pi}{3}\right) = -2$

$\tan\left(-\dfrac{4\pi}{3}\right) = -\sqrt{3}$ $\cot\left(-\dfrac{4\pi}{3}\right) = -\dfrac{\sqrt{3}}{3}$

27. 0   **29.** $-\dfrac{1}{2}$ **31.** $-\dfrac{\sqrt{3}}{2}$ **33.** $-\dfrac{\sqrt{2}}{2}$

35. (a) $-\dfrac{1}{3}$ (b) -3 **37.** (a) $-\dfrac{7}{8}$ (b) $-\dfrac{8}{7}$

39. 0.7071   **41.** -0.9900 **43.** 0.3249

45. -0.1288 **47.** 1.3940   **49.** -1.4486

51. (a) 0.25 ft   (b) 0.0177 ft   (c) -0.2475 ft

Section 2.3   *(page* 115)

WARM UP

1. $\dfrac{\pi}{6}$ **2.** $\dfrac{3\pi}{4}$ **3.** $60°$ **4.** $-270°$

5. $\sqrt{34}$ **6.** $2\sqrt{2}$ **7.** 1.24   **8.** 317.55

9. 2,785,714.29   **10.** 28.80

1. $\sin \theta = \dfrac{1}{2}$ $\csc \theta = 2$

$\cos \theta = \dfrac{\sqrt{3}}{2}$ $\sec \theta = \dfrac{2\sqrt{3}}{3}$

$\tan \theta = \dfrac{\sqrt{3}}{3}$ $\cot \theta = \sqrt{3}$

3. $\sin \theta = \dfrac{3}{5}$ $\csc \theta = \dfrac{5}{3}$

$\cos \theta = \dfrac{4}{5}$ $\sec \theta = \dfrac{5}{4}$

$\tan \theta = \dfrac{3}{4}$ $\cot \theta = \dfrac{4}{3}$

5. $\sin \theta = \dfrac{\sqrt{161}}{15}$ $\csc \theta = \dfrac{15\sqrt{161}}{161}$

$\cos \theta = \dfrac{8}{15}$ $\sec \theta = \dfrac{15}{8}$

$\tan \theta = \dfrac{\sqrt{161}}{8}$ $\cot \theta = \dfrac{8\sqrt{161}}{161}$

7. $\sin \theta = \dfrac{8\sqrt{73}}{73}$ $\csc \theta = \dfrac{\sqrt{73}}{8}$

$\cos \theta = \dfrac{3\sqrt{73}}{73}$ $\sec \theta = \dfrac{\sqrt{73}}{3}$

$\tan \theta = \dfrac{8}{3}$ $\cot \theta = \dfrac{3}{8}$

9. $\cos \theta = \dfrac{\sqrt{5}}{3}$

$\tan \theta = \dfrac{2\sqrt{5}}{5}$

$\csc \theta = \dfrac{3}{2}$

$\sec \theta = \dfrac{3\sqrt{5}}{5}$

$\cot \theta = \dfrac{\sqrt{5}}{2}$

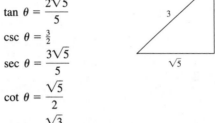

11. $\sin \theta = \dfrac{\sqrt{3}}{2}$

$\cos \theta = \dfrac{1}{2}$

$\tan \theta = \sqrt{3}$

$\csc \theta = \dfrac{2\sqrt{3}}{3}$

$\cot \theta = \dfrac{\sqrt{3}}{3}$

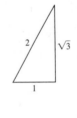

13. $\sin \theta = \dfrac{3\sqrt{10}}{10}$

$\cos \theta = \dfrac{\sqrt{10}}{10}$

$\csc \theta = \dfrac{\sqrt{10}}{3}$

$\sec \theta = \sqrt{10}$

$\cot \theta = \dfrac{1}{3}$

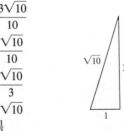

15. $\sin \theta = \dfrac{2\sqrt{13}}{13}$

$\cos \theta = \dfrac{3\sqrt{13}}{13}$

$\tan \theta = \dfrac{2}{3}$

$\csc \theta = \dfrac{\sqrt{13}}{2}$

$\sec \theta = \dfrac{\sqrt{13}}{3}$

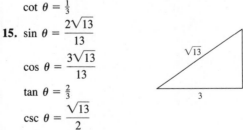

17. (a) $\sqrt{3}$ (b) $\frac{1}{2}$ (c) $\frac{\sqrt{3}}{2}$ (d) $\frac{\sqrt{3}}{3}$

19. (a) $\frac{1}{3}$ (b) $\frac{2\sqrt{2}}{3}$ (c) $\frac{\sqrt{2}}{4}$ (d) 3

21. (a) $\frac{1}{2}$ (b) $\frac{\sqrt{3}}{3}$ **23.** (a) 1 (b) $\frac{\sqrt{2}}{2}$

25. (a) 0.1736   (b) 0.1736
27. (a) 1.3499   (b) 1.3432
29. (a) 0.2815   (b) 3.5523
31. (a) 5.0273   (b) 0.1989
33. (a) 1.1884   (b) 1.1884

35. (a) $30° = \frac{\pi}{6}$ (b) $30° = \frac{\pi}{6}$

37. (a) $60° = \frac{\pi}{3}$ (b) $45° = \frac{\pi}{4}$

39. (a) $60° = \frac{\pi}{3}$ (b) $45° = \frac{\pi}{4}$

41. (a) $55° \approx 0.96$ (b) $89° \approx 1.55$
43. (a) $50° \approx 0.87$ (b) $25° \approx 0.44$

45. 57.74   **47.** $\frac{25\sqrt{3}}{3}$ **49.** 15.56

51. 9.19   **53.** 15 ft   **55.** 19.32 ft   **57.** 2145 ft

59. True, $\csc x = \dfrac{1}{\sin x}$ **61.** False, $\dfrac{\sqrt{2}}{2} + \dfrac{\sqrt{2}}{2} \neq 1$

63. False, $\dfrac{\sin 60°}{\sin 30°} = 1.7321 \neq \sin 2°$

Section 2.4   (*page* 125)

WARM UP

1. $\sin 30° = \frac{1}{2}$ **2.** $\tan 45° = 1$ **3.** $\cos \dfrac{\pi}{4} = \dfrac{\sqrt{2}}{2}$

4. $\cot \dfrac{\pi}{3} = \dfrac{\sqrt{3}}{3}$ **5.** $\sec \dfrac{\pi}{6} = \dfrac{2\sqrt{3}}{3}$ **6.** $\csc \dfrac{\pi}{4} = \sqrt{2}$

7. $\sin \theta = \dfrac{3\sqrt{13}}{13}$

$\cos \theta = \dfrac{2\sqrt{13}}{13}$

$\csc \theta = \dfrac{\sqrt{13}}{3}$

$\sec \theta = \dfrac{\sqrt{13}}{2}$

$\cot \theta = \dfrac{2}{3}$

8. $\sin \theta = \dfrac{\sqrt{5}}{3}$

$\tan \theta = \dfrac{\sqrt{5}}{2}$

$\csc \theta = \dfrac{3\sqrt{5}}{5}$

$\sec \theta = \dfrac{3}{2}$

$\cot \theta = \dfrac{2\sqrt{5}}{5}$

9. $\cos \theta = \dfrac{2\sqrt{6}}{5}$

$\tan \theta = \dfrac{\sqrt{6}}{12}$

$\csc \theta = 5$

$\sec \theta = \dfrac{5\sqrt{6}}{12}$

$\cot \theta = 2\sqrt{6}$

10. $\sin \theta = \dfrac{2\sqrt{2}}{3}$

$\cos \theta = \dfrac{1}{3}$

$\tan \theta = 2\sqrt{2}$

$\csc \theta = \dfrac{3\sqrt{2}}{4}$

$\cot \theta = \dfrac{\sqrt{2}}{4}$

1. (a) $\sin \theta = \frac{4}{5}$

$\cos \theta = \frac{3}{5}$

$\tan \theta = \frac{4}{3}$

$\csc \theta = \frac{5}{4}$

$\sec \theta = \frac{5}{3}$

$\cot \theta = \frac{3}{4}$

(b) $\sin \theta = -\frac{15}{17}$

$\cos \theta = \frac{8}{17}$

$\tan \theta = -\frac{15}{8}$

$\csc \theta = -\frac{17}{15}$

$\sec \theta = \frac{17}{8}$

$\cot \theta = -\frac{8}{15}$

3. (a) $\sin \theta = \dfrac{1}{2}$

$\cos \theta = -\dfrac{\sqrt{3}}{2}$

$\tan \theta = -\dfrac{\sqrt{3}}{3}$

$\csc \theta = 2$

$\sec \theta = -\dfrac{2\sqrt{3}}{3}$

$\cot \theta = -\sqrt{3}$

(b) $\sin \theta = -\dfrac{\sqrt{2}}{2}$

$\cos \theta = -\dfrac{\sqrt{2}}{2}$

$\tan \theta = 1$

$\csc \theta = -\sqrt{2}$

$\sec \theta = -\sqrt{2}$

$\cot \theta = 1$

5. (a) $\sin \theta = \frac{24}{25}$

$\cos \theta = \frac{7}{25}$

$\tan \theta = \frac{24}{7}$

$\csc \theta = \frac{25}{24}$

$\sec \theta = \frac{25}{7}$

$\cot \theta = \frac{7}{24}$

(b) $\sin \theta = -\frac{24}{25}$

$\cos \theta = \frac{7}{25}$

$\tan \theta = -\frac{24}{7}$

$\csc \theta = -\frac{25}{24}$

$\sec \theta = \frac{25}{7}$

$\cot \theta = -\frac{7}{24}$

7. (a) $\sin \theta = \dfrac{5\sqrt{29}}{29}$

$\cos \theta = -\dfrac{2\sqrt{29}}{29}$

$\tan \theta = -\dfrac{5}{2}$

$\csc \theta = \dfrac{\sqrt{29}}{5}$

$\sec \theta = -\dfrac{\sqrt{29}}{2}$

$\cot \theta = -\dfrac{2}{5}$

(b) $\sin \theta = -\dfrac{5\sqrt{34}}{34}$

$\cos \theta = \dfrac{3\sqrt{34}}{34}$

$\tan \theta = -\dfrac{5}{3}$

$\csc \theta = -\dfrac{\sqrt{34}}{5}$

$\sec \theta = \dfrac{\sqrt{34}}{3}$

$\cot \theta = -\dfrac{3}{5}$

9. (a) $c_1 = 5$, $b_2 = 12$, $c_2 = 15$

(b) $\sin \alpha_1 = \sin \alpha_2 = \frac{3}{5}$

$\cos \alpha_1 = \cos \alpha_2 = \frac{4}{5}$

$\tan \alpha_1 = \tan \alpha_2 = \frac{3}{4}$

$\csc \alpha_1 = \csc \alpha_2 = \frac{5}{3}$

$\sec \alpha_1 = \sec \alpha_2 = \frac{5}{4}$

$\cot \alpha_1 = \cot \alpha_2 = \frac{4}{3}$

11. (a) $b_1 = \sqrt{3}$, $a_2 = \frac{5\sqrt{3}}{3}$,

$c_2 = \frac{10\sqrt{3}}{3}$

(b) $\sin \alpha_1 = \sin \alpha_2 = \frac{1}{2}$

$\cos \alpha_1 = \cos \alpha_2 = \frac{\sqrt{3}}{2}$

$\tan \alpha_1 = \tan \alpha_2 = \frac{\sqrt{3}}{3}$

$\csc \alpha_1 = \csc \alpha_2 = 2$

$\sec \alpha_1 = \sec \alpha_2 = \frac{2\sqrt{3}}{3}$

$\cot \alpha_1 = \cot \alpha_2 = \sqrt{3}$

13. (a) Quadrant III

(b) Quadrant II

15. (a) Quadrant II

(b) Quadrant IV

17. $\sin \theta = \frac{3}{5}$ $\csc \theta = \frac{5}{3}$

$\cos \theta = -\frac{4}{5}$ $\sec \theta = -\frac{5}{4}$

$\tan \theta = -\frac{3}{4}$ $\cot \theta = -\frac{4}{3}$

19. $\sin \theta = -\frac{15}{17}$ $\csc \theta = -\frac{17}{15}$

$\cos \theta = \frac{8}{17}$ $\sec \theta = \frac{17}{8}$

$\tan \theta = -\frac{15}{8}$ $\cot \theta = -\frac{8}{15}$

21. $\sin \theta = \frac{\sqrt{3}}{2}$ $\csc \theta = \frac{2\sqrt{3}}{3}$

$\cos \theta = -\frac{1}{2}$ $\sec \theta = -2$

$\tan \theta = -\sqrt{3}$ $\cot \theta = -\frac{\sqrt{3}}{3}$

23. $\sin \theta = 0$ $\csc \theta$ is undefined.

$\cos \theta = -1$ $\sec \theta = -1$

$\tan \theta = 0$ $\cot \theta$ is undefined.

25. $\sin \theta = -\frac{2\sqrt{5}}{5}$ $\csc \theta = \frac{-\sqrt{5}}{2}$

$\cos \theta = -\frac{\sqrt{5}}{5}$ $\sec \theta = -\sqrt{5}$

$\tan \theta = 2$ $\cot \theta = \frac{1}{2}$

27. (a) $\theta' = 23°$ (b) $\theta' = 53°$

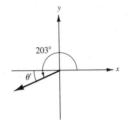 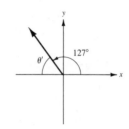

29. (a) $\theta' = 65°$ (b) $\theta' = 72°$

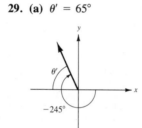 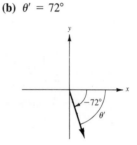

31. (a) $\theta' = \dfrac{\pi}{3}$ (b) $\theta' = \dfrac{\pi}{6}$

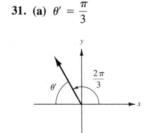 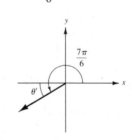

33. (a) $\theta' = 3.5 - \pi$ (b) $\theta' = 2\pi - 5.8$

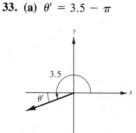 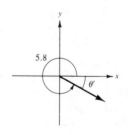

35. (a) $\sin\theta = -\dfrac{\sqrt{2}}{2}$, $\cos\theta = -\dfrac{\sqrt{2}}{2}$, $\tan\theta = 1$

    (b) $\sin\theta = \dfrac{\sqrt{2}}{2}$, $\cos\theta = -\dfrac{\sqrt{2}}{2}$, $\tan\theta = -1$

37. (a) $\sin\theta = \dfrac{1}{2}$, $\cos\theta = \dfrac{\sqrt{3}}{2}$, $\tan\theta = \dfrac{\sqrt{3}}{3}$

    (b) $\sin\theta = \dfrac{1}{2}$, $\cos\theta = -\dfrac{\sqrt{3}}{2}$, $\tan\theta = -\dfrac{\sqrt{3}}{3}$

39. (a) $\sin\theta = -\dfrac{\sqrt{3}}{2}$, $\cos\theta = -\dfrac{1}{2}$, $\tan\theta = \sqrt{3}$

    (b) $\sin\theta = \dfrac{\sqrt{3}}{2}$, $\cos\theta = -\dfrac{1}{2}$, $\tan\theta = -\sqrt{3}$

41. (a) $\sin\theta = -\dfrac{1}{2}$, $\cos\theta = \dfrac{\sqrt{3}}{2}$, $\tan\theta = -\dfrac{\sqrt{3}}{3}$

    (b) $\sin\theta = \dfrac{1}{2}$, $\cos\theta = -\dfrac{\sqrt{3}}{2}$, $\tan\theta = -\dfrac{\sqrt{3}}{3}$

43. (a) $\sin\theta = \dfrac{\sqrt{2}}{2}$, $\cos\theta = -\dfrac{\sqrt{2}}{2}$, $\tan\theta = -1$

    (b) $\sin\theta = -\dfrac{1}{2}$, $\cos\theta = \dfrac{\sqrt{3}}{2}$, $\tan\theta = -\dfrac{\sqrt{3}}{3}$

45. (a) 0.1736   **(b)** 5.7588
47. (a) 0.3640   **(b)** 0.3640
49. (a) −0.3420   **(b)** −0.3420
51. (a) 1.7321   **(b)** 1.7321
53. (a) $30° = \dfrac{\pi}{6}$, $150° = \dfrac{5\pi}{6}$

    (b) $210° = \dfrac{7\pi}{6}$, $330° = \dfrac{11\pi}{6}$

55. (a) $60° = \dfrac{\pi}{3}$, $120° = \dfrac{2\pi}{3}$

    (b) $135° = \dfrac{3\pi}{4}$, $315° = \dfrac{7\pi}{4}$

57. (a) $45° = \dfrac{\pi}{4}$, $225° = \dfrac{5\pi}{4}$

    (b) $150° = \dfrac{5\pi}{6}$, $330° = \dfrac{11\pi}{6}$

59. (a) 54.99°, 125.01°   **(b)** 195°, 345°
61. (a) 0.175, 6.109   **(b)** 2.201, 4.083
63. (a) 0.873, 4.014   **(b)** 1.693, 4.835
65. 1   **67.** 0
69. (a) 25.2°F   **(b)** 65.1°F   **(c)** 50.8°F

Section 2.5   (*page* 138)

WARM UP

1. 6π **2.** $\frac{1}{2}$ **3.** $\dfrac{\pi}{6}$ **4.** $\dfrac{7\pi}{6}$ **5.** −2
6. $-\frac{4}{3}$ **7.** 1   **8.** 0   **9.** 1   **10.** 0

1. Period: π **3.** Period: 4π
    Amplitude: 2   Amplitude: $\frac{3}{2}$
5. Period: 2   **7.** Period: 2π
    Amplitude: $\frac{1}{2}$ Amplitude: 2
9. Period: $\dfrac{\pi}{5}$ **11.** Period: 3π **13.** Period: $\frac{1}{2}$

    Amplitude: 2   Amplitude: $\frac{1}{2}$ Amplitude: 3

15. *Shift* graph of *f* π units to the right to obtain the graph of *g*.
17. *Reflect* the graph of *f* about the *x*-axis to obtain the graph of *g*.
19. The *period* of *f* is twice the period of *g*.
21. *Shift* the graph of *f* two units up to obtain the graph of *g*.

23.

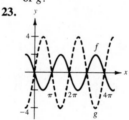

25.

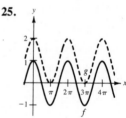

27.

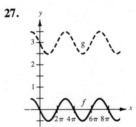

29.

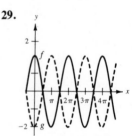

31.

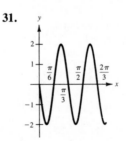

33.

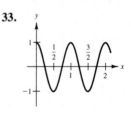

35.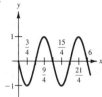

37.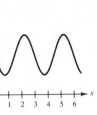

61.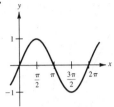

63.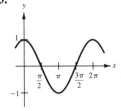

39.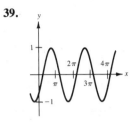

41.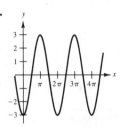

65. (a) $t = 6$ sec
(b) 10
(c)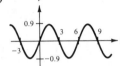

67. (a) $p = \frac{1}{440}$
(b) $f = 440$
(c)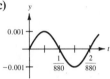

43.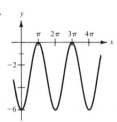

45.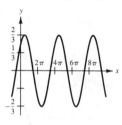

Section 2.6   (*page* 149)

WARM UP

1. 0   **2.** $\dfrac{\sqrt{2}}{2}$ **3.** 1   **4.** 0   **5.** 0   **6.** 0

7.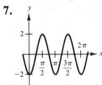

8.

47.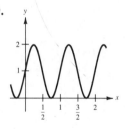

49.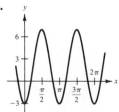

9.

10.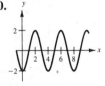

51.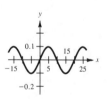

53.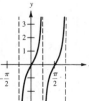

1. c, π **2.** g, $\dfrac{\pi}{3}$ **3.** e, 2π **4.** a, 4π **5.** d, 1

6. h, 2   **7.** b, 2π **8.** f, 1

9.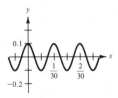

11.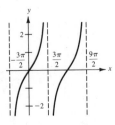

55.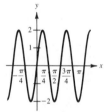

57. $y = 2 \sin 4x$

59. $y = \cos\left(2x + \dfrac{\pi}{2}\right)$

13.

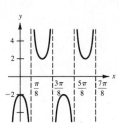

15.

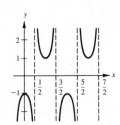

17.

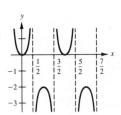

19.

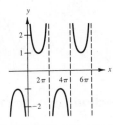

21.

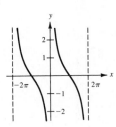

23.

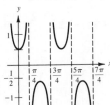

25.

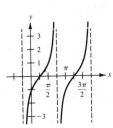

27.

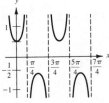

29.

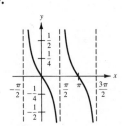

31.

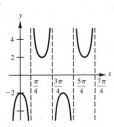

33.

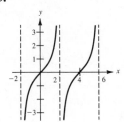

35.

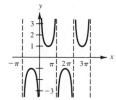

37. $d = 6 \cot x$

Section 2.7 (*page* 156)

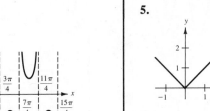

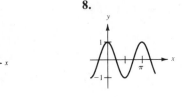

WARM UP

1. $f(x) = -1: \dfrac{3\pi}{2}$

$f(x) = 0: 0, \pi, 2\pi$

$f(x) = 1: \dfrac{\pi}{2}$

2. $f(x) = -1: \pi$

$f(x) = 0: \dfrac{\pi}{2}, \dfrac{3\pi}{2}$

$f(x) = 1: 0, 2\pi$

3. $f(x) = -1: \dfrac{3\pi}{4}, \dfrac{7\pi}{4}$

$f(x) = 0: 0, \dfrac{\pi}{2}, \pi, \dfrac{3\pi}{2}, 2\pi$

$f(x) = 1: \dfrac{\pi}{4}, \dfrac{5\pi}{4}$

4. $f(x) = -1: 2\pi$

$f(x) = 0: \pi$

$f(x) = 1: 0$

5.

6.

7.

8.

9. $0, \dfrac{\sqrt{3}\pi}{12}, \dfrac{\sqrt{2}\pi}{8}, \dfrac{\pi}{6}, 0$

10. $0, \dfrac{3+\pi}{6}, \dfrac{2\sqrt{2}+\pi}{4}, \dfrac{3\sqrt{3}+2\pi}{6}, \dfrac{\pi+2}{2}$

1.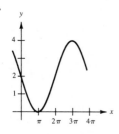

3.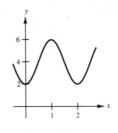

21.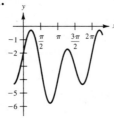

23.

5.

7.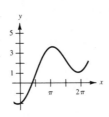

25.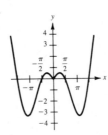

27.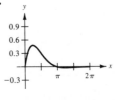

9.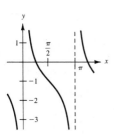

11.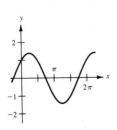

29.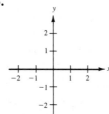

31.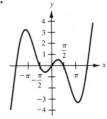

13.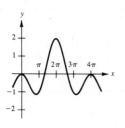

15.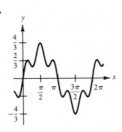

33.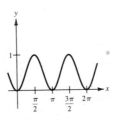

17.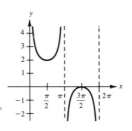

19.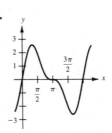

35.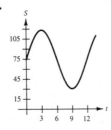

37.

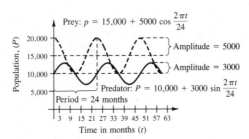

Prey: $p = 15{,}000 + 5000 \cos \frac{2\pi t}{24}$

Amplitude = 5000

Amplitude = 3000

Predator: $P = 10{,}000 + 3000 \sin \frac{2\pi t}{24}$

Period = 24 months

We can explain the cycles of this predator-prey population by noting the following cause and effect pattern:

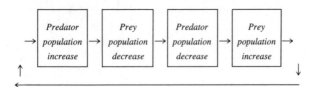

| Predator population increase | Prey population decrease | Predator population decrease | Prey population increase |

39.

x	-0.5	-0.4	-0.3	-0.2	-0.1
$\dfrac{1 - \cos x}{x}$	-0.245	-0.197	-0.149	-0.100	-0.050

x	0.1	0.2	0.3	0.4	0.5
$\dfrac{1 - \cos x}{x}$	0.050	0.100	0.149	0.197	0.245

$f(x)$ approaches 0 as x approaches 0.

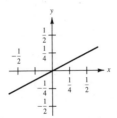

Section 2.8   (*page* 166)

WARM UP

1. -1 **2.** -1 **3.** -1 **4.** $\dfrac{\sqrt{2}}{2}$ **5.** 0

6. $\dfrac{\pi}{6}$ **7.** π **8.** $\dfrac{\pi}{4}$ **9.** 0 **10.** $\dfrac{\pi}{4}$

1. $\dfrac{\pi}{6}$ **3.** $\dfrac{\pi}{3}$ **5.** $\dfrac{\pi}{6}$ **7.** $\dfrac{5\pi}{6}$ **9.** $-\dfrac{\pi}{3}$ **11.** $\dfrac{2\pi}{3}$

13. $\dfrac{\pi}{3}$ **15.** 0 **17.** 1.29 **19.** -0.85

21. -1.11 **23.** 0.32 **25.** 1.99 **27.** 0.74

29. 0.3 **31.** -0.1 **33.** 0 **35.** $\frac{3}{5}$ **37.** $\dfrac{\sqrt{5}}{5}$

39. $\frac{12}{13}$ **41.** $\dfrac{\sqrt{34}}{5}$ **43.** $\dfrac{1}{x}$ **45.** $\sqrt{1 - 4x^2}$

47. $\sqrt{1 - x^2}$ **49.** $\dfrac{\sqrt{9 - x^2}}{x}$ **51.** $\dfrac{\sqrt{x^2 + 2}}{x}$

53. $\dfrac{9}{\sqrt{x^2 + 81}}$ **55.** $\dfrac{|x - 1|}{\sqrt{x^2 - 2x + 10}}$

57.

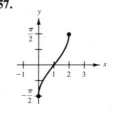

59.

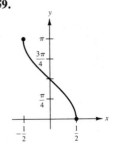

61. (a) $\beta \approx 40.6°$ **(b)** $\beta \approx 30.3°$

65. (a) $\dfrac{\pi}{4}$ **(b)** 0

Section 2.9   (*page* 176)

WARM UP

1. 8.45 **2.** 78.99 **3.** 1.06 **4.** 1.24
5. 4.88 **6.** 34.14 **7.** $4; \pi$ **8.** $\frac{1}{2}; 2$
9. $3; \frac{2}{3}$ **10.** $0.2; 8\pi$

1. $a \approx 3.64$
$c \approx 10.64$
$B = 70°$

3. $a \approx 8.26$
$c \approx 25.38$
$A = 19°$

5. $a \approx 91.34$
$b \approx 420.70$
$B = 77°45'$

7. $c = 2\sqrt{34}$
$A \approx 30.96°$
$B \approx 59.04°$

9. $a = 12\sqrt{17}$
$A \approx 72.08°$
$B \approx 17.92°$

11. 15.4 ft   **13.** 2.56 in

15. 56.3°   **17.** 12.68°   **19.** 19.9 ft

21. 508 miles north; 650 miles east   **23.** N 56.3° W

25. (a) N 58° E   **(b)** 68.8 yd   **27.** 1657.13 ft

29. 17,054 ft $\approx$ 3.23 mi   **31.** 29.389 in

33. $y = \sqrt{3}r$ **35.** $a \approx 7, c \approx 12.2$

37. (a) 4   **(b)** 4   **(c)** $t = \frac{1}{16}$

39. (a) $\frac{1}{16}$ **(b)** 60   **(c)** $t = \frac{1}{120}$

41. $\omega = 528\pi$

27. $\sin \theta = -\dfrac{3\sqrt{13}}{13}$ $\csc \theta = -\dfrac{\sqrt{13}}{3}$

$\cos \theta = -\dfrac{2\sqrt{13}}{13}$ $\sec \theta = -\dfrac{\sqrt{13}}{2}$

$\tan \theta = \frac{3}{2}$ $\cot \theta = \frac{2}{3}$

29. $\sin \theta = -\dfrac{\sqrt{11}}{6}$ $\cos \theta = \frac{5}{6}$

$\tan \theta = -\dfrac{\sqrt{11}}{5}$ $\cot \theta = -\dfrac{5\sqrt{11}}{11}$

$\csc \theta = -\dfrac{6\sqrt{11}}{11}$

31. $\cos \theta = -\dfrac{\sqrt{55}}{8}$ $\sec \theta = -\dfrac{8\sqrt{55}}{55}$

$\tan \theta = -\dfrac{3\sqrt{55}}{55}$ $\cot \theta = -\dfrac{\sqrt{55}}{3}$

$\csc \theta = \frac{8}{3}$

33. $-\dfrac{\sqrt{3}}{2}$ **35.** $-\dfrac{\sqrt{2}}{2}$ **37.** 0.65   **39.** 3.24

41. $135° = \dfrac{3\pi}{4}, 225° = \dfrac{5\pi}{4}$

43. $210° = \dfrac{7\pi}{6}, 330° = \dfrac{11\pi}{6}$

45. $57.0° \approx 0.9949, 123.0° \approx 2.1470$

47. $165.0° \approx 2.8798, 195.0° \approx 3.4034$

49. $\dfrac{\sqrt{-x^2 + 2x}}{-x^2 + 2x}$ **51.** $\dfrac{2\sqrt{4 - 2x^2}}{4 - x^2}$

53.

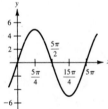

Chapter 2 Review Exercises   (*page* 180)

1. $\dfrac{3\pi}{4}, -\dfrac{5\pi}{4}$ **3.** 250°, −470°

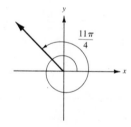

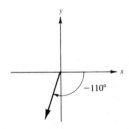

5. 135.28°   **7.** 5.38°   **9.** 135° 16′ 12″

11. −85° 9′   **13.** 128.57°   **15.** −200.54°

17. 8.3776   **19.** −0.5890   **21.** $\dfrac{\pi}{5}$ **23.** 72°

25. $\sin \theta = \dfrac{2\sqrt{53}}{53}$ $\csc \theta = \dfrac{\sqrt{53}}{2}$

$\cos \theta = -\dfrac{7\sqrt{53}}{53}$ $\sec \theta = -\dfrac{\sqrt{53}}{7}$

$\tan \theta = -\frac{2}{7}$ $\cot \theta = -\frac{7}{2}$

55.

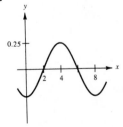

57.

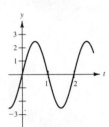

59.

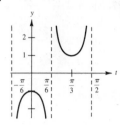

61.

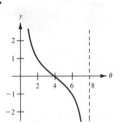

63.

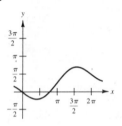

65.

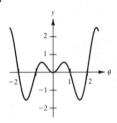

67.

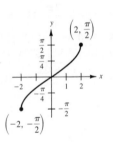

69.

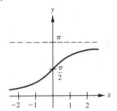

71. 1.33 mi

73. 268.8 ft

CHAPTER 3
Section 3.1 (*page* 189)

WARM UP

1. $\sin \theta = \dfrac{3\sqrt{13}}{13}$ $\sec \theta = \dfrac{\sqrt{13}}{2}$

$\cos \theta = \dfrac{2\sqrt{13}}{13}$ $\cot \theta = \frac{2}{3}$

$\csc \theta = \dfrac{\sqrt{13}}{3}$

2. $\sin \theta = \dfrac{2\sqrt{2}}{3}$ $\csc \theta = \dfrac{3\sqrt{2}}{4}$

$\cos \theta = \frac{1}{3}$ $\cot \theta = \dfrac{\sqrt{2}}{4}$

$\tan \theta = 2\sqrt{2}$

3. $\sin \theta = -\dfrac{3\sqrt{58}}{58}$ $\csc \theta = -\dfrac{\sqrt{58}}{3}$

$\cos \theta = \dfrac{7\sqrt{58}}{58}$ $\sec \theta = \dfrac{\sqrt{58}}{7}$

$\tan \theta = -\frac{3}{7}$ $\cot \theta = -\frac{7}{3}$

4. $\sin \theta = \dfrac{\sqrt{5}}{5}$ $\csc \theta = \sqrt{5}$

$\cos \theta = -\dfrac{2\sqrt{5}}{5}$ $\sec \theta = -\dfrac{\sqrt{5}}{2}$

$\tan \theta = -\frac{1}{2}$ $\cot \theta = -2$

5. $\frac{1}{2}$ **6.** $\frac{5}{4}$ **7.** $\dfrac{\sqrt{73}}{8}$ **8.** $\frac{2}{3}$

9. $\dfrac{x^2 + x + 16}{4(x + 1)}$ **10.** $\dfrac{8x - 2}{1 - x^2}$

1. $\sin x = \frac{1}{2}$ $\cot x = \sqrt{3}$

$\cos x = \dfrac{\sqrt{3}}{2}$ $\sec x = \dfrac{2\sqrt{3}}{3}$

$\tan x = \dfrac{\sqrt{3}}{3}$ $\csc x = 2$

3. $\sec \theta = \sqrt{2}$ $\tan \theta = -1$

$\sin \theta = -\dfrac{\sqrt{2}}{2}$ $\cot \theta = -1$

$\cos \theta = \dfrac{\sqrt{2}}{2}$ $\csc \theta = -\sqrt{2}$

5. $\tan x = \frac{5}{12}$ $\sin x = -\frac{5}{13}$

$\sec x = -\frac{13}{12}$ $\cot x = \frac{12}{5}$

$\cos x = -\frac{12}{13}$ $\csc x = -\frac{13}{5}$

7. $\sec \phi = -1$ $\tan \phi = 0$

$\sin \phi = 0$ $\cot \phi$ is undefined.

$\cos \phi = -1$ $\csc \phi$ is undefined.

9. $\sin x = \frac{2}{3}$ $\cot x = -\frac{\sqrt{5}}{2}$

$\tan x = -\frac{2\sqrt{5}}{5}$ $\sec x = -\frac{3\sqrt{5}}{5}$

$\cos x = -\frac{\sqrt{5}}{3}$ $\csc x = \frac{3}{2}$

11. $\tan \theta = 2$ $\sin \theta = -\frac{2\sqrt{5}}{5}$

$\sec \theta = -\sqrt{5}$ $\csc \theta = -\frac{\sqrt{5}}{2}$

$\cos \theta = -\frac{\sqrt{5}}{5}$ $\cot \theta = \frac{1}{2}$

13. $\sin \theta = -1$ $\csc \theta = -1$

$\cot \theta = 0$ $\tan \theta$ is undefined.

$\cos \theta = 0$ $\sec \theta$ is undefined.

15. d   **16.** e   **17.** a   **18.** f   **19.** b   **20.** c

21. b   **22.** c   **23.** e   **24.** a   **25.** f   **26.** d

27. $\sec \phi$ **29.** $\sin \beta$ **31.** $\cos x$ **33.** 1

35. $-\tan x$ **37.** $\tan x$ **39.** $1 + \sin y$ **41.** $\sin^2 x$

43. $\sin^2 x \tan^2 x$ **45.** $\sec^4 x$ **47.** $\sin^2 x - \cos^2 x$

49. $1 + 2 \sin x \cos x$ **51.** $\tan^2 x$ **53.** $2 \csc^2 x$

55. $2 \sec x$ **57.** $1 + \cos y$ **59.** $3(\sec x + \tan x)$

61. $5 \cos \theta$ **63.** $3 \tan \theta$ **65.** $5 \sec \theta$ **67.** $\cos \theta$

69. $27 \sec^3 \theta$ **71.** $0 \le \theta < \frac{\pi}{2}, \frac{3\pi}{2} < \theta < 2\pi$

73. $\ln |\cot \theta|$ **75.** False, $\dfrac{\sin k\theta}{\cos k\theta} = \tan k\theta$

77. True, provided $\sin \theta \ne 0$

79. (a) $\csc^2 132° - \cot^2 132° \approx 1.8107 - 0.8107 = 1$

   (b) $1.6360 - 0.6360 = 1$

81. (a) $\cos (90° - 80°) = \sin 80° \approx 0.9848$

   (b) $\cos\left(\dfrac{\pi}{2} - 0.8\right) = \sin 0.8 \approx 0.7174$

83. $\cos \theta = \pm\sqrt{1 - \sin^2 \theta}$ $\sec \theta = \pm\dfrac{1}{\sqrt{1 - \sin^2 \theta}}$

$\tan \theta = \pm\dfrac{\sin \theta}{\sqrt{1 - \sin^2 \theta}}$ $\csc \theta = \dfrac{1}{\sin \theta}$

$\cot \theta = \pm\dfrac{\sqrt{1 - \sin^2 \theta}}{\sin \theta}$

Section 3.2   (*page* 197)

WARM UP

1. (a) $x^2(1 - y^2)$ **(b)** $\sin^4 x$

2. (a) $x^2(1 + y^2)$ **(b)** 1

3. (a) $(x^2 + 1)(x^2 - 1)$ **(b)** $\sec^2 x(\tan^2 x - 1)$

4. (a) $(z + 1)(z^2 - z + 1)$

   (b) $(\tan x + 1)(\tan^2 x - \tan x + 1)$

5. (a) $(x - 1)(x^2 + 1)$ **(b)** $(\cot x - 1) \csc^2 x$

6. (a) $(x^2 - 1)^2$ **(b)** $\cos^4 x$

7. (a) $\dfrac{y^2 - x^2}{x}$ **(b)** $\tan x$

8. (a) $\dfrac{x^2 - 1}{x^2}$ **(b)** $\sin^2 x$

9. (a) $\dfrac{y^2 + (1 + z)^2}{y(1 + z)}$ **(b)** $2 \csc x$

10. (a) $\dfrac{y(1 + y) - z^2}{z(1 + y)}$ **(b)** $\dfrac{\tan x - 1}{\sec x(1 + \tan x)}$

61. $\theta = -\dfrac{\pi}{3}; -0.8660 \ne 0.8660$ **63.** $x = -\dfrac{\pi}{4}; 1 \ne -1$

Section 3.3   (*page* 207)

WARM UP

1. $\dfrac{2\pi}{3}, \dfrac{4\pi}{3}$ **2.** $\dfrac{\pi}{3}, \dfrac{2\pi}{3}$ **3.** $\dfrac{\pi}{4}, \dfrac{7\pi}{4}$

4. $\dfrac{7\pi}{4}, \dfrac{5\pi}{4}$ **5.** $\dfrac{\pi}{3}$ **6.** $\dfrac{3\pi}{4}$ **7.** $\dfrac{15}{8}$ **8.** $-3, \dfrac{5}{2}$

9. $\dfrac{2 \pm \sqrt{14}}{2}$ **10.** $-1, 3$

7. $\dfrac{2\pi}{3} + 2n\pi, \dfrac{4\pi}{3} + 2n\pi$ **9.** $\dfrac{\pi}{3} + 2n\pi, \dfrac{2\pi}{3} + 2n\pi$

11. $\dfrac{\pi}{4} + \dfrac{n\pi}{2}$ **13.** $\dfrac{\pi}{6} + n\pi, \dfrac{5\pi}{6} + n\pi$

15. $n\pi, \dfrac{\pi}{4} + n\pi$ **17.** $n\pi, \dfrac{3\pi}{2} + 2n\pi$

19. $\dfrac{\pi}{2} + n\pi, \dfrac{2\pi}{3} + 2n\pi, \dfrac{4\pi}{3} + 2n\pi$ **21.** $\dfrac{\pi}{3}, \dfrac{5\pi}{3}$

23. $\dfrac{7\pi}{6}, \dfrac{3\pi}{2}, \dfrac{11\pi}{6}$ **25.** $0, \dfrac{\pi}{2}, \pi, \dfrac{3\pi}{2}$

27. $\dfrac{\pi}{2}, \dfrac{2\pi}{3}, \dfrac{3\pi}{2}, \dfrac{4\pi}{3}$ **29.** $\dfrac{\pi}{6}, \dfrac{5\pi}{6}, \dfrac{7\pi}{6}, \dfrac{11\pi}{6}$

31. No solution   **33.** $\dfrac{2\pi}{3}, \dfrac{5\pi}{6}, \dfrac{5\pi}{3}, \dfrac{11\pi}{6}$ **35.** $\dfrac{\pi}{2}$

37. π **39.** $\dfrac{\pi}{3}, \dfrac{5\pi}{3}$

41. 0.9828, 1.7682, 4.1244, 4.9098

43. 0.3398, 0.8481, 2.2935, 2.8018   **45.** 0.8411, 5.4421

47. 1.1555, 1.3981, 4.2971, 4.5397   **49.** 0.4271, 2.7145

51. $\dfrac{\pi}{4}, \dfrac{5\pi}{4}$ **53.** 0.04, 0.43, 0.83

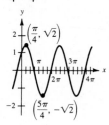

55. 37°

Section 3.4   (*page* 216)

1. $\sin 75° = \dfrac{\sqrt{2}}{4}(1 + \sqrt{3})$

$\cos 75° = \dfrac{\sqrt{2}}{4}(\sqrt{3} - 1)$

$\tan 75° = \sqrt{3} + 2$

3. $\sin 105° = \dfrac{\sqrt{2}}{4}(\sqrt{3} + 1)$

$\cos 105° = \dfrac{\sqrt{2}}{4}(1 - \sqrt{3})$

$\tan 105° = -2 - \sqrt{3}$

5. $\sin 195° = \dfrac{\sqrt{2}}{4}(1 - \sqrt{3})$

$\cos 195° = -\dfrac{\sqrt{2}}{4}(\sqrt{3} + 1)$

$\tan 195° = 2 - \sqrt{3}$

7. $\sin \dfrac{11\pi}{12} = \dfrac{\sqrt{2}}{4}(\sqrt{3} - 1)$

$\cos \dfrac{11\pi}{12} = -\dfrac{\sqrt{2}}{4}(\sqrt{3} + 1)$

$\tan \dfrac{11\pi}{12} = -2 + \sqrt{3}$

9. $\sin \dfrac{17\pi}{12} = -\dfrac{\sqrt{2}}{4}(\sqrt{3} + 1)$ **11.** $\cos 40°$

$\cos \dfrac{17\pi}{12} = \dfrac{\sqrt{2}}{4}(1 - \sqrt{3})$

$\tan \dfrac{17\pi}{12} = 2 + \sqrt{3}$

13. $\sin 200°$ **15.** $\tan 239°$ **17.** $\sin 1.8$

19. $\tan 3x$ **21.** $\dfrac{33}{65}$ **23.** $-\dfrac{56}{65}$ **25.** $-\dfrac{3}{5}$ **27.** $\dfrac{44}{125}$

49. (a) $\sqrt{2} \sin\left(\theta + \dfrac{\pi}{4}\right)$ (b) $\sqrt{2} \cos\left(\theta - \dfrac{\pi}{4}\right)$

51. (a) $13 \sin(3\theta + 0.3948)$ (b) $13 \cos(3\theta - 1.1760)$

53. $\sqrt{2} \sin \theta + \sqrt{2} \cos \theta$ **55.** 1   **57.** $\dfrac{\pi}{2}$

59. $\dfrac{\pi}{4}, \dfrac{7\pi}{4}$ **61.** $0, \dfrac{\pi}{3}, \pi, \dfrac{5\pi}{3}$

Section 3.5   (*page* 226)

1. $\sin 90° = 1$ **3.** $\sin 60° = \dfrac{\sqrt{3}}{2}$

$\cos 90° = 0$ $\cos 60° = \dfrac{1}{2}$

$\tan 90°$ is undefined. $\tan 60° = \sqrt{3}$

5. $\sin \dfrac{2\pi}{3} = \dfrac{\sqrt{3}}{2}$ **7.** $\sin 2u = \dfrac{24}{25}$

$\cos \dfrac{2\pi}{3} = -\dfrac{1}{2}$ $\cos 2u = \dfrac{7}{25}$

$\tan \dfrac{2\pi}{3} = -\sqrt{3}$ $\tan 2u = \dfrac{24}{7}$

9. $\sin 2u = \dfrac{4}{5}$

$\cos 2u = \dfrac{3}{5}$

$\tan 2u = \dfrac{4}{3}$

11. $\sin 2u = -\dfrac{4\sqrt{21}}{25}$

$\cos 2u = -\dfrac{17}{25}$

$\tan 2u = \dfrac{4\sqrt{21}}{17}$

13. $\sin 105° = \dfrac{1}{2}\sqrt{2 + \sqrt{3}}$

$\cos 105° = -\dfrac{1}{2}\sqrt{2 - \sqrt{3}}$

$\tan 105° = -2 - \sqrt{3}$

15. $\sin 112° \, 30' = \dfrac{1}{2}\sqrt{2 + \sqrt{2}}$

$\cos 112° \, 30' = -\dfrac{1}{2}\sqrt{2 - \sqrt{2}}$

$\tan 112° \, 30' = -1 - \sqrt{2}$

17. $\sin \dfrac{\pi}{8} = \dfrac{1}{2}\sqrt{2 - \sqrt{2}}$

$\cos \dfrac{\pi}{8} = \dfrac{1}{2}\sqrt{2 + \sqrt{2}}$

$\tan \dfrac{\pi}{8} = \sqrt{2} - 1$

19. $\sin \dfrac{u}{2} = \dfrac{5\sqrt{26}}{26}$

$\cos \dfrac{u}{2} = \dfrac{\sqrt{26}}{26}$

$\tan \dfrac{u}{2} = 5$

21. $\sin \dfrac{u}{2} = \sqrt{\dfrac{\sqrt{89} - 8}{2\sqrt{89}}}$

$\cos \dfrac{u}{2} = -\sqrt{\dfrac{\sqrt{89} + 8}{2\sqrt{89}}}$

$\tan \dfrac{u}{2} = \dfrac{8 - \sqrt{89}}{5}$

23. $\sin \dfrac{u}{2} = \dfrac{3\sqrt{10}}{10}$

$\cos \dfrac{u}{2} = -\dfrac{\sqrt{10}}{10}$

$\tan \dfrac{u}{2} = -3$

25. $\sin 3x$ **27.** $-\tan 4x$

29. (a) $\frac{1}{8}(3 + 4 \cos 2x + \cos 4x)$

(b) $\frac{1}{32}(2 + \cos 2x - 2 \cos 4x - \cos 6x)$

31. $3(\sin \dfrac{\pi}{2} + \sin 0)$ **33.** $\frac{1}{2}(\sin 8\theta + \sin 2\theta)$

35. $\frac{5}{2}(\cos 8\beta + \cos 2\beta)$

37. $\frac{1}{2}(\cos 2y - \cos 2x)$

39. $\frac{1}{2}[\sin 2\theta + \sin 2\pi]$ **41.** $2 \sin 45° \cos 15°$

43. $-2 \sin \dfrac{\pi}{2} \sin \dfrac{\pi}{4}$ **45.** $2 \cos 4x \cos 2x$

47. $2 \cos \alpha \sin \beta$ **49.** $2 \cos(\phi + \pi) \cos \pi$

73. $\dfrac{\pi}{12}, \dfrac{5\pi}{12}, \dfrac{13\pi}{12}, \dfrac{17\pi}{12}$ **75.** $0, \dfrac{2\pi}{3}, \dfrac{4\pi}{3}$

77. $0, \dfrac{\pi}{2}, \pi, \dfrac{3\pi}{2}$ **79.** $0, \dfrac{\pi}{4}, \dfrac{\pi}{2}, \dfrac{3\pi}{4}, \pi, \dfrac{5\pi}{4}, \dfrac{3\pi}{2}, \dfrac{7\pi}{4}$

81. $0, \dfrac{\pi}{4}, \dfrac{\pi}{2}, \dfrac{3\pi}{4}, \pi, \dfrac{5\pi}{4}, \dfrac{3\pi}{2}, \dfrac{7\pi}{4}$ **83.** $\dfrac{\pi}{6}, \dfrac{5\pi}{6}$

85.

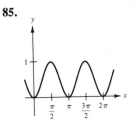

87. $2x\sqrt{1 - x^2}$

Chapter 3 Review Exercises (*page* 228)

1. $\sin^2 x$ **3.** $1 + \cot \alpha$ **5.** 1 **7.** 1

9. $\cos^2 2x$ **41.** $\dfrac{\sqrt{2}}{4}(\sqrt{3} + 1)$ **43.** $-\frac{1}{2}\sqrt{2 + \sqrt{2}}$

45. $-\frac{3}{52}(5 + 4\sqrt{7})$ **47.** $\frac{1}{52}(36 + 5\sqrt{7})$

49. $\frac{1}{4}\sqrt{2(4 - \sqrt{7})}$ **51.** $0, \pi$ **53.** $\dfrac{\pi}{3}, \dfrac{5\pi}{3}$

55. $0, \dfrac{3\pi}{4}, \pi, \dfrac{5\pi}{4}$ **57.** $0, \dfrac{\pi}{2}, \pi$ **59.** $\dfrac{\pi}{4}, \dfrac{5\pi}{4}$

61. $2 \cos \dfrac{5\theta}{2} \cos \dfrac{\theta}{2}$ **63.** $\frac{1}{2}(\cos \alpha - \cos 5\alpha)$

CHAPTER 4

Section 4.1 *(page 238)*

1. $C = 105°$, $b \approx 14.1$, $c \approx 19.3$
3. $C = 110°$, $b \approx 22.4$, $c \approx 24.4$
5. $B \approx 21.6°$, $C \approx 122.4°$, $c \approx 11.5$
7. $B = 10°$, $b \approx 69.5$, $c \approx 136.8$
9. $B = 42° \, 4'$, $a \approx 22.1$, $b \approx 14.9$
11. $A \approx 10° \, 11'$, $C \approx 154° \, 19'$, $c \approx 11.0$
13. $A \approx 25.6°$, $B \approx 9.4°$, $a \approx 10.5$
15. $B \approx 18° \, 13'$, $C \approx 51° \, 32'$, $c \approx 40.1$
17. No solution
19. Two solutions $B \approx 70.4°$, $C \approx 51.6°$, $c \approx 4.16$; $B \approx 109.6°$, $C \approx 12.4°$, $c \approx 1.14$
21. No solution
23. (a) $b \le 5$, $b = \dfrac{5}{\sin 36°}$ (b) $5 < b < \dfrac{5}{\sin 36°}$
    (c) $b > \dfrac{5}{\sin 36°}$ 25. 10.4   27. 1675.2
29. 474.9   31. 6.0 units   33. 77 yd   35. 5 mi
37. 26.1 mi, 15.9 mi   39. 4.55 mi   41. No

Section 4.2 *(page 245)*

1. $A \approx 27.7°$, $B \approx 40.5°$, $C \approx 111.8°$
3. $B \approx 23.8°$, $C \approx 126.2°$, $a \approx 12.4$
5. $A \approx 36.9°$, $B \approx 53.1°$, $C = 90°$
7. $A \approx 92.9°$, $B \approx 43.55°$, $C \approx 43.55°$

9. $a \approx 11.79$, $B \approx 12.73°$, $C \approx 47.27°$
11. $A \approx 158°36'$, $C \approx 12°38'$, $b \approx 10.4$
13. $A = 27°10'$, $B = 27°10'$, $c \approx 56.9$ 15. 16.25
17. 54   19. 96.82   21. S 52°37' E, S 25°20' W
23. 43.3 mi   25. 116.35 ft, 133.09 ft   27. 114.95°
29. $\overline{PQ} \approx 9.4$ ft, $\overline{QS} \approx 5.0$ ft, $\overline{RS} \approx 12.8$ ft
31. (a) 63.7 ft   (b) 47.6 ft
33. (a) N 58.3° W   (b) S 81.6° W
35. (a) 570.60   (b) 5910.08   (c) 177.09

Section 4.3 *(page 258)*

1.
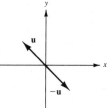

3.
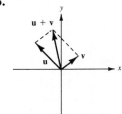

5.
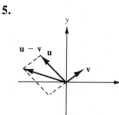

7. $\mathbf{v} = \langle 3, 4 \rangle$, $\|\mathbf{v}\| = 5$
9. $\mathbf{v} = \langle -3, 2 \rangle$, $\|\mathbf{v}\| = \sqrt{13}$
11. $\mathbf{v} = \langle 0, 5 \rangle$, $\|\mathbf{v}\| = 5$

13. $\mathbf{v} = \langle 16, -3 \rangle$, $\|\mathbf{v}\| = \sqrt{265}$
15. $\mathbf{v} = \langle 8, 4 \rangle$, $\|\mathbf{v}\| = 4\sqrt{5}$
17. (a) $\langle 4, 3 \rangle$ (b) $\langle -2, 1 \rangle$ (c) $\langle -7, 1 \rangle$
19. (a) $\langle -4, 4 \rangle$ (b) $\langle 0, 2 \rangle$ (c) $\langle 2, 3 \rangle$
21. (a) $\langle 4, -2 \rangle$ (b) $\langle 4, -2 \rangle$ (c) $\langle 8, -4 \rangle$
23. (a) $3\mathbf{i} - 2\mathbf{j}$ (b) $-\mathbf{i} + 4\mathbf{j}$ (c) $-4\mathbf{i} + 11\mathbf{j}$
25. (a) $2\mathbf{i} + \mathbf{j}$ (b) $2\mathbf{i} - \mathbf{j}$ (c) $4\mathbf{i} - 3\mathbf{j}$

27. $\mathbf{v} = \langle 3, 0 \rangle$

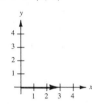

29. $\mathbf{v} = \left\langle -\dfrac{\sqrt{3}}{2}, \dfrac{1}{2} \right\rangle$

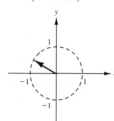

31. $\mathbf{v} = \left\langle -\dfrac{3\sqrt{6}}{2}, \dfrac{3\sqrt{2}}{2} \right\rangle$

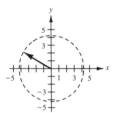

33. $\mathbf{v} = \left\langle \dfrac{\sqrt{10}}{5}, \dfrac{3\sqrt{10}}{5} \right\rangle$

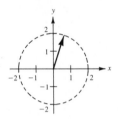

35. $\mathbf{v} = \left\langle 3, -\dfrac{3}{2} \right\rangle$

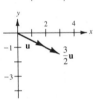

37. $\mathbf{v} = \langle 4, 3 \rangle$

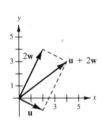

39. $\mathbf{v} = \left\langle \dfrac{7}{2}, -\dfrac{1}{2} \right\rangle$

41. $\langle 5, 5 \rangle$

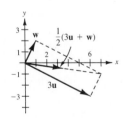

43. $\langle (10\sqrt{2} - 50), 10\sqrt{2} \rangle$　　**45.** $\frac{4}{5}\mathbf{i} - \frac{3}{5}\mathbf{j}$　　**47.** $\mathbf{j}$
49. $90°$　　**51.** $63.4°$　　**53.** 82.2 lb
55. $71.3°, 228.5$ lb　　**57.** 425 ft-lb
59. N $52.1°$ W, 569.5 mph　　**61.** N $25.2°$ E, 82.8 mph
63. $62.7°$

Section 4.4　(*page* 270)

WARM UP

1. $\sqrt{17}$　　**2.** $\sqrt{10}$　　**3.** $\langle 6, 15 \rangle$　　**4.** $\langle -2, -3 \rangle$
5. $\langle 4, 8 \rangle$　　**6.** $\langle 1, 0 \rangle$　　**7.** $\langle 7, -6 \rangle$　　**8.** $\langle 11, 13 \rangle$
9. $\mathbf{i} - 2\mathbf{j}$　　**10.** $2\mathbf{i} - 24\mathbf{j}$

1. **(a)** -6　　**(b)** 25　　**(c)** 25
　　(d) $\langle -12, 18 \rangle$　　**(e)** -12
3. **(a)** -18　　**(b)** 13　　**(c)** 13
　　(d) $\langle 0, -108 \rangle$　　**(e)** -36
5. **(a)** 3　　**(b)** 5　　**(c)** 5　　**(d)** $3\mathbf{i} - 3\mathbf{j}$　　**(e)** 6
7. $90°$　　**9.** $\arccos\left(-\dfrac{\sqrt{2}}{10} \right) \approx 98.1°$
11. $\arccos\left(-\frac{4}{5} \right) \approx 143.1°$　　**13.** $45°$　　**15.** Neither
17. Orthogonal　　**19.** Neither　　**21.** Orthogonal
23. **(a)** $\langle \frac{5}{2}, \frac{1}{2} \rangle$　　**(b)** $\langle -\frac{1}{2}, \frac{5}{2} \rangle$
25. **(a)** $\langle 0, 1 \rangle$　　**(b)** $\langle 2, 0 \rangle$
27. **(a)** $\langle \frac{6}{5}, \frac{3}{5} \rangle$　　**(b)** $\langle -\frac{1}{5}, \frac{2}{5} \rangle$
29. **(a)** $\langle 5, 0 \rangle$　　**(b)** $\langle 0, -4 \rangle$
31. **(a)** 8282.2 lb　　**(b)** $30{,}909.6$ lb　　**33.** 425 ft·lb
35. 32　　**37.** $\$11{,}905.30$

39. **(a)** $\theta = \dfrac{\pi}{2}$　　**(b)** $0 < \theta < \dfrac{\pi}{2}$　　**(c)** $\dfrac{\pi}{2} < \theta < \pi$

Chapter 4 Review Exercises　(*page* 271)

1. $A \approx 29.7°, B \approx 52.4°, C \approx 97.9°$
3. $C = 110°, b \approx 20.4, c \approx 22.6$
5. $A = 35°, C = 35°, b \approx 6.6$　　**7.** No solution
9. $A \approx 25.9°, C \approx 39.1°, c \approx 10.1$
11. $B \approx 31.2°, C \approx 133.8°, c \approx 13.9$
　　$B \approx 148.8°, C \approx 16.2°, c \approx 5.38$
13. $A \approx 9.9°, C \approx 20.1°, b \approx 29.1$
15. $A \approx 40.9°, C \approx 114.1°, c \approx 8.6$
　　$A \approx 139.1°, C \approx 15.9°, c \approx 2.6$
17. $4\sqrt{6}$　　**19.** 9.08　　**21.** 31 ft　　**23.** 31.1 m
25. 1135 mi　　**27.** $7\mathbf{i} - 7\mathbf{j}$　　**29.** $-4\mathbf{i} + 4\sqrt{3}\mathbf{j}$

31. $\dfrac{1}{\sqrt{61}}(6\mathbf{i} - 5\mathbf{j})$　　**33.** $-26\mathbf{i} - 35\mathbf{j}$

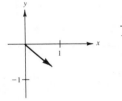

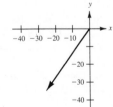

35. 133.92 lb   **37.** 92.3°, 117.0
39. 460.3 mph, N 32.2° E   **41.** $-\frac{3}{10}(\mathbf{i} - 3\mathbf{j})$ **45.** No
47. (a) $\mathbf{i} - 3\mathbf{j}$ (b) $-\mathbf{i} + 3\mathbf{j}$ (c) $3\mathbf{i} + \mathbf{j}$

CHAPTER 5

Section 5.1 *(page 281)*

```
┌─────────────────────────────────────────┐
│  WARM UP                                 │
```

1. $2\sqrt{3}$ **2.** $10\sqrt{5}$ **3.** $\sqrt{5}$ **4.** $-6\sqrt{3}$

5. 12   **6.** 48   **7.** $\dfrac{\sqrt{3}}{3}$ **8.** $\sqrt{2}$

9. $-\dfrac{1}{2} \pm \dfrac{\sqrt{5}}{2}$ **10.** $-1 \pm \sqrt{2}$

```
└─────────────────────────────────────────┘
```

1. $i, -1, -i, 1, i, -1, -i, 1,$
$i, -1, -i, 1, i, -1, -i, 1$
3. $a = -10, b = 6$ **5.** $a = 6, b = 5$
7. $4 + 3i, 4 - 3i$ **9.** $2 - 3\sqrt{3}i, 2 + 3\sqrt{3}i$
11. $5\sqrt{3}i, -5\sqrt{3}i$ **13.** $-1 - 6i, -1 + 6i$
15. $-5i, 5i$ **17.** $8, 8$ **19.** $11 - i$
21. 4 **23.** $3 - 3\sqrt{2}i$ **25.** $\frac{1}{6} + \frac{7}{6}i$
27. $-2\sqrt{3}$ **29.** -10 **31.** $5 + i$ **33.** 41
35. $12 + 30i$ **37.** $-40 + 16i$ **39.** 24
41. $-9 + 40i$ **43.** $\frac{16}{41} + \frac{20}{41}i$ **45.** $\frac{3}{5} + \frac{4}{5}i$
47. $-7 - 6i$ **49.** $\frac{1}{8}i$ **51.** $-\frac{5}{4} - \frac{5}{4}i$
53. $\frac{35}{29} + \frac{595}{29}i$ **55.** -10 **57.** $1 \pm i$ **59.** $-2 \pm \frac{1}{2}i$
61. $-\frac{3}{2}, -\frac{5}{2}$ **63.** $\frac{1}{8} \pm \frac{\sqrt{11}}{8}i$

Section 5.2 *(page 287)*

```
┌─────────────────────────────────────────┐
│  WARM UP                                 │
```

1. $4 - \sqrt{29}i, 4 + \sqrt{29}i$
2. $-5 - 12i, -5 + 12i$
3. $-1 + 4\sqrt{2}i, -1 - 4\sqrt{2}i$ **4.** $6 + \frac{1}{2}i, 6 - \frac{1}{2}i$
5. $-13 + 9i$ **6.** $12 + 16i$ **7.** $26 + 22i$
8. 29 **9.** i **10.** $-9 + 46i$

```
└─────────────────────────────────────────┘
```

1. One   **3.** Two   **5.** None   **7.** Two
9. $\pm\sqrt{5}$ **11.** $-5 \pm \sqrt{6}$ **13.** 4 **15.** $-1 \pm 2i$

17. $\frac{1}{2} \pm i$ **19.** $20 \pm 2\sqrt{215}$
21. $x^3 - x^2 + 25x - 25$ **23.** $x^3 - 10x^2 + 33x - 34$
25. $x^4 + 37x^2 + 36$
27. $x^4 + 8x^3 + 9x^2 - 10x + 100$
29. $16x^4 + 36x^3 + 16x^2 + x - 30$
31. $\pm 5i; (x + 5i)(x - 5i)$
33. $2 \pm \sqrt{3}; (x - 2 - \sqrt{3})(x - 2 + \sqrt{3})$
35. $\pm 3, \pm 3i; (x + 3)(x - 3)(x + 3i)(x - 3i)$
37. $1 \pm i; (z - 1 + i)(z - 1 - i)$
39. $2, 2 \pm i; (x - 2)(x - 2 + i)(x - 2 - i)$
41. $-5, 4 \pm 3i; (t + 5)(t - 4 + 3i)(t - 4 - 3i)$
43. $-10, -7 \pm 5i; (x + 10)(x + 7 - 5i)(x + 7 + 5i)$
45. $-\frac{3}{4}, 1 \pm \frac{1}{2}i; (4x + 3)(2x - 2 + i)(2x - 2 - i)$
47. $-2, 1 \pm \sqrt{2}i; (x + 2)(x - 1 + \sqrt{2}i)(x - 1 - \sqrt{2}i)$
49. $-\frac{1}{5}, 1 \pm \sqrt{5}i; (5x + 1)(x - 1 + \sqrt{5}i)(x - 1 - \sqrt{5}i)$
51. $2, \pm 2i; (x - 2)^2(x + 2i)(x - 2i)$
53. $\pm i, \pm 3i; (x + i)(x - i)(x + 3i)(x - 3i)$
55. $-2, -\frac{1}{2}, \pm i; (x + 2)(2x + 1)(x + i)(x - i)$
57. $\pm 3i, \pm\sqrt{3}; (x + 3i)(x - 3i)(x + \sqrt{3})(x - \sqrt{3})$
59. $1 \pm \sqrt{3}, 1 \pm \sqrt{2}i;$
$(x - 1 + \sqrt{3})(x - 1 - \sqrt{3})(x - 1 + \sqrt{2}i)(x - 1 - \sqrt{2}i)$
61. $\pm 5i, -\frac{3}{2}$ **63.** $\pm 2i, 1, -\frac{1}{2}$ **65.** $-3 \pm i, \frac{1}{4}$
67. $1, 2, -3 \pm \sqrt{2}i$ **69.** $\frac{3}{4}, \frac{1}{2} \pm \frac{\sqrt{5}}{2}i$ **71.** $x^2 + b$

Section 5.3 *(page 293)*

```
┌─────────────────────────────────────────┐
│  WARM UP                                 │
```

1. $-5 - 10i$ **2.** $7 + 3\sqrt{6}i$ **3.** $-1 - 4i$
4. $-3i$ **5.** $6 - 14i$ **6.** $6 + 4\sqrt{2}i$
7. $-22 + 16i$ **8.** 13 **9.** $-\frac{3}{2} + \frac{5}{2}i$
10. $-\frac{5}{2} - \frac{3}{2}i$

```
└─────────────────────────────────────────┘
```

1. $4\left(\cos\dfrac{\pi}{2} + i\sin\dfrac{\pi}{2}\right)$ **3.** $3\sqrt{2}\left(\cos\dfrac{5\pi}{4} + i\sin\dfrac{5\pi}{4}\right)$

5. $3\sqrt{2}\left(\cos\dfrac{7\pi}{4} + i\sin\dfrac{7\pi}{4}\right)$ **7.** $2\left(\cos\dfrac{\pi}{6} + i\sin\dfrac{\pi}{6}\right)$

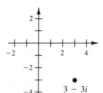

$3 - 3i$

$\sqrt{3} + i$

9. $4\left(\cos \dfrac{4\pi}{3} + i \sin \dfrac{4\pi}{3}\right)$ **11.** $6\left(\cos \dfrac{\pi}{2} + i \sin \dfrac{\pi}{2}\right)$

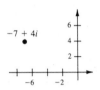

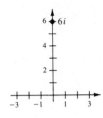

13. $\sqrt{65}(\cos 2.62 + i \sin 2.62)$

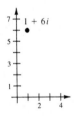

15. $7(\cos 0 + i \sin 0)$

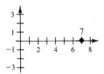

17. $\sqrt{37}(\cos 1.41 + i \sin 1.41)$

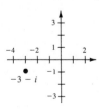

19. $\sqrt{10}(\cos 3.46 + i \sin 3.46)$

21. $-\sqrt{3} + i$ **23.** $\dfrac{3}{4} - \dfrac{3\sqrt{3}}{4}i$

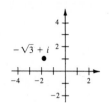

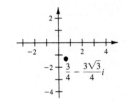

25. $\dfrac{-15\sqrt{2}}{8} + \dfrac{15\sqrt{2}}{8}i$ **27.** $-4i$

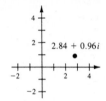

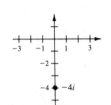

29. $2.8408 + 0.9643i$

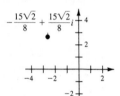

31. $12(\cos 90° + i \sin 90°)$
33. $\dfrac{10}{9}(\cos 200° + i \sin 200°)$
35. $0.27(\cos 150° + i \sin 150°)$
39. $\cos \dfrac{2\pi}{3} + i \sin \dfrac{2\pi}{3}$
41. $4[\cos(-58°) + i \sin(-58°)]$
43. $4(\cos 0° + i \sin 0°) = 4$
45. $2\sqrt{2}\left(\cos \dfrac{7\pi}{4} + i \sin \dfrac{7\pi}{4}\right) = 2 - 2i$
47. $\dfrac{5\sqrt{13}}{13}[\cos(-56.3°) + i \sin(-56.3°)] \approx \dfrac{5}{13}(2 - 3i)$
51. (a) r^2 (b) $\cos 2\theta + i \sin 2\theta$
53.

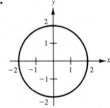

Section 5.4   (*page* 300)

WARM UP

1. $3\sqrt[3]{2}$ **2.** $2\sqrt{2}$

3. $5\sqrt{2}(\cos 135° + i \sin 135°)$

4. $3(\cos 270° + i \sin 270°)$

5. $12(\cos 180° + i \sin 180°)$

6. $12(\cos 0° + i \sin 0°)$

7. $\cos \dfrac{3\pi}{4} + i \sin \dfrac{3\pi}{4}$ **8.** $\cos \dfrac{11\pi}{12} + i \sin \dfrac{11\pi}{12}$

9. $2\left(\cos \dfrac{\pi}{2} + i \sin \dfrac{\pi}{2}\right)$ **10.** $\frac{2}{3}(\cos 45° + i \sin 45°)$

1. $-4 - 4i$ **3.** $-32i$ **5.** $-128\sqrt{3} - 128i$

7. $\dfrac{125}{2} + \dfrac{125\sqrt{3}}{2}i$ **9.** i

11. $608.02 + 144.69i$

13. $3\left(\cos \dfrac{\pi}{3} + i \sin \dfrac{\pi}{3}\right) = \dfrac{3}{2} + \dfrac{3\sqrt{3}}{2}i$

$3\left(\cos \dfrac{4\pi}{3} + i \sin \dfrac{4\pi}{3}\right) = -\dfrac{3}{2} - \dfrac{3\sqrt{3}}{2}i$

15. $2\left(\cos \dfrac{\pi}{3} + i \sin \dfrac{\pi}{3}\right) = 1 + \sqrt{3}i$

$2\left(\cos \dfrac{5\pi}{6} + i \sin \dfrac{5\pi}{6}\right) = -\sqrt{3} + i$

$2\left(\cos \dfrac{4\pi}{3} + i \sin \dfrac{4\pi}{3}\right) = -1 - \sqrt{3}i$

$2\left(\cos \dfrac{11\pi}{6} + i \sin \dfrac{11\pi}{6}\right) = \sqrt{3} - i$

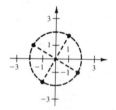

17. $5\left(\cos \dfrac{3\pi}{4} + i \sin \dfrac{3\pi}{4}\right) = -\dfrac{5\sqrt{2}}{2} + \dfrac{5\sqrt{2}}{2}i$

$5\left(\cos \dfrac{7\pi}{4} + i \sin \dfrac{7\pi}{4}\right) = \dfrac{5\sqrt{2}}{2} - \dfrac{5\sqrt{2}}{2}i$

19. $5\left(\cos \dfrac{4\pi}{9} + i \sin \dfrac{4\pi}{9}\right) = 0.8682 + 4.9240i$

$5\left(\cos \dfrac{10\pi}{9} + i \sin \dfrac{10\pi}{9}\right) = -4.6985 - 1.7101i$

$5\left(\cos \dfrac{16\pi}{9} + i \sin \dfrac{16\pi}{9}\right) = 3.8302 - 3.2139i$

21. $2(\cos 0 + i \sin 0) = 2$

$2\left(\cos \dfrac{2\pi}{3} + i \sin \dfrac{2\pi}{3}\right) = -1 + \sqrt{3}i$

$2\left(\cos \dfrac{4\pi}{3} + i \sin \dfrac{4\pi}{3}\right) = -1 - \sqrt{3}i$

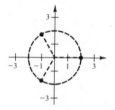

23. $\cos 0 + i \sin 0 = 1$

$$\cos \frac{2\pi}{5} + i \sin \frac{2\pi}{5} = 0.3090 + 0.9510i$$

$$\cos \frac{4\pi}{5} + i \sin \frac{4\pi}{5} = -0.8090 + 0.5878i$$

$$\cos \frac{6\pi}{5} + i \sin \frac{6\pi}{5} = -0.8090 - 0.5878i$$

$$\cos \frac{8\pi}{5} + i \sin \frac{8\pi}{5} = 0.3090 - 0.9510i$$

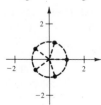

25. $\cos \dfrac{\pi}{8} + i \sin \dfrac{\pi}{8}$

$\cos \dfrac{5\pi}{8} + i \sin \dfrac{5\pi}{8}$

$\cos \dfrac{9\pi}{8} + i \sin \dfrac{9\pi}{8}$

$\cos \dfrac{13\pi}{8} + i \sin \dfrac{13\pi}{8}$

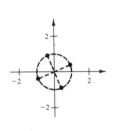

27. $3\left(\cos \dfrac{\pi}{5} + i \sin \dfrac{\pi}{5}\right)$

$3\left(\cos \dfrac{3\pi}{5} + i \sin \dfrac{3\pi}{5}\right)$

$3(\cos \pi + i \sin \pi)$

$3\left(\cos \dfrac{7\pi}{5} + i \sin \dfrac{7\pi}{5}\right)$

$3\left(\cos \dfrac{9\pi}{5} + i \sin \dfrac{9\pi}{5}\right)$

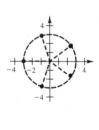

29. $4\left(\cos \dfrac{\pi}{2} + i \sin \dfrac{\pi}{2}\right)$

$4\left(\cos \dfrac{7\pi}{6} + i \sin \dfrac{7\pi}{6}\right)$

$4\left(\cos \dfrac{11\pi}{6} + i \sin \dfrac{11\pi}{6}\right)$

31. $\sqrt[6]{2}(\cos 105° + i \sin 105°)$

$\sqrt[6]{2}(\cos 225° + i \sin 225°)$

$\sqrt[6]{2}(\cos 345° + i \sin 345°)$

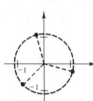

Chapter 5 Review Exercises (*page* 301)

1. $3 + 7i$ **3.** $-\sqrt{2}i$ **5.** $40 + 65i$ **7.** $-4 - 46i$

9. $1 - 6i$ **11.** $\frac{4}{3}i$ **13.** Two   **15.** One

17. None   **19.** Two

21. $6x^4 + 13x^3 + 7x^2 - x - 1$ **23.** $-8, 2$

25. $4, 7 \pm i$ **27.** $-5, 1 \pm 2i$

29. $-2, 3, -3 \pm \sqrt{5}i$ **31.** $1, -3 \pm i, 2 \pm 3i$

33. $5\sqrt{2}(\cos 315° + i \sin 315°)$

35. $13(\cos 67.38° + i \sin 67.38°)$ **37.** $-50 - 50\sqrt{3}i$

39. (a) $z_1 = 6(\cos \pi + i \sin \pi)$, $z_2 = 5\left(\cos \dfrac{\pi}{2} + i \sin \dfrac{\pi}{2}\right)$

(b) $z_1 z_2 = 30\left(\cos \dfrac{3\pi}{2} + i \sin \dfrac{3\pi}{2}\right)$,

$\dfrac{z_1}{z_2} = \dfrac{6}{5}\left(\cos \dfrac{\pi}{2} + i \sin \dfrac{\pi}{2}\right)$

41. (a) $z_1 = 3\sqrt{2}\left(\cos \dfrac{5\pi}{4} + i \sin \dfrac{5\pi}{4}\right)$,

$z_2 = 4\left(\cos \dfrac{\pi}{6} + i \sin \dfrac{\pi}{6}\right)$

(b) $z_1 z_2 = 12\sqrt{2}\left(\cos \dfrac{17\pi}{12} + i \sin \dfrac{17\pi}{12}\right)$,

$\dfrac{z_1}{z_2} = \dfrac{3\sqrt{2}}{4}\left(\cos \dfrac{13\pi}{12} + i \sin \dfrac{13\pi}{12}\right)$

43. $\dfrac{5^4}{2} + \dfrac{5^4\sqrt{3}}{2}i$ **45.** $2035 - 828i$

47. $3\left(\cos \dfrac{\pi}{4} + i \sin \dfrac{\pi}{4}\right)$ $3\left(\cos \dfrac{5\pi}{4} + i \sin \dfrac{5\pi}{4}\right)$

$3\left(\cos \dfrac{7\pi}{12} + i \sin \dfrac{7\pi}{12}\right)$ $3\left(\cos \dfrac{19\pi}{12} + i \sin \dfrac{19\pi}{12}\right)$

$3\left(\cos \dfrac{11\pi}{12} + i \sin \dfrac{11\pi}{12}\right)$ $3\left(\cos \dfrac{23\pi}{12} + i \sin \dfrac{23\pi}{12}\right)$

49. $\cos \dfrac{\pi}{3} + i \sin \dfrac{\pi}{3} = \dfrac{1}{2} + \dfrac{\sqrt{3}}{2}i$

$\cos \pi + i \sin \pi = -1$

$\cos \dfrac{5\pi}{3} + i \sin \dfrac{5\pi}{3} = \dfrac{1}{2} - \dfrac{\sqrt{3}}{2}i$

CHAPTER 6

Section 6.1 (*page* 313)

WARM UP

1. 5^x **2.** 3^{2x} **3.** 4^{3x} **4.** 10^x **5.** 4^{2x}

6. 4^{10x} **7.** $\left(\frac{3}{2}\right)^x$ **8.** 4^{3x} **9.** 2^{-x} **10.** $16^{x/4}$

1. 946.852   **3.** 747.258   **5.** 5.256
7. 472,369.379   **9.** 673.639   **11.** 7.389
13. 0.472   **15.** g   **16.** e   **17.** b   **18.** h
19. d   **20.** a   **21.** f   **22.** c

23.

25.

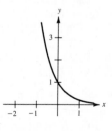

39.

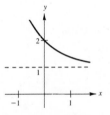

41.

n	1	2	4
A	$7,764.62	$8,017.84	$8,155.09

n	12	365	Continuous compounding
A	$8,250.97	$8,298.66	$8,300.29

27.

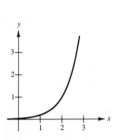

29.

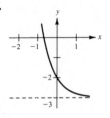

43.

n	1	2	4
A	$24,115.73	$25,714.29	$26,602.23

n	12	365	Continuous compounding
A	$27,231.38	$27,547.07	$27,557.94

31.

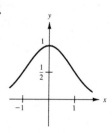

33.

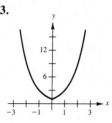

45.

t	1	10	20
P	$91,393.12	$40,656.97	$16,529.89

t	30	40	50
P	$6,720.55	$2,732.37	$1,110.90

35.

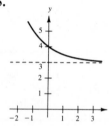

37.

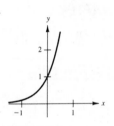

47.

t	1	10	20
P	$90,521.24	$36,940.70	$13,646.15

t	30	40	50
P	$5,040.98	$1,862.17	$687.90

49. (a) $472.70   **(b)** $298.29
51. (a) 100   **(b)** 300   **(c)** 900

Section 6.2   *(page* 323)

1. 4   **3.** −2   **5.** $\frac{1}{2}$ **7.** 0   **9.** −2   **11.** 3
13. −2   **15.** 2   **17.** $\log_5 125 = 3$ **19.** $\log_{81} 3 = \frac{1}{4}$
21. $\log_6 \frac{1}{36} = -2$ **23.** $\ln 20.0855 \approx 3$ **25.** $\ln 4 = x$
27. 2.538   **29.** 2.913   **31.** 1.005
33.   **35.**

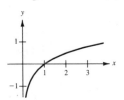

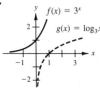

37. d   **38.** e   **39.** a   **40.** c   **41.** f   **42.** b
43. Domain: $(0, \infty)$
     Vertical asymptote: $x = 0$
     Intercept: $(1, 0)$

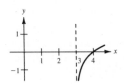

45. Domain: $(3, \infty)$
     Vertical asymptote: $x = 3$
     Intercept: $(4, 0)$

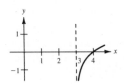

47. Domain: $(2, \infty)$
     Vertical asymptote: $x = 2$
     Intercept: $(3, 0)$

49. Domain: $(-\infty, 0)$
     Vertical asymptote: $x = 0$
     Intercept: $(-1, 0)$

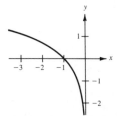

51. $\dfrac{\log_{10} 5}{\log_{10} 3}$ **53.** $\dfrac{\log_{10} x}{\log_{10} 2}$ **55.** $\dfrac{\ln 5}{\ln 3}$ **57.** $\dfrac{\ln x}{\ln 2}$
59. 1.771   **61.** −2.000   **63.** −0.417   **65.** 2.633
67. (a) 80.0   **(b)** 68.1   **(c)** 62.3   **69.** 23.68 yr

71.

r	0.005	0.010	0.015
t	138.6 yr	69.3 yr	46.2 yr

r	0.020	0.025	0.030
t	34.7 yr	27.7 yr	23.1 yr

73. (a)

x	1	5	10
$f(x)$	0	0.322	0.230

x	10^2	10^4	10^6
$f(x)$	0.046	0.00092	0.0000138

(b) 0

Section 6.3   (*page* 330)

WARM UP

1. 2   **2.** −5   **3.** −2   **4.** −3   **5.** e^5

6. $\dfrac{1}{e}$ **7.** e^6 **8.** 1   **9.** x^{-2} **10.** $x^{1/2}$

1. $\log_2 5 + \log_2 x$ **3.** $\log_3 5 - \log_3 x$ **5.** $4 \log_8 x$
7. $\frac{1}{2} \ln z$ **9.** $\log_2 x + \log_2 y + \log_2 z$
11. $\frac{1}{2} \ln(a - 1)$ **13.** $\ln z + 2 \ln(z - 1)$
15. $2 \log_b x - 2 \log_b y - 3 \log_b z$ **17.** $\frac{1}{3} \ln x - \frac{1}{3} \ln y$
19. $4 \log_9 x + \frac{1}{2} \log_9 y - 5 \log_9 z$ **21.** $\ln 2x$

23. $\log_4 \dfrac{z}{y}$ **25.** $\log_2(x + 4)^2$

27. $\ln \dfrac{x}{(x + 1)^3}$ **29.** $\log_3 \sqrt[3]{5x}$ **31.** $\log_3 \dfrac{x - 2}{x + 2}$

33. $\ln \dfrac{x}{(x^2 - 4)^2}$ **35.** $\ln \sqrt[3]{\dfrac{x(x + 3)^2}{x^2 - 1}}$

37. $\ln \dfrac{\sqrt[3]{y(y + 4)^2}}{y - 1}$ **39.** $\ln \dfrac{9}{\sqrt{x^2 + 1}}$ **41.** 0.9208

43. 0.2084   **45.** 1.6542   **47.** 0.1781   **49.** 1.8957
51. −0.7124   **53.** 0.9136   **55.** 2.0367   **57.** 2
59. 2.4   **61.** 4.5   **63.** $\frac{3}{2}$ **65.** $\frac{1}{2} + \frac{1}{2} \log_7 10$
67. $-3 - \log_5 2$ **69.** $6 + \ln 5$
73.

x	y	$\dfrac{\ln x}{\ln y}$	$\ln \dfrac{x}{y}$	$\ln x - \ln y$
1	2	0	−0.6931	−0.6931
3	4	0.7925	−0.2877	−0.2877
10	5	1.4307	0.6931	0.6931
4	0.5	−2.000	2.0794	2.0794

Section 6.4   (*page* 338)

WARM UP

1. $\dfrac{\ln 3}{\ln 2}$ **2.** $1 + \dfrac{2}{\ln 4}$ **3.** $\dfrac{e}{2}$ **4.** $2e$
5. $2 \pm i$ **6.** $\frac{1}{2}, 1$ **7.** $2x$ **8.** $3x$ **9.** $2x$
10. $-x^2$

1. 2   **3.** −2   **5.** 3   **7.** 64   **9.** $\frac{1}{10}$ **11.** x^2
13. $5x + 2$ **15.** x^2 **17.** $\log_{10} 42$ **19.** $\frac{1}{2} \log_{10} 36$
21. $1 + \log_{10} \frac{2}{3}$ **23.** $\ln 10$ **25.** $\ln \frac{5}{3}$ **27.** $\frac{2}{3} \ln \frac{40}{3}$
29. $-\frac{1}{2} + \frac{1}{2} \ln \frac{962}{25}$ **31.** $\dfrac{\ln 3}{4 \ln\left(1 + \frac{0.09}{4}\right)}$ **33.** $\dfrac{\ln 3}{0.09}$
35. $\dfrac{\ln 2}{12 \ln\left(1 + \frac{0.10}{12}\right)}$ **37.** $-5 \ln \frac{4}{19}$ **39.** $-\dfrac{\ln 0.2247}{\ln 1.0775}$
41. $\dfrac{7 - \log_{10} 5}{1 + \log_{10} 5}$ **43.** $\frac{1}{2} \ln \frac{1}{3}$ **45.** 0   **47.** $\frac{1}{2} \ln 3$
49. e^5 **51.** $e^{7/2}$ **53.** $\frac{1}{4}$ **55.** 103
57. $1 + \sqrt{1 + e}$ **59.** $-\dfrac{1}{2} + \dfrac{\sqrt{17}}{2}$
61. $-\dfrac{3}{2} + \dfrac{\sqrt{9 + 4e}}{2}$ **63.** $1, e^2$ **65.** 8.155 yr
67. (a) 1426 units
(b) 1498 units
69. (a) 29.33 yr
(b) 39.79 yr

Section 6.5   (*page* 347)

1.

Initial investment	Annual % rate	Effective yield	Time to double	Amount after 10 years
$1000	12%	12.75%	5.78 yr	$3,320.12

3.

Initial investment	Annual % rate	Effective yield	Time to double	Amount after 10 years
$750	8.94%	9.35%	7.75 yr	$1,833.67

5.

Initial investment	Annual % rate	Effective yield	Time to double	Amount after 10 years
$500	9.5%	9.97%	7.30 yr	$1,292.85

7.

Initial investment	Annual % rate	Effective yield	Time to double	Amount after 10 years
$6,392.79	11%	11.63%	6.30 yr	$19,205.00

9.

Initial investment	Annual % rate	Effective yield	Time to double	Amount after 10 years
$5000	8%	8.33%	8.66 yr	$11,127.70

11. $112,087.09   **13. (a)** 6.642 yr   **(b)** 6.330 yr
(c) 6.302 yr   **(d)** 6.301 yr

15.

r	2%	4%	6%
t	54.93 yr	27.47 yr	18.31 yr

r	8%	10%	12%
t	13.73 yr	10.99 yr	9.16 yr

17. $26,111.12   **19.** $640,501.62

21. 2009   **23.** $\dfrac{\ln 2}{r}, \dfrac{\ln 3}{r}$

25.

Half-life (in years)	Initial quantity	Amount after 1000 years	Amount after 10,000 years
1620	10 g	6.52 g	0.14 g

27.

Half-life (in years)	Initial quantity	Amount after 1000 years	Amount after 10,000 years
5730	6.71 g	5.95 g	2 g

29.

Half-life (in years)	Initial quantity	Amount after 1000 years	Amount after 10,000 years
24,360	2.16 g	2.1 g	1.62 g

31. 95.8%   **33.** $\frac{1}{4}\ln 10 \approx 0.5756$
35. $\frac{1}{4}\ln\frac{1}{4} \approx -0.3466$
37. (a) $S(t) = 100(1 - e^{-0.1625t})$ **(b)** 55,625
39. (a) 2137   **(b)** 4.6 months
41. (a) 20   **(b)** 70   **(c)** 95   **(d)** 120
43. (a) 7.91   **(b)** 7.68
45. 4.64   **47.** 1.58×10^{-6} moles per liter
49. (a) $T(t) = 70 + 280e^{-0.02784t}$ **(b)** 122.7°
(c) 119.7 min

Chapter 6 Review Exercises (*page* 349)

1.
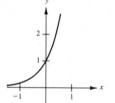

3.
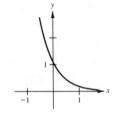

5.
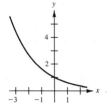

7.
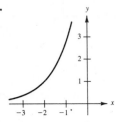

9.
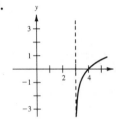

11.
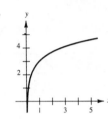

13.
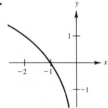

15.
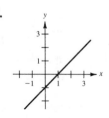

17. $1 + 2\log_5 x$ **19.** $\log_{10} 5 + \frac{1}{2}\log_{10} y - 2\log_{10} x$
21. $\ln(x^2 + 1) - \ln|x|$ **23.** $\ln(x^2 + 1) + \ln(x - 1)$
25. $\log_{10}\dfrac{5}{(x + 4)^2}$ **27.** $\ln\dfrac{\sqrt{|2x - 1|}}{(x + 1)^2}$ **29.** $\ln x^{7/3}$
31. $\ln\dfrac{3\sqrt[3]{4 - x^2}}{x}$ **33.** False   **35.** False   **37.** True
39. True   **41.** 229.2 units per milliliter
43. (a) 15.3%   **(b)** 48.7%   **(c)** 81.1%
45.

Speed	50	55	60	65	70
Miles per gallon	28	26.4	24.8	23.4	22.0

47. $y = 2e^{0.1014t}$ **49.** $y = 4e^{-0.4159t}$

51. (a) 1151 units   **(b)** 1325 units

53. (a) 8.94%   **(b)** $1,834.37   **(c)** 9.36%

55.

x	2	5
e^{-x}	1.3534×10^{-1}	6.7379×10^{-3}

x	10	15
e^{-x}	4.5399×10^{-5}	3.0590×10^{-7}

57. (a) Sum: 2.7083, Calculator: 2.7183

(b) Sum: 1.6484, Calculator: 1.6487

(c) Sum: 0.6068, Calculator: 0.6065

CHAPTER 7

Section 7.1   (*page* 358)

WARM UP

1. $-\frac{9}{2}$ **2.** $-\frac{13}{3}$ **3.** 0   **4.** 0

5. $y = \frac{2}{3}x - \frac{5}{3}$ **6.** $y = -2x$

7. $y = 3x - 1$ **8.** $y = \frac{2}{3}x + 5$

9. $y = -2x + 7$ **10.** $y = x + 3$

1. 1   **3.** 0   **5.** -3

7.

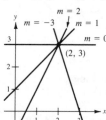

9. $m = 2$

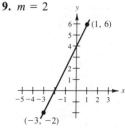

11. m is undefined

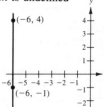

13. $m = \frac{4}{3}$

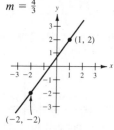

15. $(0, 1)$, $(3, 1)$, $(-1, 1)$

17. $(6, -5)$, $(7, -4)$, $(8, -3)$

19. $(-8, 0)$, $(-8, 2)$, $(-8, 3)$ **21.** $m = 5$, $(0, 3)$

23. m is undefined, no y-intercept

25. $m = -\frac{7}{6}$, $(0, 5)$

27. $3x + 5y - 10 = 0$ **29.** $x + 2y - 3 = 0$

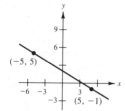

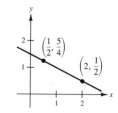

31. $x + 8 = 0$ **33.** $2x - 5y + 1 = 0$

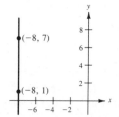

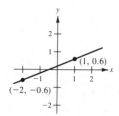

35. $3x - y - 2 = 0$ **37.** $2x + y = 0$

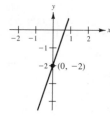

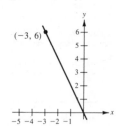

39. $x + 3y - 4 = 0$ **41.** $x - 6 = 0$

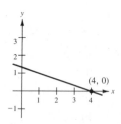

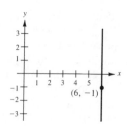

43. $8x - 6y - 17 = 0$ **47.** $3x + 2y - 6 = 0$

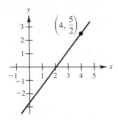

49. $12x + 3y + 2 = 0$ **51.** $x + y - 3 = 0$
53. $F = \frac{9}{5}C + 32$ **55.** $I = 95t$
57. $S = 0.85L$ **59.** $W = 0.07S + 2500$
61. (a) $C = 16.75t + 36,500$ **(b)** $R = 27t$
     (c) $P = 10.25t - 36,500$ **(d)** $t \approx 3561$ hr

Section 7.2   (page 369)

┌───┐
│ **WARM UP** │

1. 0   **2.** -2 **3.** $-\frac{1}{2}$ **4.** 2   **5.** 4   **6.** $\frac{1}{2}$
7. $-\frac{5}{4}$ **8.** $\frac{1}{2}$ **9.** $38.66°$ **10.** $66.80°$
└───┘

1. Perpendicular   **3.** Parallel
5. (a) $2x - y - 3 = 0$ **(b)** $x + 2y - 4 = 0$
7. (a) $3x + 4y + 2 = 0$ **(b)** $4x - 3y + 36 = 0$
9. (a) $y = 0$ **(b)** $x + 1 = 0$
11. Right triangle   **13.** Right triangle
15. Right triangle   **17.** Collinear   **19.** Not collinear
21. 2   **23.** $\frac{5\sqrt{2}}{2}$ **25.** $\frac{4}{3}$ **27.** 0   **29.** $2\sqrt{2}$
31. $30°$ **33.** $45°$ **35.** $121.0°$ **37.** $172.4°$
39. $45°$ **41.** $\arctan \frac{7}{24} \approx 16.3°$ **43.** $0°$
45. $y - \sqrt{3}x = 6$ **47.** $y + 6.162x = -17.49$

Section 7.3   (page 376)

┌───┐
│ **WARM UP** │

1. $x^2 - 10x + 5$ **2.** $x^2 + 6x + 8$
3. $-x^2 - 8x - 6$ **4.** $-x^2 + 4x$
5. $(x + 3)^2 - 1$ **6.** $(x - 5)^2 - 4$
7. $2 - (x - 1)^2$ **8.** $-2(x - 1)^2$
9. $2x + 3y - 20 = 0$ **10.** $3x - 4y - 17 = 0$
└───┘

1. e   **2.** f   **3.** a   **4.** c   **5.** d   **6.** b
7. Vertex: $(0, 0)$ **9.** Vertex: $(0, 0)$
  Focus: $(0, \frac{1}{16})$ Focus: $(-\frac{3}{2}, 0)$
  Directrix: $y = -\frac{1}{16}$ Directrix: $x = \frac{3}{2}$
11. Vertex: $(0, 0)$ **13.** Vertex: $(1, -2)$
  Focus: $(0, -2)$ Focus: $(1, -4)$
  Directrix: $y = 2$ Directrix: $y = 0$
15. Vertex: $(5, -\frac{1}{2})$ **17.** Vertex: $(1, 1)$
  Focus: $(\frac{11}{2}, -\frac{1}{2})$ Focus: $(1, 2)$
  Directrix: $x = \frac{9}{2}$ Directrix: $y = 0$
19. Vertex: $(8, -1)$ **21.** Vertex: $(-2, -3)$
  Focus: $(9, -1)$ Focus: $(-4, -3)$
  Directrix: $x = 7$ Directrix: $x = 0$
23. Vertex: $(-1, 2)$ **25.** Vertex: $(-2, 2)$
  Focus: $(0, 2)$ Focus: $(-2, 1)$
  Directrix: $x = -2$ Directrix: $y = 3$
27. $x^2 + 6y = 0$ **29.** $y^2 + 8x = 0$
31. $x^2 - 4y = 0$ **33.** $y^2 - 4y + 8x - 20 = 0$
35. $x^2 - 8y + 32 = 0$ **37.** $x^2 + 8y - 16 = 0$
39. $5x^2 - 14x - 3y + 9 = 0$ **41.** $x^2 + y - 4 = 0$
43. $x^2 - 800y = 0$ **45.** $x^2 - 12y = 0$
47. $4x - y - 8 = 0$ **49.** $4x - y + 2 = 0$
51. (a) $4x^2 + 25y - 1200 = 0$
     (b) $10\sqrt{3} \approx 17.32$ ft from the point directly below the
     end of the pipe.
53. (a) $x^2 + 64y - 4800 = 0$ **(b)** $40\sqrt{3} \approx 69.28$ ft

Section 7.4   (page 384)

┌───┐
│ **WARM UP** │

1. 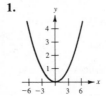   **2.**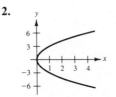

3.   **4.**
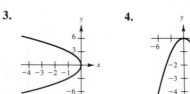
└───┘

5. $c = 12$ **6.** $b = 1$ **7.** $a = 10$
8. $c = 2\sqrt{6}$ **9.** $4x^2 + 3y^2 = 1$
10. $\dfrac{9(x-1)^2}{4} + 9(y+2)^2 = 1$

1. e **2.** a **3.** c **4.** b **5.** f **6.** d

7. Center: (0, 0)
Foci: (± 3, 0)
Vertices: (± 5, 0)
$e = \frac{3}{5}$

9. Center: (0, 0)
Foci: (0, ± 3)
Vertices: (0, ± 5)
$e = \frac{3}{5}$

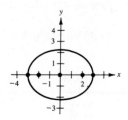

11. Center: (0, 0)
Foci: (± 2, 0)
Vertices: (± 3, 0)
$e = \frac{2}{3}$

13. Center: (0, 0)
Foci: ($\pm\sqrt{3}$, 0)
Vertices: (± 2, 0)
$e = \dfrac{\sqrt{3}}{2}$

15. Center: (0, 0)
Foci: (0, ± 1)
Vertices: (0, $\pm\sqrt{3}$)
$e = \dfrac{\sqrt{3}}{3}$

17. Center: (0, 0)
Foci: $\left(0, \pm\dfrac{\sqrt{3}}{2}\right)$
Vertices: (0, ± 1)
$e = \dfrac{\sqrt{3}}{2}$

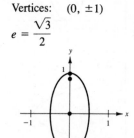

19. Center: (1, 5)
Foci: (1, 9), (1, 1)
Vertices: (1, 10), (1, 0)
$e = \frac{4}{5}$

21. Center: (-2, 3)
Foci: (-2, $3 \pm \sqrt{5}$)
Vertices: (-2, 6), (-2, 0)
$e = \dfrac{\sqrt{5}}{3}$

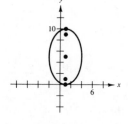

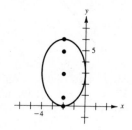

23. Center: (1, -1)
Foci: $\left(\frac{1}{4}, -1\right), \left(\frac{7}{4}, -1\right)$
Vertices: $\left(-\frac{1}{4}, -1\right), \left(\frac{9}{4}, -1\right)$
$e = \frac{3}{5}$

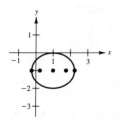

25. Center: $\left(\frac{1}{2}, -1\right)$
Foci: $\left(\frac{1}{2} \pm \sqrt{2}, -1\right)$
Vertices: $\left(\frac{1}{2} \pm \sqrt{5}, -1\right)$
$e = \dfrac{\sqrt{10}}{5}$

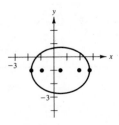

27. $\dfrac{x^2}{36} + \dfrac{y^2}{11} = 1$ **29.** $\dfrac{x^2}{25} + \dfrac{y^2}{21} = 1$

31. $\dfrac{x^2}{1} + \dfrac{y^2}{4} = 1$ **33.** $\dfrac{(x-2)^2}{4} + \dfrac{(y-2)^2}{1} = 1$

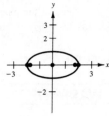

35. $\dfrac{x^2}{48} + \dfrac{(y-4)^2}{64} = 1$ **37.** $\dfrac{21x^2}{400} + \dfrac{y^2}{25} = 1$

39. $\dfrac{(x-3)^2}{36} + \dfrac{(y-2)^2}{32} = 1$ **41.** $\dfrac{x^2}{25} + \dfrac{y^2}{16} = 1$

43. $\frac{3}{2}$ ft from the center, 5 ft

45. **(a)**

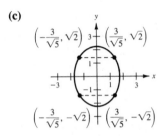

$\left(-\sqrt{3}, \frac{1}{2}\right)$ $\left(\sqrt{3}, \frac{1}{2}\right)$
$\left(-\sqrt{3}, -\frac{1}{2}\right)$ $\left(\sqrt{3}, -\frac{1}{2}\right)$

(b)

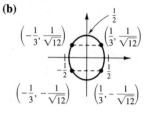

$\left(-\frac{1}{3}, \frac{1}{\sqrt{12}}\right)$ $\left(\frac{1}{3}, \frac{1}{\sqrt{12}}\right)$
$\left(-\frac{1}{3}, -\frac{1}{\sqrt{12}}\right)$ $\left(\frac{1}{3}, -\frac{1}{\sqrt{12}}\right)$

(c)

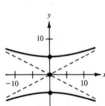

$\left(-\frac{3}{\sqrt{5}}, \sqrt{2}\right)$ $\left(\frac{3}{\sqrt{5}}, \sqrt{2}\right)$
$\left(-\frac{3}{\sqrt{5}}, -\sqrt{2}\right)$ $\left(\frac{3}{\sqrt{5}}, -\sqrt{2}\right)$

47. 91,419,000 mi, 94,581,000 mi   **49.** $e \approx 0.052$

Section 7.5   *(page* 392)

WARM UP

1. $\sqrt{61}$ **2.** $\sqrt{65}$

3.

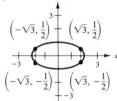

4.

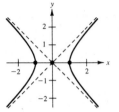

5.

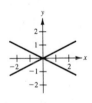

6.

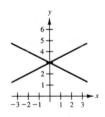

7. Parabola   **8.** Ellipse   **9.** Circle
10. Parabola

1. e   **2.** a   **3.** f   **4.** c   **5.** d   **6.** b
7. Center: (0, 0)
Vertices: $(\pm1, 0)$
Foci: $(\pm\sqrt{2}, 0)$

9. Center: (0, 0)
Vertices: $(0, \pm1)$
Foci: $(0, \pm\sqrt{5})$

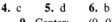

11. Center: (0, 0)
Vertices: $(0, \pm5)$
Foci: $(0, \pm13)$

13. Center: (0, 0)
Vertices: $(\pm\sqrt{3}, 0)$
Foci: $(\pm\sqrt{5}, 0)$

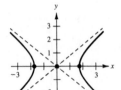

15. Center: (0, 0)
Vertices: $(0, \pm2)$
Foci: $(0, \pm3)$

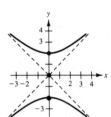

17. Center: $(1, -2)$
Vertices: $(-1, -2), (3, -2)$
Foci: $(1 \pm\sqrt{5}, -2)$

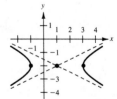

19. Center: (2, −6)
    Vertices: (2, −5), (2, −7)
    Foci: (2, −6 ± √2)

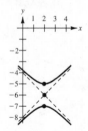

21. Center: (2, −3)
    Vertices: (1, −3), (3, −3)
    Foci: (2 ± √10, −3)

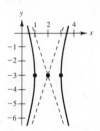

23. Center: (1, −3)
    Vertices: (1, −3 ± √2)
    Foci: (1, −3 ± 2√5)

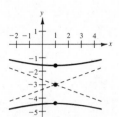

25. Degenerate hyperbola: graph is two intersecting lines
    with center at (−1, −3)

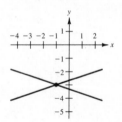

27. $\dfrac{y^2}{4} - \dfrac{x^2}{12} = 1$ **29.** $\dfrac{x^2}{1} - \dfrac{y^2}{9} = 1$

31. $\dfrac{(x-4)^2}{4} - \dfrac{y^2}{12} = 1$

33. $\dfrac{(y-5)^2}{16} - \dfrac{(x-4)^2}{9} = 1$

35. $\dfrac{y^2}{9} - \dfrac{4x^2}{9} = 1$ **37.** $\dfrac{x^2}{4} - \dfrac{3(y-1)^2}{4} = 1$

39. $\dfrac{(x-3)^2}{9} - \dfrac{(y-2)^2}{4} = 1$

41. (4400, −4290) **43.** Circle **45.** Hyperbola
47. Ellipse **49.** Parabola

Section 7.6 (*page* 402)

WARM UP

1. b **2.** e **3.** d **4.** a **5.** f **6.** c

7. $\dfrac{1}{2}x - \dfrac{\sqrt{3}}{2}y$ **8.** $-\dfrac{1}{2}x + \dfrac{\sqrt{3}}{2}y$

9. $\dfrac{4x^2 - 12xy + 9y^2}{13}$ **10.** $\dfrac{x^2 - 2\sqrt{2}xy + 2y^2}{3}$

1. $\dfrac{(y')^2}{2} - \dfrac{(x')^2}{2} = 1$ **3.** $y' = \dfrac{(x')^2}{6} - \dfrac{x'}{3}$

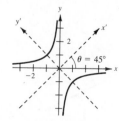

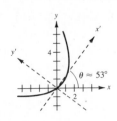

5. $\dfrac{(x')^2}{1/4} - \dfrac{(y')^2}{1/6} = 1$

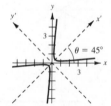

7. $\dfrac{(x' - 3\sqrt{2})^2}{16} - \dfrac{(y' - \sqrt{2})^2}{16} = 1$

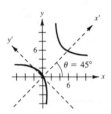

9. $\dfrac{(x')^2}{3} + \dfrac{(y')^2}{2} = 1$ **11.** $x' = -(y')^2$

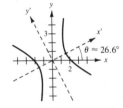

13. $\dfrac{(x')^2}{3} - \dfrac{(y')^2}{5} = 1$ **15.** $\dfrac{(x')^2}{1.096} - \dfrac{(y')^2}{6.153} = 1$

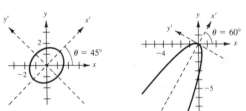

17. Parabola   **19.** Ellipse   **21.** Hyperbola
23. Parabola

Chapter 7 Review Exercises   *(page 402)*

1. $x = 0$ **3.** $5x - 12y + 2 = 0$
5. $2x - 7y + 2 = 0$ **7.** $t = \frac{7}{3}$ **9.** $t = 3$
15. a   **16.** c   **17.** d   **18.** h   **19.** g   **20.** f
21. e   **22.** b
23. Circle   **25.** Hyperbola

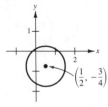

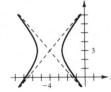

27. Ellipse

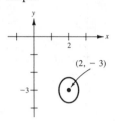

29. Parabola

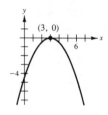

31. Parabola

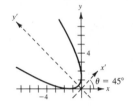

33. $(x - 4)^2 = -8(y - 2)$
35. $y^2 - 12x - 4y + 4 = 0$ **37.** $\dfrac{(x - 2)^2}{25} + \dfrac{y^2}{21} = 1$
39. $\dfrac{2x^2}{9} + \dfrac{y^2}{36} = 1$ **41.** $\dfrac{y^2}{1} - \dfrac{x^2}{8} = 1$
43. $\dfrac{5(x - 4)^2}{16} - \dfrac{5y^2}{64} = 1$ **45.** $(0, 50)$
47. $-2x + 3y = 25$ **49.** $\frac{2}{3}x - \sqrt{3}y = 1$
51. $(3, 4), (0, 5)$ **53.** $(\pm\sqrt{15}, 5)$

CHAPTER 8
Section 8.1   *(page 411)*

WARM UP

1. $\dfrac{3\pi}{4}$ **2.** $\dfrac{7\pi}{6}$ **3.** $\sin \theta = \dfrac{\sqrt{5}}{5}$

$\cos \theta = \dfrac{2\sqrt{5}}{5}$

4. $\sin \theta = -\frac{3}{5}$ **5.** $\dfrac{3\pi}{4}$ **6.** 0.5880 **7.** $-\dfrac{\sqrt{3}}{2}$
$\cos \theta = \frac{4}{5}$

8. $-\dfrac{\sqrt{2}}{2}$ **9.** -0.3090 **10.** 0.9735

1.

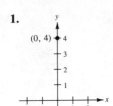

3.

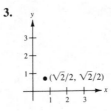

5.

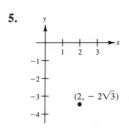

7.

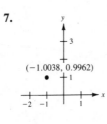

9. $\left(\sqrt{2}, \dfrac{\pi}{4}\right), \left(-\sqrt{2}, \dfrac{5\pi}{4}\right)$

11. $(5, 2.214), (-5, 5.356)$

13. $\left(\sqrt{6}, \dfrac{5\pi}{4}\right), \left(-\sqrt{6}, \dfrac{\pi}{4}\right)$

15. $(2\sqrt{13}, 0.983), (-2\sqrt{13}, 4.124)$

17. $r = 3$ **19.** $r = 2a \cos \theta$ **21.** $r = 4 \csc \theta$

23. $r = 10 \sec \theta$ **25.** $r = \dfrac{-2}{3 \cos \theta - \sin \theta}$

27. $r^2 = 8 \csc 2\theta$ **29.** $r^2 = 9 \cos(2\theta)$

31. $x^2 + y^2 - 4y = 0$ **33.** $\sqrt{3}x - 3y = 0$

35. $y = 2$ **37.** $(x^2 + y^2)^2 = 2y(3x^2 - y^2)$

39. $4x^2 - 5y^2 - 36y - 36 = 0$

Section 8.2 *(page* 418)

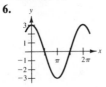
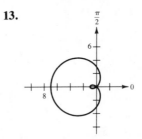
7.

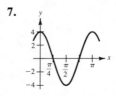

8.

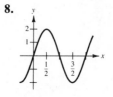

9. $\dfrac{1}{2}(\sqrt{3} \sin x - \cos x)$ **10.** $\dfrac{\sqrt{2}}{2}(\cos x - \sin x)$

1.

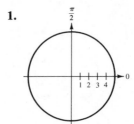

3.

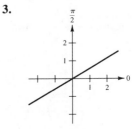

5.

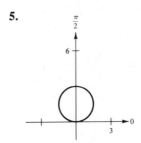

7.

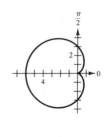

9.

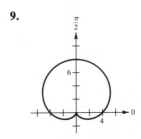

11.

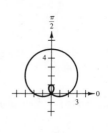

13.

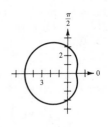

15.

17.

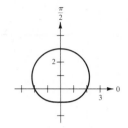

19.

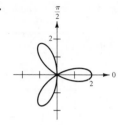

21.

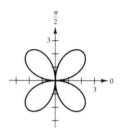

23.

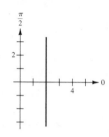

Section 8.3 (*page* 425)

WARM UP

1. $\left(\dfrac{3\sqrt{2}}{2}, -\dfrac{3\sqrt{2}}{2}\right)$ **2.** $(-2, -2\sqrt{3})$

3. $\left(3, \dfrac{3\pi}{2}\right), \left(-3, \dfrac{\pi}{2}\right)$

4. $(13, 1.9656), (-13, 5.1072)$ **5.** $r = 5$

6. $r^3 = 4 \sec^2 \theta \csc \theta$ **7.** $y = -4$

8. $x^2 + y^2 - 4x = 0$

9.

10.

25.

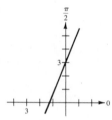

27.

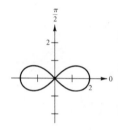

29.

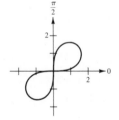

1. c   **2.** f   **3.** a   **4.** e   **5.** b   **6.** d

7. Parabola   **9.** Parabola

33. (a) $r = 2 - \sin\left(\theta - \dfrac{\pi}{4}\right) = 2 - \dfrac{\sqrt{2}}{2}(\sin\theta - \cos\theta)$

(b) $r = 2 + \cos\theta$ **(c)** $r = 2 + \sin\theta$

(d) $r = 2 - \cos\theta$

35. (a)

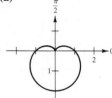

(b)

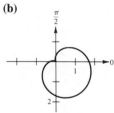

11. Ellipse   **13.** Ellipse

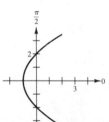

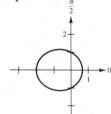

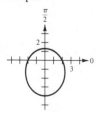

15. Hyperbola

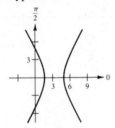

17. Parabola

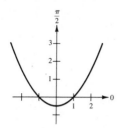

19. Hyperbola

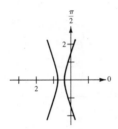

21. $r = \dfrac{1}{1 - \cos \theta}$

23. $r = \dfrac{1}{2 + \sin \theta}$

25. $r = \dfrac{2}{1 + 2 \cos \theta}$

27. $r = \dfrac{2}{1 - \sin \theta}$ or $r = \dfrac{-2}{1 + \sin \theta}$

29. $r = \dfrac{8}{3 + \sin \theta}$ or $r = \dfrac{-8}{3 - \sin \theta}$

31. $r = \dfrac{-9}{4 + 5 \sin \theta}$ or $r = \dfrac{9}{4 - 5 \sin \theta}$

33. $r = \dfrac{10}{2 + 3 \cos \theta}$ or $r = \dfrac{-10}{2 - 3 \cos \theta}$

35. $r = \dfrac{10}{1 - \cos \theta}$ or $r = \dfrac{-10}{1 + \cos \theta}$

39. $r^2 = \dfrac{24{,}336}{169 - 25 \cos^2 \theta}$

41. $r^2 = \dfrac{144}{25 \cos^2 \theta - 9}$

43. $r^2 = \dfrac{144}{25 \sin^2 \theta - 16}$

45. $r = \dfrac{7975.8}{1 + 0.937 \cos \theta}$

47. $r = \dfrac{8200}{1 + \sin \theta}$

Section 8.4   (*page* 432)

WARM UP

1.

2.
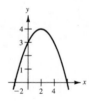

3.

4.

5.

6.

7. 10   **8.** 5 tan² θ   **9.** sec² x + tan² x
10. ½ sin θ

1. $2x - 3y + 5 = 0$

3. $y = (x - 1)^2$

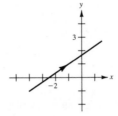

5. $y = 1 - x^2,\ x \geq 0$

7. $y = \frac{1}{2}x^{2/3}$

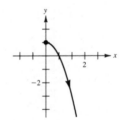

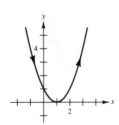

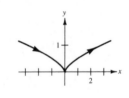

9. $y = \dfrac{x+1}{x}$

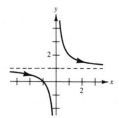

11. $x^2 + y^2 = 9$

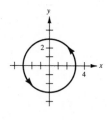

25. $y = \dfrac{1}{x^3}, x > 0$

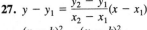

13. $\dfrac{x^2}{16} + \dfrac{y^2}{4} = 1$

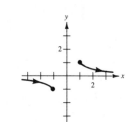

15. $y = 2 - 2x^2, -1 \le x \le 1$

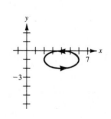

27. $y - y_1 = \dfrac{y_2 - y_1}{x_2 - x_1}(x - x_1)$

29. $\dfrac{(x - h)^2}{a^2} + \dfrac{(y - k)^2}{b^2} = 1$

31. $x = 5t$
$y = -2t$

33. $x = 2 + 4\cos\theta$
$y = 1 + 4\sin\theta$

35. $x = 5\cos\theta$
$y = 3\sin\theta$

37. $x = 4\sec\theta$
$y = 3\tan\theta$

39. $x = t, y = t^3$
$x = \sqrt[3]{t}, y = t$

41.

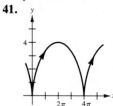

17. $y = \dfrac{1}{x}, |x| \ge 1$

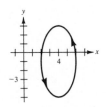

19. $\dfrac{(x - 4)^2}{4} + \dfrac{(y + 1)^2}{1} = 1$

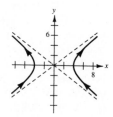

43.

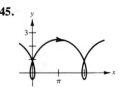

45.

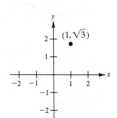

47. $x = a\theta - b\sin\theta$
$y = a - b\cos\theta$

21. $\dfrac{(x - 4)^2}{4} + \dfrac{(y + 1)^2}{16} = 1$ **23.** $\dfrac{x^2}{16} - \dfrac{y^2}{9} = 1$

Chapter 8 Review Exercises   *(page* 434)

1. $(1, \sqrt{3})$

3. $\left(\dfrac{5\sqrt{2}}{2}, -\dfrac{5\sqrt{2}}{2}\right)$

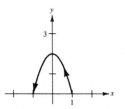

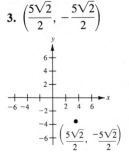

5. $\left(8, \dfrac{\pi}{2}\right), \left(-8, \dfrac{3\pi}{2}\right)$

7. $(\sqrt{5}, 2.0344), (-\sqrt{5}, 5.1760)$

9. Circle

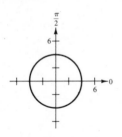

11. Rose curve

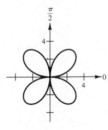

13. Cardioid

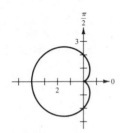

15. Limaçon

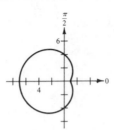

17. Rose curve

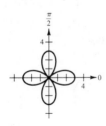

19. Rose curve

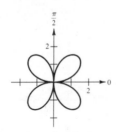

21. Line

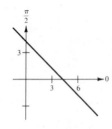

23. Parabola

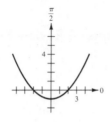

25. $x^2 + y^2 - 3x = 0$ **27.** $x^2 + 4y - 4 = 0$

29. $(x^2 + y^2)^2 - x^2 + y^2 = 0$ **31.** $r = a \cos^2 \theta \sin \theta$

33. $r = \dfrac{4}{1 - \cos \theta}$ **35.** $r = \dfrac{7}{3 + 4 \cos \theta}$

37. $r = 10 \sin \theta$

39. $y = 2x$

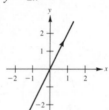

41. $3x + 4y = 11$

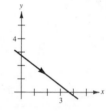

43. $xy = 1, x > 0, y > 0$

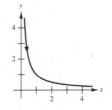

45. $x^2 + y^2 = 36$

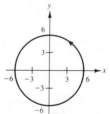

47. $y = \dfrac{1}{x^2}$

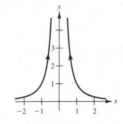

49. $x^{2/3} + \left(\dfrac{y}{4}\right)^{2/3} = 1$

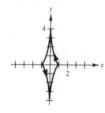

51. $x = -3 + 4 \cos \theta, y = 4 + 3 \sin \theta$

Appendix A (*page* A7)

1. (a) 2.6201 **(b)** -1.3799

3. (a) 3.7993 **(b)** 3

5. (a) -0.6081 **(b)** 3.8959

7. (a) 0.7715 **(b)** 4.2520

9. (a) 25,000 **(b)** 0.025

11. (a) 1,360,000 **(b)** 0.0136

13. (a) 3.6420 **(b)** -0.2176

15. (a) 414,500 **(b)** 0.007075 **17.** 18.10

19. 4.42 **21.** 901.5

23. (a) 1.8310 **(b)** 2.2565

25. (a) 7.46 **(b)** 4.23

27. (a) 33.115 **(b)** 0.0302

29. (a) 0.8290 **(b)** 1.483

31. (a) 0.8660 **(b)** 1.732

33. (a) 25°, 155° **(b)** 25°, 205°

35. (a) 64° 58′, 294° 2′ **(b)** 14° 2′, 194° 2′

Index